Food Science and Food Microbiology: Applied Nanotechnology

Food Science and Food Microbiology: Applied Nanotechnology

Editor: Dorothy Green

STATES
ACADEMIC PRESS
www.statesacademicpress.com

States Academic Press,
109 South 5th Street,
Brooklyn, NY 11249, USA

Visit us on the World Wide Web at:
www.statesacademicpress.com

ISBN: 978-1-63989-208-2 (Hardback)

Trademark Notice: Registered trademark of products or corporate names are used only for explanation and identification without intent to infringe.

Cataloging-in-Publication Data

Food science and food microbiology : applied nanotechnology / edited by Dorothy Green.
    p. cm.
Includes bibliographical references and index.
ISBN 978-1-63989-208-2
1. Food science. 2. Food--Microbiology. 3. Nanotechnology. I. Green, Dorothy.
TP370.5 .F66 2022
664--dc23

Table of Contents

Preface

The world is advancing at a fast pace like never before. Therefore, the need is to keep up with the latest developments. This book was an idea that came to fruition when the specialists in the area realized the need to coordinate together and document essential themes in the subject. That's when I was requested to be the editor. Editing this book has been an honour as it brings together diverse authors researching on different streams of the field. The book collates essential materials contributed by veterans in the area which can be utilized by students and researchers alike.

Food science is a multi-disciplinary field which involves chemistry, microbiology, biochemistry, nutrition and engineering to solve real life problems associated with food systems. The study of organisms which inhibit, contaminate or create food is termed as food microbiology. Nanotechnology is an upcoming field of engineering which deals with creation of functional systems at a molecular level. It has a wide variety of applications in the fields of food science and food microbiology. A few of its applications are targeted delivery systems, nanosensors and metal oxide nanoparticles. Targeted delivery systems using nanoencapsulation are found to increase the bioavailability of bioactive compounds. Metal oxide nanoparticles are used to increase the shelf life of food as they have antimicrobial properties. Nanosensors are being used extensively for detection of pathogens and other contaminants. Nanotechnology is also employed to replace non-biodegradable plastic packaging materials with environment friendly solutions. This book traces the progress of this field and highlights some of its key concepts and applications. From theories to research to practical applications, case studies related to all contemporary topics of relevance to food science and food microbiology have been included in it. A number of latest researches have been included in this book to keep the readers up-to-date with the global concepts in this area of study.

Each chapter is a sole-standing publication that reflects each author´s interpretation. Thus, the book displays a multi-facetted picture of our current understanding of application, resources and aspects of the field. I would like to thank the contributors of this book and my family for their endless support.

Editor

Cationic Nanostructures against Foodborne Pathogens

Letícia Dias de Melo Carrasco [1,2], Ronaldo Bertolucci Jr. [1], Rodrigo T. Ribeiro [1],
Jorge L. M. Sampaio [2] and Ana M. Carmona-Ribeiro [1,2]*

[1] Laboratório de Biocolóides, Departamento de Bioquímica, Instituto de Química, Universidade de São Paulo, São Paulo, Brazil, [2] Laboratório de Microbiologia, Departamento de Análises Clínicas e Toxicológicas, Faculdade de Ciências Farmacêuticas, Universidade de São Paulo, São Paulo, Brazil

Keywords: gramicidin D, dioctadecyldimethylammonium bromide, polydiallyldimethyl ammonium chloride, polymethylmethacrylate, Escherichia coli, Staphylococcus aureus, Salmonella enterica, Listeria monocytogenes

Edited by:
Spiros Paramithiotis,
Agricultural University of Athens,
Greece

Reviewed by:
Carmen Losasso,
Istituto Zooprofilattico Sperimentale
delle Venezie, Italy
Yalong Bai,
Shanghai Academy of Agricultural
Sciences, China

*Correspondence:
Ana M. Carmona-Ribeiro
amcr@usp.br

In food microbiology, novel strategies to fight foodborne pathogens are certainly welcome. In this data report, cationic nanostructures built from combinations of nanoparticles, antimicrobial peptide and cationic lipid are evaluated against important foodborne pathogens such as *Escherichia coli*, *Salmonella enterica* subsp. serovar Typhymurium, *Staphylococcus aureus* and *Listeria monocytogenes*.

The cationic lipid dioctadecyldimethylammonium bromide (DODAB), the antimicrobial peptide gramicidin D (Gr), the antimicrobial cationic polymer poly (diallyldimethylammonium chloride) (PDDA) and the biocompatible polymer poly (methyl methacrylate) (PMMA) can be combined to yield a variety of antimicrobial cationic nanostructures as previously described by our group (Carmona Ribeiro and Chaimovich, 1983; Martins et al., 1997; Lincopan et al., 2003, 2005; Pereira et al., 2008; Melo et al., 2010, 2011; Carvalho et al., 2012; Naves et al., 2013; Ragioto et al., 2014; Carrasco et al., 2015; Sanches et al., 2015). However, these nanostructures were not specifically evaluated against foodborne pathogens before. This data report aims at filling up this gap.

The cationic lipid (DODAB) and the cationic polymer PDDA bear quaternary antimicrobial nitrogens and form a variety of cationic nanostructures as the closed or open bilayers; the hybrid polymeric nanoparticles NPs and the DODAB/Gr combinations. Schemes, physical properties and antimicrobial activity for the cationic assemblies against the foodborne pathogens are on the data set (https://www.researchgate.net/publication/308140571_September_15_2016_data_set_on_cationic_assemblies_against_food_pathogens.)

DODAB, Gr, PDDA, PMMA, ethanol, 2,2,2-trifluoroethanol (TFE) and NaCl were from Sigma-Aldrich (St. Louis, MO, USA). DODAB LV were obtained by hydrating and vortexing the DODAB powder in 1 mM NaCl aqueous solution, at 60°C at 2 mM DODAB (Carmona Ribeiro and Chaimovich, 1983). For obtaining the DODAB BF, LV were ultrasonically disrupted with a macrotip (85 W/15 min/70°C) before centrifuging (10,000 rpm/60 min/4°C) and collecting the supernatant (Carmona-Ribeiro, 2006). DODAB analysis was via microtitration of its bromide counterion (Schales and Schales, 1941).

A Gr stock solution (6.4 mM Gr) in TFE was added to previously prepared LV or BF at a 1:10 Gr:DODAB molar ratio. DODAB/Gr dispersions were prepared from DODAB LV incubated for 1 h/60°C with Gr (Ragioto et al., 2014). DODAB LV/ Gr sonicated with macrotip (85 W/15 min/70°C) and centrifuged (10,000 rpm/60 min/4°C) yield the DODAB BF/Gr.

Abbreviations: ATCC, American Type Culture Collection; BF, Bilayer fragments; CFU, Colony forming unit; CTAB, cetyltrimethyl ammonium bromide; DLS, Dynamic light scattering; DODAB, Dioctadecyldimethylammonium bromide; Dz, zeta-average diameter; Gr, Gramicidin D; LV, Large vesicles; MBC, minimal bactericidal concentration; MIC, Minimal inhibitory concentration; MHA, Mueller-Hinton agar; NP, nanoparticle; P, Polydispersity; PDDA, poly (diallyldimethylammonium) chloride; PMMA, poly (methylmethacrylate); TFE, 2,2,2-trifluoroethanol (TFE); ζ, zeta-potencial.

PMMA/DODAB and PMMA/PDDA NPs were obtained as previously described (Naves et al., 2013; Sanches et al., 2015).

All nanostructures were characterized for sizing (zeta-average diameter or Dz), zeta-potential (ζ) and polydispersity (P) using a Zeta Plus-Zeta Potential Analyzer (Brookhaven Instruments Corporation, Holtsville, NY, USA) equipped with a laser (677 nm) for DLS with measurements at 90° (Grabowski and Morrison, 1983). ζ values were calculated from the electrophoretic mobility (μ) and Smoluchovski equation ($\zeta = \mu\eta / \varepsilon$, where η is the medium viscosity and ε, the dielectric constant).

Food pathogens from American Type Culture Collection (ATCC) (Manassas, VA, USA) were S. aureus ATCC 29213, E. coli ATCC 25922, L. monocytogenes ATCC 19111 and S. enterica ATCC 14028. After reactivation from frozen stocks in MHA, strains' cultures incubated in MHA (37°C/18–48 h depending on the pathogen) had some colonies transferred to a 1 mM NaCl solution and turbidity was adjusted to 0.5 McFarland (Chapin and Lauderdale, 2007). After 1 h interaction between nanostructures and bacteria in 1 mM NaCl over a range of DODAB, Gr or PDDA concentrations in the nanostructures, mixtures were diluted up to 100,000 before plating 0.1 mL onto MHA surface in triplicate. Controls were bacteria only in 1 mM NaCl (plated after 1 h). After incubation (37°C/24–48 h) and CFU/mL counting, MBC is the lowest concentration yielding the minimal CFU counting.

Gr insertion in DODAB LV bilayer reduced Dz and increased the positive ζ-potential. Gr tryptophans anchoring the peptide at the bilayer-water interface sterically stabilized DODAB/Gr. Disrupting DODAB LV/Gr led to cationic bilayers with Gr molecules inserted as dimeric channels so that the packing of the cationic lipids in the bilayer and the ζ-potentials increased. Other assemblies also tested in this work against the food pathogens had DODAB embedded in PMMA or PDDA making an outer layer (shell) in core-shell PMMA/PDDA positively charged NPs.

DODAB not only carried Gr but also displayed antimicrobial activity and reduced the MBC values against most strains tested. Table 1 shows MBCs in mM or mg/mL and the total reduction in viability caused by the antimicrobials. The Gr peptide was effective against the two Gram-positive bacteria.

DODAB BF or LV affected all bacteria tested with exception of S. enterica. Mostly DODAB BF was more efficacious against the bacteria than LV. The Gr peptide in DODAB BF reduced MBC values against three bacteria strains (lines 2 and 3, Table 1). This effect was important due to the toxicity of the cationic lipid and the antimicrobial peptide. S. enterica was the most refractory strain to the cationic agents alone or in combinations with exception of PDDA or PMMA/PDDA NPs (Table 1). In particular, the PMMA/PDDA NPs (last line on Table 1) were very efficient against S. enterica. DODAB in the PMMA/DODAB NPs displayed a reduced antimicrobial activity whereas PDDA exposure as an outer shell on the PMMA/PDDA NPs increased the antimicrobial activity (Table 1). The log of viability reduction at MBC against S. enterica of DODAB BF/Gr and DODAB LV/Gr was slightly lower than the one for DODAB BF and DODAB LV possibly due to the bulky nature of Gr tryptophans located at the bilayer/water interface, which prevented the close electrostatic attraction between the cationic moieties of the nanostructures and the anionic moieties of the bacteria.

Antimicrobial activity can be determined as inhibition of growth or minimal inhibitory concentration (MIC) or as cells survival in log (CFU/mL) (MBC). Here MBC determinations and reduction in log(CFU/mL) at MBC properly quantified the antimicrobial effect of the cationic nanostructures (Table 1). As concentrations required for inhibition are smaller than those for death, the consistency of the results can be checked: MIC for Gr against S. aureus was 2.5 μM. (Wang et al., 2012) and MBC for Gr against S. aureus was 10 μM (Table 1), a value consistently higher than the MIC value. Gr displayed a high toxicity against mammalian (Sorochkina et al., 2012; Wang et al., 2012) and eukaryotic cells such as S. cerevisae seen as 50% of cell viability at 1 μM Gr (Ragioto et al., 2014). However, in formulations with DODAB, Gr toxicity decreased against S. cerevisae (Ragioto et al., 2014). Reductions in MBC for DODAB in the combinations with Gr mean reduction in Gr doses since Gr concentration is always 10% of the DODAB concentration in each combination. Against mammalian cells, 0.5 mM DODAB killed 50% of fibroblasts in culture (Carmona-Ribeiro et al., 1997). Despite the DODAB relative toxicity in vitro, there were instances of good activity

TABLE 1 | MBC (mM; mg/mL) and log of viability reduction at MBC for cationic nanostructures against food pathogens.

Assembly	MBC in mM; mg/mL/ Reduction in log(CFU/mL)			
	E. coli	S. enterica	S. aureus	L. monocytogenes
Gr	0.010; 0.019/ 0.3	0.010; 0.019/ 0.5	0.010; 0.019/ 2.1	0.005; 0.009/ 7.6
DODAB BF	0.063; 0.039/ 7.6	0.500; 0.316/ 1.3	0.063; 0.039/ 3.4	0.125; 0.079/ 7.8
DODAB BF/Gr	0.031; 0.019/ 7.5	0.250; 0.158/ 0.9	0.015; 0.010/ 3.8	0.125; 0.079/ 8.0
DODAB LV	0.015; 0.010/ 4.5	0.500; 0.316/ 0.7	0.015; 0.010/ 2.9	0.250; 0.158/ 5.7
DODAB LV/Gr	0.015; 0.010/ 4.6	0.500; 0.316/ 0.4	0.031; 0.019/ 2.7	0.063; 0.039/ 6.0
PMMA/DODAB	------; 2.500/ 2.2	------; 1.250/ 0.1	------; 5.000/ 3.1	------; 5.000/ 1.5
PDDA	------; 0.005/ 7.5	------; 4.810/ 3.3	------; 0.010/ 5.8	------; 0.048/ 0.5
PMMA/PDDA	------; 0.009/ 7.5	------; 0.940/ 6.9	------; 0.940/ 7.1	------; 0.940/ 5.1

For DODAB/Gr combinations, the molar ratio is [Gr] = 0.1[DODAB].

for DODAB formulations *in vivo*. For example, DODAB could be used as an effective immunoadjuvant in combination with peptides or proteins for vaccines (Tsuruta et al., 1997; Carmona-Ribeiro, 2014) or could incorporate amphotericin B against systemic candidiasis in mice inducing about 100% of mice survival after treatment in absence of nephrotoxicity (Lincopan et al., 2003, 2005).

DODAB and DODAB/Gr interacted with bacteria driven by the electrostatic attraction and their mechanism of action involved lysis of the bacteria with leakage of intracellular compounds to the external medium and distortions in cell morphology (Martins et al., 1997; Ragioto et al., 2014). Gr required insertion in the bacterial cell membrane in order to act as a channel for permeation of cations across the membrane; this disturbed the ionic balance and ultimately led to the observed Gr antibiotic activity (Harold and Baarda, 1967; Clement and Gould, 1981; Hamada et al., 2010). Thus, for the DODAB/Gr combinations, the mechanism involved would include both the lytic aspects of DODAB interaction with the bacteria and the Gr effects on membrane function and selectivity in the transport of ions and nutrients and ion distribution in the cell.

DODAB could be incorporated in a polymeric biocompatible network of PMMA (Pereira et al., 2008) but displayed limited antimicrobial activity therein (**Table 1**) in contrast to the one of the more mobile CTAB surfactant which readily diffused across the polymeric PMMA network, reached attached or free bacteria and displayed good antimicrobial activity (Melo et al., 2011). Therefore, the good miscibility of DODAB lipid in the polymeric network of PDDA hampered DODAB diffusion to the outer medium where DODAB would act against the bacteria.

L. monocytogenes was very sensitive to the cationic lipid DODAB and the antimicrobial neutral peptide Gr (**Table 1**). Lysozyme and cationic peptides targeting the *L. monocytogenes* cell wall to promoted bacterial lysis. The introduction of specific modifications in components of the cell envelope as a strategy developed by bacteria rendered them undetectable to both immune recognition and to the bacteriolytic activity of host defense enzymes such as lysozyme and cationic antimicrobial peptides (Davis and Weiser, 2011; Carvalho et al., 2014). It seems that *L. monocytogenes* did not develop yet any mechanism against DODAB or Gr so that these might be advantageously employed in anti- *L. monocytogenes* coatings. On the other hand, the cationic antimicrobial polymer PDDA, similarly to cationic peptides did not affect this pathogen (**Table 1**).This is understandable from the already disclosed *L. monocytogenes* mechanisms to fight the cationic antimicrobial peptides. Curiously, the spherical assembly of PDDA as an outer shell of a PMMA/PDDA NP exhibits a reduction of 5 logs against *L. monocytogenes* (**Table 1**), suggesting that this bacterium is not prepared against this cationic NP and this also may become an asset in the fight against the pathogen.

Alternating layers of branched polyethylenimine and styrene maleic anhydride copolymer were applied onto the surface of polypropylene yielding coatings with low surface energy and enhanced antimicrobial character due to the presence of both cationic and N-halamine moieties; the coating inactivated *L. monocytogenes* by ~3 logarithmic cycles whereas in the form of N-halamines there was more than 5 logarithmic cycles in the viable cells counting (Bastarrachea and Goddard, 2015). In this respect, it seemed advantageous to introduce PMMA/PDDA NPs as efficient assemblies to reduce *L. monocytogenes* cell viability by 5 logarithmic cycles (**Table 1**).

S. enterica is one of the most important foodborne pathogens, leading to millions of cases of enteric diseases, thousands of hospitalizations and deaths worldwide each year (Hur et al., 2011). These bacteria were not sensitive to the majority of the cationic assemblies tested (**Table 1**) with exception of PDDA (3 logs reduction in viability) or PMMA/PDDA (5 logs reduction in viability) (**Table 1**). Although the antibacterial effect of antimicrobial peptides and polymers was mediated by membrane disruption with leakage of intracellular compounds (Carrasco et al., 2015), it was not clear how they reached the bacterial cytoplasmic membrane, crossing barriers such as the external membrane of Gram-negative bacteria and the cell wall of Gram-positive bacteria. Possibly the peptide or polymer first targets the outer cell wall and then undergoes a self-promoted uptake (Hancock, 1997; Yaron et al., 2003). In this respect, our results suggested that only PDDA and PMMA/PDDA NPs targeted the cytoplasmic membrane of *S. enterica* causing lysis and death. In particular, the activity of the NPs was higher than the one of the free polymer (**Table 1**), suggesting that they were more effective in inducing membrane disruption than the free polymer.

AUTHOR CONTRIBUTIONS

LC and RB performed all the experiments, analyzed the results and helped writing the manuscript; RR provided technical assistance and helped discussing the manuscript; JS provided all bacterial strains and important advice for growing them; AC designed the study, interpreted the results and wrote the manuscript.

FUNDING

Financial support was from Conselho Nacional de Desenvolvimento Científico e Tecnológico (CNPq 302352/2014-7).

ACKNOWLEDGMENTS

LC and RB were recipients of fellowships from FAPESP (2012/24534-1) and Programa Unificado de Bolsas (Pro-Reitoria de Graduação- Universidade de São Paulo), respectively. AC thanks CNPq for a 1A research fellowship (302352/2014-7).

REFERENCES

Bastarrachea, L. J., and Goddard, J. M. (2015). Antimicrobial coatings with dual cationic and N-halamine character: characterization and biocidal efficacy. *J. Agric. Food Chem.* 63, 4243–4251. doi: 10.1021/acs.jafc.5b00445

Carmona-Ribeiro, A. M. (2006). Lipid bilayer fragments and disks in drug delivery. *Curr. Med. Chem.* 13, 1359–1370. doi: 10.2174/092986706776872925

Carmona-Ribeiro, A. M. (2014). *Cationic Nanostructures for Vaccines, Immune Response Activation.* InTech. doi: 10.5772/57543. Available online at: http://www.intechopen.com/books/immune-response-activation/cationic-nanostructures-for-vaccines

Carmona Ribeiro, A. M., and Chaimovich, H. (1983). Preparation and characterization of large dioctadecyldimethylammonium chloride liposomes and comparison with small sonicated vesicles. *Biochim. Biophys. Acta* 733, 172–179. doi: 10.1016/0005-2736(83)90103-7

Carmona-Ribeiro, A. M., Ortis, F., Schumacher, R. I., and Armelin, M. C. S. (1997). Interactions between cationic vesicles and cultured mammalian cells. *Langmuir* 13, 2215–2218. doi: 10.1021/la960759h

Carrasco, L. D. M., Sampaio, J. L., and Carmona-Ribeiro, A. M. (2015). Supramolecular cationic assemblies against multidrug-resistant microorganisms: activity and mechanism of action. *Int. J. Mol. Sci.* 16, 6337–6352. doi: 10.3390/ijms16036337

Carvalho, C. A., Olivares-Ortega, C., Soto-Arriaza, M. A., and Carmona-Ribeiro, A. M. (2012). Interaction of gramicidin with DPPC/DODAB bilayer fragments. *Biochim. Biophys Acta.* 1818, 3064–3071. doi: 10.1016/j.bbamem.2012.08.008

Carvalho, F., Sousa, S., and Cabanes, D. (2014). How *Listeria monocytogenes* organizes its surface for virulence. *Front. Cell. Infect. Microbiol.* 4:48. doi: 10.3389/fcimb.2014.00048

Chapin, K. C., and Lauderdale, T. (2007). "Reagents, stains, and media: bacteriology," in *Manual of Clinical Microbiology,* eds P. R. Murray, E. J. Baron, J. H. Jorgensen, M. L. Landry and M. A. Pfaller (Washington, DC: ASM Press), 334–364.

Clement, N. R., and Gould, J. M. (1981). Kinetics for development of gramicidin-induced ion permeability in unilamellar phospholipid vesicles. *Biochemistry* 20, 1544–1548. doi: 10.1021/bi00509a021

Davis, K. M., and Weiser, J. N. (2011). Modifications to the peptidoglycan backbone help bacteria to establish infection. *Infect. Immun.* 79, 562–570. doi: 10.1128/IAI.00651-10

Grabowski, E., and Morrison, I. (1983). "Particle size distribution from analysis of quasielastic light scattering data," in *Measurements of Suspended Particles by Quasielastic Light Scattering,* ed B. Dahneke (New York, NY: Wiley-Interscience), 199–236.

Hamada, T., Matsunaga, S., Fujiwara, M., Fujita, K., Hirota, H., Schmucki, R., et al. (2010). Solution structure of polytheonamide B, a highly cytotoxic nonribosomal polypeptide from marine sponge. *J. Am. Chem. Soc.* 132, 12941–12945. doi: 10.1021/ja104616z

Hancock, R. E. W. (1997). Peptide antibiotics. *Lancet* 349, 418–422. doi: 10.1016/S0140-6736(97)80051-7

Harold, F. M., and Baarda, J. R. (1967). Gramicidin, valinomycin, and cation permeability of *Streptococcus faecalis. J. Bacteriol.* 94, 53–60.

Hur, J., Jawale, C., and Lee, J. H. (2011). Antimicrobial resistance of Salmonella isolated from food animals: a review. *Food Res. Int.* 45, 819–830. doi: 10.1016/j.foodres.2011.05.014

Lincopan, N., Mamizuka, E. M., and Carmona-Ribeiro, A. M. (2003). *In vivo* activity of a novel amphotericin B formulation with synthetic cationic bilayer fragments. *J. Antimicrob. Chemother.* 52, 412–418. doi: 10.1093/jac/dkg383

Lincopan, N., Mamizuka, E. M., and Carmona-Ribeiro, A. M. (2005). Low nephrotoxicity of an effective amphotericin B formulation with cationic bilayer fragments. *J. Antimicrob. Chemother.* 55, 727–734. doi: 10.1093/jac/dki064

Martins, L. M. S., Mamizuka, E. M., and Carmona-Ribeiro, A. M. (1997). Cationic vesicles as bactericides. *Langmuir* 13, 5583–5587. doi: 10.1021/la970353k

Melo, L. D. M., Mamizuka, E. M., and Carmona-Ribeiro, A. M. (2010). Antimicrobial particles from cationic lipid and polyelectrolytes. *Langmuir* 26, 12300–12306. doi: 10.1021/la101500s

Melo, L. D. M., Palombo, R. R., Petri, D. F. S., Bruns, M., Pereira, E. M. A., and Carmona-Ribeiro, A. M. (2011). Structure-activity relationship for quaternary

ammonium compounds hybridized with poly(methyl methacrylate). *ACS Appl. Mater. Interfaces* 3, 1933–1939. doi: 10.1021/am200150t

Naves, A. F., Palombo, R. R., Carrasco, L. D., and Carmona-Ribeiro, A. M. (2013). Antimicrobial particles from emulsion polymerization of methyl methacrylate in the presence of quaternary ammonium surfactants. *Langmuir* 29, 9677–9684. doi: 10.1021/la401527j

Pereira, E. M. A., Kosaka, P. M., Rosa, H., Vieira, D. B., Kawano, Y., Petri, D. F. S., et al. (2008). Hybrid materials from intermolecular associations between cationic lipid and polymers. *J. Phys. Chem. B* 112, 9301–9310. doi: 10.1021/jp801297t

Ragioto, D. A., Carrasco, L. D., and Carmona-Ribeiro, A. M. (2014). Novel gramicidin formulations in cationic lipid as broad-spectrum microbicidal agents. *Int. J. Nanomed.* 9, 3183–3192. doi: 10.2147/IJN.S65289

Sanches, L. M., Petri, D. F., de Melo Carrasco, L. D., and Carmona-Ribeiro, A. M. (2015). The antimicrobial activity of free and immobilized poly (diallyldimethylammonium) chloride in nanoparticles of poly (methylmethacrylate). *J. Nanobiotechnol.* 13, 58. doi: 10.1186/s12951-015-0123-3

Schales, O., and Schales, S. S. (1941). A simple and accurate method for the determination of chloride in biological fluids. *J. Biol. Chem.* 140, 879–884.

Sorochkina, A. I., Plotnikov, E. Y., Rokitskaya, T. I., Kovalchuk, S. I., Kotova, E. A., Sychev, S. V., et al. (2012). N-terminally glutamate-substituted analogue of gramicidin A as a protonophore and selective mitochondrial uncoupler. *PLoS ONE* 7:41919. doi: 10.1371/journal.pone.0041919

Tsuruta, L. R., Quintilio, W., Costa, M. H. B., and Carmona-Ribeiro, A. M. (1997). Interactions between cationic liposomes and an antigenic protein: the physical chemistry of the immunoadjuvant action. *J. Lipid Res.* 38, 2003–2011.

Wang, F., Qin, L., Pace, C. J., Wong, P., Malonis, R., and Gao, J. (2012). Solubilized gramicidin A as potential systemic antibiotics. *Chem. Bio. Chem.* 13, 51–55. doi: 10.1002/cbic.201100671

Yaron, S., Rydlo, T., Shachar, D., and Mor, A. (2003). Activity of dermaseptin K4-S4 against foodborne pathogens. *Peptides* 24, 1815–1821. doi: 10.1016/j.peptides.2003.09.016

Conflict of Interest Statement: The authors declare that the research was conducted in the absence of any commercial or financial relationships that could be construed as a potential conflict of interest.

2

All New Faces of Diatoms: Potential Source of Nanomaterials and Beyond

Meerambika Mishra¹, Ananta P. Arukha², Tufail Bashir³, Dhananjay Yadav⁴* and G. B. K. S. Prasad⁵*

¹ School of Life Sciences, Sambalpur University, Burla, India, ² Department of Infectious Diseases and Pathology, University of Florida, Gainesville, FL, United States, ³ School of Biotechnology, Yeungnam University, Gyeongsan, South Korea, ⁴ Department of Medical Biotechnology, Yeungnam University, Gyeongsan, South Korea, ⁵ School of Biochemistry, Jiwaji University, Gwalior, India

Edited by:
Spiros Paramithiotis,
Agricultural University of Athens,
Greece

Reviewed by:
Qazi Mohd Rizwanul Haq,
Jamia Millia Islamia, India
Abdul Qader Abbady,
Department of Molecular Biology
and Biotechnology, Atomic Energy
Commission of Syria, Syria

*Correspondence:
G. B. K. S. Prasad
gbksprasad@gmail.com
Dhananjay Yadav
dhanyadav16481@gmail.com

Nature's silicon marvel, the diatoms have lately astounded the scientific community with its intricate designs and lasting durability. Diatoms are a major group of phytoplanktons involved in the biogeochemical cycling of silica and are virtually inherent in every environment ranging from water to ice to soil. The usage of diatoms has proved prudently cost effective and its handling neither requires costly materials nor sophisticated instruments. Diatoms can easily be acquired from the environment, their culture requires ambient condition and does not involve any costly media or expensive instruments, besides, they can be transported in small quantities and proliferated to a desirable confluence from that scratch, thus are excellent cost effective industrial raw material. Naturally occurring diatom frustules are a source of nanomaterials. Their silica bio-shells have raised curiosity among nanotechnologists who hope that diatoms will facilitate tailoring minuscule structures which are beyond the capabilities of material scientists. Additionally, there is a colossal diversity in the dimensions of diatoms as the frustule shape differs from species to species; this provides a scope for the choice of a particular species of diatom to be tailored to an exacting requisite, thus paving the way to create desired three dimensional nanocomposites. The present article explores the use of diatoms in various arenas of science, may it be in nanotechnology, biotechnology, environmental science, biophysics or biochemistry and summarizes facets of diatom biology under one umbrella. Special emphasis has been given to biosilicification, biomineralization and use of diatoms as nanomaterials', drug delivery vehicles, optical and immune-biosensors, filters, immunodiagnostics, aquaculture feeds, lab-on-a-chip, metabolites, and biofuels.

Keywords: biosensors, diatoms, drug delivery, nanomaterials, nanocomposites, diatom nanotechnology

INTRODUCTION

Diatoms are unicellular algae (~1–500 mm length) belonging to Class Bacillariophyceae, division Bacillariophyta, either of order centrales or pennales owing to their morphology or habitat. These phytoplanktons are further categorized into centric diatoms (Coscinodiscophyceae), pennate diatoms (Fragilariophyceae; no raphe), and pennate diatoms (Bacillariophyceae; with raphe), they exist either as unicellular or colonies, filaments, ribbons (Fragilaria), fans (Meridion), zigzags (Tabellaria), or stellate (Asterionella). Diatoms are producers within the food chain;

globally contributing to almost 25% of primary productivity (Scala and Bowler, 2001). Asexual reproduction in diatoms: cell division produces two daughter cells each inheriting one parental valve, subsequently grows another smaller valve within. Owing to this size reduction division, with every generation the size of the diatom cell reduces but upon reaching a minimal size; they invert the scenario by forming an auxospore which subsequently grows larger and then undergoes size-diminishing divisions.

Diatoms can easily be acquired from the environment and transported in small quantities and proliferated to a desirable confluence. They uptake silicon from the environment and deposit it in their cell walls forming frustules which are intricate, homogenous, regularly spaced, mesoporous, siliceous nanostructures and further allow genetic modification to tailor frustules shape and pore size according to requirement. Diatoms can incorporate desired material into their frustules enhancing their use in making hybrid biosensors, bioreactors and in biotechnology, nanomedicine, photonic devices, and microfluidics. Intact frustules can be obtained from live diatoms with minimal abrasive treatment; these nanomaterials can then be further processed according to their final goal. They have been successfully used as templates for the synthesis of advanced nanostructured bio-hybrids (Nassif and Livage, 2011). Understanding and modifying the processes of biomineralization in diatoms would further accentuate its applicability in nanotechnology.

In this review, attempt to conscientiously compile the multidisciplinary applicability of diatoms in the field of nanotechnology, and biotechnology, especially in biosensor design, drug delivery, immunodiagnostics, metabolite production has been done.

BIOSILICIFICATION OF DIATOMS

Nature has blessed diatoms with an innate ability to uptake silicon from the environment and deposit in their cell walls; thereby generating silica shells which pose as nanomaterials with multifaceted applicability. Silicon is absorbed from the surroundings at low concentration (<1 µM) and is actively transported across membranes, as silicic acid through silicic acid transporters (SITs), leading to an internal soluble silicon pool, which subsequently makes insoluble silicon for incorporation into cell walls (Martin-Jézéquel et al., 2000; Knight et al., 2016). The biogenic silica for forming frustules is manufactured intracellularly by the polymerization of silicic acid monomers. Comparatively, low molecular weight amorphous silica is transported to the edge of Silica Deposition Vesicle (SDV) by silica-transport vesicles (STVs). Upon release into interior of the SDV, these particles diffuse till they come across the part of the breeding aggregate, unto which they stick. The surface consists of silanol groups [Si $(OH)_2$ or Si–OH], facilitating them to disseminate over the surface of aggregate in a pH and temperature dependent process called 'sintering.' Relocation permits the molecules to restructure themselves to attain a thermodynamic stability, typically resulting in a smoothening of the aggregate surface. Silica structure formation in diatoms

is normally categorized into three distinct scales progressing from the nano to the meso and finally to the microscale (Hildebrand et al., 2006, 2007). The microscale is the overall shaping of the valve and girdle bands within the SDV through active and passive molding and involves cytoskeleton, actin, and microtubules (Round et al., 1990; Van De Meene and Pickett-Heaps, 2002; Tesson and Hildebrand, 2010a; Knight et al., 2016). The organic components required for biosilica polymerization (Kröger and Wetherbee, 2000) are LCPAs (long-chain polyamines, a component of biosilica) and silaffins (Kröger et al., 2002; Poulsen and Kröger, 2004; Tesson and Hildebrand, 2010b).

High variability in shell shape from sparse skeletons of criss-crossing bars to barrels, pods, stars, triangles, and elaborate disks that look like flying saucers is evident. During replication, the two diatom halves (epitheca and hypotheca) and girdle bands separate and new ones are synthesized intracellularly inside the SDVs. Girdle bands may be split rings or continuous, encircling the cell or scale-like (Round et al., 1990; Hildebrand et al., 2009). Although the girdle bands are less ornate than valves they still have a structure that appears to be species dependent and are synthesized within SDV (Kröger and Wetherbee, 2000). In centric diatoms (**Figure 1**), initial valve formation occurs by the deposition of linear ribs that radiate out from the center (Round et al., 1990; Taylor et al., 2007; Hildebrand et al., 2009). Although, the basic ribbed structure of centric diatoms appears to be conserved, that being a reasonably flat ribbed structure radiating out from the center, there are variations in the nanoscale structure.

MULTIPURPOSE USES OF DIATOMS

Both live diatoms and their modified frustules have innumerable uses. Diatoms have evolved by secondary endocytobiosis, possessing atypical cell biology and genetic makeup. Advances in molecular biology and genetic engineering will unravel usage of diatoms in nanotechnology and biotechnology (Kroth, 2007). In nature, they potently remove carbon-dioxide from the atmosphere and are largely used for environmental reconstruction and audit, forensic investigation of drowning victims and water quality monitoring. The various properties leading to the use of diatoms and their frustules in different areas of technology has been summarized in **Table 1**.

As a Source of Nanomaterials
Diatoms can self-replicate and can further be engineered to provide cost-effective and programmable industrialized system. Efforts to substitute silicon with metal oxides of established optical, electrical, thermal, biological, and chemical properties as germanium, titanium; even zinc have paid off bountifully (Rorrer et al., 2005; Jeffryes et al., 2008; Jaccard et al., 2009). Rorrer et al. (2005) have used diatom to controllably fabricate semiconductor titanium dioxide nanostructured by a bottom-up self-assembly course on a massively parallel scale. They metabolically inserted nano-structured TiO_2, forming a nano-composite of titanium and silicon in the diatom *Pinnularia* sp., by cultivating the

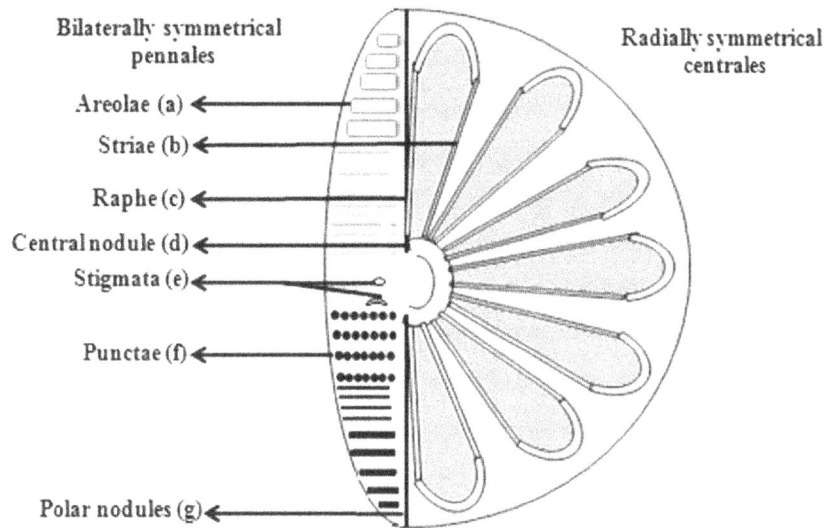

FIGURE 1 | The intricate structures of the diatom. Diatom encompasses (a). Areolae (hexagonal or polygonal boxlike perforation with a sieve present on the surface of diatom, b). Striae (pores, punctae, spots or dots in a line on the surface, c). Raphe (slit in the valves, d). Central nodule (thickening of wall at the midpoint of raphe, e). Stigmata (holes through valve surface which looks rounded externally but with a slit like internal, f). Punctae (spots or small perforations on the surface, g). Polar nodules (thickening of wall at the distal ends of the raphe) diagram modified from Taylor et al. (2007).

diatom in a controlled two-stage bioreactor process. Greatly useful in dye-sensitized solar cells designed for improved light trapping efficiency and structured photocatalysts for the superior breakdown of toxic chemicals. Lang et al. (2013) have used live diatom cells to formulate organo-silica assemblies without any loss in the intricate frustule patterning. Addition of various metals to the already existant silica frustues improves their durability and usability in various nanotechnological purposes.

As Filterant in Water Purification

Diatomaceous earth (DE) is a heterogeneous concoction of the fossil residue of dead diatoms with filtration capability. The use of diatoms over DE is advantageous because; usage of a single culture will ensure homogenous permeability and fixed pore size (Hildebrand, 2008). They can be transported cost-effectively in small numbers and cultured to desired confluence, ideal for industrial processes (Lobo et al., 1991).

As Biodevices

Diatom cells have been grown on self-assembled monolayers. The surface of glass was activated with the addition of trifluoromethyl, methyl, carboxyl, and amino groups by the self-assembled monolayers (SAM) process following which diatom was cultured on the modified glass surface. Upon rinsing post adhesion, diatoms had formed a 2D array, thus aggrandizing their use in bio-devices development (Umemura et al., 2001). Freshwater diatoms have been used to make biosensors for water quality assessment using alternating current dielectrophoresis to chain live diatom cells in order to create a 2D array (Siebman et al., 2017).

INDUSTRIAL APPLICATIONS

Metabolite Production

Diatoms are artificially cultivated for their intracellular metabolites like eicosapentaenoic acid (EPA), essential lipids, and amino acids for pharmaceutical and cosmetic purposes (Lebeau and Robert, 2003; Hemaiswarya et al., 2011). Live diatoms as *Chaetoceros* and *Thalassiosira* species are used as larval feed (Spolaore et al., 2006), *Tetrasel missuecica*, *Thalassiosira pseudonana*, *Pavlova lutheri*, *Isochrysis galbana*, and *Skeletonema costatum* are used to feed bivalve molluscs (Hemaiswarya et al., 2011). The extracellular metabolites are used as chicken and fish feeds. *P. tricornutum* and *Nitzschia laevis* have been cultivated in various photobioreactors like perfusion cell bleeding, helical tubular photobioreactor, glass tank and glass tube outdoors photobioreactor for EPA production (Lebeau et al., 2002), used to thwart coronary heart disease, hyper-triglyceridemia, blood platelet aggregation and reduction in blood cholesterol level, preventing risk of arteriosclerosis and inflammation. EPA from more popular sources like fish oil products possess poor taste, instability and higher purification cost (Abedi and Sahari, 2014). Predominantly, *Nitzschia inconspicia* (1.9–4.7% dw EPA), *Nitzschia laevis* (2.5–2.76% dw EPA), *Navicula saprophila* and *Phaeodactylum tricornutum* (2.2–3.9% dw EPA) are cultured for EPA (Wen and Chen, 2001a,b; Lebeau and Robert, 2003; Abedi and Sahari, 2014; Wah et al., 2015). *Nitzschia inconspicia* has been reported to produce arachidonic acid around 0.6–4.7% total fatty acids (Chu et al., 1994; Lebeau and Robert, 2003). Aspartic acid and isoleucine are synthesized by *Chaetoceros calcitrans* and *S. costatum*, while leucine is synthesized only by *C. calcitrans*, ornithine by *S. costatum*, serine, glutamic acid and tyrosine

TABLE 1 | Properties of diatoms which make them suitable for various uses.

Uses	Property	References
Nanotechnology and material science	• Cell wall of pectin drenched with high amount of silica. • Reproducibility of the three-dimensional structures • Ability to self-replicate • Possibility of genetic engineering and low cost of production • Intricate pore sizes which can be modified • Natural variability of design includes costae (rib-like structure further longitudinal rib and axial rib), canaliculi (tube like channels), areolae (box-like), punctae (pore-like). • Heat-resistant insulation favorable for use in boilers and blast furnaces. • Very hard hence used as abrasives	Sandhage et al., 2002, 2005; Gordon and Parkinson, 2005; Hildebrand et al., 2006, 2007; Losic et al., 2006; Jeffryes et al., 2008; Mock et al., 2008; Lang et al., 2013; Rorrer and Wang, 2016
Biosensor and Forensic limnology	• Micron sized and homogenous spaced with striae • Possibility of decreasing striae width further • Prospect to cheaply create thousands of channels on a single silicon chip • Low-cost and naturally available material • Limited dispersion through ecosystems thus give identity of their environment • Frustules vary according to species and environment hence generate flora profiles for positive identification in crime scenes, drowning victims, and time of death estimation	Dempsey et al., 1997; De Stefano et al., 2009; Gordon et al., 2009; Verma, 2013
Immunoisolation, Immunodiagnostics and Immunosensors	• High sensitivity and option to chemically modify the surface to attach bioactive molecules • Filtration and encapsulation properties of diatom frustules • Probability of controlling pore size • Evades complements of the immune system	Colton, 1995; Desai et al., 1998; Townley et al., 2008; Rorrer and Wang, 2016
Filtration and water purification	• Filters micro-organisms • Homogeneous permeability and fixed pore size • Transport in small numbers • Easy multiplication post transport • Cost effective • USEPA approved	Lobo et al., 1991; Fulton, 2000
Aquaculture feed	• Lipid and amino acid rich algal content • Anti-proliferative blue green pigment • Abundantly found in nature	Duerr et al., 1998; Lebeau et al., 1999, 2000, 2002; Turpin et al., 1999
Metabolite and biofuel production, solar panel	• EPA production • Reserve food is oil, volutin, and chrysolaminarin • Production of anti-bacterial, anti-fungal, and anti-tumoral peptides • Manufacture of neutral lipids that are lipid-fuel precursors • Production of more oil under nutrient deprivation • Photosynthetic (chlorophyll a, chlorophyll c along with xanthophylls like fucoxanthin, diatoxanthin, and diadinoxanthin) and possibility of desirable engineering	Lincoln et al., 1990; Pesando, 1990; Alonso et al., 1996; Dunahay et al., 1996; Carbonnelle et al., 1998; Ramachandra et al., 2009
Bioremediation	• Heavy metal resistance due to phytochelatin synthesis or competition for metal uptake • Efficient removal of ammonium, cadmium, phosphorous, and orthophosphate • Can be re-administered to bivalves as feed • Non-invasive as are already present in the environment	Lefebvre et al., 1996; Pistocchi et al., 2000; Schmitt et al., 2001; Medarević et al., 2016
Drug delivery	• Uniform nanoscale pore structure • Chemically inert and biocompatible • Sustained release of drugs • Filtration property • Non-toxic • Species dependent drug delivery rate	Curnow et al., 2012; Zhang et al., 2013; Milovic et al., 2014; Rea et al., 2014; Vasani et al., 2015

by *Thalassiosira* (Derrien et al., 1998; Hildebrand et al., 2012). A strong neuroexcitatory adversary of glutamate, domoic acid is also produced by *Nitzschia navisvaringica* with about 1.7 pg cell^{-1} (Kotaki et al., 2000; Martin-Jézéquel et al., 2015). Domoic acid is also established as anti-helminthic and insecticidal (Lincoln et al., 1990; Lebeau and Robert, 2003). Antibacterial and antifungal activities of diatoms are attributed to a complex of fatty acids (Pesando, 1990; Thillairajasekar et al., 2009). *S. costatum* inhibits growth of *Vibrio* in aquaculture (Naviner et al., 1999). Organic extracts of *S. costatum* (Bergé et al., 1996)

and aqueous extract of *Haslea ostrearia* (Rowland et al., 2001) are anti-tumoral, effective against human lung cancer and HIV (Hildebrand et al., 2012). A C_{25} highly branched isoprenoidpolyenes which are polyunsaturated sesterpenes oils or haslenes are responsible for anti-tumoral activities (Lebeau and Robert, 2003; Hildebrand et al., 2012).

Biofuels

Oil as food reserve is produced by diatoms during vegetative phase which keeps them afloat while awaiting favorable conditions. Using these oils glands they also produce neutral lipids which are lipid-fuel precursors; yield a lot more oil than soybean, oil seeds and palm. Ramachandra et al. (2009) professed that diatom substantially produces more oil under stress as lesser silica or nitrogen in the culture. Micro spectrometry comparative analysis of diatom oil compared with known crude oil revealed that the former has 60–70% more saturated fatty acid than the latter. A lion's share of the existent petrol has arisen from the fossilized diatoms. Diatoms imbibe CO_2 and sink on the ocean floor, gets preserved to yield petroleum (Ramachandra et al., 2009; Vinayak et al., 2015).

Ramachandra et al. (2009) also established a time-saving method of producing diatom oil which reduces the production time. They have successfully modified diatom to secrete oil as contrary to storage, which facilitates daily extraction of oil. Diatoms are adhered to a solar panel on an angiosperm leaf wherein the photosynthetic diatom substitutes mesophyll. Thus stomata facilitate gaseous exchange and leaf provides a humid growth environment for diatom while it photosynthesizes. Subsequently, they have genetically engineered diatoms to directly secrete gasoline which averts additional processing (Ramachandra et al., 2009). Diatom fuels may substitute fossil fuels thus substantially reducing greenhouse gases burden. *Cyclotella cryptica* has been genetically engineered for biodiesel production (Dunahay et al., 1996). *Phaeodactylum tricornutum* Bohlin UTEX 640 was mutated to exhibit 44% higher EPA production (Alonso et al., 1996; Lebeau and Robert, 2003).

NANOMEDICINE AND MEDICAL APPLICATIONS

Nanomedicine employs nanomaterials, nanoelectric biosensors and molecular nanotechnology with drug delivery vehicles, diagnostic devices and physical therapy applications being equally pivotal in it. However, the major shortcoming faced by nanomedicine is toxicity, biodegradability, and environmental impact. Using diatoms or their derived frustules instead provides intricate homogeneity while also surpassing the shortcomings as they are non-toxic, biodegradable, and readily available in the environment (Bradbury, 2004; Dolatabadi and de la Guardia, 2011; Jamali et al., 2012; Li et al., 2016).

Biosensors

The striae (**Figure 1**) in pennales are microscopic and are constantly spaced which can further be decreased using the compustat approach. The possibility of cheaply making such arrays of channels leading to Lab-on-a-chip (numerous channels on a single silicon chip) and the filtration ability of diatoms are favorable for numerous biosensor designs (Dempsey et al., 1997; Gordon et al., 2009; Siebman et al., 2017). These sensitive devices possess a biological molecular recognition constituent allied to a transducer, proficient of inducing a signal relative to the changing concentration of the molecule being sensed (Collings and Caruso, 1997). The flaw in extant biosensors is interference due to clustering of biomolecules in the circumference of the sensor. Frustules can filter; pore size is controllable, thus by

incorporating a specific frustule in specific sensing chamber of biosensor, selective trafficking of the molecule can be achieved. Due to their extremely refractive nature, frustules amplify signal and thus can be used as fluorescent probe.

Immunodiagnostics

Immunoisolating bio-encapsulation benefits from the filtration and encapsulation features of frustules. Lately, a biocapsule competent of selectively immune-isolating transplants was fashioned. The researchers used UV lithography, silicon thin film deposition and selective etching techniques (Desai et al., 1998). These capsules are adept in shielding its enclosure from defensive components of the immune system while concomitantly permitting the ample inflow of nutrients and oxygen to the transplanted tissue. Since frustules are naturally mesoporous, they are ideal vehicles for transporting nutrients to the girdled cells. In order to armor the frustules to filter immunoglobulins and complement system apparatus, the pore size is constrained in dimensions (30 nm) impenetrable to C1q and IgM (Colton, 1995). Furthermore, controlling the dimension of the pores, overall dimensions of frustule can also be altered so that hefty biocapsules adroit of enclosing several mammalian cells can be designed.

The diatom frustule can be chemically tailored for artificially tethering antibodies and bioactive molecules to it. The attached antibodies or molecules retain their inherent biological activity. These customized structures are crucial in antibody arrays and also form the basis of immunodiagnostics. As diatom biosilica requires only light and nominal nutrients hence they spawn an outstandingly low-priced and renewable starting matter (Townley et al., 2008).

Optical Biosensors

The frustules of the central *Coscinodiscus concinnus Wm. Smith* have been chemically modified to bind to an exceedingly selective bio-probe as an antibody. Measuring the photoluminescence emission of these modified diatoms frustules, reveal the degree of antibody–ligand interaction. Diatom frustules are nanostructured, inexpensive, abundantly available naturally and also exhibit extreme sensitivity, therefore, are the ultimate entrant for the lab-on-a-chip applications (De Stefano et al., 2009).

Drug Delivery

Homogenous pore size, constant spacing of striae, hard biosilica, genetically modifiable, chemically inert and biocompatibility are the decisive features facilitating the use of frustules as drug delivery vehicles. Pore size and rate at which the drug would be released from the diatom frustules is species-specific which gives investigators ample choices. Drug-laden diatoms can be directed to the site of release by integrating ferromagnetic elements into the frustules and then using a magnet. Currently, diatom nanotechnology is an exceedingly interdisciplinary yet a rapidly growing research front with extremely divergent applicability (Gordon and Parkinson, 2005). High-resolution imaging techniques establish a baseline for investigating biomineralization in diatoms that ultimately impact device manufacturing capabilities. Zhang et al. (2013) have efficiently used diatom for the oral delivery of drugs for gastrointestinal diseases. Usage of diatom microparticles has no toxicity rather effectively enhanced the permeability of prednisone and mesalamine while also enabling their sustained release. The use of diatom as a solid carrier for BCS Class II drugs notorious for their low water solubility for oral administration through self-emulsifying drug delivery system (SEDDS) has been reported. Two approaches using diverse self-emulsifying phospholipid suspension of carbamazepine (CBZ) first by directly mixing with diatoms, second by dispersing diatoms into its ethanolic preparation was employed. While the physical mixture procedure was more efficient, mixing with the ethanolic extract deemed faster. Both processes, however, showed prolonged longevity (Milovic et al., 2014). Diatom has also been used for transport of siRNA into tumor cells (Rea et al., 2014). Besides, diatom frustules have also been used for antibiotic delivery (Vasani et al., 2015). The genome sequences of two diatom species, *Thalassiosira pseudonana* and *Phaeodactylum tricornutum*, has already been deciphered, works on others is in progress (Armbrust et al., 2004; Bowler et al., 2008; Hildebrand et al., 2012) to effectively identify the proteins involved in fabrication of diatom skeleton features enhancing expression or direct production of desired products.

FUTURE PROSPECTS

Diatoms make gargantuan variety of shapes. Some of these structures are dependent on microtubules and possibly are sensitive to microgravity. The NASA Single Loop for Cell Culture (SLCC) for culturing and observing microbes authorizes economical, low labor in-space experiments. Three diatom species were sent to the International Space Station, together with the huge (6 mm length) diatoms of Antarctica and the exclusive colonial diatom, *Bacillaria paradoxa*. The cells of *Bacillaria* moved next to each other in partial but opposite synchrony by a microfluidics method. Swift, directed evolution is achievable by using the SLCC as a compustat. Since the structural details are well conserved in hard silica, the development of normal and deviant morphogenesis can be achieved by drying the samples on a moving diatom filter paper. Owing to the massive biodiversity of diatoms, its nanotechnology will present a condensed and portable diatom nanotechnology toolkit for space exploration (Gordon and Parkinson, 2005).

Diatoms pose a novel example of a natural enigma which has been unfolded recently. There are still many unanswered questions, as the equation amid the genotype and phenotype of diatom, its further manipulation without breaking the balance of its 3D shape and pattern, methods of genetic engineering applicable. Other speculations are about the limits for diatom evolution, how can we make the most out of them and in what other fields can diatoms find use. As our comprehension of genetic composition of diatoms gets enlightened, the possibility of designing molecularly explicit architectures of large (mm) and minute (nm) dimensions would be more feasible. Genetically engineered diatoms are employed as vectors for vaccine delivery and used for enhancing the nutritional quality of the feedstuff for crustaceans and aqua-cultured fish, few diatom based vaccines have been successfully used and patented as well (Gladue and Maxey, 1994; Hempel et al., 2011; Corbeil et al., 2015; Doron et al., 2016). Various researches have been structured to find novel diatoms even in unconventional places to decipher these siliceous mysteries (Amspoker, 2016; Noga et al., 2016). The future harbors promising challenges endowed with great rewards for diatomists and nanotechnologists eventually as the research on diatoms gets more illumined.

AUTHOR CONTRIBUTIONS

All authors listed have made a substantial, direct and intellectual contribution to the work, and approved it for publication.

ACKNOWLEDGMENT

The authors are thankful to Dr. Nanda Kishore Mishra (Department of English, Gangadhar Meher University, Odisha) for improving the English language quality.

REFERENCES

Abedi, E., and Sahari, M. A. (2014). Long-chain polyunsaturated fatty acid sources and evaluation of their nutritional and functional properties. *Food Sci. Nutr.* 2, 443–463. doi: 10.1002/fsn3.121

Alonso, D. L., Segura del Castillo, C. I., Grima, E. M., and Cohen, Z. (1996). First insights into improvement of eicosapentaenoic acid content in *Phaeodactylum tricornutum* (Bacillariophyceae) by induced mutagenesis. *J. Phycol.* 32, 339–345. doi: 10.1111/j.0022-3646.1996.00339.x

Amspoker, M. C. (2016). *Eunotogramma litorale* sp. nov., a marine epipsammic diatom from Southern California, USA. *Diatom Res.* 31, 389–395. doi: 10.1080/0269249X.2016.1256350

Armbrust, E. V., Berges, J. A., Bowler, C., Green, B. R., Martinez, D., Putnam, N. H., et al. (2004). The genome of the diatom *Thalassiosira pseudonana*: ecology, evolution, and metabolism. *Science* 306, 79–86. doi: 10.1126/science.1101156

Bergé, J., Bourgougnon, N., Carbonnelle, D., Le Bert, V., Tomasoni, C., Durand, P., et al. (1996). Antiproliferative effects of an organic extract from the marine diatom *Skeletonema costatum* (Grev.) Cleve. Against a non-small-cell bronchopulmonary carcinoma line (NSCLC-N6). *Anticancer Res.* 17, 2115–2120.

Bowler, C., Allen, A. E., Badger, J. H., Grimwood, J., Jabbari, K., Kuo, A., et al. (2008). The *Phaeodactylum* genome reveals the evolutionary history of diatom genomes. *Nature* 456, 239–244. doi: 10.1038/nature07410

Bradbury, J. (2004). Nature's nanotechnologists: unveiling the secrets of diatoms. *PLoS Biol.* 2:e306. doi: 10.1371/journal.pbio.0020306

Carbonnelle, D., Pondaven, P., Morançais, M., Massé, G., Bosch, S., Jacquot, C., et al. (1998). Antitumor and antiproliferative effects of an aqueous extract from the marine diatom *Haslea ostrearia* (Simonsen) against solid tumors: lung carcinoma (NSCLC-N6), kidney carcinoma (E39) and melanoma (M96) cell lines. *Anticancer Res.* 19, 621–624.

Chu, W.-L., Phang, S.-M., and Goh, S.-H. (1994). "Studies on the production of useful chemicals, especially fatty acids in the marine diatom *Nitzschia conspicua* Grunow," in *Ecology and Conservation of Southeast Asian Marine and Freshwater Environments including Wetlands*, eds A. Sasekumar, N. Marshall, and D. J. Macintosh (Berlin: Springer), 33–40.

Collings, A., and Caruso, F. (1997). Biosensors: recent advances. *Rep. Prog. Phys.* 60, 1397–1445. doi: 10.1088/0034-4885/60/11/005

Colton, C. K. (1995). Implantable biohybrid artificial organs. *Cell Transplant.* 4, 415–436. doi: 10.1016/0963-6897(95)00025-S

Corbeil, L. B., Hildebrand, M., Shrestha, R., Davis, A., Schrier, R., Oyler, G. A., et al. (2015). Diatom-based vaccines. U.S. Patent 20150037370 A1.

Curnow, P., Senior, L., Knight, M. J., Thamatrakoln, K., Hildebrand, M., and Booth, P. J. (2012). Expression, purification, and reconstitution of a diatom silicon transporter. *Biochemistry* 51, 3776–3785. doi: 10.1021/bi3000484

De Stefano, L., Rotiroti, L., De Stefano, M., Lamberti, A., Lettieri, S., Setaro, A., et al. (2009). Marine diatoms as optical biosensors. *Biosens. Bioelectron.* 24, 1580–1584. doi: 10.1016/j.bios.2008.08.016

Dempsey, E., Diamond, D., Smyth, M. R., Urban, G., Jobst, G., Moser, I., et al. (1997). Design and development of a miniaturised total chemical analysis system for on-line lactate and glucose monitoring in biological samples. *Anal. Chim. Acta* 346, 341–349. doi: 10.1016/S0003-2670(97)90075-1

Derrien, A., Coiffard, L. J., Coiffard, C., and De Roeck-Holtzhauer, Y. (1998). Free amino acid analysis of five microalgae. *J. Appl. Phycol.* 10, 131–134. doi: 10.1023/A:1008003016458

Desai, T. A., Chu, W. H., Tu, J. K., Beattie, G. M., Hayek, A., and Ferrari, M. (1998). Microfabricated immunoisolating biocapsules. *Biotechnol. Bioeng.* 57, 118–120. doi: 10.1002/(SICI)1097-0290(19980105)57:1<118::AID-BIT14>3.0.CO;2-G

Dolatabadi, J. E. N., and de la Guardia, M. (2011). Applications of diatoms and silica nanotechnology in biosensing, drug and gene delivery, and formation of complex metal nanostructures. *TrAC Trends Anal. Chem.* 30, 1538–1548. doi: 10.1016/j.trac.2011.04.015

Doron, L., Segal, N. A., and Shapira, M. (2016). Transgene expression in microalgae—from tools to applications. *Front. Plant Sci.* 7:505. doi: 10.3389/fpls.2016.00505

Duerr, E. O., Molnar, A., and Sato, V. (1998). Cultured microalgae as aquaculture feeds. *J. Mar. Biotechnol.* 6, 65–70.

Dunahay, T. G., Jarvis, E. E., Dais, S. S., and Roessler, P. G. (1996). "Manipulation of microalgal lipid production using genetic engineering," in *Proceedings of the Seventeenth Symposium on Biotechnology for Fuels and Chemicals* (Berlin: Springer), 223–231. doi: 10.1007/978-1-4612-0223-3_20

Fulton, G. P. (2000). *Diatomaceous Earth Filtration for Safe Drinking Water.* Reston, VA: American Society of Civil Engineers. doi: 10.1061/9780784404294

Gladue, R. M., and Maxey, J. E. (1994). Microalgal feeds for aquaculture. *J. Appl. Phycol.* 6, 131–141. doi: 10.1007/BF02186067

Gordon, R., Losic, D., Tiffany, M. A., Nagy, S. S., and Sterrenburg, F. A. (2009). The glass menagerie: diatoms for novel applications in nanotechnology. *Trends Biotechnol.* 27, 116–127. doi: 10.1016/j.tibtech.2008.11.003

Gordon, R., and Parkinson, J. (2005). Potential roles for diatomists in nanotechnology. *J. Nanosci. Nanotechnol.* 5, 35–40. doi: 10.1166/jnn.2005.002

Hemaiswarya, S., Raja, R., Kumar, R. R., Ganesan, V., and Anbazhagan, C. (2011). Microalgae: a sustainable feed source for aquaculture. *World J. Microbiol. Biotechnol.* 27, 1737–1746. doi: 10.1007/s11274-010-0632-z

Hempel, F., Bozarth, A. S., Lindenkamp, N., Klingl, A., Zauner, S., Linne, U., et al. (2011). Microalgae as bioreactors for bioplastic production. *Microb. Cell Fact.* 10:81. doi: 10.1186/1475-2859-10-81

Hildebrand, M. (2008). Diatoms, biomineralization processes, and genomics. *Chem. Rev.* 108, 4855–4874. doi: 10.1021/cr078253z

Hildebrand, M., Davis, A. K., Smith, S. R., Traller, J. C., and Abbriano, R. (2012). The place of diatoms in the biofuels industry. *Biofuels* 3, 221–240. doi: 10.4155/bfs.11.157

Hildebrand, M., Frigeri, L. G., and Davis, A. K. (2007). Synchronized growth of *Thalassiosira pseudonana* (Bacillariophyceae) provides novel insights into cell-wall synthesis processes in relation to the cell cycle. *J Phycol.* 43, 730–740. doi: 10.1111/j.1529-8817.2007.00361.x

Hildebrand, M., Kim, S., Shi, D., Scott, K., and Subramaniam, S. (2009). 3D imaging of diatoms with ion-abrasion scanning electron microscopy. *J. Struct. Biol.* 166, 316–328. doi: 10.1016/j.jsb.2009.02.014

Hildebrand, M., York, E., Kelz, J. I., Davis, A. K., Frigeri, L. G., Allison, D. P., et al. (2006). Nanoscale control of silica morphology and three-dimensional structure during diatom cell wall formation. *J. Mater. Res.* 21, 2689–2698. doi: 10.1557/jmr.2006.0333

Jaccard, T., Ariztegui, D., and Wilkinson, K. J. (2009). Incorporation of zinc into the frustule of the freshwater diatom *Stephanodiscus hantzschii*. *Chem. Geol.* 265, 381–386. doi: 10.1016/j.chemgeo.2009.04.016

Jamali, A. A., Akbari, F., Ghorakhlu, M. M., De La Guardia, M., and Yari Khosroushahi, A. (2012). Applications of diatoms as potential microalgae in nanobiotechnology. *Bioimpacts* 2, 83–89. doi: 10.5681/bi.2012.012

Jeffryes, C., Gutu, T., Jiao, J., and Rorrer, G. L. (2008). Metabolic insertion of nanostructured TiO_2 into the patterned biosilica of the diatom *Pinnularia* sp. by a two-stage bioreactor cultivation process. *ACS Nano* 2, 2103–2112. doi: 10.1021/nn800470x

Knight, M. J., Senior, L., Nancolas, B., Ratcliffe, S., and Curnow, P. (2016). Direct evidence of the molecular basis for biological silicon transport. *Nat. Commun.* 7:11926. doi: 10.1038/ncomms11926

Kotaki, Y., Koike, K., Yoshida, M., Van Thuoc, C., Huyen, N. T. M., Hoi, N. C., et al. (2000). Domoic acid production in *Nitzschia* sp. (Bacillariophyceae) isolated from a shrimp-culture pond in Do Son, Vietnam. *J. Phycol.* 36, 1057–1060. doi: 10.1046/j.1529-8817.2000.99209.x

Kröger, N., Lorenz, S., Brunner, E., and Sumper, M. (2002). Self-assembly of highly phosphorylated silaffins and their function in biosilica morphogenesis. *Science* 298, 584–586. doi: 10.1126/science.1076221

Kröger, N., and Wetherbee, R. (2000). Pleuralins are involved in theca differentiation in the diatom *Cylindrotheca fusiformis*. *Protist* 151, 263–273. doi: 10.1078/1434-4610-00024

Kroth, P. G. (2007). Genetic transformation: a tool to study protein targeting in diatoms. *Methods Mol. Biol.* 390, 257–267. doi: 10.1007/978-1-59745-466-7_17

Lang, Y., Del Monte, F., Collins, L., Rodriguez, B. J., Thompson, K., Dockery, P., et al. (2013). Functionalization of the living diatom *Thalassiosira weissflogii* with thiol moieties. *Nat. Commun.* 4:2683. doi: 10.1038/ncomms3683

Lebeau, T., Gaudin, P., Junter, G.-A., Mignot, L., and Robert, J.-M. (2000). Continuous marennin production by agar-entrapped *Haslea ostrearia* using a tubular photobioreactor with internal illumination. *Appl. Microbiol. Biotechnol.* 54, 634–640. doi: 10.1007/s002530000380

Lebeau, T., Gaudin, P., Moan, R., and Robert, J.-M. (2002). A new photobioreactor for continuous marennin production with a marine diatom: influence of the light intensity and the immobilised-cell matrix (alginate beads or agar layer). *Appl. Microbiol. Biotechnol.* 59, 153–159. doi: 10.1007/s00253-002-0993-9

Lebeau, T., Junter, G.-A., Jouenne, T., and Robert, J.-M. (1999). Marennine production by agar-entrapped *Haslea ostrearia* Simonsen. *Bioresour. Technol.* 67, 13–17. doi: 10.1016/S0960-8524(99)00096-6

Lebeau, T., and Robert, J. M. (2003). Diatom cultivation and biotechnologically relevant products. Part I: cultivation at various scales. *Appl. Microbiol. Biotechnol.* 60, 612–623. doi: 10.1007/s00253-002-1176-4

Lefebvre, S., Hussenot, J., and Brossard, N. (1996). Water treatment of land-based fish farm effluents by outdoor culture of marine diatoms. *J. Appl. Phycol.* 8, 193–200. doi: 10.1007/BF02184971

Li, A., Zhang, W., Ghaffarivardavagh, R., Wang, X., Anderson, S. W., and Zhang, X. (2016). Towards uniformly oriented diatom frustule monolayers: experimental and theoretical analyses. *Microsyst. Nanoeng.* 2:16064. doi: 10.1038/micronano.2016.64

Lincoln, R. A., Strupinski, K., and Walker, J. M. (1990). Biologically active compounds from diatoms. *Diatom Res.* 5, 337–349. doi: 10.1080/0269249X.1990.9705124

Lobo, E. A., Oliveira, M. A., Neves, M., and Schuler, S. (1991). Caracterização de ambientes de terras úmidas, no Estado do Rio Grande do Sul, onde ocorrem espécies de anatídeos com valor cinegético. *Acta Biol. Leopoldensia* 13, 19–60.

Losic, D., Triani, G., Evans, P. J., Atanacio, A., Mitchell, J. G., and Voelcker, N. H. (2006). Controlled pore structure modification of diatoms by atomic layer deposition of TiO_2. *J. Mater. Chem.* 16, 4029–4034. doi: 10.1039/b610188g

Martin-Jézéquel, V., Calu, G., Candela, L., Amzil, Z., Jauffrais, T., Séchet, V., et al. (2015). Effects of organic and inorganic nitrogen on the growth and production of domoic acid by *Pseudo-nitzschia multiseries* and *P. australis* (Bacillariophyceae) in Culture. *Mar. Drugs* 13, 7067–7086. doi: 10.3390/md13127055

Martin-Jézéquel, V., Hildebrand, M., and Brzezinski, M. A. (2000). Silicon metabolism in diatoms: implications for growth. *J. Phycol.* 36, 821–840. doi: 10.1046/j.1529-8817.2000.00019.x

Medarević, D. P., Lošić, D., and Ibrić, S. R. (2016). Diatoms-nature materials with great potential for bioapplications. *Hemijska Industrija* 70, 613–627. doi: 10.2298/HEMIND150708069M

Milovic, M., Simovic, S., Losic, D., Dashevskiy, A., and Ibric, S. (2014). Solid self-emulsifying phospholipid suspension (SSEPS) with diatom as a drug carrier. *Eur. J. Pharm. Sci.* 63, 226–232. doi: 10.1016/j.ejps.2014.07.010

Mock, T., Samanta, M. P., Iverson, V., Berthiaume, C., Robison, M., Holtermann, K., et al. (2008). Whole-genome expression profiling of the marine diatom *Thalassiosira pseudonana* identifies genes involved in silicon bioprocesses. *Proc. Natl. Acad. Sci. U.S.A.* 105, 1579–1584. doi: 10.1073/pnas. 0707946105

Nassif, N., and Livage, J. (2011). From diatoms to silica-based biohybrids. *Chem. Soc. Rev.* 40, 849–859. doi: 10.1039/c0cs00122h

Naviner, M., Bergé, J.-P., Durand, P., and Le Bris, H. (1999). Antibacterial activity of the marine diatom *Skeletonema costatum* against aquacultural pathogens. *Aquaculture* 174, 15–24. doi: 10.1016/S0044-8486(98)00513-4

Noga, T., Stanek-Tarkowska, J., Kochman-Kędziora, N., Pajączek, A., and Peszek, Ł (2016). The inside of a dam as an unusual habitat for two rare species of *Gomphosphenia-G. fontinalis* and *G. holmquistii*. *Diatom Res.* 31, 379–387. doi: 10.1080/0269249X.2016.1247019

Pesando, D. (1990). "Antibacterial and antifungal activities of marine algae," in *Introduction to Applied Phycology*, ed. I. Akatsuka (The Hague: SPB Academic Publishing B.V), 3–26.

Pistocchi, R., Mormile, M., Guerrini, F., Isani, G., and Boni, L. (2000). Increased production of extra-and intracellular metal-ligands in phytoplankton exposed to copper and cadmium. *J. Appl. Phycol.* 12, 469–477. doi: 10.1023/A: 1008162812651

Poulsen, N., and Kröger, N. (2004). Silica morphogenesis by alternative processing of silaffins in the diatom *Thalassiosira pseudonana*. *J. Biol. Chem.* 279, 42993–42999. doi: 10.1074/jbc.M407734200

Ramachandra, T. V., Mahapatra, D. M., and Gordon, R. (2009). Milking diatoms for sustainable energy: biochemical engineering versus gasoline-secreting diatom solar panels. *Ind. Eng. Chem. Res.* 48, 8769–8788. doi: 10.1021/ie900 044j

Rea, I., Martucci, N. M., De Stefano, L., Ruggiero, I., Terracciano, M., Dardano, P., et al. (2014). Diatomite biosilica nanocarriers for siRNA transport inside cancer cells. *Biochim. Biophys. Acta* 1840, 3393–3403. doi: 10.1016/j.bbagen.2014. 09.009

Rorrer, G. L., Chang, C.-H., Liu, S.-H., Jeffryes, C., Jiao, J., and Hedberg, J. A. (2005). Biosynthesis of silicon–germanium oxide nanocomposites by the marine diatom *Nitzschia frustulum*. *J. Nanosci. Nanotechnol.* 5, 41–49. doi: 10.1166/jnn.2005.005

Rorrer, G. L., and Wang, A. X. (2016). Nanostructured diatom frustule immunosensors. *Front. Nanosci. Nanotechnol.* 2:128–130. doi: 10.15761/FNN. 1000122

Round, F. E., Crawford, R. M., and Mann, D. G. (1990). *Diatoms: Biology and Morphology of the Genera*. New York, NY: Cambridge University Press.

Rowland, S., Belt, S., Wraige, E., Massé, G., Roussakis, C., and Robert, J.-M. (2001). Effects of temperature on polyunsaturation in cytostatic lipids of *Haslea ostrearia*. *Phytochemistry* 56, 597–602. doi: 10.1016/S0031-9422(00)00434-9

Sandhage, K. H., Allan, S. M., Dickerson, M. B., Gaddis, C. S., Shian, S., Weatherspoon, M. R., et al. (2005). Merging biological self-assembly with synthetic chemical tailoring: the potential for 3-D genetically engineered micro/nano-devices (3-D GEMS). *Int. J. Appl. Ceram. Technol.* 2, 317–326. doi: 10.1111/j.1744-7402.2005.02035.x

Sandhage, K. H., Dickerson, M. B., Huseman, P. M., Caranna, M. A., Clifton, J. D., Bull, T. A., et al. (2002). Novel, bioclastic route to self-assembled, 3D, chemically tailored meso/nanostructures: shape-preserving reactive conversion of biosilica (diatom) microshells. *Adv. Mater.* 14, 429–433. doi: 10.1002/1521-4095(20020318)14:6<429::AID-ADMA429>3.0.CO;2-C

Scala, S., and Bowler, C. (2001). Molecular insights into the novel aspects of diatom biology. *Cell. Mol. Life Sci.* 58, 1666–1673. doi: 10.1007/PL00000804

Schmitt, D., Müller, A., Csögör, Z., Frimmel, F. H., and Posten, C. (2001). The adsorption kinetics of metal ions onto different microalgae and siliceous earth. *Water Res.* 35, 779–785. doi: 10.1016/S0043-1354(00)00317-1

Siebman, C., Velev, O. D., and Slaveykova, V. I. (2017). Alternating current-dielectrophoresis collection and chaining of phytoplankton on chip: comparison of individual species and artificial communities. *Biosensors* 7:4. doi: 10.3390/bios7010004

Spolaore, P., Joannis-Cassan, C., Duran, E., and Isambert, A. (2006). Commercial applications of microalgae. *J. Biosci. Bioeng.* 101, 87–96. doi: 10.1263/jbb.101.87

Taylor, J. C., Harding, W. R., and Archibald, C. (2007). *An Illustrated Guide to Some Common Diatom Species from South Africa*. Gezina: Water Research Commission.

Tesson, B., and Hildebrand, M. (2010a). Dynamics of silica cell wall morphogenesis in the diatom *Cyclotella cryptica*: substructure formation and the role of microfilaments. *J. Struct. Biol.* 169, 62–74. doi: 10.1016/j.jsb.2009.08.013

Tesson, B., and Hildebrand, M. (2010b). Extensive and intimate association of the cytoskeleton with forming silica in diatoms: control over patterning on the meso-and micro-scale. *PLoS ONE* 5:e14300. doi: 10.1371/journal.pone.0014300

Thillairajasekar, K., Duraipandiyan, V., Perumal, P., and Ignacimuthu, S. (2009). Antimicrobial activity of *Trichodesmium erythraeum* (Ehr)(microalga) from south East coast of Tamil Nadu, India. *Int. J. Integr. Biol.* 5, 167–170.

Townley, H. E., Parker, A. R., and White-Cooper, H. (2008). Exploitation of diatom frustules for nanotechnology: tethering active biomolecules. *Adv. Funct. Mater.* 18, 369–374. doi: 10.1002/adfm.200700609

Turpin, V., Robert, J.-M., and Goulletquer, P. (1999). Limiting nutrients of oyster pond seawaters in the Marennes-Oléron region for *Haslea ostrearia*: applications to the mass production of the diatom in mesocosm experiments. *Aquat. Living Resour.* 12, 335–342. doi: 10.1016/S0990-7440(99)00114-X

Umemura, K., Ishikawa, M., and Kuroda, R. (2001). Controlled immobilization of DNA molecules using chemical modification of mica surfaces for atomic force microscopy: characterization in air. *Anal. Biochem.* 290, 232–237. doi: 10.1006/abio.2001.4996

Van De Meene, A. M., and Pickett-Heaps, J. D. (2002). Valve morphogenesis in the centric diatom *Proboscia alata* Sundstrom. *J. Phycol.* 38, 351–363. doi: 10.1046/j.1529-8817.2002.01124.x

Vasani, R., Losic, D., Cavallaro, A., and Voelcker, N. (2015). Fabrication of stimulus-responsive diatom biosilica microcapsules for antibiotic drug delivery. *J. Mater. Chem. B* 3, 4325–4329. doi: 10.1039/C5TB00648A

Verma, K. (2013). Role of diatoms in the world of forensic science. *J. Forensic Res.* 4:181. doi: 10.4172/2157-7145.1000181

Vinayak, V., Manoylov, K. M., Gateau, H., Blanckaert, V., Hérault, J., Pencréac'h, G., et al. (2015). Diatom milking: a review and new approaches. *Mar. Drugs* 13, 2629–2665. doi: 10.3390/md13052629

Wah, N. B., Ahmad, A. L. B., Chieh, D. C. J., and Hwai, A. T. S. (2015). Changes in lipid profiles of a tropical benthic diatom in different cultivation temperature. *Asian J. Appl. Sci. Eng.* 4, 91–101.

Wen, Z.-Y., and Chen, F. (2001a). A perfusion–cell bleeding culture strategy for enhancing the productivity of eicosapentaenoic acid by *Nitzschia laevis*. *Appl. Microbiol. Biotechnol.* 57, 316–322.

Wen, Z. Y., and Chen, F. (2001b). Application of statistically-based experimental designs for the optimization of eicosapentaenoic acid production by the diatom *Nitzschia laevis*. *Biotechnol. Bioeng.* 75, 159–169.

Zhang, H., Shahbazi, M. A., Makila, E. M., Da Silva, T. H., Reis, R. L., Salonen, J. J., et al. (2013). Diatom silica microparticles for sustained release and permeation enhancement following oral delivery of prednisone and mesalamine. *Biomaterials* 34, 9210–9219. doi: 10.1016/j.biomaterials.2013. 08.035

Conflict of Interest Statement: The authors declare that the research was conducted in the absence of any commercial or financial relationships that could be construed as a potential conflict of interest.

3

Gut Microbiota Modulation and its Relationship with Obesity using Prebiotic Fibers and Probiotics

Dinesh K. Dahiya[1], Renuka[2], Monica Puniya[3], Umesh K. Shandilya[4], Tejpal Dhewa[5], Nikhil Kumar[6], Sanjeev Kumar[7], Anil K. Puniya[8,9]* and Pratyoosh Shukla[10]*

[1] Advanced Milk Testing Research Laboratory, Post Graduate Institute of Veterinary Education and Research – Rajasthan University of Veterinary and Animal Sciences at Bikaner, Jaipur, India, [2] Department of Biochemistry, Basic Medical Science, South Campus, Panjab University, Chandigarh, India, [3] Food Safety Management System Division, Food Safety and Standards Authority of India, New Delhi, India, [4] Animal Biotechnology Division, National Bureau of Animal Genetic Resources, Karnal, India, [5] Department of Nutrition Biology, Central University of Haryana, Mahendergarh, India, [6] Department of Life Sciences, Shri Venkateshwara University, JP Nagar, India, [7] Department of Life Science, Central Assam University, Silchar, India, [8] College of Dairy Science and Technology, Guru Angad Dev Veterinary and Animal Sciences University, Ludhiana, India, [9] Dairy Microbiology Division, ICAR-National Dairy Research Institute, Karnal, India, [10] Enzyme Technology and Protein Bioinformatics Laboratory, Department of Microbiology, Maharshi Dayanand University, Rohtak, India

Edited by:
Jayanta Kumar Patra,
Dongguk University Seoul,
South Korea

Reviewed by:
Manoj Kumar Rout,
Haritoshan, Australia
Young Min Kwon,
University of Arkansas, USA
Andrea Masotti,
Bambino Gesù Ospedale Pediatrico
(IRCCS), Italy

*Correspondence:
Pratyoosh Shukla
pratyoosh.shukla@gmail.com
Anil K. Puniya
akpuniya@gmail.com

In the present world scenario, obesity has almost attained the level of a pandemic and is progressing at a rapid rate. This disease is the mother of all other metabolic disorders, which apart from placing an added financial burden on the concerned patient also has a negative impact on his/her well-being and health in the society. Among the various plausible factors for the development of obesity, the role of gut microbiota is very crucial. In general, the gut of an individual is inhabited by trillions of microbes that play a significant role in host energy homeostasis by their symbiotic interactions. Dysbiosis in gut microbiota causes disequilibrium in energy homeostasis that ultimately leads to obesity. Numerous mechanisms have been reported by which gut microbiota induces obesity in experimental models. However, which microbial community is directly linked to obesity is still unknown due to the complex nature of gut microbiota. Prebiotics and probiotics are the safer and effective dietary substances available, which can therapeutically alter the gut microbiota of the host. In this review, an effort was made to discuss the current mechanisms through which gut microbiota interacts with host energy metabolism in the context of obesity. Further, the therapeutic approaches (prebiotics/probiotics) that helped in positively altering the gut microbiota were discussed by taking experimental evidence from animal and human studies. In the closing statement, the challenges and future tasks within the field were discussed.

Keywords: gut microbiota, prebiotic, probiotics, obesity, nanotechnology

INTRODUCTION

Obesity is a pathological state marked by the accumulation of excess body mass in the abdominal region as a result of disequilibrium between energy intake and its consumption. It is a metabolic disorder that is on the rise globally and if allowed to spread unchecked would assume the proportions of a pandemic. Obesity is the mother of many other deadly diseases, particularly diabetes, cardiovascular, non-alcoholic fatty liver disease (NAFLD) and some form of cancers (Kopelman, 2007; Nikolopoulou and Kadoglou, 2012; Vucenik and Stains, 2012). Obesity not only affects the well-being of a person, but also places an unwanted economic burden on the society (Wang et al., 2011; Withrow and Alter, 2011). According to a report, more than 500 million people across the world are living with the stigma of obesity, that shows the severity of the disease and the challenges confronting health practitioners (Swinburn et al., 2011). Several factors such as host genetics, metabolism, lifestyle, and diet have been pinpointed as the key etiological agents responsible for the progression of obesity. However, the in-depth mechanisms that lead to the development of obesity are yet to be disclosed. The most recent studies have speculated that the gut microbiota present in the human gastrointestinal tract (GIT) have a paramount role in the onset and establishment of obesity. The adhered gut microbiota affects the host's nutrients acquisition and energy homeostasis by influencing the number of effector molecules that finally decide the fat storage in adipocytes (Rosenbaum et al., 2015). Nonetheless, there is growing evidence that some dietary substances, especially probiotics and prebiotics can modulate the gut microbiota of the host in a positive way and are therefore considered as important assets in the management of obesity. Various approaches such as omics methods, systems biology and metabolic engineering enable us to understand and optimize the metabolic processes (Yadav et al., 2016a,b). The major objectives of this review are to provide an overview of how prebiotics and probiotics modulate the gut microbiota in context of prevention or treatment of obesity. Before we progress further, we elaborate our current understanding of how gut microbiota are predisposed toward obesity.

RELATIONSHIP BETWEEN "GUT MICROBIOTA AND OBESITY"

Human Gut Microbiome, the "Unforeseen Organ"

It is believed that the gut of a fetus during the intrauterine period is deprived of any bacterial communities, i.e., it is nearly sterile; however, some microbes before birth and during parturition transit from the mother to the fetus gut and constitute the rudimentary microbiota (Aagaard et al., 2014). The gut composition of a child varies widely during the first few years of life due to factors like changes in gut physiology, introduction of solid foods, use of therapeutic drugs, host genotype and proximity to adult microbiota (Koenig et al., 2011). During adolescence, however, the gut microbiota is nearly consistent and predominated by a few colonizers. Thereafter, it changes during old age when the host physiology and dietary habits change dramatically. Nevertheless, the dynamics and structure of an individual's gut microbiota is unique and people can actually be identified on the basis of microbiota "fingerprints" alone, with the help of the metagenomics approach (Franzosa et al., 2015). The gut harbors a trillion microbes, thereby constituting a complex microbial community that is approximately comprised of 1000–1100 different bacterial species altogether representing 10^{14}–10^{15} microbes. This population is 10 times the number of cells present in a eukaryotic host (Qin et al., 2010) and resemble a "world within a world." The collective genes of these different microbial species are termed as "microbiome," while a combination of microbiome and host genes is called "metagenome" (Quigley, 2011). Before the advent of sophisticated sequencing techniques, the gut remained a neglected organ because of the limitations of culturing methods, but it is now considered to be a vital organ as it helps in various metabolic functions of the host that would otherwise not be possible (Sommer and Bäckhed, 2013). An earlier study inferred that the gut of an adult human being is mainly inhabited by bacteria from three major divisions, the *Firmicutes* (Gram-positive), *Bacteroidetes* (Gram-negative) and *Actinobacteria* (Gram-positive), which together make up more than 90% of total bacteria presented in the gut. In case of Archaea, one species *Methanobrevibacter smithii* predominates over others (Eckburg et al., 2005). However, obtaining an accurate picture of the gut is very difficult as several factors such as availability of oxygen, diet, and physiochemical properties of the gut (e.g., pH, bile) rapidly influence its composition.

Arumugam et al. (2011) made an attempt to understand the variation in species composition and gene pools within the human population from the previously available data and found the existence of three main distinct bacterial communities or "enterotypes" – *Bacteroides*, *Prevotella*, and *Ruminococcus*-based on their abundance (Arumugam et al., 2011). Later studies reduced the concept of three enterotypes to two – *Bacteroides* and *Prevotella* (Koren et al., 2013; Knights et al., 2014). From above studies, it can be inferred that gut microbiota have occupied a significant position in human biology that interplays with the metabolic physiology and influences the health status.

Evidence that Gut Microbiota Have a Role in Obesity and Dysbiosis

The pioneering evidence that linked gut microbiota to the development of obesity came from the findings of Bäckhed et al. (2004), when they transplanted the microbiota from normally grown mice to germ free (GF) mice. The latter, consequently, gained more fat pad mass and body weight despite reduction in food consumption. Increased body weight led to insulin resistance, along with higher glucose and leptin levels in blood. The authors postulated that the transplanted microbiota helped GF mice in harvesting excess energy from the diet. Further, they advocated that microbiota increases the expression of key transcriptional factors to enhance lipogenesis in the liver and promoted lipoprotein lipase (LPL) activity to store triglyceride (TG) in adipocytes (Bäckhed et al., 2004). Surprisingly, when GF mice were maintained on a high fat diet (HFD), they were protected from the development of obesity. Interesting evidence in this context emerged from the effect of antibiotic experiments on body weight. Antibiotics have been used in the livestock sector for decades to promote the growth and body weight of animals, which indirectly indicate that role of the gut microbiota in weight modulation. Evidence from mice has shown that early exposure to antibiotics had altered their gut microbiota, increased fat mass, and negatively modulated hepatic metabolism and associated hormones, which predisposed them toward adiposity (Cho et al., 2012; Cox et al., 2014). The effect of early administration of antibiotics on human adiposity has also been seriously reviewed over the past few years (Mueller et al., 2014; Turta and Rautava, 2016; Podolsky, 2017) and there is growing consensus that their increased use maybe a reason for the obesity explosion we are witnessing today.

If microbiota have a crucial role in the development of obesity, then it is obvious that the obese phenotype should have a microbial composition distinct from lean individuals. Ley et al. (2005) during the analysis of the gut microbiota from *ob/ob* mice, lean *ob/+* and wild-type counterparts, found that genetically obese mice have more of *Firmicutes* and less of *Bacteroidetes* compared with lean mice (Ley et al., 2005). These *Firmicutes* help the obese mice to draw more calories from the ingested diet, leading to obesity (Turnbaugh et al., 2006). Upon transplantation of microbiota from obese mice to GF mice, the obese phenotype is transferred. Similar findings were observed with obese people who had less of *Bacteroidetes* and more of *Firmicutes* in their gut. The proportion of *Bacteroidetes* increased with the initiation of a low calorie diet (Ley et al., 2006). In another study, obese children were found to have more of *Firmicutes* and less of *Bacteroidetes* in their gut. In fact, they also had higher short chain fatty acids (SCFAs) that were correlated with the development of obesity (Riva et al., 2017). Overall, obese people have less microbial diversity in comparison with lean ones (Le Chatelier et al., 2013) and dietary intervention may improve the microbial richness and associated clinical phenotypes (Cotillard et al., 2013).

Alterations in the gut microbial population also occurred at genus and species level, but these results were not consistent, especially in case of *lactobacilli*. In some findings, increase in the population of *lactobacilli* was observed in obese subjects and correlated with its pro-obesity effects (Armougom et al., 2009; Million et al., 2012b). In contrast, several studies have documented their anti-obesity effects as discussed elsewhere in a review (Arora et al., 2013). This mystery was resolved with the help of a meta-analysis study which depicted that anti-obesity activity of *lactobacilli* is species-specific attribute and is not a common feature of whole genera (Million et al., 2012a). Likewise, the population of *bifidobacteria* is negatively correlated with obesity, and its supplementation provided anti-obesity effects in some findings (Yin et al., 2010; An et al., 2011). In addition, *Faecalibacterium prausnitzii* and *Akkermansia muciniphila* were also found to be significantly linked with obesity. In general, *F. prausnitzii* found abundant in healthy adults and its supplementation in mice have colitis preventive effects (Miquel et al., 2013). However, there is inconsistency in *F. prausnitzii* population among obese human subjects. As in one case study their population was found to be increased in obese subjects (Balamurugan et al., 2010) while in a recent finding, opposite results were obtained (Dao et al., 2016). Whereas, Feng et al. (2014) in reported non-significant results in their findings. Similarly, *A. muciniphila* is negatively correlated with obesity (Schneeberger et al., 2015; Remely et al., 2016) and its administration has weight lowering effects (Everard et al., 2013; Dao et al., 2016).

The above findings clearly indicate that gut microbiota have a crucial role in the etiology of obesity and offer an opportunity to prevent or treat obesity by its therapeutic modulation. However, it is still a matter of debate to define which "indicator" microbial group is responsible for causing obesity as there are many contradictory findings with regard to the presence or absence of a particular microbiota in obesity. The discrepancies observed in the findings might be due to genetic background of host, age, sex, gut transit time, geographical location, and the diverse nature of gut microbiota. We believe that an in-depth study of gut microbiota at functional levels, i.e., metagenomics studies, along with focus on meta-transcriptomics and meta-proteomics, would provide an improved view of the picture by correlating the interlinked mechanisms. The outcomes will definitely help in understanding the known as well as unknown metabolic functions adhered by the gut microbiota of the host in leading to or preventing obesity.

Gut Microbiota Link with Obesity: Mechanistic Insight

Gut microbiota play several crucial roles in host physiology such as immune modulation, digestion of indigestible food materials, and production of vitamins, bile acids, bioactive compounds [conjugated linoleic acid (CLA), bacteriocins]. They are also known to be involved in the degradation of toxins, carcinogens, inhibition of enteric pathogens, and maintenance of intestinal epithelia, all of which the host cannot achieve alone (Cani et al., 2013). It is proved that dysbiosis (imbalance in microbial community due to pathological state) of gut microbiota leads to the progression of several diseases in human beings such as obesity, diabetes, NAFLD), certain form of cancers, and even anxiety and depression (Luna and Foster, 2015; Leung et al., 2016; Perez-Chanona and Trinchieri, 2016). Therefore, understanding the relationship between host physiology and gut microbiota would pave new therapeutic opportunities. In the next section, we will describe the various mechanisms by which gut microbes influence host physiology, metabolism and energy storage, thereby making it susceptible to obesity. Yet, the interplay of these mechanisms and how they affect the overall metabolic status of an individual is not fully understood.

Gut Microbiota in Energy Harvesting from Indigestible Food

As our digestive system is deprived of enzymes to digest higher polysaccharides such as cellulose, xylan and pectin, upon ingestion, they reach the distal gut where these are fermented by the action of microbiota lying there. Actual digestion depends upon the type of microbial composition. *Bacteroides* are the dominating anaerobes there, which digest these polysaccharides, and in this context the starch hydrolytic system of *Bacteroides thetaiotaomicron* has been studied extensively. The simple sugars released after the fermentation of complex polysaccharides were influxed into glycolysis to generate ATP (adenosine triphosphate). Further hydrolysis of these biological molecules, which are produced by different microbial fermentation pathways, lead to the generation of more ATPs and simple carbon molecules. Of the SCFAs, acetate, propionate and butyrate are the most important end products of gut-situated microbial species (Koh et al., 2016) and absorbed in the body by passive diffusion and the action of mono-carboxylic acid transporters (MCT). Nearly 10% of the daily energy requirement by the host colonic epithelial cells and more than 70% of energy for cellular respiration is obtained from SCFAs. Among SCFAs, butyrate is the most liked source of energy for colonic epithelial cells (Kasubuchi et al., 2015). Persistent acquisition of energy from SCFAs leads to extra fat deposition in the body, which leads to obesity. However, the human diet varies greatly in fiber composition and that significantly alters the SCFA production. Studies of obese animal models showed an increased presence of SCFAs in the fecal material and similar findings was observed in human subjects. A reduced butyrate level was recorded in the fecal material of obese human subjects, who received varied carbohydrate content as part of their diet. Besides, a significant reduction in the population of *Roseburia/Eubacterium* rectal was also observed, which signified the important role of this group in butyrate formation (Louis and Flint, 2009). However, there lies a controversy over this matter as production of SCFAs from indigestible material depend on several factors in the gut environment such as availability of substrate, mucosal absorption, transit time of food, and interactions between different gut microbial species (Duncan et al., 2007). In addition to their role in providing energy, SCFAs also reduce the pH of the gut, thereby altering the composition of microbiota. An increase in pH from 5.5 to 6.5 reduces the abundance of butyrate producers and simultaneously increases the population of propionate producers. At a slightly acidic pH (at 5.5), proportions of *Firmicutes* was found to be predominated that is responsible for butyrate production. Whereas at pH 6.5, the population was predominated by *B. thetaiotaomicron*, which produced propionate as fermentation product (Duncan et al., 2009). These findings suggest that a particular microbial group outclasses another group/species for carbohydrates' utilization at a specific luminal pH. However, these studies are confounding in nature and exact mechanisms are yet to be established.

Gut Microbiota Influence Fatty Acid Oxidation

Adenosine monophosphate kinase (AMPK), which is an important enzyme expressed mainly in the liver and skeletal muscles, plays a crucial role in cellular energy homeostasis. Drugs that increase the expression of AMPK lead to increase in fatty acid oxidation in liver and muscle tissues, incites energy loss, and disfavor obesity (Kim et al., 2017). Activation of AMPK eventually triggers carnitine palmitoyltransferase-1 (Cpt-1) *via* acyl-CoA carboxylase (Acc) activity, which in turn enhances mitochondrial fatty acid oxidation and inhibition of anabolic pathways such as glycogen storage and improved insulin sensitivity (Angin et al., 2016). Inhibition of AMPK by gut microbiota negatively influences fatty acid oxidation in target organs and tissues, promotes the synthesis of cholesterol and TG, and favor lipogenesis, which leads to excess fat storage and obesity (Boulangé et al., 2016). The fact was well understood by an experiment in which GF mice on a Western type diet had higher levels of phosphorylated AMPK, ACC and CPT-1 in the liver and skeletal muscles in comparison with conventionally raised mice.

These elevated levels result in increased fatty acid oxidation in target tissues (Bäckhed et al., 2007). From here, it is inferred that gut microbiota have a suppressive effect on AMPK activity, which in turn affect fatty acid oxidation and make the host susceptible to obesity.

Gut Microbiota Influences Fasting Induced Adipose Factor (FIAF)

Fasting induced adipose factor, also called Angiopoietin-like 4 protein (ANGPTL4), is produced by adipose tissue, liver, skeletal muscle and intestine in response to fasting. It is also a powerful metabolism and a adiposity regulator (Dutton and Trayhurn, 2008). It is the main site of action for Peroxisome proliferator-activated receptor proteins (PPARs). Its main role is the inhibition of LPL, which in turn restricts TG accumulation in adipocytes (Wang and Eckel, 2009). Bäckhed et al. (2004) found that when GF mice were transplanted with the distal gut microbiota of conventionally grown mice, a 60% increase in the epididymal body fat was determined. They proposed that the transferred gut microbiota suppressed the FIAF expression in intestinal epithelium that in turn caused enhanced fatty acid uptake by adipocytes *via* increased LPL activity (Bäckhed et al., 2004). Further, the same group reported that GF *Fiaf*−/− mice were not protected from the development of obesity in comparison with their normal GF littermates fed on the same HFD. They concluded that the gut microbiota in wild-type GF mice suppressed the expression of *FIAF*, thereby increasing LPL activity and fat storage in adipocytes. In addition, the authors highlighted that Fiaf might modulate fatty acid oxidation in gastrocnemius muscle by means of controlling the expression of peroxisomal proliferator activated receptor co-activator 1α, which (Pgc1α) is accountable for coactivating every recognized nuclear receptors as well as many other transcription factors involved in mitochondrial fatty acid oxidation, including Cpt1 and medium-chain acyl-CoA dehydrogenase (Bäckhed et al., 2007). Thus, gut microbiota induces obesity with the help of the above-explained mechanisms. However, there lies a piece of evidence, which suggests gut microbiota are not able to provide resistance against obesity development or modulation in circulation of Fiaf/Angptl 4 levels. When GF mice and conventional mice were raised on HFD and Western type diet, then more weight gain was observed in GF mice on both the diets in comparison with their conventional littermates. The important thing was that this weight gain in GF was associated with increased intestinal mRNA levels of fasting-induced Fiaf/Angptl4, but not with circulating Fiaf/Angptl4. The population of gut microbiota was also found changed among conventional mice fed on HFD and wild-type diets. Thus, the study found that diet modulates the type of gut microbiota, and intestinal Fiaf/Angptl 4 does not have a crucial role in adipocytes' fat storage as suggested by others (Fleissner et al., 2010). Therefore, the matter concerning the gut microbiota influence on *Fiaf* levels in obesity is still open for debate.

Gut Microbiota Influences Bile Acids

Bile acids are significant physiological molecules that facilitate digestion and absorption of fats in the small intestine and aid in the removal of lipids and toxic metabolites in the feces. Cholic acid (CA) and chenodeoxycholic acid (CDCA) are the main primary bile acids synthesized in the liver from cholesterol and are conjugated with taurine or glycine to form bile salts prior to secretion in bile. After their secretion into the intestinal lumen, these are converted into secondary bile acids deoxycholic acid and lithocholic acid by the dehydroxylation activity of bacteria. Subsequently, these bile acids are reabsorbed from ileum via ileal bile acid transporter (IBAT) through active transport and passive diffusion into the upper small intestine and colon. They are then transported back to the liver *via* blood circulation for re-secretion and feedback inhibition of bile acid synthesis in a process known as enterohepatic circulation. In this way, the bile acids affect intestinal absorption of fats, lipogenesis and ultimately metabolic homeostasis. Swann et al. (2011) demonstrated that mice having a distinct microbial structure in the gut possess different bile acid metabolites in their organs and hence have a divergent energy metabolism (Swann et al., 2011). Although, the underlying molecular mechanism of bile acid feedback inhibition is still not clear, but it has been suggested that nuclear receptor farnesoid X receptor (FXR) plays an important role in this regulation. FXR negatively regulates the expression of two key genes, namely, cholesterol 7a-hydroxylase (*CYP7A1*) and *CYP27A1*. *CYP7A1* is required for the initiation of classic pathways of bile synthesis while *CYP27A1* is required for the alternative pathway (Chiang, 2009). Recent studies have shown that intestinal FXR regulates hepatic CYP7A1 with the help of a fibroblast growth factor 15 (FGF15)-dependent mechanism (Zimmer et al., 2012). Sayin et al. (2013) in their re-derivation study of *FXR−/−* mice to GF showed that gut microbiota regulate expression of *FGF15* and *CYP7A1* by FXR-dependent mechanisms. The outcomes from this study suggest that the gut microbiota inhibits bile acid synthesis in the liver by alleviating the levels of FXR in the ileum (Sayin et al., 2013). Another mechanism by which bile acids regulate energy metabolism is by activating the G-protein-coupled bile acid receptor 1 (GPBAR1) or TGR5. This protein gets activated by interacting with secondary bile acids, as ligands, present in the intestinal lumen, thereby aiding in glucose homeostasis by activating secretion of glucagon-like peptide 1 (GLP1; Aron-Wisnewsky et al., 2013). Thus, in this manner, gut microbiota modulate bile acid metabolism by influencing FXR/TGR5 signaling and indirectly contributing toward the development of obesity. In addition, it is well known that bile acids exert an antimicrobial effect on gut microbiota by damaging the cell membrane integrity and thus its pool size and composition are considered as significant factors in the gut microbial community structure regulation. Composite and important alterations in the microbiome structure of animals were noticed when they were administered with bile acids (Ridlon et al., 2014). From these studies, it can be inferred that decrease in the levels of bile acids in the gut favors the population of gram-negative members, including some important pathogens. Conversely, an increase in bile acid amounts in the gut seem to promote gram-positive members of the *Firmicutes*, which include those bacteria that convert host primary bile acids to toxic secondary bile acids by 7α-dehydroxylation (Ridlon et al., 2014).

Gut Microbiota Influences Satiety

Apart from the role of SCFAs as substrate in energy metabolism, they also function as ligands for some receptors. Of those receptors, G-protein-coupled receptors; GPR41 (now called as FFAR3) and GPR43 (now called as FFAR2) are important target receptors. FFAR3 is expressed by the host immune cells, adipose tissue, spleen, bone marrow, large intestine, liver, and skeletal muscle (Le Poul et al., 2003; Regard et al., 2008). FFAR3 is mainly triggered by the presence of propionate, followed by butyrate and acetate, whereas FFAR2 is stimulated by all three SCFAs at the same rate (Brown et al., 2003). Notably, the presence of these receptors in different peripheral tissues clearly indicates that these SCFAs can directly influence several different functions such as satiety and host metabolism. One of the underlying mechanisms by which SCFAs regulate food intake, and satiety are *via* modulation of intestinal enteroendocrine L cells derived peptides, mainly GLP1 and peptide YY (PYY). These cells are found in abundance in the ileum and colon (De Silva and Bloom, 2012). The function of PYY is to reduce appetite by acting upon neuropeptide Y (NPY), thereby inhibiting gastric motility and reducing food intake (Karra et al., 2009). Likewise, the functions of GLP1, an incretin, are to regulate appetite, inhibit gastric emptying, and at the same time stimulate insulin secretion (Steinert et al., 2016). Nøhr et al. (2013) demonstrated that SCFAs activate GLP1 and PYY *via* stimulation of FFAR3 and FFAR2 present on L cells. These findings let us postulate that SCFAs produced from dietary polysaccharides, as a result of gut microbial fermentation, have direct influence on L cells, which in turn results in the rise of intestinal and plasma GLP 1 level. It is well documented in animal and human studies that ingestion of indigestible polysaccharides upregulates total GLP1 and PYY levels through SCFAs (Zhou et al., 2008; Tarini and Wolever, 2010). Tolhurst et al. (2012) reported that *FFAR2* or *FFAR3* knockout mice had reduced levels of GLP-1 and impaired glucose tolerance *in vitro* and *in vivo* at the same time due to lack of interaction with SCFA ligands. In a different gene knockout study, the authors revealed that mice lacking *FFAR2* gene became obese even after receiving a normal diet, while mice overexpressing *FFAR2* in adipose tissue stayed lean even after receiving a HFD. In addition, FFAR2 also suppresses insulin-mediated fat accumulation, which in turn regulates the energy balance by inhibiting the deposition of excess energy and inducing fat consumption (Kimura et al., 2013). Another mechanism, by which gut microbiota modulate energy homeostasis *via* SCFAs is their effect on leptin secretion from adipocytes through GPR41/43 dependent process. Thus, SCFAs and GPR41/43 interplay the role of significant messengers amidst gut microbiota and host metabolism (Xiong et al., 2004; Zaibi et al., 2010).

Gut Microbiota Influences Lipogenesis

The first experimental evidence that demonstrated that gut microbiota promote *de novo* hepatic lipogenesis came from the study of Bäckhed et al. (2004) on GF mice. In their pioneering research, the authors observed that transplantation of gut microbiota from normally raised mice to GF mice helps in inducing excess body fat storage and insulin resistance within the first 2 weeks despite reduced food intake. In subsequent years, another group studied the influence of gut microbiota on energy and lipid metabolism of host by comparing the serum metabolome and the lipidomes of serum, adipose tissue, and liver of conventionally raised and GF mice with the help of the MS-based metabolomics approach. Conventionally raised mice had an increased number of energy metabolites (e.g., pyruvic acid and citric acid) in their serum while, the levels of cholesterol and fatty acids were reduced. Moreover, they found that microbiota altered a number of lipid species in serum, adipose tissue, and liver, with the effect, mainly visible on TG and phosphatidylcholine species (Velagapudi et al., 2010). Enhanced TG synthesis observed was associated with an increase in the expression of the lipogenic genes, mainly acetyl-CoA carboxylase (Acc1) and fatty acid synthase (Fas). Both Acc1 and Fas are transcriptional sites of two key transcription factors, sterol response element binding protein 1c (SREBP-1c) and carbohydrate response element binding protein (ChREBP), required for lipogenesis in liver in response to insulin and glucose (Bäckhed et al., 2004). In conventionally raised mice, a significant enhancement in the levels of ChREBP was found in the liver as well as in the nucleus after its nuclear translocation, followed by its dephosphorylation by PP2A. Noticeably, PP2A was successively activated by xylulose-5-phosphate (Xu5P), an intermediate in the hexose monophosphate shunt. Conventionally raised mice reported to have higher levels of liver Xu5P compared with their GF littermates, suggesting that enhanced levels of this hexose monophosphate shunt intermediate further promote the liver ChREBP levels and consequently, liver lipogenesis.

These findings suggest that with an increase in fermentation of dietary polysaccharides, with the help of microbes, in conventionally raised mice, there is an increased supply of monosaccharides to the liver, which subsequently increases the activation of lipogenic enzymes by ChREBP and perhaps SREBP-1. The liver has two ways to tackle this increased influx of calories: it either increases the inefficient metabolism (futile cycles) or stores these surplus calories as fat in peripheral tissues (Bäckhed et al., 2004). Further, another research group demonstrated that gut microbiota induces *de novo* lipogenesis and TG synthesis in HcpG2 cells by production of t10,c12 CLA. They found that treating cells with t10,c12 CLA increased lipid deposition *via* increased incorporation of acetate, palmitate, oleate, and 2-deoxyglucose into TG. CLA treatment also led to upregulate the mRNA expression as well as protein levels of lipogenic genes, including *SREBP1*, *ACC1*, *FASN*, *ELOVL6*, *GPAT1*, and *DGAT1*, thereby presenting a potential mechanism by which CLA increased lipid accumulation. Most importantly, CLA treatment also increased the phosphorylation of mTOR, S6K, and S6. Together, the authors concluded that t10,c12 CLA production by gut microbiota induces liver *de novo* lipogenesis and TG synthesis is linked with the activation of the mTOR/SREBP1 pathway that consequently, leads to increased lipid incorporation in HepG2 cells (Go et al., 2013).

Gut Microbiota and Innate Immunity

Toll-like receptors (TLRs) are groups of proteins that play an important role in the innate immune system. They are membrane-spanning, non-catalytic receptors normally expressed on sentinel cells that recognize structurally conserved motifs of microbes called pathogen-associated molecular patterns (PAMPS) (Medzhitov, 2001). The interaction of these PAMPS with host TLRs induces several antimicrobial immune responses through the activation of inflammatory signaling pathways that are necessary for the effective immune response. Therefore, there is no doubt about the fact that the microbiota we harbor in our gut, and which interacts with epithelium TLRs at the luminal interface, is vital for maintaining the immune homeostasis (Peterson et al., 2015). Of the various PAMPS of bacteria, TLR5 mainly detects bacterial flagellin from invading bacteria and are found highly expressed in the intestinal mucosa. Vijay-Kumar et al. (2010) elucidated the role of TLR5 receptor in adiposity progression and associated metabolic syndrome. They found that TLR5 deficient mice (TLR5KO) exhibited many features of metabolic syndrome such as hyperphagia, hyperlipidemia, hypertension, hypercholesterolemia, high blood pressure, insulin resistance, and enhanced fat deposition in comparison with normal counterparts. They demonstrated that these changes were associated with an increase in adipocytes secretion of proinflammatory cytokines IL-1β and INF-γ (Vijay-Kumar et al., 2010). Next, the authors examined whether changes in the gut microbiota, resulting from loss of TLR5, helped in the development of metabolic syndrome. In order to do so, they placed TLR5KO mice and wild-type littermates on antibiotics and found that destruction of gut microbiota in TLR5KO mice ameliorated metabolic syndrome similar to wild-type mice. UniFrac analysis showed that the gut microbiota composition of TLR5KO, and wild-type littermate mice was remarkably different. Besides marked inter-individual differences in species diversity, they observed 116 bacterial phylotypes from various phyla to be consistently enriched or reduced in TLR5KO mice in comparison to wild-type mice (Vijay-Kumar et al., 2010). To further assess whether alteration in the gut microbiota was a factor responsible for the development of metabolic syndrome in TLR5KO mice, they transplanted the microbiota from TLR5KO mice to wild-type, GF mice. They found that the transplanted microbiota conferred many phenotypic effects of TLR5KO to wild-type mice. The authors concluded that the gut microbiota play a crucial role in the development of metabolic diseases and opined that dysfunction of the innate immune system may be one factor that favor their development. However, there is one study in which TLR5KO mice from two different animal colonies, neither exhibited evidence of metabolic abnormalities nor showed enhanced basal intestinal inflammation (Letran et al., 2011). Therefore, the authors concluded that basal inflammatory phenotype is not a consistent feature of TLR5-deficient mice.

Gut Microbiota, Metabolic Endotoxinemia and the Endocannabinoid System

The progression of obesity is associated with the activation of low grade inflammatory signaling molecules from adipose tissue such as TNF-α, IL-1, IL-6, which disrupt normal metabolic processes and mediate insulin resistance (Hotamisligil, 2006; Ouchi et al., 2011). The adverse effects of insulin resistance lead to hyperinsulinemia, and excessive hepatic and adipose tissue storage of fat. For a long time, however, the inflammation triggering molecules in HFD-induced obesity remained unknown and it was Cani et al. (2007a) who first proposed that a Gram-negative bacterial outer membrane component known as lipopolysaccharide (LPS) was responsible for early onset of inflammation, insulin resistance, obesity and diabetes (Cani et al., 2007a). The authors found that supplementation of HFD in mice for 4 weeks chronically increased plasma LPS levels 2- to 3-fold than those of control animals and called it "metabolic endotoxemia." Notably, increased LPS levels in the HFD group were associated with a decreased abundance of *Bacteroides*, *Eubacterium rectale-Clostridium coccoides* group and *Bifidobacterium* species. In a subsequent set of experiments, the authors subcutaneously infused LPS in GF mice for 4 weeks and found that changes in body weight, metabolic physiology, and endotoxemia were similar to the ones earlier seen with HFD. However, the effect of LPS-induced metabolic changes was diminished when the mice were made deficient in the genes *cd14* and *tlr4* (Cani et al., 2007a; Davis et al., 2008; Vijay-Kumar et al., 2010). This signifies that LPS induces systemic inflammation *via* these markers. Next, to assess whether modulating the gut microbiota could control the occurrence of metabolic endotoxemia and the resultant metabolic diseases, the authors made use of antibiotics on intestinal microbiota of HFD and genetically obese (*ob/ob*) mice. As a result, a decrease in inflammation, obesity-related bio-markers and endotoxemia levels were observed. Noticeably, high fat feeding also increased intestinal permeability and simultaneously reduced the expression of genes coding for two tight junction proteins ZO-1 and occludin (Cani et al., 2008). HFD dramatically decreased the population of *Lactobacillus* spp., *Bifidobacterium* spp., and *Bacteroides–Prevotella* spp. Interestingly, feeding of *bifidobacteria* reversed metabolic endotoxemia, and improved gut integrity and associated metabolic changes in mice (Wang et al., 2006; Cani et al., 2007c). However, no relationship was found between endotoxemia and other bacterial groups *E. rectale–C. coccoides, lactobacilli–enterococci, Bacteroides*, and sulfate-reducing bacteria (Cani et al., 2007c). Until this point, no information was available concerning molecular mechanism that linked how modulation in gut microbiota improved metabolic endotoxemia, tight junction integrity, obesity-related hepatic and metabolic disorders. Therefore, to decipher the underlying mechanism, Cani et al. (2009b) performed three different sets of experiments on genetically obese mice (*ob/ob*) using different strategies. In the end, they found that selective modulation of gut microbiota by probiotic supplementation regulates and enhances the endogenous production of intestinotrophic GLP-2, which in turn improves gut barrier integrity and functions by way of a GLP-2-dependent mechanism during obesity and diabetes (Cani et al., 2009b). In addition, they advocated the role of the endocannabinoid (eCB) system in gut barrier integrity and obesity. The eCB system consists of eCBs, their receptors, and enzymes that synthesize and degrade eCBs (Mackie, 2008). Cannabinoid receptor type 1 (CB1) and type 2 (CB2) are two important G-protein-coupled cannabinoid receptors activated by the eCB system. Two eCBs, namely anandamide

and 2-Arachidonoylglycerol (2-AG), play a significant role in adipogenesis by activating their receptors. Anandamide activates CB1 while 2-AG activates both cannabinoid receptors. The eCB system was found hyperactive (greater system tone) in case of obesity and type 2 diabetes. It has been seen in several studies that there is a close connection between LPS and the eCB system. In fact, some *in vitro* and *in vivo* studies reflect that LPS regulates the synthesis of eCBs *via* LPS receptor-mediated signaling pathways (Muccioli et al., 2010). But the influence of gut microbiota on eCB signaling was yet to be understood. Muccioli et al. (2010) found that gut microbiota modulate the intestinal eCB system tone, which regulates gut permeability and plasma LPS levels. Besides, they also showed that LPS plays a central role in adipose tissue metabolism both under *in vivo* and *in vitro* by blocking cannabinoid-driven adipogenesis. From their study, it could be figured that gut microbiota regulate adipogenesis through LPS–eCB system loop (Muccioli et al., 2010).

In subsequent studies, the same research group tried to investigate the effect of eCB, LPS, and the gut microbiota in the regulation of apelin and APJ expression in adipose tissue (Geurts et al., 2011). Apelin and APJ are found widely expressed in mammalian tissues and deploy their functional effects both in the central nervous system and in the peripheral nervous system. Apelin is the endogenous ligand for the G-protein-coupled receptor known as APJ receptor. Apelin was found to play a significant role in the cardiovascular system by acting on heart contractility, blood pressure, fluid homeostasis, vessel formation, and cell proliferation (Maenhaut and Van de Voorde, 2011). Interestingly, apelin also acted on glucose homeostasis via AMP-kinase- and nitric oxide (NO)-dependent mechanisms (Dray et al., 2008). At the end of the study, the investigators came up with the inferences that apelin and APJ expressions were suppressed by the eCB system in physiological conditions and increased by LPS in pathological situations such as obesity and diabetes (Geurts et al., 2011).

Thus, it seems that several factors play important roles in the regulation of gut permeability and adiposity, among which the role of LPS is visualized as central to all these mechanisms. All the above proposed mechanisms are represented in a pictorial manner in **Figure 1**.

MODULATION OF GUT MICROBIOTA BY DIETARY APPROACHES FOR THERAPEUTIC EFFECTS

From the aforementioned studies, it can be inferred that gut microbiota plays a central role in host physiology in obesity. Therefore, it is quite feasible to hypothesize that its positive modulation by external approaches may provide beneficial effects to the host. Out of the available intervention approaches (diet, antibiotics, surgery), dietary strategy is much preferred by medical practitioners due to associated lesser cost and safety issues. Future therapeutic strategies can be formulated by understanding which dietary substance has a positive modulatory effect. Probiotics and prebiotics are promising because of their

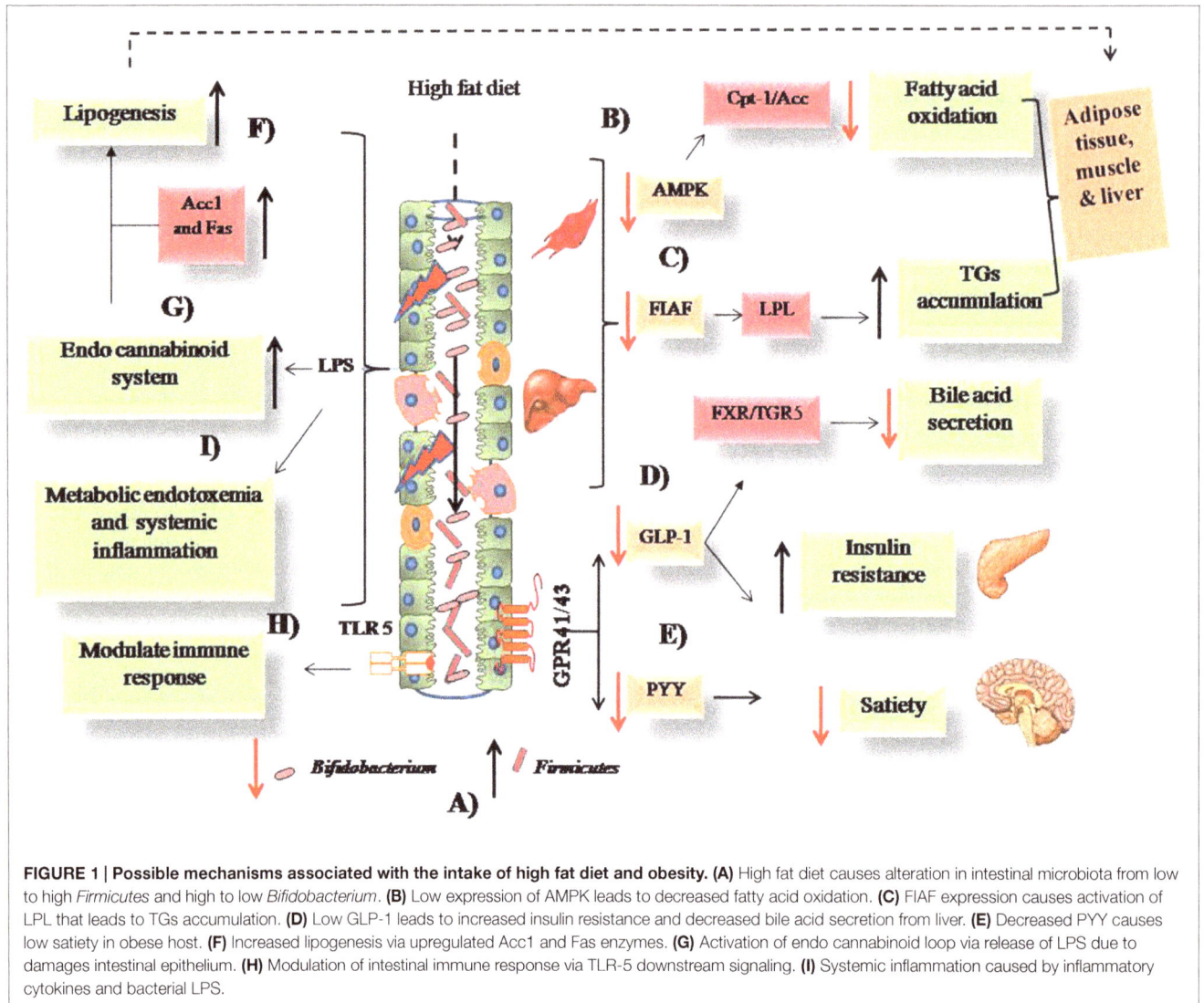

FIGURE 1 | Possible mechanisms associated with the intake of high fat diet and obesity. (A) High fat diet causes alteration in intestinal microbiota from low to high *Firmicutes* and high to low *Bifidobacterium*. **(B)** Low expression of AMPK leads to decreased fatty acid oxidation. **(C)** FIAF expression causes activation of LPL that leads to TGs accumulation. **(D)** Low GLP-1 leads to increased insulin resistance and decreased bile acid secretion from liver. **(E)** Decreased PYY causes low satiety in obese host. **(F)** Increased lipogenesis via upregulated Acc1 and Fas enzymes. **(G)** Activation of endo cannabinoid loop via release of LPS due to damages intestinal epithelium. **(H)** Modulation of intestinal immune response via TLR-5 downstream signaling. **(I)** Systemic inflammation caused by inflammatory cytokines and bacterial LPS.

direct influence on the gut microbiota. In the coming sections, we have described the effect of prebiotics and probiotics on the gut microbiota and their outcomes in experimental settings (animal and human). However, in the past few decades, fecal transplantation of the gut microbiota has also gained momentum, but this practice is only limited to some countries or to certain research laboratories/institutions and are not discussed here in this review.

Prebiotics in Modulation of Gut Microbiota in Context to Obesity
Evidence from Animal Studies

Prebiotics are the indigestible dietary polysaccharides that promote the growth of inherited gut microbes or probiotics when supplied externally. The most commonly used prebiotics in practice are fructooligosaccharides, galactooligosaccharides, lactulose, and non-digestible carbohydrates inulin, cellulose, resistant starches, hemicelluloses, gums, and pectins because they fulfill the criterion as suggested by Gibson et al. (2004).

The science of using prebiotics for therapeutic applications is not new as they were used by our elders to assist in restoring people back to health when diseases struck. But the science picked up pace during the past few decades, when Cani et al. (2004) found that inulin type dietary fructans (ITF) [oligofructose (OFS) and Synergy 1] have the potential to increase intestinal proglucagon and GLP-1 levels, and simultaneously decrease the expression of ghrelin in the treated Wistar rat than the control. These gut hormones are critically involved in the regulation of appetite and body weight in human and animal models (Cani et al., 2004). A similar hypothesis was tested among HFD fed Wistar rats administered with OFS as prebiotic. Consequently, feeding of OFS provides obesity ameliorating effect in rat due to modulation in the expression of gut situated peptides as described in a previous finding except for GLP-2. However, the exact mechanism how these prebiotic fibers made changes in secretion of orexigenic- and anorexigenic peptides, and thereby changes in the energy homeostasis was not elusive, but proposed due to activity of SCFAs that promoted the production of these peptides from endocrine L cells (Cani et al., 2004, 2007b). Later on, the concept incepted that these prebiotic fibers modulate the microbial community upon ingestion in gut in particular of *bifidobacteria* and *lactobacilli* (Everard et al., 2011; Neyrinck et al., 2011; Gérard, 2016). In a meta-analysis, review concerning the modulation of gut microbiota by prebiotics and probiotics to counter obesity, the authors found that in most of the studies, *bifidobacteria* plays a central role in ameliorating obesity by promoting its growth in presence of prebiotics (da Silva et al., 2013). However, there is a study which has shown that the stimulating effect of prebiotics is not only restricted to *bifidobacteria, lactobacilli,* and *F. prausnitzii,* but also influences other bacterial taxa that play an important role in obesity (Respondek et al., 2013; Everard et al., 2014). Notably, more often, this alteration in gut microbiota by prebiotic induction provides obesity reducing effects by indirectly acting on various pathological sites responsible for the development of obesity.

Cani et al. (2006b, 2009b) found that feeding OFS to HFD mice led to a considerable increase in the bifidobacterial count, which in turn decreased the inflammatory markers by way of reduced LPS production. Decreased LPS production improves gut permeability and reduces adiposity. Later on it was elucidated that low metabolic endotoxemia resulted because of the bifidogenic effect of prebiotic. As these fibers increase the expression of gut hormones GLP-1 and GLP-2 from L cells and also modulate the eCB system; these modulations in-turn alleviate inflammation and insulin resistance in mice (Cani et al., 2006b, 2009b; Muccioli et al., 2010). In addition to *Bifidobacteriaceae,* the impact of prebiotic feeding on other gut microbiota was also revealed with the help of molecular biology approaches. Prebiotic feeding in genetically obese mice led to a decrease in *Firmicutes,* while registering an increase in *Bacteroidetes.* Change in proportion of more than 100 distinct taxa was also revealed, out of which 16 taxa displayed more than a 10-fold change. This led to the identification of *A. muciniphila,* whose population in the gut is negatively correlated with obesity (Everard et al., 2011).They hypothesized that *A. muciniphila* has a positive role in obesity, that was validated by a recent finding wherein feeding of bacteria to dietary HFD mice provided alleviation of pathophysiological parameters and reduction in body weight (Schneeberger et al., 2015). In addition to modulation of gut microbiota, prebiotic feeding also increases the number of L cells and positively modulates the various parameters (GLP-1, fat mass development, oxidative stress, etc.) responsible for the development of metabolic syndromes. The researchers unraveled a new mechanism linking gut microbiota-mediated change in metabolism of genetically obese mice in which feeding of prebiotics had improved leptin sensitivity (Everard et al., 2011). Subsequently, the mechanism by which *A. muciniphila* plays an important role in the amelioration of obesity, and related disorders was elucidated by Everard et al. (2013). Prebiotic feeding stimulates the growth of *A. muciniphila* that concomitantly increases the intestinal levels of eCB, which regulates inflammation, gut permeability, and anorexigenic peptide. However, only viable cells of *A. muciniphila* can address these effects.

Evidence from Human Studies

If we talk about the impact of prebiotic (inulin type) supplementation on healthy human physiology, then they have been reported to induce satiety, increase breath-hydrogen excretion, modulate gut peptides involved in appetite regulation (Cani et al., 2006a, 2009a; Parnell and Reimer, 2009), and prompted the growth of *bifidobacteria* and *lactobacilli* (Gibson et al., 2004). Whether these prebiotic (inulin) stimulated the growth of whole *bifidobacteria* genus or a particular species or other members of human gut microbiota was unknown. Ramirez-Farias et al. (2009) found that inulin ingestion specifically stimulated the growth of *B. adolescentis* among other analyzed species. Besides, *F. prausnitzii* was found as bacterial species other than lactic acid bacteria that was stimulated because of inulin ingestion. However, the study is not elusive because of the involvement of only a few volunteers in the study. In a similar finding, Joossens et al. (2011) reported that ingestion of OFS-enriched inulin to 17 human volunteers led to the significant increase in *B. longum* and *B. adolescentis* species. A later prebiotic intervention study in obese women provided an insight of the effect of this treatment on the gut microbiota. Inulin type prebiotics promoted growth of *Firmicutes* and *Actinobacteria,* and inhibition of *Bacteroidetes.* A deeper analysis revealed that there was an increase in the population of *Bifidobacterium* and *F. prausnitzii,* while a decrease was noticed in *Bacteroides intestinalis* and *B. vulgatus,* after prebiotic treatment. Despite that increase in the population of *lactobacilli* was also observed after prebiotic treatment. From the correlation analysis between prebiotic treatment and host metabolism, it could be speculated that *Bifidobacterium* and *F. prausnitzii* were negatively correlated with serum LPS levels, while changes in *B. intestinalis* and *B. vulgatus* and *Propionibacterium* were positively correlated with changes in body composition and glucose homeostasis (Dewulf et al., 2012). In conclusion, the authors suggested that treatment with ITF prebiotics alleviated host obesity related mechanism *via* selective modulation in the gut microbial signatories of obese women. In a subsequent study, the investigator tries to establish a correlation between *Bifidobacterium* species, SCFAs, and key metabolic markers of host physiology. Ingestion of ITF by obese women led to an increase in the population of *B. longum, B. pseudocatenulatum,* and *B. adolescentis.* Modulation in numbers of *B. bifidum* and *B. adolescentis* was inversely linked with fat mass percentage, while *B. breve* was negatively correlated with serum cholesterols. Strikingly, *B. longum* was negatively linked to serum LPS. The levels of SCFAs (acetate and propionate) were also found to be low in treatment groups compared with control ones. In summary, the authors affirmed that ingestion of ITF prebiotics in obese women led to an increase in the population of *Bifidobacterium* species and a decrease in the production of SCFAs, which ultimately reduce the host metabolic parameters associated with obesity (Salazar et al., 2015).

However, it is predicted that instead of SCFA other metabolites (bile acids, choline, vitamins, polyamines, and lipids) produced by gut microbiota under influence of prebiotics also have a significant role in the host physiology. It is reflected from a finding wherein authors fed a HFD and prebiotic rich diet (ITF or arabinoxylans) to mice and found an increase in the rumenic acid (cis-9, trans-11-18 : 2 CLA) content in both the caecal and liver tissues compared with the control group. Of the two prebiotics tested, only arabinoxylans were able to increase the rumenic acid content because their prebiotic fibers might have provided high fat-binding capacity which provides more substrates for bacterial metabolism to differentially modulate the gut microbiota. Rumenic acid is produced from linoleic acid by gut microbes by their biohydrogenation activity during a detoxifying mechanism. A similar effect was also observed with gut isolated microbes when they were subjected to substrate linoleic acid during in vitro studies. In conclusion, the authors suggested that the CLA-producing bacteria could be a responsible for addressing the metabolic effects in both HFD feeding and prebiotic supplementation (Druart et al., 2013).

Altogether, prebiotics manage obesity by lowering the production of LPS by modulating the gut microbiota that ultimately hinders the process of low grade inflammation and modulates the eCB system. They also reported to induce satiety via promotion of satiety peptides from L cells in the gut.

Probiotics in Modulation of Gut Microbiota in Context to Obesity
Evidence from Animal Studies

Apart from prebiotics, there lies another alternative dietary approach in which probiotics are used to modulate gut microbiota. This method led to a rise in anti-obesity effects across animal and human studies. Probiotics are the live microorganisms which, when fed in adequate amount, confer health promoting effects on the host (Sanders, 2008). Members of lactic acid bacteria, namely Lactobacillus spp. and Bifidobacterium spp. are the two extensively studied probiotics that have provided anti-obesity effects in animal models and human beings (**Tables 1, 2**). However, these days only those strains that pass the prescribed probiotic and functional tests are used for animal and human use (Dahiya and Puniya, 2015). The proposed mechanism of action includes alteration in the gut microbial community, production of bioactive compounds by probiotic strains, reduction in fat storage, alterations in serum lipid profiles, induction in fatty acid oxidation genes, interaction of probiotics with host TLRs, reduced expression of pro-inflammatory cytokines, and stimulating the production of satiety-inducing peptides (Stanton et al., 2005; Tsai et al., 2014; Villena and Kitazawa, 2014; Dahiya and Puniya, 2017).

In most of the accomplished in vivo studies, gut microbiota was not studied, although modulation of gut microbiota by probiotic feeding presented an interesting therapeutic approach. Yadav et al. (2013) demonstrated that feeding of probiotic VSL#3 consortiums attenuate obesity and diabetes in mouse models via modulation of the gut flora. Deeper investigation revealed that VSL#3 stimulated the production of GLP-1 via butyrate production from altered gut microbiota, which addressed reduced food intake, improved glucose tolerance, and reduced adiposity. In another study, oral feeding of L. curvatus HY7601 and L. plantarum KY1032 to HFD mice significantly shifted the microbial communities, which ultimately reduced obesity in mice. The comparative abundance of four species belonging to the Ruminococcaceae and Lachnospiraceae families of the order Clostridiales and phylum Firmicutes decreased in the high fat control group and increased among the probiotics-administered mice. This microbial shift was accompanied with anti-obesity effects in mice that were probably due to induced positive influence on the expression of inflammatory and lipid oxidation markers situated in the liver and adipose tissue.

Murphy et al. (2013) demonstrated that feeding bacteriocin producing probiotic L. salivarius UCC118Bac+ to mice had the potential to alter their gut microbiota. The feeding of this strain to mice results in a relative increase in Bacteroidetes and Proteobacteria, decrease in Actinobacteria, but no effect on Firmicutes in comparison with non-bacteriocin producing strain. However, this strain was unable to address any change in the metabolic physiology of mice. In their subsequent investigation, the same group showed interest in elucidating the time dependent effect of feeding the L. salivarius UCC118Bac+ and a shift in the gut microbial composition. Initial treatment resulted in a significant increase in amount of Peptococcaceae and decrease amount of Rikenellaceae and Porphyromonadaceae in comparison with the gut microbiota of control mice. The findings highlighted the ability of gut microbiota to recover its shape after a period of time and require long term probiotic treatment to undergo sustained modification (Clarke et al., 2013).

Toral et al. (2014) showed that administration of L. coryniformis CECT5711 reduces gut dysbiosis that improves metabolic endotoxemia by lowering LPS levels and improving gut permeability, which thereby improves obesity in mice. Another study found that feeding of probiotic dahi, which contains L. casei NCDC 19, led to a reduction in epididymal fat weights, blood glucose, plasma lipids, leptin levels, and body weight among HFD mice (Rather et al., 2014). These observed effects were correlated with an increase in the population of bifidobacteria. Kim and co-workers found that administrating L. brevis OK56 to HFD mice abrogated the adverse effect of diet on gut microbiota. Despite the increase in population of bifidobacteria, OK56 supplementation suppressed colonic and plasmatic LPS and decreased production of H_2 breath gas. The authors suggested that the anti-obesity effect exerted by OK56 was due to inhibition of LPS production by modulation of gut microbiota and suppression of other inflammatory pathways (Kim et al., 2015). Similar results were observed by Lim et al. (2016) who found that feeding L. sakei OK67 to HFD mice helped in ameliorating obesity by reducing production of LPS, which was possibly due to modulation of gut microbiota. They also opined that probiotic feeding induces the expression of tight junction proteins, which are responsible for maintaining gut integrity. In a recent finding, the authors found that feeding diabetic rats L. rhamnosus NCDC17 increases the population of bifidobacteria and lactobacilli in the cecum, although it also resulted in attenuation of other biomarkers responsible for development of obesity (Singh et al., 2016). Similar findings were also conducted by others. Alard et al. (2016) showed that adiposity dampens the effect of probiotics, which are linked to the improvement of dysbiotic

TABLE 1 | Effects of probiotics on gut microbiota of animals and their physiological outcomes.

Probiotics	Animal model	Influence on gut microbiota	Metabolic outcome	Reference
VSL#3	C57J/B67 HFD and ob/ob mice	↓ Firmicutes, ↑Bacteroidetes, and bifidobacteria. * NC on lactobacilli	↓ Body weight (BW), food intake and adiposity, ↑ insulin sensitivity, glucose tolerance, production of GLP-1 and SCFA butyrate	Yadav et al., 2013
L. curvatus HY7601 and L. Plantarum KY1034	C57BL/6J HFD mice	NC in Firmicutes and Bacteroidetes. Phylum Verrucomicrobia was absent and ↓ Proteobacteria. ↑ Four Species belonging to the Ruminococcaceae and Lachnospiraceae families of the order Clostridiales and phylum Firmicutes. ↑ endogenous B. pseudolongum was found higher	↓ BW and fat accumulation, plasma insulin, leptin, total cholesterol (TC) and liver toxicity biomarkers and adipose tissue, ↓ pro-inflammatory genes in adipose tissue while ↑ fatty acid oxidation genes in liver	Park et al., 2013
L. salivarius UCC118	C57BL/J6 HFD mice	NC in Firmicutes, ↑ Bacteroidetes and Proteobacteria and ↓ Actinobacteria	No alteration in metabolic profiles	Murphy et al., 2013
L. salivarius JCC118	C57BL/J6 HFD mice	↑ In proportions of Bacteroidetes and Peptococcaceae and significantly ↓ in population of Rikenellaceae and Porphyromonadaceae	↓ BW at appropriate time	Clarke et al., 2013
L. coryniformis CECT5711	C57BL/6J HFD mice	↑ Lactobacillus spp. cluster	↓ Plasma glycaemia, insulin resistance and LPS levels and improves gut permeability.	Toral et al., 2014
L. casei NCDC 19	C57BL/6J HFD mice	NC in total bacterial counts, Eubacterium rectale–Clostridium coccoides and lactobacilli–Enterococcus (LAB) groups. Significant ↑ in population of bifidobacteria was observed	↓ BW, epididymal fat, blood glucose, plasma lipids and leptin; ↑ adiponectin	Rather et al., 2014
L. brevis OK56	C57BL/6J HFD mice	↑ Bifidobacteria population	↓ Body and epididymal fat, ↓ NF-κB activation and LPS production. ↓ Pro-inflammatory cytokines	Kim et al., 2015
L. rhamnosus NCDC17	High fat diet fed and streptozotocin treated Wistar rats	↑ Bifidobacteria and lactobacilli proportions	↑ Glucose tolerance, blood glucose, plasma insulin, glycosylated hemoglobin, FFA, TGs, serum lipids and cholesterol, oxidative stress. ↑ GLP1 and adiponectin, ↓ pro-inflammatory cytokines and propionate in caecum	Singh et al., 2016
Probiotic mixture (L. salivarius 33, L. rhamnosus LMG S-28148, B. animalis subsp. lactis LMG P-28149)	C57BL/6J HFD mice	↑ Bacteroidetes, Restores the abundance of Rikenellaceae ↓ Lactobacillaceae. Probiotic treatment induced a modification in the bifidobacterial population: B. pseudolongum was no longer detected that was present in HFD group. ↓ Lactobacillus–Leuconostoc–Pediococcus group. Restoration of A. muciniphila after treatment	↓ BW and adiposity with improvement in insulin resistance, ↓ adipose tissue inflammation, ↑ dyslipidemia through adipose tissue immune cell-remodeling	Alard et al., 2016
L. plantarum HAC01	C57BL/6J HFD mice	↑ Lachnospiraceae, ↓ Deferribacteraceae, Mucispirillum and Lactobacillaceae with NC in proportion of Firmicutes and Bacteroidetes	↓ BW gain and mesenteric fat weight, blood glucose, TC and triacylglycerol, ↑ lipid oxidative genes	Park et al., 2017

*NC, No significant change.

TABLE 2 | Effect of probiotics supplementation on human gut microbiota and their metabolic outcomes.

Probiotic and subject	Study Design	Influence on gut microbiota	outcome	Reference
L. rhamnosus GG, ATCC 53103 pregnant women and children	Double-blind placebo controlled Mothers: 4 weeks before deliver Children: 6 months after birth	No clear cut study on gut/fecal microbiota	The authors proposed that the reduction in BW was due to positive modulation of gut microbiota by probiotic during the critical development period.	Luoto et al., 2010
L. salivarius Ls-33 obese adolescents	Double-blind placebo controlled 12 weeks	↑ *Bacteroides–Prevotella–Porphyromonas* group to *Firmicutes* belonging bacteria, including *Clostridium* cluster XIV, *Blautia coccoides_Eubacteria rectale* and *Roseburia intestinalis* group. NC of ingestion on *Clostridium* cluster I/IV, *F. prausnitzii*, *Enterobacteriaceae, Enterococcus, Lactobacillus,* and *Bifidobacterium* group	Feeding strain might have modified the fecal microbiota in obese adolescents by a mechanism that is not associated to metabolic syndrome.	Larsen et al., 2013
Probiotic (*S. thermophiles*, *L. plantarum, L. acidophilus*, *L. rhamnosus, B. lactis, B. longum* and *B. breve*) and *Bofutsushosan* herb, subjects having BMI > 25 kg/m² and waist circumference > 85 cm were only included in the study	Double-blind, placebo controlled 8 weeks	↑ *B. breve, B. lactis, L. rhamnosus,* and *L. plantarum*. ↓ *Firmicutes/Bacteroidetes* ratio was also observed upon treatment.	↓ Weight, waist circumference and ↑ in HDL-cholesterol. Change in body composition is positively related to levels of LPS and *L. plantarum*	Lee et al., 2014
L. salivarius UBL S22 and Prebiotic Fructoligosaccharide healthy human volunteers	Double-blind, placebo controlled 6 weeks	↑ *lactobacilli* and ↓ *E. coli* after intervention period.	Improvement in serum TGs and lipid profile with ↓ in inflammatory cytokines	Rajkumar et al., 2015

gut microbiota. They observed that feeding a probiotic strain restores the abundance of *A. muciniphila* and *Rikenellaceae* and decreases *Lactobacillaceae*. These gut-associated alterations are linked with improvement in other pathological parameters and obesity (Alard et al., 2016). Recently, Park et al. (2017) demonstrated that feeding of a probiotic strain *L. plantarum* HAC01 to HFD mice resulted in reduction of body weight, fat mesenteric fat, and other biomarkers associated with obesity. In spite of these changes, significant alterations in several bacterial taxa, both on family and genus level, were observed, as revealed in metagenomics' studies. HAC01 feeding led to a significantly increase in the abundance of the family *Lachnospiraceae* (phylum *Firmicutes*), while decreasing the population of *Deferribacteraceae*. The decrease in abundance was due to a significant reduction in genus *Mucispirillum* numbers. Interestingly, administration of HAC01 also resulted in a decrease in the population of *Lactobacillaceae*. Moreover, no remarkable change in the relative proportion of *Firmicutes* and *Bacteroidetes* was observed post treatment. Finally, the authors suggested that administration of probiotic strain induces modulations in the gut microbiota, which in turn influences the regulation of genes associated with lipid metabolism. These series of changes may consequently, abrogate the fat storage and alleviation of host metabolism (Park et al., 2017).

The above studies made us understand that probiotics address obesity, at least in animals, *via* modulation of gut microbiota. But clear-cut studies are still missing. Also, in most of the studies, only few bacteria tax or phylum or family or genus were studied and that too with the help of conventional techniques. Only one studied provided deeper inside of modulated gut microbiota under influence of probiotic treatment. The discrepancy between the studies may be due to differently inherited gut microbiota of the host of varied genetic background, age and diet. So, here we emphasized on the application of advanced "omics"-based techniques to study the changes in the gut microbiota in probiotic intervention studies, whereby particular "indicator" taxa instead of phyla can be linked with anti-obesity potential. Besides, mechanistic studies are also warranted that decipher how particular "indicator" taxa crosstalk with probiotic strains during reversal of obesity. Moreover, the effects of probiotics are strain specific, so exploring the impact of a single strain on gut microbiota modulation further improves our understanding in the context of host metabolism.

Evidence from Human Clinical Trials

Most of the human studies concerning the impact of probiotics on body weight were restricted to the analysis of biochemical (inflammatory markers) and physical parameters related to metabolic disorders (reviewed by Sanz et al., 2013). Only few studies have evaluated their effect on gut microbiota in the context of obesity and associated disorders (**Table 2**). Luoto et al. (2010) evaluated the impact of perinatal probiotic (*L. Rhamnosus* GG, ATCC 53103) feeding on childhood growth and development patterns up to a period of 10 years in follow-up study. The results signified that probiotic feeding during the early years modulated the gut microbiota of children, which in turn changed growth patterns by way of restraining excessive weight gain (Luoto et al., 2010). In a subsequent clinical trial, the authors studied the impact of *L. salivarius* Ls-33 supplementation on the fecal microbiota of obese adolescents. The administration significantly increased the ratio of *Bacteroides–Prevotella–Porphyromonas* group to *Firmicutes* belonging bacteria. The population of *Lactobacillus* spp. and *Bifidobacterium* spp. changed remarkably post feeding. Also, no change in the production of SCFAs was observed between treatment and placebo group. The authors concluded that the probiotic modulated fecal microbiota by a method not related to metabolic syndrome (Larsen et al., 2013). A later study was designed to assess the combined effect of probiotic capsules (*L. plantarum*, *L. acidophilus*, *L. rhamnosus*, *B. lactis*, *B. longum*, *B. breve*, and *Streptococcus thermophilus*) and herbal medicine in the treatment of obesity among patients having BMI > 25 kg/m^2 and waist circumference >85 cm. The results demonstrated a major reduction in body weight and waist circumference, but no remarkable differences in body composition and metabolic biomarkers were noticed. When they correlated the change in body composition with LPS level and the population of gut *L. plantarum*, a positive relation was revealed. A positive correlation was also documented for Gram-negative bacteria with alterations in body composition and total cholesterol level. A negative correlation was found between *B. breve* population and LPS level. The conclusion corroborated the fact that probiotics play a significant role in deterring obesity by a reduction in LPS production through altered gut microbiota (Lee et al., 2014). In a subsequent study, the individual or symbiotic effect of probiotic *L. salivarius* UBLS22 and prebiotic (fructo-oligosaccharide) supplementation on the various biomarkers of obesity and gut microbiota in healthy young volunteers was examined. After treatment, significant positive alterations in the serum lipid profile were observed in the probiotic as well as symbiotic groups. The serum concentrations of inflammatory cytokines were also reduced in the two treatment groups. They observed a noteworthy boost in the population of *lactobacilli*, and a decrease in total coliforms and *Escherichia coli* across both groups. However, a more pronounced effect was observed in the symbiotic group than the individual one. The authors advocated that the symbiotic mixture could be used for the treatment of obesity by modulating the serum lipid profiles, inflammatory cytokines, and gut microbiota (Rajkumar et al., 2015).

From the aforementioned clinical trials, it can be inferred that gut microbiota display crucial alterations during probiotic intervention, but none of these trials have clearly stated that these alterations are solely responsible for reduction in body weight or obesity. Also, the effects of probiotic supplementation on gut microbiota modulation in context with gut permeability, satiety hormones, eCB system are needed to be studied in detail. . .. In addition, the comparative effect of different strains was also not studied, even though probiotic effects varied among individuals. These gaps in our current understanding open the platform for future research. Further research is required to prove their beneficial effects on humans to gain an insight into the mechanisms through which live bacterial organisms improve the human gut barrier function.

CONCLUSIONS AND FUTURE PROSPECTS

Obesity and related diseases have enormously increased in society and considered the biggest plausible factor for disturbing well-being and health. Studies performed in animal models, and human subjects have clearly indicated that dysbiosis of gut microbiota predisposed them toward obesity and other associated disorders. Gut microbiota influences obesity by acting on the various mechanisms that are central to energy homeostasis and development. In most of the studies, LPS stimulated low grade inflammation was understood as the prime mechanism by which gut microbiota induces obesity. Supplementation of prebiotics and probiotics addressed therapeutic effect on the altered gut microbiota that provides us an opportunity to prevent or treat obesity. However, the discrepancy observed in some studies, in the context of gut microbiota, might be due to the adoption of different sequencing techniques, intra-individual strain differences, age and genotype of individuals. In a nutshell, we believe that the science of prebiotics and probiotics have the potential to tackle obesity and associated metabolic disorders.

But before that, several problems need to be seriously addressed. Till date, it is not clear which microbial community contributes more to the obesity etiology. In some studies, a particular species was positively influenced, while contrasting results were obtained in other studies. This might be due to the complex nature of gut microbiota. The next challenge is to figure, what would be the appropriate dose of these dietary modulators for improving health. Whether they should be same for all age groups, is a major point to discuss. None of the studies have analyzed the comparative effect of different strains with regard to their anti-obesity potential, so the issue certainly requires further research. The biggest one is the safety issue of probiotics, although they are known to be safe for human consumption, but at the same time we cannot deny from the fact that they may spread antibiotic resistance (Gad et al., 2014) Likewise, some probiotics could also cause gastrointestinal disorders as previously discussed in a review (Marteau and Seksik, 2004). This reflects the need for a stricter regulatory framework globally. Products containing probiotics should be analyzed for safety risks before sale in the market.

Analyzing the crosstalk between probiotics and gut microbiota would be one of the most important future research tasks in broadening our understanding on the topic. One of the main aspects to be studied in this area would be to understand how probiotics make genetic communication with the intestinal microbiota by means of genetic material exchange. If we transform the genetic properties of probiotic bacteria in some of the intestinal bacteria, then it is possible to confer few beneficial traits to the host. There is an emerging need to look for those strategies that would not only positively modify the gut microbiota, but also be safe for use.

AUTHOR CONTRIBUTIONS

DD conceived the idea for the article, prepared and edited the final manuscript. R prepared the figure, tables and edited the final manuscript. MP, US, TD, NK, SK help in final editing of the manuscript. AP and PS participated in developing the idea and critically revised the manuscript. All authors approved it for publication.

ACKNOWLEDGMENTS

The authors (MP and R) would like to acknowledge Department of Biotechnology and Council of Scientific and Industrial Research, Govt of India, for funding in the form of a BioCARe Women Scientist Award/2013-16 and CSIR-SRF (09/135(0645)/2011-EMR-I), respectively.

REFERENCES

Aagaard, K., Ma, J., Antony, K. M., Ganu, R., Petrosino, J., and Versalovic, J. (2014). The placenta harbors a unique microbiome. *Sci. Trans. Med.* 6, 237ra265. doi: 10.1126/scitranslmed.3008599

Alard, J., Lehrter, V., Rhimi, M., Mangin, I., Peucelle, V., Abraham, A. L., et al. (2016). Beneficial metabolic effects of selected probiotics on diet-induced obesity and insulin resistance in mice are associated with improvement of dysbiotic gut microbiota. *Environ. Microbiol.* 18, 1484–1497. doi: 10.1111/1462-2920.13181

An, H. M., Park, S. Y., Lee, D. K., Kim, J. R., Cha, M. K., Lee, S. W., et al. (2011). Antiobesity and lipid-lowering effects of *Bifidobacterium* spp. in high fat diet-induced obese rats. *Lipids Health Dis.* 10:116. doi: 10.1186/1476-511X-10-116

Angin, Y., Beauloye, C., Horman, S., and Bertrand, L. (2016). "Regulation of carbohydrate metabolism, lipid metabolism, and protein metabolism by AMPK," in *AMP-Activated Protein Kinase*, (Berlin: Springer), 23–43. doi: 10.1007/978-3-319-43589-3_2

Armougom, F., Henry, M., Vialettes, B., Raccah, D., and Raoult, D. (2009). Monitoring bacterial community of human gut microbiota reveals an increase in *Lactobacillus* in obese patients and Methanogens in anorexic patients. *PLoS ONE* 4:e7125. doi: 10.1371/journal.pone.0007125

Aron-Wisnewsky, J., Gaborit, B., Dutour, A., and Clement, K. (2013). Gut microbiota and non-alcoholic fatty liver disease: new insights. *Clin. Microbiol. Infect* 19, 338–348. doi: 10.1111/1469-0691.12140

Arora, T., Singh, S., and Sharma, R. K. (2013). Probiotics: interaction with gut microbiome and antiobesity potential. *Nutrition* 29, 591–596. doi: 10.1016/j.nut.2012.07.017

Arumugam, M., Raes, J., Pelletier, E., Le Paslier, D., Yamada, T., Mende, D. R., et al. (2011). Enterotypes of the human gut microbiome. *Nature* 473, 174–180. doi: 10.1038/nature09944

Bäckhed, F., Ding, H., Wang, T., Hooper, L. V., Koh, G. Y., Nagy, A., et al. (2004). The gut microbiota as an environmental factor that regulates fat storage. *Proc. Natl. Acad. Sci. U.S.A.* 101, 15718–15723. doi: 10.1073/pnas.0407076101

Bäckhed, F., Manchester, J. K., Semenkovich, C. F., and Gordon, J. I. (2007). Mechanisms underlying the resistance to diet-induced obesity in germ-free mice. *Proc. Natl. Acad. Sci. U.S.A.* 104, 979–984. doi: 10.1073/pnas.0605374104

Balamurugan, R., George, G., Kabeerdoss, J., Hepsiba, J., Chandragunasekaran, A. M., and Ramakrishna, B. S. (2010). Quantitative differences in intestinal *Faecalibacterium prausnitzii* in obese Indian children. *Br. J. Nutr.* 103, 335–338. doi: 10.1017/S0007114509992182

Boulangé, C. L., Neves, A. L., Chilloux, J., Nicholson, J. K., and Dumas, M.-E. (2016). Impact of the gut microbiota on inflammation, obesity, and metabolic disease. *Genome Med.* 8:42. doi: 10.1186/s13073-016-0303-2

Brown, A. J., Goldsworthy, S. M., Barnes, A. A., Eilert, M. M., Tcheang, L., Daniels, D., et al. (2003). The Orphan G protein-coupled receptors GPR41 and GPR43 are activated by propionate and other short chain carboxylic acids. *J. Biol. Chem.* 278, 11312–11319. doi: 10.1074/jbc.M211609200

Cani, P. D., Amar, J., Iglesias, M. A., Poggi, M., Knauf, C., Bastelica, D., et al. (2007a). Metabolic endotoxemia initiates obesity and insulin resistance. *Diabetes Metab. Res. Rev.* 56, 1761–1772. doi: 10.2337/db06-1491

Cani, P. D., Bibiloni, R., Knauf, C., Waget, A., Neyrinck, A. M., Delzenne, N. M., et al. (2008). Changes in gut microbiota control metabolic endotoxemia-induced inflammation in high-fat diet–induced obesity and diabetes in mice. *Diabetes Metab. Res. Rev.* 57, 1470–1481. doi: 10.2337/db07-1403

Cani, P. D., Dewever, C., and Delzenne, N. M. (2004). Inulin-type fructans modulate gastrointestinal peptides involved in appetite regulation (glucagon-like peptide-1 and ghrelin) in rats. *Br. J. Nutr.* 92, 521–526. doi: 10.1079/BJN20041225

Cani, P. D., Everard, A., and Duparc, T. (2013). Gut microbiota, enteroendocrine functions and metabolism. *Curr. Opin. Pharmacol.* 13, 935–940. doi: 10.1016/j.coph.2013.09.008

Cani, P. D., Hoste, S., Guiot, Y., and Delzenne, N. M. (2007b). Dietary non-digestible carbohydrates promote L-cell differentiation in the proximal colon of rats. *Br. J. Nutr.* 98, 32–37.

Cani, P. D., Joly, E., Horsmans, Y., and Delzenne, N. (2006a). Oligofructose promotes satiety in healthy human: a pilot study. *Eur. J. Clin. Nutr.* 60, 567–572.

Cani, P. D., Knauf, C., Iglesias, M. A., Drucker, D. J., Delzenne, N. M., and Burcelin, R. (2006b). Improvement of glucose tolerance and hepatic insulin sensitivity by oligofructose requires a functional glucagon-like peptide 1 receptor. *Diabetes Metab. Res. Rev.* 55, 1484–1490.

Cani, P. D., Lecourt, E., Dewulf, E. M., Sohet, F. M., Pachikian, B. D., Naslain, D., et al. (2009a). Gut microbiota fermentation of prebiotics increases satietogenic and incretin gut peptide production with consequences for appetite sensation and glucose response after a meal. *Am. J. Clin. Nutr.* 90, 1236–1243. doi: 10.3945/ajcn.2009.28095

Cani, P. D., Neyrinck, A., Fava, F., Knauf, C., Burcelin, R., Tuohy, K., et al. (2007c). Selective increases of bifidobacteria in gut microflora improve high-fat-diet-induced diabetes in mice through a mechanism associated with endotoxaemia. *Diabetologia* 50, 2374–2383.

Cani, P. D., Possemiers, S., Van de Wiele, T., Guiot, Y., Everard, A., Rottier, O., et al. (2009b). Changes in gut microbiota control inflammation in obese mice through a mechanism involving GLP-2-driven improvement of gut permeability. *Gut* 58, 1091–1103. doi: 10.1136/gut.2008.165886

Chiang, J. Y. (2009). Bile acids: regulation of synthesis. *J. Lipid Res.* 50, 1955–1966. doi: 10.1194/jlr.R900010-JLR200

Cho, I., Yamanishi, S., Cox, L., Methé, B. A., Zavadil, J., Li, K., et al. (2012). Antibiotics in early life alter the murine colonic microbiome and adiposity. *Nature* 488, 621–626. doi: 10.1038/nature11400

Clarke, S. F., Murphy, E. F., O'Sullivan, O., Ross, R. P., O'Toole, P. W., Shanahan, F., et al. (2013). Targeting the microbiota to address diet-induced obesity: a time dependent challenge. *PLoS ONE* 8:e65790. doi: 10.1371/journal.pone.0065790

Cotillard, A., Kennedy, S. P., Kong, L. C., Prifti, E., Pons, N., Le Chatelier, E., et al. (2013). Dietary intervention impact on gut microbial gene richness. *Nature* 500, 585–588. doi: 10.1038/nature12480

Cox, L. M., Yamanishi, S., Sohn, J., Alekseyenko, A. V., Leung, J. M., Cho, I., et al. (2014). Altering the intestinal microbiota during a critical developmental window has lasting metabolic consequences. *Cell* 158, 705–721. doi: 10.1016/j.cell.2014.05.052

da Silva, S. T., dos Santos, C. A., and Bressan, J. (2013). Intestinal microbiota; relevance to obesity and modulation by prebiotics and probiotics. *Nutr. Hosp.* 28, 1039–1048. doi: 10.3305/nh.2013.28.4.6525

Dahiya, D. K., and Puniya, A. K. (2015). Evaluation of survival, free radical scavenging and human enterocyte adherence potential of Lactobacilli with anti-obesity and anti-inflammatory CLA isomer-producing attributes. *J. Food Process. Preserv.* 36, 2866–2877. doi: 10.1111/jfpp.12538

Dahiya, D. K., and Puniya, A. K. (2017). Isolation, molecular characterization and screening of indigenous lactobacilli for their abilities to produce bioactive conjugated linoleic acid (CLA). *J. Food Sci. Technol.* 54, 792–801. doi: 10.1007/s13197-017-2523-x

Dao, M. C., Everard, A., Aron-Wisnewsky, J., Sokolovska, N., Prifti, E., Verger, E. O., et al. (2016). *Akkermansia muciniphila* and improved metabolic health during a dietary intervention in obesity: relationship with gut microbiome richness and ecology. *Gut* 65, 426–436. doi: 10.1136/gutjnl-2014-308778

Davis, J. E., Gabler, N. K., Walker-Daniels, J., and Spurlock, M. E. (2008). Tlr-4 deficiency selectively protects against obesity induced by diets high in saturated fat. *Obesity* 16, 1248–1255. doi: 10.1038/oby.2008.210

De Silva, A., and Bloom, S. R. (2012). Gut hormones and appetite control: a focus on PYY and GLP-1 as therapeutic targets in obesity. *Gut Liver* 6, 10–20. doi: 10.5009/gnl.2012.6.1.10

Dewulf, E. M., Cani, P. D., Claus, S. P., Fuentes, S., Puylaert, P. G., Neyrinck, A. M., et al. (2012). Insight into the prebiotic concept: lessons from an exploratory, double blind intervention study with inulin-type fructans in obese women. *Gut* 62, 1112–1121. doi: 10.1136/gutjnl-2012-303304

Dray, C., Knauf, C., Daviaud, D., Waget, A., Boucher, J., Buléon, M., et al. (2008). Apelin stimulates glucose utilization in normal and obese insulin-resistant mice. *Cell Metab.* 8, 437–445. doi: 10.1016/j.cmet.2008.10.003

Druart, C., Neyrinck, A. M., Dewulf, E. M., De Backer, F. C., Possemiers, S., Van de Wiele, T., et al. (2013). Implication of fermentable carbohydrates targeting the gut microbiota on conjugated linoleic acid production in high-fat-fed mice. *Br. J. Nutr.* 110, 998–1011. doi: 10.1017/S0007114513000123

Duncan, S. H., Belenguer, A., Holtrop, G., Johnstone, A. M., Flint, H. J., and Lobley, G. E. (2007). Reduced dietary intake of carbohydrates by obese subjects results in decreased concentrations of butyrate and butyrate-producing bacteria in feces. *Appl. Environ. Microbiol.* 73, 1073–1078. doi: 10.1128/AEM.02340-06

Duncan, S. H., Louis, P., Thomson, J. M., and Flint, H. J. (2009). The role of pH in determining the species composition of the human colonic microbiota. *Environ. Microbiol.* 11, 2112–2222. doi: 10.1111/j.1462-2920.2009.01931.x

Dutton, S., and Trayhurn, P. (2008). Regulation of angiopoietin-like protein 4/fasting-induced adipose factor (Angptl4/FIAF) expression in mouse white adipose tissue and 3T3-L1 adipocytes. *Br. J. Nutr.* 100, 18–26. doi: 10.1017/S0007114507882961

Eckburg, P. B., Bik, E. M., Bernstein, C. N., Purdom, E., Dethlefsen, L., Sargent, M., et al. (2005). Diversity of the human intestinal microbial flora. *Science* 308, 1635–1638. doi: 10.1126/science.1110591

Everard, A., Belzer, C., Geurts, L., Ouwerkerk, J. P., Druart, C., Bindels, L. B., et al. (2013). Cross-talk between *Akkermansia muciniphila* and intestinal epithelium controls diet-induced obesity. *Proc. Natl. Acad. Sci. U.S.A.* 110, 9066–9071. doi: 10.1073/pnas.1219451110

Everard, A., Lazarevic, V., Derrien, M., Girard, M., Muccioli, G. G., Neyrinck, A. M., et al. (2011). Responses of gut microbiota and glucose and lipid metabolism to prebiotics in genetic obese and diet-induced leptin-resistant mice. *Diabetes Metab. Res. Rev.* 60, 2775–2786. doi: 10.2337/db11-0227

Everard, A., Lazarevic, V., Gaïa, N., Johansson, M., Ståhlman, M., Backhed, F., et al. (2014). Microbiome of prebiotic-treated mice reveals novel targets involved in host response during obesity. *ISME J.* 8, 2116–2130. doi: 10.1038/ismej.2014.45

Feng, J., Tang, H., Li, M., Pang, X., Wang, L., Zhang, M., et al. (2014). The abundance of fecal *Faecalibacterium prausnitzii* in relation to obesity and gender in Chinese adults. *Arch. Microbiol.* 196, 73–77. doi: 10.1007/s00203-013-0942-2

Fleissner, C. K., Huebel, N., El-Bary, M. M. A., Loh, G., Klaus, S., and Blaut, M. (2010). Absence of intestinal microbiota does not protect mice from diet-induced obesity. *Br. J. Nutr.* 104, 919–929. doi: 10.1017/S0007114510001303

Franzosa, E. A., Huang, K., Meadow, J. F., Gevers, D., Lemon, K. P., Bohannan, B. J., et al. (2015). Identifying personal microbiomes using metagenomic codes. *Proc. Natl. Acad. Sci. U.S.A.* 112, E2930–E2938. doi: 10.1073/pnas.1423854112

Gad, G. F. M., Abdel-Hamid, A. M., and Farag, Z. S. H. (2014). Antibiotic resistance in lactic acid bacteria isolated from some pharmaceutical and dairy products. *Braz. J. Microbiol.* 45, 25–33. doi: 10.1590/S1517-83822014000100005

Gérard, P. (2016). Gut microbiota and obesity. *Cell. Mol. Life Sci.* 73, 147–162. doi: 10.1007/s00018-015-2061-5

Geurts, L., Lazarevic, V., Derrien, M., Everard, A., Van Roye, M., Knauf, C., et al. (2011). Altered gut microbiota and endocannabinoid system tone in obese and diabetic leptin-resistant mice: impact on apelin regulation in adipose tissue. *Front. Microbiol.* 2:149. doi: 10.3389/fmicb.2011.00149

Gibson, G. R., Probert, H. M., Van Loo, J., Rastall, R. A., and Roberfroid, M. B. (2004). Dietary modulation of the human colonic microbiota: updating the concept of prebiotics. *Nutr. Res. Rev.* 17, 259–275. doi: 10.1079/NRR200479

Go, G.-W., Oh, S., Park, M., Gang, G., McLean, D., Yang, H.-S., et al. (2013). t10, c12 conjugated linoleic acid upregulates hepatic de novo lipogenesis and triglyceride synthesis via mTOR pathway activation. *J. Microbiol. Biotechnol.* 23, 1569–1576. doi: 10.4014/jmb.1308.08008

Hotamisligil, G. S. (2006). Inflammation and metabolic disorders. *Nature* 444, 860–867. doi: 10.1038/nature05485

Joossens, M., Huys, G., Van Steen, K., Cnockaert, M., Vermeire, S., Rutgeerts, P., et al. (2011). High-throughput method for comparative analysis of denaturing gradient gel electrophoresis profiles from human fecal samples reveals significant increases in two bifidobacterial species after inulin-type prebiotic intake. *FEMS Microbiol. Ecol.* 75, 343–349. doi: 10.1111/j.1574-6941.2010.01008.x

Karra, E., Chandarana, K., and Batterham, R. L. (2009). The role of peptide YY in appetite regulation and obesity. *J. Physiol.* 587, 19–25. doi: 10.1113/jphysiol.2008.164269

Kasubuchi, M., Hasegawa, S., Hiramatsu, T., Ichimura, A., and Kimura, I. (2015). Dietary gut microbial metabolites, short-chain fatty acids, and host metabolic regulation. *Nutrients* 7, 2839–2849. doi: 10.3390/nu7042839

Kim, J., Yang, G., and Ha, J. (2017). Targeting of AMP-activated protein kinase: prospects for computer-aided drug design. *Expert Opin. Drug Dis.* 12, 47–59. doi: 10.1080/17460441.2017.1255194

Kim, K.-A., Jeong, J.-J., and Kim, D.-H. (2015). *Lactobacillus brevis* OK56 ameliorates high-fat diet-induced obesity in mice by inhibiting NF-κB activation and gut microbial LPS production. *J. Funct. Foods* 13, 183–191. doi: 10.1016/j.jff.2014.12.045

Kimura, I., Ozawa, K., Inoue, D., Imamura, T., Kimura, K., Maeda, T., et al. (2013). The gut microbiota suppresses insulin-mediated fat accumulation via the short-chain fatty acid receptor GPR43. *Nat. Commun.* 4, 1829. doi: 10.1038/ncomms2852

Knights, D., Ward, T. L., McKinlay, C. E., Miller, H., Gonzalez, A., McDonald, D., et al. (2014). Rethinking "enterotypes". *Cell Host Microbe* 16, 433–437. doi: 10.1016/j.chom.2014.09.013

Koenig, J. E., Spor, A., Scalfone, N., Fricker, A. D., Stombaugh, J., Knight, R., et al. (2011). Succession of microbial consortia in the developing infant gut microbiome. *Proc. Natl. Acad. Sci. U.S.A.* 108(Suppl. 1), 4578–4585. doi: 10.1073/pnas.1000081107

Koh, A., De Vadder, F., Kovatcheva-Datchary, P., and Bäckhed, F. (2016). From dietary fiber to host physiology: short-chain fatty acids as key bacterial metabolites. *Cell* 165, 1332–1345. doi: 10.1016/j.cell.2016.05.041

Kopelman, P. (2007). Health risks associated with overweight and obesity. *Obes. Rev.* 8, 13–17. doi: 10.1111/j.1467-789X.2007.00311.x

Koren, O., Knights, D., Gonzalez, A., Waldron, L., Segata, N., Knight, R., et al. (2013). A guide to enterotypes across the human body: meta-analysis of microbial community structures in human microbiome datasets. *PLoS Comput. Biol.* 9:e1002863. doi: 10.1371/journal.pcbi.1002863

Larsen, N., Vogensen, F. K., Gøbel, R. J., Michaelsen, K. F., Forssten, S. D., Lahtinen, S. J., et al. (2013). Effect of *Lactobacillus salivarius* Ls-33 on fecal microbiota in obese adolescents. *Clin. Nutr.* 32, 935–940. doi: 10.1016/j.clnu.2013.02.007

Le Chatelier, E., Nielsen, T., Qin, J., Prifti, E., Hildebrand, F., Falony, G., et al. (2013). Richness of human gut microbiome correlates with metabolic markers. *Nature* 500, 541–546. doi: 10.1038/nature12506

Le Poul, E., Loison, C., Struyf, S., Springael, J.-Y., Lannoy, V., Decobecq, M.-E., et al. (2003). Functional characterization of human receptors for short chain fatty acids and their role in polymorphonuclear cell activation. *J. Biol. Chem.* 278, 25481–25489. doi: 10.1074/jbc.M301403200

Lee, S. J., Bose, S., Seo, J.-G., Chung, W.-S., Lim, C.-Y., and Kim, H. (2014). The effects of co-administration of probiotics with herbal medicine on obesity, metabolic endotoxemia and dysbiosis: a randomized double-blind controlled clinical trial. *Clin. Nutr.* 33, 973–981. doi: 10.1016/j.clnu.2013.12.006

Letran, S. E., Lee, S.-J., Atif, S. M., Flores-Langarica, A., Uematsu, S., Akira, S., et al. (2011). TLR5-deficient mice lack basal inflammatory and metabolic defects but exhibit impaired CD4 T cell responses to a flagellated pathogen. *J. Immunol.* 186, 5406–5412. doi: 10.4049/jimmunol.1003576

Leung, C., Rivera, L., Furness, J. B., and Angus, P. W. (2016). The role of the gut microbiota in NAFLD. *Nat. Rev. Gastroenterol. Hepatol.* 13, 412–425. doi: 10.1038/nrgastro.2016.85

Ley, R. E., Bäckhed, F., Turnbaugh, P., Lozupone, C. A., Knight, R. D., and Gordon, J. I. (2005). Obesity alters gut microbial ecology. *Proc. Natl. Acad. Sci. U.S.A.* 102, 11070–11075. doi: 10.1073/pnas.0504978102

Ley, R. E., Turnbaugh, P. J., Klein, S., and Gordon, J. I. (2006). Microbial ecology: human gut microbes associated with obesity. *Nature* 444, 1022–1023. doi: 10.1038/4441022a

Lim, S.-M., Jeong, J.-J., Woo, K. H., Han, M. J., and Kim, D.-H. (2016). Lactobacillus sakei OK67 ameliorates high-fat diet–induced blood glucose intolerance and obesity in mice by inhibiting gut microbiota lipopolysaccharide production and inducing colon tight junction protein expression. *Nutr. Res.* 36, 337–348. doi: 10.1016/j.nutres.2015.12.001

Louis, P., and Flint, H. J. (2009). Diversity, metabolism and microbial ecology of butyrate-producing bacteria from the human large intestine. *FEMS Microbiol. Lett.* 294, 1–8. doi: 10.1111/j.1574-6968.2009.01514.x

Luna, R. A., and Foster, J. A. (2015). Gut brain axis: diet microbiota interactions and implications for modulation of anxiety and depression. *Curr. Opin. Biotechnol.* 32, 35–41. doi: 10.1016/j.copbio.2014.10.007

Luoto, R., Kalliomäki, M., Laitinen, K., and Isolauri, E. (2010). The impact of perinatal probiotic intervention on the development of overweight and obesity: follow-up study from birth to 10 years. *Int. J. Obes.* 34, 1531–1537. doi: 10.1038/ijo.2010.50

Mackie, K. (2008). Cannabinoid receptors: where they are and what they do. *J. Neuroendocrinol.* 20, 10–14. doi: 10.1111/j.1365-2826.2008.01671.x

Maenhaut, N., and Van de Voorde, J. (2011). Regulation of vascular tone by adipocytes. *BMC Med.* 9:25. doi: 10.1186/1741-7015-9-25

Marteau, P., and Seksik, P. (2004). Tolerance of probiotics and prebiotics. *J. Clin. Gastroenterol.* 38, S67–S69. doi: 10.1097/01.mcg.0000128929.37156.a7

Medzhitov, R. (2001). Toll-like receptors and innate immunity. *Nat. Rev. Immunol.* 1, 135–145. doi: 10.1038/35100529

Million, M., Angelakis, E., Paul, M., Armougom, F., Leibovici, L., and Raoult, D. (2012a). Comparative meta-analysis of the effect of *Lactobacillus* species on weight gain in humans and animals. *Microb. Pathog.* 53, 100–108. doi: 10.1016/j.micpath.2012.05.007

Million, M., Maraninchi, M., Henry, M., Armougom, F., Richet, H., Carrieri, P., et al. (2012b). Obesity-associated gut microbiota is enriched in *Lactobacillus reuteri* and depleted in *Bifidobacterium animalis* and *Methanobrevibacter smithii*. *Int. J. Obes.* 36, 817–825. doi: 10.1038/ijo.2011.153

Miquel, S., Martín, R., Rossi, O., Bermúdez-Humarán, L. G., Chatel, J. M., Sokol, H., et al. (2013). *Faecalibacterium prausnitzii* and human intestinal health. *Curr. Opin. Microbiol.* 16, 255–261. doi: 10.1016/j.mib.2013.06.003

Muccioli, G. G., Naslain, D., Bäckhed, F., Reigstad, C. S., Lambert, D. M., Delzenne, N. M., et al. (2010). The endocannabinoid system links gut microbiota to adipogenesis. *Mol. Syst. Biol.* 6, 392. doi: 10.1038/msb.2010.46

Mueller, N. T., Whyatt, R., Hoepner, L., Oberfield, S., Dominguez-Bello, M. G., Widen, E., et al. (2014). Prenatal exposure to antibiotics, cesarean section and risk of childhood obesity. *Int. J. Obes.* 39, 665–670. doi: 10.1038/ijo.2014.180

Murphy, E. F., Cotter, P. D., Hogan, A., O'sullivan, O., Joyce, A., Fouhy, F., et al. (2013). Divergent metabolic outcomes arising from targeted manipulation of the gut microbiota in diet-induced obesity. *Gut* 62, 220–226. doi: 10.1136/gutjnl-2011-300705

Neyrinck, A. M., Possemiers, S., Druart, C., Van de Wiele, T., De Backer, F., Cani, P. D., et al. (2011). Prebiotic effects of wheat arabinoxylan related to the increase in bifidobacteria, *Roseburia* and *Bacteroides/Prevotella* in diet-induced obese mice. *PLoS ONE* 6:e20944. doi: 10.1371/journal.pone.0020944

Nikolopoulou, A., and Kadoglou, N. P. (2012). Obesity and metabolic syndrome as related to cardiovascular disease. *Expert Rev. Cardiovasc. Ther.* 10, 933–939. doi: 10.1586/erc.12.74

Nøhr, M. K., Pedersen, M. H., Gille, A., Egerod, K. L., Engelstoft, M. S., Husted, A. S., et al. (2013). GPR41/FFAR3 and GPR43/FFAR2 as cosensors for short-chain fatty acids in enteroendocrine cells vs FFAR3 in enteric neurons and FFAR2 in enteric leukocytes. *Endocrinology* 154, 3552–3564. doi: 10.1210/en.2013-1142

Ouchi, N., Parker, J. L., Lugus, J. J., and Walsh, K. (2011). Adipokines in inflammation and metabolic disease. *Nat. Rev. Immunol.* 11, 85–97. doi: 10.1038/nri2921

Park, D. Y., Ahn, Y. T., Park, S. H., Huh, C. S., Yoo, S. R., Yu, R., et al. (2013). Supplementation of *Lactobacillus curvatus* HY7601 and *Lactobacillus plantarum* KY1032 in diet-induced obese mice is associated with gut microbial changes and reduction in obesity. *PLoS ONE* 8:e59470. doi: 10.1371/journal.pone.0059470

Park, S., Ji, Y., Jung, H.-Y., Park, H., Kang, J., Choi, S.-H., et al. (2017). *Lactobacillus plantarum* HAC01 regulates gut microbiota and adipose tissue accumulation in a diet-induced obesity murine model. *Appl. Microbiol. Biotechnol.* 101, 1605–1614. doi: 10.1007/s00253-016-7953-2

Parnell, J. A., and Reimer, R. A. (2009). Weight loss during oligofructose supplementation is associated with decreased ghrelin and increased peptide YY in overweight and obese adults. *Am. J. Clin. Nutr.* 89, 1751–1759. doi: 10.3945/ajcn.2009.27465

Perez-Chanona, E., and Trinchieri, G. (2016). The role of microbiota in cancer therapy. *Curr. Opin. Immunol.* 39, 75–81. doi: 10.1016/j.coi.2016.01.003

Peterson, C., Sharma, V., Elmén, L., and Peterson, S. (2015). Immune homeostasis, dysbiosis and therapeutic modulation of the gut microbiota. *Clin. Exp. Immunol.* 179, 363–377. doi: 10.1111/cei.12474

Podolsky, S. H. (2017). Historical perspective on the rise and fall and rise of antibiotics and human weight gainhistorical perspective on antibiotics and human weight gain. *Ann. Intern. Med.* 166, 133–138. doi: 10.7326/M16-1855

Qin, J., Li, R., Raes, J., Arumugam, M., Burgdorf, K. S., Manichanh, C., et al. (2010). A human gut microbial gene catalogue established by metagenomic sequencing. *Nature* 464, 59–65. doi: 10.1038/nature08821

Quigley, E. M. (2011). Gut microbiota and the role of probiotics in therapy. *Curr. Opin. Pharmacol.* 11, 593–603. doi: 10.1016/j.coph.2011.09.010

Rajkumar, H., Kumar, M., Das, N., Kumar, S. N., Challa, H. R., and Nagpal, R. (2015). Effect of probiotic *Lactobacillus salivarius* UBL S22 and prebiotic fructo-oligosaccharide on serum lipids, inflammatory markers, insulin sensitivity, and gut bacteria in healthy young volunteers: a randomized controlled single-blind pilot study. *J. Cardiovasc. Pharmacol. Ther.* 20, 289–298. doi: 10.1177/1074248414555004

Ramirez-Farias, C., Slezak, K., Fuller, Z., Duncan, A., Holtrop, G., and Louis, P. (2009). Effect of inulin on the human gut microbiota: stimulation of *Bifidobacterium adolescentis* and *Faecalibacterium prausnitzii*. *Br. J. Nutr.* 101, 541–550. doi: 10.1017/S0007114508019880

Rather, S. A., Pothuraju, R., Sharma, R. K., De, S., Mir, N. A., and Jangra, S. (2014). Anti-obesity effect of feeding probiotic dahi containing *Lactobacillus casei* NCDC 19 in high fat diet-induced obese mice. *Int. J. Dairy Technol.* 67, 504–509. doi: 10.1111/1471-0307.12154

Regard, J. B., Sato, I. T., and Coughlin, S. R. (2008). Anatomical profiling of G protein-coupled receptor expression. *Cell* 135, 561–571. doi: 10.1016/j.cell.2008.08.040

Remely, M., Hippe, B., Zanner, J., Aumueller, E., Brath, H., and Haslberger, A. G. (2016). Gut microbiota of obese, type 2 diabetic individuals is enriched in *Faecalibacterium prausnitzii*, *Akkermansia muciniphila* and *Peptostreptococcus anaerobius* after weight loss. *Endocr. Metab. Immune Disord. Drug Targets* 16, 99–106. doi: 10.2174/1871530316666160831093813

Respondek, F., Gerard, P., Bossis, M., Boschat, L., Bruneau, A., Rabot, S., et al. (2013). Short-chain fructo-oligosaccharides modulate intestinal microbiota and metabolic parameters of humanized gnotobiotic diet induced obesity mice. *PLoS ONE* 8:e71026. doi: 10.1371/journal.pone.0071026

Ridlon, J. M., Kang, D. J., Hylemon, P. B., and Bajaj, J. S. (2014). Bile acids and the gut microbiome. *Curr. Opin. Gastroenterol.* 30, 332–338. doi: 10.1097/MOG.0000000000000057

Riva, A., Borgo, F., Lassandro, C., Verduci, E., Morace, G., Borghi, E., et al. (2017). Pediatric obesity is associated with an altered gut microbiota and discordant shifts in Firmicutes populations. *Environ. Microbiol.* 19, 95–105. doi: 10.1111/1462-2920.13463

Rosenbaum, M., Knight, R., and Leibel, R. L. (2015). The gut microbiota in human energy homeostasis and obesity. *Trends Endocrinol. Metab.* 26, 493–501. doi: 10.1016/j.tem.2015.07.002

Salazar, N., Dewulf, E. M., Neyrinck, A. M., Bindels, L. B., Cani, P. D., Mahillon, J., et al. (2015). Inulin-type fructans modulate intestinal *Bifidobacterium* species populations and decrease fecal short-chain fatty acids in obese women. *Clin. Nutr.* 34, 501–507. doi: 10.1016/j.clnu.2014.06.001

Sanders, M. E. (2008). Probiotics: definition, sources, selection, and uses. *Clin. Infect. Dis.* 46(Suppl. 2), S58–S61. doi: 10.1086/523341

Sanz, Y., Rastmanesh, R., and Agostonic, C. (2013). Understanding the role of gut microbes and probiotics in obesity: how far are we? *Pharmacol. Res.* 69, 144–155. doi: 10.1016/j.phrs.2012.10.021

Sayin, S. I., Wahlström, A., Felin, J., Jäntti, S., Marschall, H.-U., Bamberg, K., et al. (2013). Gut microbiota regulates bile acid metabolism by reducing the levels of tauro-beta-muricholic acid, a naturally occurring FXR antagonist. Cell Metab. 17, 225–235. doi: 10.1016/j.cmet.2013.01.003

Schneeberger, M., Everard, A., Gómez-Valadés, A. G., Matamoros, S., Ramírez, S., Delzenne, N. M., et al. (2015). Akkermansia muciniphila inversely correlates with the onset of inflammation, altered adipose tissue metabolism and metabolic disorders during obesity in mice. Sci. Rep. 5:16643. doi: 10.1038/srep16643

Singh, S., Sharma, R., Malhotra, S., Pothuraju, R., and Shandilya, U. (2016). Lactobacillus rhamnosus NCDC17 ameliorates type-2 diabetes by improving gut function, oxidative stress and inflammation in high-fat-diet fed and streptozotocintreated rats. Benef. Microbes doi: 10.3920/BM2016.0090 [Epub ahead of print].

Sommer, F., and Bäckhed, F. (2013). The gut microbiota—masters of host development and physiology. Nat. Rev. Microbiol. 11, 227–238. doi: 10.1038/nrmicro2974

Stanton, C., Ross, R. P., Fitzgerald, G. F., and Van Sinderen, D. (2005). Fermented functional foods based on probiotics and their biogenic metabolites. Curr. Opin. Biotechnol. 16, 198–203. doi: 10.1038/nrmicro2974

Steinert, R., Beglinger, C., and Langhans, W. (2016). Intestinal GLP-1 and satiation: from man to rodents and back. Int. J. Obes. 40, 198–205. doi: 10.1038/ijo.2015.172

Swann, J. R., Want, E. J., Geier, F. M., Spagou, K., Wilson, I. D., Sidaway, J. E., et al. (2011). Systemic gut microbial modulation of bile acid metabolism in host tissue compartments. Proc. Natl. Acad. Sci. U.S.A. 108(Suppl. 1), 4523–4530. doi: 10.1073/pnas.1006734107

Swinburn, B. A., Sacks, G., Hall, K. D., McPherson, K., Finegood, D. T., Moodie, M. L., et al. (2011). The global obesity pandemic: shaped by global drivers and local environments. Lancet 378, 804–814. doi: 10.1016/S0140-6736(11)60813-1

Tarini, J., and Wolever, T. M. (2010). The fermentable fibre inulin increases postprandial serum short-chain fatty acids and reduces free-fatty acids and ghrelin in healthy subjects. Appl. Physiol. Nutr. Metab. 35, 9–16. doi: 10.1139/H09-119

Tolhurst, G., Heffron, H., Lam, Y. S., Parker, H. E., Habib, A. M., Diakogiannaki, E., et al. (2012). Short-chain fatty acids stimulate glucagon-like peptide-1 secretion via the G-protein–coupled receptor FFAR2. Diabetes Metab. Res. Rev. 61, 364–371. doi: 10.2337/db11-1019

Toral, M., Gómez-Guzmán, M., Jiménez, R., Romero, M., Sánchez, M., Utrilla, M. P., et al. (2014). The probiotic Lactobacillus coryniformis CECT5711 reduces the vascular pro-oxidant and pro-inflammatory status in obese mice. Clin. Sci. 127, 33–45. doi: 10.1042/CS20130339

Tsai, Y.-T., Cheng, P.-C., and Pan, T.-M. (2014). Anti-obesity effects of gut microbiota are associated with lactic acid bacteria. Appl. Microbiol. Biotechnol. 98, 1–10. doi: 10.1042/CS20130339

Turnbaugh, P. J., Ley, R. E., Mahowald, M. A., Magrini, V., Mardis, E. R., and Gordon, J. I. (2006). An obesity-associated gut microbiome with increased capacity for energy harvest. Nature 444, 1027–1131. doi: 10.1007/s00253-013-5346-3

Turta, O., and Rautava, S. (2016). Antibiotics, obesity and the link to microbes-what are we doing to our children? BMC Med. 14:57. doi: 10.1186/s12916-016-0605-7

Velagapudi, V. R., Hezaveh, R., Reigstad, C. S., Gopalacharyulu, P., Yetukuri, L., Islam, S., et al. (2010). The gut microbiota modulates host energy and lipid metabolism in mice. J. Lipid Res. 51, 1101–1112. doi: 10.1194/jlr.M002774

Vijay-Kumar, M., Aitken, J. D., Carvalho, F. A., Cullender, T. C., Mwangi, S., Srinivasan, S., et al. (2010). Metabolic syndrome and altered gut microbiota in mice lacking Toll-like receptor 5. Science 328, 228–231. doi: 10.1126/science.1179721

Villena, J., and Kitazawa, H. (2014). Modulation of intestinal TLR4-inflammatory signaling pathways by probiotic microorganisms: lessons learned from Lactobacillus jensenii TL2937. Front. Immunol. 4:512. doi: 10.3389/fimmu.2013.00512

Vucenik, I., and Stains, J. P. (2012). Obesity and cancer risk: evidence, mechanisms, and recommendations. Ann. N. Y. Acad. Sci. 1271, 37–43. doi: 10.1111/j.1749-6632.2012.06750.x

Wang, H., and Eckel, R. H. (2009). Lipoprotein lipase: from gene to obesity. Am. J. Physiol. Endocrinol. Metab. 297, E271–E288. doi: 10.1152/ajpendo.90920.2008

Wang, Y. C., McPherson, K., Marsh, T., Gortmaker, S. L., and Brown, M. (2011). Health and economic burden of the projected obesity trends in the USA and the UK. Lancet 378, 815–825. doi: 10.1016/S0140-6736(11)60814-3

Wang, Z., Xiao, G., Yao, Y., Guo, S., Lu, K., and Sheng, Z. (2006). The role of bifidobacteria in gut barrier function after thermal injury in rats. J. Trauma Acute Care Surg. 61, 650–657. doi: 10.1016/S0140-6736(11)60814-3

Withrow, D., and Alter, D. (2011). The economic burden of obesity worldwide: a systematic review of the direct costs of obesity. Obes. Rev. 12, 131–141. doi: 10.1111/j.1467-789X.2009.00712.x

Xiong, Y., Miyamoto, N., Shibata, K., Valasek, M. A., Motoike, T., Kedzierski, R. M., et al. (2004). Short-chain fatty acids stimulate leptin production in adipocytes through the G protein-coupled receptor GPR41. Proc. Natl. Acad. Sci. U.S.A. 101, 1045–1050. doi: 10.1111/j.1467-789X.2009.00712.x

Yadav, H., Lee, J. H., Lloyd, J., Walter, P., and Rane, S. G. (2013). Beneficial metabolic effects of a probiotic via butyrate-induced GLP-1 hormone secretion. J. Biol. Chem. 288, 25088–25097. doi: 10.1074/jbc.M113.452516

Yadav, R., Singh, P. K., Puniya, A. K., and Shukla, P. (2016a). Catalytic interactions andmolecular docking of bile salt hydrolase (BSH) from L. plantarum RYPR1 and itsprebiotic utilization. Front. Microbiol. 7:2116. doi: 10.3389/fmicb.2016.02116

Yadav, R., Singh, P. K., and Shukla, P. (2016b). Metabolic engineering for probiotics and their genome-wide expression profiling. Curr. Protein Pept. Sci. doi: 10.2174/1389203718666161111130157 [Epub ahead of print].

Yin, Y.-N., Yu, Q.-F., Fu, N., Liu, X.-W., and Lu, F.-G. (2010). Effects of four Bifidobacteria on obesity in high-fat diet induced rats. World J. Gastroenterol. 16, 3394–3401. doi: 10.3748/wjg.v16.i27.3394

Zaibi, M. S., Stocker, C. J., O'Dowd, J., Davies, A., Bellahcene, M., Cawthorne, M. A., et al. (2010). Roles of GPR41 and GPR43 in leptin secretory responses of murine adipocytes to short chain fatty acids. FEBS Lett. 584, 2381–2386. doi: 10.1016/j.febslet.2010.04.027

Zhou, J., Martin, R. J., Tulley, R. T., Raggio, A. M., McCutcheon, K. L., Shen, L., et al. (2008). Dietary resistant starch upregulates total GLP-1 and PYY in a sustained day-long manner through fermentation in rodents. Am. J. Physiol. Endocrinol. Metab. 295, E1160–E1166. doi: 10.1152/ajpendo.90637.2008

Zimmer, J., Lange, B., Frick, J.-S., Sauer, H., Zimmermann, K., Schwiertz, A., et al. (2012). A vegan or vegetarian diet substantially alters the human colonic faecal microbiota. Eur. J. Clin. Nutr. 66, 53–60. doi: 10.1038/ejcn.2011.141

Conflict of Interest Statement: The authors declare that the research was conducted in the absence of any commercial or financial relationships that could be construed as a potential conflict of interest.

4

Current Demands for Food-Approved Liposome Nanoparticles in Food and Safety Sector

Shruti Shukla[1], Yuvaraj Haldorai[2], Seung Kyu Hwang[3], Vivek K. Bajpai[1]*, Yun Suk Huh[3]* and Young-Kyu Han[1]*

[1] *Department of Energy and Materials Engineering, Dongguk University, Seoul, South Korea,* [2] *Department of Nanoscience and Technology, Bharathiar University, Coimbatore, India,* [3] *Department of Biological Engineering, Biohybrid Systems Research Center (BSRC), World Class Smart Lab (WCSL), Inha University, Incheon, South Korea*

Edited by:
Giovanna Suzzi,
Università di Teramo, Italy

Reviewed by:
Eleonore Fröhlich,
Medical University of Graz, Austria
Lorenzo Siroli,
Università di Bologna, Italy

*Correspondence:
Yun Suk Huh
yunsuk.huh@inha.ac.kr
Vivek K. Bajpai
vbajpai04@yahoo.com
Young-Kyu Han
ykenergy@dongguk.edu

Safety of food is a noteworthy issue for consumers and the food industry. A number of complex challenges associated with food engineering and food industries, including quality food production and safety of the food through effective and feasible means can be explained by nanotechnology. However, nanoparticles have unique physicochemical properties compared to normal macroparticles of the same composition and thus could interact with living system in surprising ways to induce toxicity. Further, few toxicological/safety assessments have been performed on nanoparticles, thereby necessitating further research on oral exposure risk prior to their application to food. Liposome nanoparticles are viewed as attractive novel materials by the food and medical industries. For example, nanoencapsulation of bioactive food compounds is an emerging application of nanotechnology. In several food industrial practices, liposome nanoparticles have been utilized to improve flavoring and nutritional properties of food, and they have been examined for their capacity to encapsulate natural metabolites that may help to protect the food from spoilage and degradation. This review focuses on ongoing advancements in the application of liposomes for food and pharma sector.

Keywords: nanotechnology, liposomes, food, agriculture, nanosensors

INTRODUCTION

Food is a natural nano-structured substance. A simple step of boiling an egg can change its sub-nanometer level, as proteins tangle together to form a solid egg white. For decades, a number of processed foods have relied on processes that are now described as components of nanotechnology (Rashidi and Khosravi-Darani, 2011). For instance, tomato ketchup is composed of very small particles dispersed in the water, whereas the fatty grains of creamer coffee powder are coated with silica nanoparticles to prevent them sticking together. Current nanotechnology enables researchers to study what happens at this scale, and these studies can aid the design of new nanostructures that improve food quality (Kampers, 2016).

Nanotechnology combined with other technologies and scientific fields such as biotechnology, chemistry, physics, and engineering has increased transformative potential. Nanotechnology has shown several advantages in a variety of fields including, civil, mechanical, electronics and electrical engineering as well as new products for medicine, wastewater, potable water treatment, biology, biochemistry, agriculture, and food processing (Chellaram et al., 2014). In the food and agricultural fields, nanotechnology offers improved food security and processing, flavor, nutrition, delivery methods, pathogen detection, food functionality, environment protection, cost-effectiveness of storage and distribution, and improved abilities of plants to absorb nutrients (He and Hwang, 2016).

A longer time span of usability, insightful packaging, and more advantageous or functional-food nourishments containing dietary supplements are among the potential outcomes offered by nanotechnology. In 2016, the USFDA (US Food and Drug Administration) issued a draft directive on the utilization of nanotechnology in foods and nourishment related items (Food Safety Authority of Ireland [FSAI], 2008). The vulnerabilities associated with nanotechnology in foods are numerous, and the FDA requires venture owners to discuss them before marketing these types of products. This move has been welcomed by health and environment campaigners, as stated by George Kimbrell of the Campaign for Food Safety: "The agency is no longer ignoring scientific consensus that these nanomaterials have the capacity to be fundamentally different, and can create new and novel risks, necessitating new testing." Nanotechnology is likewise seen by the food industry as a method for improving the food security and supplement bioavailability, and a few instances of these products are already being marketed in some countries (Food Safety Authority of Ireland [FSAI], 2008).

Liposomes have been utilized in immunoassays for decades in various formats supported by visible spectrophotometry. Liposomes are minor spheres extending in diameter from 50 nm to a few microns (Bangham and Horne, 1964) as well as vesicles composed of bilayers of phospholipids or other similar amphipathic lipids. Liposomes are used commercially as drug delivery systems, carriers for medical diagnostics, signal enhancers in analytical biochemistry, medicinal vehicles, solubilizers for various food ingredients to kill pathogens and detect toxins, and as penetration enhancers in the cosmetic industry (Papahadjopoulos, 1978; Kaes and Sackmann, 1991).

There are two types of liposomes, one called conventional and the other ones having modified surface properties. Conventional liposomes are first-generation products and include several lipid compositions with altered physicochemical properties. However, their biological properties do not change after intravenous administration (Umalkar et al., 2011). Liposomes tagged with bio-recognition agents such as DNA probe, aptamer, ganglioside, antibody, and streptavidin are used as signaling reagents for enzymes, fluorophores, latex beads, colloidal gold, and others (Edwards and Baeumner, 2013). The most commonly used recognition agents are antibodies, and antibody-sensitized liposomes (immunoliposomes) have produced encouraging results in various diagnostic applications. Immunoliposomes are prepared by tagging antibodies onto the outer surfaces of liposomes, which bind specifically to target antigens (Shin and Kim, 2008).

Immunoassays are biological analytical techniques in which the quantitation of an analyte relies on the specificity of the reaction between an antigen (analyte) and an antibody. Immunoassay methods are important pharmaceutical analytical tools for diagnosing diseases, detecting toxins, therapeutic drug monitoring, as well as clinical pharmacokinetics and bioequivalence studies in drug discovery and pharmaceutical development (Findlay et al., 2000). Immunoassays results have extremely high sensitivities and low limits of detection (Pieniaszek et al., 2003; Suzuki et al., 2003). Although antibodies and other signal generating labels are used for immunoassay development, antibodies are undoubtedly the key reagents (Darwish, 2006). Many types of signal generating labels are used in immunoassays, such as enzymes, fluorescent probes, metals and chelates, and liposomes. Based on review, the most common labels used in immunoassays are enzymes and liposomes due to their high signal amplifications and sensitivities. Immunoliposome-based immunoassays have been applied in food chemistry, clinical microbiology, food microbiology, nanobiotechnology, and diagnostics for the effective detection of food-borne pathogens, toxins, and hazardous elements in foods or the human body (Shukla et al., 2016c). Liposome immunoassays (LIA), which are based on enzyme immunoassays, utilize a liposome-encapsulated marker (SRB dye) prepared as phospholipid compositions coupled with either an analyte or antibody using standardized procedures (Shukla et al., 2012). Detection limits of these LIAs depend on liposome lysis as well as the release of encapsulated markers (Shukla et al., 2016c). Shin and Kim (2008) observed higher fluorescence signals as liposome size increased due to the greater number of SRB molecules encapsulated in larger liposomes, resulting in improved detection limits. Recent developments in immunoliposome assays include immunoliposome-coupled enzyme-linked immunosorbent assay (ELISA), immunoliposome-coupled magnetic separation, and liposome-based immunochromatographic strip assay (Shukla et al., 2016c).

Until now, there have only been a few studies published on the use of liposomes nanoparticles in various food systems. Further, the interactions of liposomes with food matrices are poorly understood with regards to potential food applications. Interest in liposome nanoparticles has increased due to their reported functional properties, including their physicochemical properties, kinetic stability, efficient encapsulation capacity, biocompatibility with food constituents, low cost of raw materials used for manufacturing, and their feasibility as nutraceuticals, antimicrobials, and flood flavoring agents (Hsieh et al., 2002; Laridi et al., 2003; Were et al., 2003; Taylor et al., 2005). The main advantage of liposome nanoparticles is their completely natural compositions, which can eliminate or even reduce issues related to their inclusion in various food systems (Taylor et al., 2005).

The main aim of this review is to provide updated information on the field of liposome nanoparticles from a research perspective relevant to food scientists. Information of liposome properties, manufacturing procedures, and their major applications in food and non-food systems such as medical systems and diagnostic strategies are discussed. The updated information and research strategies provided here may assist manufacturers and scientists dealing in food and/or food associated research making broader use of liposome-based promising technology in order to improve the quality and value of food products.

Additionally, we undertook the present review to provide an update on consumer requirements, the food industry in terms of safe food production, and approved legislation concerning the presence of hazardous toxins and food pathogens. In addition, common nanotechniques and liposome nanoparticle types, food safety concerns, and research progress are discussed.

NANOSTRUCTURES IN FOOD

Food proteins are often globular structures with a size of 1–10 nm. Most polysaccharides (sugars) and lipids (fats) are direct polymers with sub-nanometer thickness, and the functionalities of numerous crude materials and the effective handling of food emerge from the presence, alteration, and production of self-assembled nano-structures (Chen et al., 2006; Ravichandran, 2010). For instance, gelatinization and benefits of nutritional influences are controlled by crystal structure of starch, starch-based processed food, and cellulose-fibril-based planer assemblies of plant cell wall, whereas fibrous structures control gel melting, setting, and texture. Oil–water and air–water interface originated two-dimensional (2D)-nanostructures control stabilities of food-foam and food-emulsion (Rudolph, 2004). Interpretation of chemical nature of food-based nanostructures can give new insights on selecting better raw materials that may enhance the quality of the food. Various analytical and innovative approaches such as atomic force microscopy, including electron microscopy have been used to determine the nature of these nanostructures (Chau et al., 2007; Ravichandran, 2010).

STATISTICS REGARDING INCREASING INTEREST OF NANOTECHNOLOGY AND NANOPARTICLES

According to statistics published on the StatNano website, ~137,500 nanotechnology articles were indexed in the Web of Science (WoS) Database in 2016, which represents 9.5% of all articles indexed in the database. Of these 137,500 articles, 34% were published in China and 16% in the United States (Stat Nano, 2017). Nine of the 20 top countries that contributed are located in Asia and Oceania, and more than half of the nanotechnology articles were published in Asia and Oceania. Brazil was among top 20 Latin American countries with nanotechnology articles. Egypt was the top African country with 1,423 nanotechnology articles and ranked 25th, and South Africa was the second ranked African country (Stat Nano, 2017). In brief, in current years, China, United States, and India were ranked 1st, 2nd, and 3rd for publishing research in the related field of nanotechnology, thus showing increasing interest in the similar topics in the field of nanotechnology by the researchers of these countries.

The number of publications on nanotechnology has soared during the last 10~11 years (2006–2017), as depicted by the expanding number of scientific articles every year retrieved from the PubMed database by searching the term "nanotechnology," "liposome," "nanoparticles," "gold nanoparticles," and "food" in titles, abstracts, and keywords (**Figure 1**). Further, outputs probably underestimate the number of academic journal articles on nanotechnology topics unrelated to the food sector. As

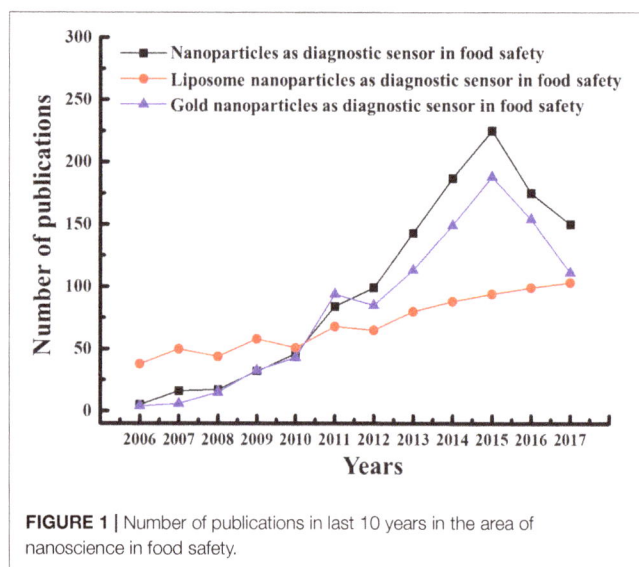

FIGURE 1 | Number of publications in last 10 years in the area of nanoscience in food safety.

shown in **Figure 1**, the food safety sector has shown increasing interest in liposome nanoparticles, which can be used as alternatives for detecting various microbial pathogens and chemical contaminants in a variety of food materials and food products. As demonstrated, although the number of publications for gold nanoparticles (GNPs) are higher than liposome nanoparticles but decreased drastically after the year 2015, while in case of liposome nanoparticles, number of publications are also in increasing order after the year 2015 (**Figure 1**).

CHARACTERISTICS OF NANOPARTICLES USEFUL IN THE FOOD SECTOR

Nanoparticles are invisible to the human eye. With a diameter of 100 nm or less, nanoparticles are 1,000 and 100-times smaller in diameter as compared to an average thickness of a book page and human hair, respectively. The various structures of nanoparticles can possibly be useful in different sectors of the food and pharmaceutical industries in food science-based research programs with proper processing and regulatory approval.

Nanoparticles incorporate huge range of classes associated with traditional materials, that include all types of metal, polymer, ceramic, and biomaterial. The development of ever smaller particles represents major challenges, mainly due to difficulties associated with characterizing the reliability, comparability, and reproducibility of materials on this scale. In light these challenges, a new sub-discipline of metrology is starting to evolve. Nanometrology, as it has become known, is the science of performing measurements at sub-100 nm dimensions as well as the characterization of these nanomaterials. However, without a righteous method of estimation, it is difficult to decide human or animal exposures via food or feed (Food Safety

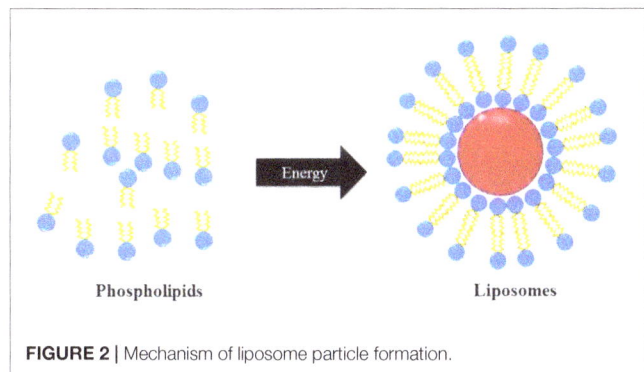

FIGURE 2 | Mechanism of liposome particle formation.

Authority of Ireland [FSAI], 2008). Of particular significance, any potential regulatory system requires assurance of the physiochemical properties of nanoparticles and estimation of their levels in food and feed. These properties incorporate size of the particle, distribution of the particle, surface range, surface charge, topography, purity and composition of particle, hydrophobicity, solubility, bioactivity, particle reactivity, and degrees of scattering.

MECHANISM OF LIPOSOME FORMATION

The detailed mechanistic phenomenon of liposome formation is not yet well known. However, predominantly there are two distinct ways of vesicle formation in the lipid-in-water suspension, which include bilayer fragmentation with subsequent self-closure of bilayered fragments, and separation of sibling vesicle from parental liposome. In simple terms, when phospholipids are placed in water and sufficient energy is provided by sonication, heating, homogenization, or other methods, bilayer vesicles are formed (**Figure 2**) (Mozafari et al., 2008). This phenomenon is probably associated with the critical micelle concentration, defined as the concentration of lipids in water above which lipids form micelles or bilayer structures rather than remaining in solution as monomers (Sharma et al., 2010; Moghimipour and Handali, 2013) of phospholipids in water.

APPLICATIONS OF LIPOSOME PARTICLES IN FOOD, AGRICULTURAL, AND OTHER SECTORS

Liposome-based micro-encapsulation techniques using edible biocompatible materials, including biopolymer matrices composed of sugar, starch, gum, protein, synthetics, dextrins, and alginates, are currently used in the food and medicine industries (Reineccius, 1995; Gibbs et al., 1999). Based on the successful use of liposome nanoparticles in biomedical and the pharmaceutical industries (cancer, and drug delivery, etc.), scientists have now started using liposomes for the controlled-delivery of functional constituents such as protein,

polysaccharide, enzyme, vitamin, and flavor in a diverse range of food practices (Aishwarya et al., 2015). **Table 1** summarizes several commercialized liposome products that have been approved for different applications including drug development, food supplements and as food preservative. Nevertheless, despite their improved stabilities and enhanced circulation times, these polymer-stabilized liposomes lack selectivity for disease targets as well as controlled drug delivery. To overcome these limitations, several engineering strategies involving the modification of various liposome components have been employed, such as surface functionalization, to improve their performances and develop various liposome detection strategies (**Figure 3**). Few other recent developments and updates related to use of liposomes in vaccine development, and toxin detection are listed in **Table 2**. Further, there are very few food applications of liposome nanoparticles, which are discussed in the following sections.

LIPOSOME NANOPARTICLE-ENCAPSULATED ENZYMES IN DAIRY FOODS

In recent decades, liposome nanoparticles have attracted huge interest for their use in the dairy food industry. During cheese ripening and hardening, time and costs can be reduced by the addition of edible enzymes such as proteinases prior to isolation of curds. However, addition of enzymes has some drawbacks such as rapid proteolysis of casein. As a solution to this problem, Law and King (1985) observed that addition of proteinases encapsulated in liposome nanoparticles to cheese-curd resulted in the formation of firm cheese through inhibiting the proteolytic process of β-casein and maintained the curd structure by preventing any enzymatic offense. Use of lipase for improving the production of cheese has been recently reviewed. Kheadr et al. (2002) reported that use of liposome-encapsulated lipase was able to reduce cheese firmness and increased the elasticity and cohesiveness of the cheddar cheese. In addition, addition of proteinases and lipases encapsulated in liposome nanoparticles to cheddar items should be attentively controlled for improved ripening and acceleration process in order to avoid negative effect on overall food quality. Composition of liposome is one of the key regulatory factors that must be considered by the manufacturers so as to keep the advancement of off-flavors and assure the payload release in a predictable way (Taylor et al., 2005).

As a Food Fortifier

Liposome nanoparticles also have potential in the food fortification field. Due to low fish consumption in many countries, dietary supplementation with fish oil capsules appears to provide a straightforward means of increasing omega-3 LC PUFA intake (Kolanowski, 2010; Bender et al., 2014). Consumers have shown considerable interest in fortified foods containing different micronutrients (Siro et al., 2008). Nevertheless, the developmental challenges related to fish oil are substantial and include instability and undesirable flavors/odors.

TABLE 1 | Approved selected liposome-based commercial products.

Liposomal product	Company	Role in human health	
Liposomal turmeric with fulvic acid	Purathrive	Anti-inflammatory	In dietary supplement
Liposomal curcumin syrup	Company Health	Anti-inflammatory	
Oral curcumin liposome syrup	Nutra Ingredients	Anti-inflammatory	
Liposomal vitamin C tablets	Dr. Mercola	Against vitamin C deficiency	
Liposomal glutathione capsules	Pure Encapsulations	Antioxidant support for liver	
Hemp liposomal syrup	Lipolife	Antioxidant, clinical strength	
Nisin-loaded liposomes	Various companies	For controlling spoilage and pathogenic bacteria in food	As a food preservative

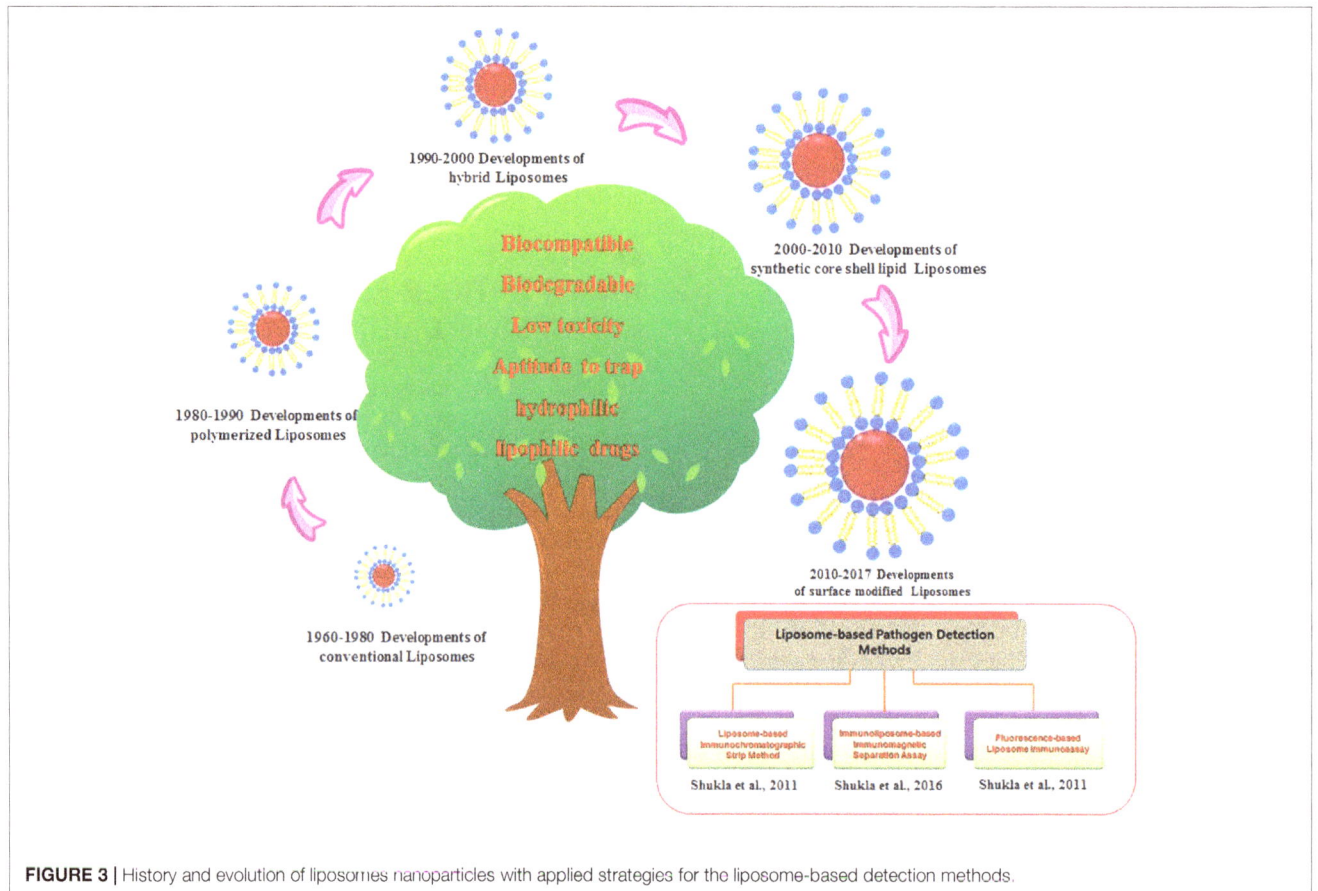

FIGURE 3 | History and evolution of liposomes nanoparticles with applied strategies for the liposome-based detection methods.

Further, their strong odors and rapid deterioration limit their applications in food formulations. Highly refined, odorless, or microencapsulated fish oil may provide a means of resolving undesirable sensory characteristics (Iafelice et al., 2008). For these reasons, Iafelice et al. (2008) used liposomes to nanoencapsulate fish oil for the fortification of yogurt. Ghorbanzade et al. (2017) also used liposome-encapsulated fish oil to fortify yogurt and observed that the sensory characteristics were closer to that of the control than yogurt fortified with free fish oil.

In recent years, several researchers have focused on encapsulation of various essential oils in liposome nanoparticles for overcoming the drawback of their being unstable and easy to degrade in natural environment since they can be easily affected by oxygen, light and temperature. Liposomes can protect the fluidity of essential oils and are stable at 4–5°C at least for 6 months (Sherry et al., 2013). Recently, Khosravi-Darani et al. (2016) reported increased antimicrobial activity of Zataria multiflora essential oil when encapsulated in liposomes. This can emphasize the potential use of these liposome nanoparticles while using natural products as potential conservative agents for using in the food and pharma industries. In addition, Sebaaly et al. (2015) synthesized clove essential oil (eugenol as a main bioactive compound) loaded liposomes, and found that these liposomes exhibited nanometric oligolamellar and spherical shaped vesicles and protected eugenol from degradation induced by UV exposure. Furthermore, these liposomes also maintained the DPPH-scavenging activity of free eugenol. These all studies proved liposome nanoparticles as a suitable system for encapsulation of volatile unstable essential oils for their various practical applicability.

TABLE 2 | Updates on liposome nanoparticles used for vaccine developments and for detection techniques.

	Liposome type/formulation/test method	Target against	Reference
For vaccine developments	Lipopeptide vaccine	Group A *Streptococcus*	Ghaffar et al., 2016
	Mannosylated ionic liposome vaccine	HIV DNA vaccine	
	PEB modified liposome with CRX-601	Influenza vaccine	Oberoi et al., 2016
	Peptide-based liposome	Antitumor vaccine	Kakhi et al., 2015
	Liposome constituting microneedles	Oral mucosal vaccination	Zhen et al., 2015
	Liposome containing E-protein vaccine	Tembusu virus vaccine in duck	Ma et al., 2016
	Ophiopogon polysaccharide liposome	Parvovirus vaccine	Fan et al., 2016
For detection strategies	Liposome-encapsulated spin-trap (LEST)	Nitric oxide	Hirsh et al., 2016
	Liposome-based microcapillary immunosensor	*E. coli*	Ho et al., 2004
	GM incorporated liposome piezoelectric agglutination	Cholera toxin	Yu et al., 2011
	Secretory proteins encapsulated liposome-based detection	*Mycobacterium tuberculosis*	Tiwari et al., 2015
	Immunoliposomes-based assays	*Salmonella, Cronobacter*	Shukla et al., 2012, 2016c

In addition, liposome nanoparticles have been used for the fortification of dairy food products by using vitamins to enhance nutritional quality and shelf-life (Taylor et al., 2005). Banville et al. (2000) also observed a significant difference in the recovery of vitamin D from cheeses containing vitamin D encapsulated in liposome nanoparticles compared to commercially prepared vitamin D-containing cheese samples, confirming protection of vitamin D via liposomal encapsulation. Other dairy products have also shown increased use of liposome nanoparticles for inducing low rate of lactose digestion, which aids digestion of dairy foods by lactose intolerant β-galactosidase entrapped in soy-phosphatidylcholine (PC)/cholesterol liposome nanoparticles (Taylor et al., 2005). Hsieh et al. (2002) reported that liposome incorporating stearic acid and alpha-tocopherol were found effective in protecting entrapped alpha-amylase from pepsin assault, storage of cold temperature and outrageous pH condition.

LIPOSOME NANOPARTICLES FOR STABILIZING FOOD COMPONENTS AGAINST DEGRADATION

In recent years, liposome nanoparticles have received interest for various food applications. Liposomes have been shown to preserve and protect vitamins in a variety of food practices. Previously, Kirby (1991) described that the antioxidant efficacy of liposome encapsulating ascorbic acid was dramatically increased when compared with free vitamin C solution. Liposome-encapsulated ascorbic acid retained >50% activity after several days of storage in the refrigerator, whereas free ascorbic acid lost all activity after 1 week of storage (Taylor et al., 2005). Additionally, other studies reported that the rates of light- or heat-induced degradation of retinol decreased upon encapsulation within liposomes regardless of the light source or temperature in comparison with free retinol (Lee S.C. et al., 2002). Further, Lee S.K. et al. (2002) also tested the antioxidant activity of free phosvitin and liposome-entrapped phosvitin in a model meat system (pork muscle homogenate and ground pork) and found that entrapped phosvitin was most effective than free phosvitin

in inhibiting lipid oxidation in unsalted, uncooked port samples and least effective in inhibiting lipid oxidation in salted cooked pork samples (Lee S.K. et al., 2002).

Overall, above findings demonstrate potent efficacy of liposomes in food system to protect and maintain nutrient bioactivity from degradation in addition to the ability of liposome-entrapped vitamins to fortify the food in order to enhance overall food quality and health benefits

ENCAPSULATION OF LIPOSOME NANOPARTICLES WITH BACTERIOCIN

Liposome nanoparticles have been utilized to encapsulate natural and synthetic bacteriocins. This represents an alternative approach to overcome the problems associated with direct application of bacteriocins to food products, such as proteolytic degradation or interactions with food components. Laridi et al. (2003) previously investigated the stability and effects of nisin Z entrapped in liposomes during cheddar cheese manufacturing. Although encapsulated nisin Z lowered the bacterial cell numbers of Lactococci, fermentation process of cheddar cheese formation was not severely hampered by encapsulated nisin Z, and it showed stability throughout the processing steps. In further support of the promising application of bacteriocin-loaded liposomes to cheese manufacturing (da Silva Malheiros et al., 2010). Benech et al. (2003) reported that nisin when encapsulated in the liposome had no adversary effect on proteolytic, rheological, and sensory properties of cheese, while direct incorporation of nisin producing strain into the cheese starter culture significantly altered proteolytic and lipolysis properties of cheese. However, no significant effect of nisinogenic strain was observed on rheological properties of cheese (da Silva Malheiros et al., 2010).

Further, da Silva Malheiros et al. (2010) encapsulated commercial nisin A into liposome nanoparticles made of partially purified PC of soybean as a cheap and readily available commodity. The prepared nisin-loaded liposomes presented high encapsulation efficiency and displayed enhanced antimicrobial activity. Overall, these investigations on liposome encapsulating antimicrobial peptides are advantageous in terms of comparing efficacy of liposome-entrapped molecule and free bacteriocin compound. These innovative research outcomes have encouraged storming demand on liposome-based research in order to improve food applications, shelf-life, and safety of food.

LIPOSOME NANOPARTICLES AS DETECTION SENSOR IN FOOD SECTOR

In Food Pathogen Detection

Kim et al. (2003) previously developed a procedure involving immunomagnetic separation as well as a test strip immunoliposome immunoassay for the sensitive detection of *Escherichia coli* O157:H7. Later, we tried to develop a rapid detection method for *Salmonella* spp. using an immunomagnetic separation/immunoliposome technique (Shin and Kim, 2008) and succeeded in developing a liposome immunosorbent immunoassay system using anti-*Salmonella* IgG (antibody)-tagged liposome (immunoliposome) to detect *Salmonella* Typhimurium (Shukla et al., 2012). More recently, we developed an immunoliposome-based chromatographic test strip assay for the detection of *S.* Typhimurium based on non-specific binding (Shukla et al., 2011). In these studies, immunoliposomes were developed by tagging with specific developed antibodies against the target pathogen. According to Chen et al. (2005), universal protein G-liposomal nanovesicles were easily conjugated by the antibodies within 30 min, and these conjugates (protein G-immunoliposomes) showed great detection ability against *E. coli* O157:H7 when incorporated in the immunomagnetic bead (IMB) assay. Further, Chen and Durst (2006) developed an array-based immunosorbent assay and confirmed the detection of *E. coli* O157:H7, *Salmonella* spp., and *Listeria monocytogenes* in pure and mixed cultures using protein G-liposomal nanovesicles. These findings confirmed the practical application of protein G-liposomal nanovesicles which can be used in various immunoassays to simultaneously detect foodborne pathogens as well as demonstrated their effectiveness as universal immunoassay reagents. Connelly et al. (2008) developed a bioanalytical detection method specific to viable human pathogenic *Cryptosporidium* spp. using lateral flow sandwich assays and liposome-encapsulated dye as a signal amplification system. These assays are inexpensive, rapid, specific, and straightforward to perform. Zaytseva et al. (2005) developed a microfluidic biosensor module based on DNA/RNA hybridization and liposome signal amplification using a fluorescence detector for the identification of pathogenic organisms and viruses. This biosensor module was designed for easy integration into micro total analysis systems that combine sample preparation and detection steps in a single chip. Park et al. (2004) also developed an immunoliposome sandwich fluorometric assay to detect *E. coli* O157:H7. These findings confirmed usefulness and practical application of immunoliposome-encapsulated fluorophores employing micro-titer plates in the automated and rapid detection of molecules at multivalent antigenic sites and demonstrated that immunoliposomes have wide applications for pathogen

detection. Baeumner et al. (2003) also developed a nucleic acid coupled liposome-based RNA biosensor assay for the rapid detection of *E. coli* in drinking water. The developed biosensor, which used a membrane-based DNA and RNA hybridization system combined with a liposome amplification process, was shown to be specific for *E. coli* and did not produce false signals due to other microorganisms. Further, DeCory et al. (2005) developed an IMB-immunoliposome fluorescence assay for the rapid detection of *E. coli* O157:H7 in aqueous samples. Their results demonstrated the usefulness of immunoliposomes for detecting *E. coli* O157:H7 in aqueous sample incorporating IMBs and sulforhodamine B. Patricia and Durst (2000) also proposed the application of an LIA to the agricultural field. They developed a liposome immunomigration, liquid-phase competition strip immunoassay for the determination of potato glycoalkaloids. In this assay, polyclonal anti-solanine antibodies were raised and used. For this format, similar cross-reactivities were measured between the glycoalkaloids R-solanine and R-chaconine. This assay has been commercialized for the quantification of total glycoalkaloids in potatoes.

Previously, Zhao et al. (2006) reported liposome-doped nanocomposites as an artificial cell-based biosensor for the sensitive detection of a pore forming hemolysin, listeriolysin O (LLO) from bacterial origin. During pore formation and membrane insertion by LLO, immobilized liposome acted as cellular surrogates. Fluorescence quenching and leaching assays were used for measuring the integrity of liposomes in solid and solid-gel glass states. Although slower but a similar kinetics was observed by LLO in liposome-doped silica composites during pore formation. Further, insertion of LLO into immobilized-liposomes was found to be pH-dependent. Also, liposome doped composites did not show any increase in the membrane permeability in the presence of LLO at pH 7.4 (Gutierrez et al., 2009). In this study, immobilized liposomes detected LLO within ~1.5 h and 30 min when steady state and kinetic calibrations were used, respectively. These liposome silica composites could potentially be used to detect hemolysin-producing *L. monocytogenes* and other bacteria that produce pore-forming toxins.

During the last decade, a variety of liposome-based assays have been developed to detect pathogens in various food matrices. Zhang et al. (2014) introduced a carbon-nanotube-based multiple-cycled liposome signal amplification assay for the detection of *E. coli* O157:H7 with a lowest detectable concentration (LOD) of 10^2 CFU/mL. Bui et al. (2015) developed a liposome-amplified plasmonic immunoassay for the visual detection of single-digit, and live pathogens, including *Salmonella, Listeria,* and *E. coli* O157:H7, in water and food samples. This approach involved the integration of a cysteine-loaded nanoliposomes into a conventional ELISA as a signal amplifier as well as the release of cysteine from nanoliposomes to cause aggregation of plasmonic GNPs as a signal-amplified response. The lowest analyte concentration analyzed and detected with a visible color shift was found to be 6.7 attomolar, which is the lowest naked eye LOD reported without use of enzymes or visualization equipment. Although signal amplification approaches have shown highly significant potential for the detection of bacterial pathogens, still they face substantial challenges as far as their large-scale manufacturing repeatability is concerned. Thus, a simple, reproducible, and sensitive means of detecting low levels of *E. coli* O157:H7 is still required.

Rapid and cheap virus particle detection using liposomes is also a research topic of considerable interest. Reichert et al. (1995) developed sialic acid-conjugated liposomes that mimic cell surface molecular recognition and signal transduction for the colorimetric detection of influenza viruses. Virus recognition is based on binding between sialic acid and hemagglutinin lectin on the viral surface. Upon binding to influenza virus, liposome particles exhibited a color change from blue to pink/orange, which was quantitative spectroscopically (Shinde et al., 2012). Charych et al. (1993) previously optimized the colorimetric detection of influenza virus particles using polymerized liposomes containing sialic acid, and Charych et al. (1996) reported a LOD of ~1 HAU in 250 μL (~4,000 viruses per μL) using this method. Hidari et al. (2007) reported LODs as low as ~0.1 pM (~6×10^4 viruses per μL) using a method based on surface plasmon resonance and immobilized sialic acid-containing liposomes. A similar LOD was obtained for influenza A virus using SPR based on a chip with immobilized bovine brain lipid-containing sialoglycolipids (Critchley and Dimmock, 2004). Egashira et al. (2008) developed a rapid and highly sensitive detection method by combining electrochemiluminescence with an immunoliposome-encapsulated Ru complex. Under optimum measurement conditions, hemagglutinin concentrations in influenza virus were measured in the concentration range of 3×10^{-13} to 4×10^{-11} g/mL, suggesting that 6×10^{-19} mol/50 μL could be considered as the LOD of this method for viral hemagglutinin. Notably, these findings showed that a method with high detection sensitivity at the attomolar level could be devised for detecting trace amounts of proteins in influenza virus. Damhorst et al. (2013) also developed an ELISA-inspired lab-on-a-chip strategy using an electrical sensing technique to detect biological entities tagged with liposome-encapsulated ion nanoparticles based on ion release impedance spectroscopy measurement. According to this method, ion-containing dipalmitoylphosphatidylcholine liposomes functionalized with antibodies were stable in deionized water but became permeable and released ions upon heating, suggesting that these liposomes could be optimal representatives for biosensing of surface-immobilized antigens electrically.

The authors further demonstrated the potential of these liposomes in quantifying viral counts using real-time impedance measurements as well as their feasibility as a reliable platform for the determination of viral loads with an ability to detect pathogenic microbes and other bio-fragments (Damhorst et al., 2013). In addition, these liposome-based methods for pathogen detection have been investigated by others (Shukla et al., 2016a,b; Song et al., 2016). Recently, attempts have been made to develop advanced liposome-based detection methods to achieve multiplexing and gain approval for commercial applications. Shukla et al. (2016a) reported limitations due to lack of multiplexing formats. However, since all liposome-based detection methods are based on fluorescence signals that differ

according to various pathogenic bacteria, single-step multiple detection is a feasible target. Possible research result formats based on our earlier research were published by Shukla et al. (2016a,b).

Pesticide Detection

A general awareness has become vibrant regarding extensive applications of pesticides and their deleterious impact on the global environment. Nowadays, there are huge applications of pesticides in food and agriculture industries including pest control. Although a trace amount of pesticide compound has ability to affect the central nervous system and can cause respiratory, myocardial, and neuromuscular malfunctioning. Further, due to their widespread use, pesticide residues can be found in the soil, atmosphere, agricultural products, and groundwater. For these reasons, methods are needed that can detect pesticides directly at low concentrations in foods and drinking water (Vamvakaki and Chaniotakis, 2007).

Pesticides are usually detected using analytical methods such as LC, GC (Hall et al., 1997), and MS (Yao et al., 1991). However, they require longer time for analysis, highly qualified technical hands for operation, and are highly expensive, and cannot be used in the field trials. Much effort has been expended to develop detection methods for pesticides, but few efforts have been made to develop pesticide detection methods based on liposome nanoparticles. Immunoliposome assays are used for various environmental monitoring purposes, including field determination of pesticides and other micro-pollutants in food and water samples.

Liposome-encapsulated ascorbic acid has been used in competitive assay formats (both lateral and horizontal flow formats) with amperometric detection following Triton X-100-induced lysis for the detection of the pesticide atrazine (Baumner and Schmid, 1998). Vamvakaki and Chaniotakis (2007) described a unique approach to the development of an AChE-based inhibitor nanobiosensor for pesticide analysis. A fluorescence detection scheme was chosen as a transduction scheme since it generally provides higher sensitivity, lower LODs, and wider detection ranges. The inherently unstable enzyme AChE from *Drosophila melanogaster* was previously encapsulated by liposomes, which showed great ability to improve enzyme stabilization against unfolding, denaturation, and dilution effect (Colletier et al., 2002; Chaize et al., 2003). Further, substrate release by liposomes has been shown to be facilitated by porins in the liposome membrane (Vamvakaki et al., 2005). In the method devised by Vamvakaki and Chaniotakis (2007), biosensor response was achieved using a pH-sensitive fluorescent indicator (pyranine; pKa 7.3) with a pKa close to the operational pH of AChE (pH 7.0). These stable nanobiosensors were used to directly detect of two widely used organophosphorus pesticides, dichlorvos and paraoxon, as well as to determine the total toxicities of drinking water samples.

Toxin Detection

Wen et al. (2005) developed a novel method based on an Ara h1-tagged liposome-based lateral flow assay for the extraction and detection of peanut allergenic proteins in chocolate. Gangliosides, as natural toxin receptors of bacteria and viruses, are expressed on the surface of the cells, and several assays have been established for their specific targeting (Shinde et al., 2012). Singh et al. (2000) assembled GT1b or GM1 ganglioside-bearing liposomes to detect bacterial toxins of tetanus, botulinum, and cholera. These engineered liposomes acted as a cell and showed ability for the recognition of targeted toxins, thereby constituting a basis for the detection of specific toxins (Shinde et al., 2012). Further, a fluorescent rhodamine marker dye was used for the labeling of liposomes and a sandwich fluoroimmunoassay was applied on antibody-coated microtiter plate. Similarly, Ahn-Yoon et al. (2003) reported detection of cholera toxin employing GM1 ganglioside-containing liposome assay. In this bioassay, cholera toxin was detected as a colored band on nitrocellulose membrane strips, implying cholera toxin bound to liposomes was captured by immobilized antibodies. Ahn-Yoon et al. (2003, 2004) developed a ganglioside-LIA using the strip assay format for the ultrasensitive detection of cholera toxin and botulinum toxin and observed no any cross-reactivity. Ho and Wauchope (2002) developed a strip LIA for the detection of aflatoxin B1 at levels as low as 20 ng, thereby providing a potentially rapid means of visually screening agricultural and food samples. Hirsh et al. (2016) described a liposome-encapsulated spin-trap (LEST) method for the capture and in situ detection of NO (nitric oxide) by electron paramagnetic resonance and obtained a linear response for [NO] > 4 mM with a detection limit of [NO] > 40 nM in a 500 mL sample (>20 pmol). The LEST method determined time-dependent NO production kinetics with no inhibitory effect on the activity of inducible NO synthase or nitrate reductase. Also, the method showed minimal abiotic production of NO with nitrite and NADH. Further, the method could detect nitrate reductase-like activity in cell lysates of the coccolithophore Emiliania huxleyi, and this activity was shown to be elevated in virus-infected cultures. The LEST exhibited potential ability on the detection of NO in cell lysate and for the preparation of crude NO-producing tissue. Chu and Wen (2013) also described an advanced IMB-based liposomal fluorescence immunoassay for the detection of gliadin toxic and allergic protein in gluten-free food products. The assay incorporated anti-gliadin antibody-conjugated IMBs and fluorescent dye-loaded immunoliposomal nanovesicles (IMLNs) to capture gliadin in the sample and to produce and/or enhance the detection of the signals, respectively. The complex formed is an "IMB–gliadin–IMLN" sandwich. Based on the experimental results, the developed immunoassay exhibited good detection sensitivity of gliadin with a lowest gliadin detection limit of 0.6 μg/mL, although the polyclonal antibody used exhibited slight cross-reactivity with barley and rye.

A number of immunoassays utilizing antibody-coupled liposomes (immunoliposomes) were performed homo- and heterogeneously with an aim to detect antigens with multiple binding sites (Singh et al., 1995). In many of these homogeneous immunoassays, immunoliposomes are designed to undergo lysis upon exposure to targeted antigens that bind to antibodies on the surfaces of liposomes (Singh et al., 1995). On the other hand, most of the heterogeneous immunoassays devised with immunoliposomes use a sandwich format (Singh et al.,

1995). Singh et al. (1995) utilized bifunctional vesicles tagged with enzymes and antibodies to detect d-dimer in human plasma and observed a nine-times lower detection limit than traditional ELISAs. Since the method employs increasing time of incubation to enable enzyme–substrate reactions for signal amplification and is thus unsuitable for routine automated assays. Rongen et al. (1995) developed a sandwich format of biotinylated liposome-encapsulated carboxyfluorescein employing a microtiter plate to measure the comparative sensitivities to that of the best colorimetric immunoassay for the detection of human interferon-γ. The assay system consisting a pump and auto-sampler was also automatized via placing a microtiter plate into a flow injection system. In the findings of Emanuel et al. (1996), it was found that polyethylene glycols steric barrier significantly improved the specific binding between immunoliposomes and targeted cells on the surface of the liposomes by preventing the non-specific binding on the surface of the cells. Currently, intensive efforts are being made on the use of immunodiagnostic assays employing these liposomes for the analysis of a wide range of analytes.

Research is also being conducted on other types of nanoparticle-based sensing devices for bacterial pathogens, virus particles, hazardous chemicals, toxic proteins, peptides, and other entities (Kim et al., 2016a,b). **Figure 3** represents the evolution of various kinds of liposome nanoparticles for their potential commercial usage. Our research group also has tried to develop applications and novel strategies for liposome-based detection of major food contaminants (**Figure 3**).

Liposome nanoparticles are noted down with several potent advantages, including low cost, easy biocompatibility and biodegradability, and easy surface modification nature of phospholipid bilayer as compared to conventional and other alternative nanoparticles such as GNPs, silver nanoparticles (SNPs), magnetic nanoparticles (MNPs), and silica nanoparticles (Shukla et al., 2011; Sung et al., 2016). Especially for diagnostic detections, liposomes can encapsulate hundreds of thousands of bioluminescence, fluorescent dyes or other signals which provide strong signal amplification. This makes liposomes very sensitive in biosensing assays as compared to other nanoparticles. Prior publications also show that liposome nanoparticles have received increasing research interest as diagnostic sensors due to their natural biocompatible properties than GNPs (**Figure 1**).

Although GNPs have proved as suitable candidates used for the diagnostic detection of foodborne pathogens, pesticides and toxins in food industries (Priyadarshini and Pradhan, 2017), apart from the positive factors, a majorly noted negative factor with liposome nanoparticles is associated with their low stability and shelf-life during their storage. Therefore, in order to improve the stability and long storage ability of liposome nanoparticles, researchers are finding the alternative ways to prepare stable liposome nanoparticles by re-constituting their composition. In a recent research, Peng et al. (2017) prepared hybrid liposomes, composed of amphiphilic chitosan and phospholipid and speculated that hydrophobic force between amphiphilic chitosan and phospholipid would help to stabilize liposomal membrane that may provide better detection strategies than

GNPs and other detection strategies and may offer improved applications in food sectors.

PROS AND CONS OF INDUSTRIAL APPLICABILITY OF LIPOSOMES

Liposomes are composed of phospholipids, which in water form lipid bilayer spheres enclosing aqueous cores. Phospholipids are amphiphilic, and this characteristic is responsible for the formation of vesicles in aqueous solutions caused by hydrogen bonding, van der Waals forces, and other electrostatic interactions (Pattni et al., 2015). This vesicle structure means that liposomes can encapsulate hydrophilic or lipophilic molecules and determine the solubilities and *in vivo* fates of these encapsulated entities. The advantages offered by liposomes are as follows: (i) they improve the solubilities of encapsulated targets, (ii) they prevent the chemical and biological degradation of encapsulated materials under storage conditions, (iii) they reduce the side effects and toxicities for human health and thus improve efficacies and therapeutic indices, (iv) they can be chemically modified with specific surface ligands for targeting purposes, and (v) they are compatible with biodegradable and non-toxic materials.

According to a recent literature survey, liposome nanoparticles were used as a biological sensor for microbial pathogens, food toxins, and pesticides. Although a few nanobiosensors have been developed for various detection strategies, several other nanomaterials are also commonly used as an alternative to liposome nanoparticles for the detection of foodborne pathogens and toxins, including GNPs, gold nanorods, MNPs, quantum dots (QDs), SNPs, and silica nanoparticles (Leonard et al., 2003; Valdes et al., 2009; Lopez and Merkoci, 2011; Inbaraj and Chen, 2016). Generally, although the detection of foodborne pathogens and toxins is performed by exploiting the optical (optical sensors) or electronic (electrochemical sensors) properties of the nanomaterials as their value added characteristics (Lopez and Merkoci, 2011; Inbaraj and Chen, 2016), liposome nanoparticles have potential advantages over other kinds of nanoparticles used for various different approaches, including sensing approach because of their biocompatibility and biodegradability (Hsieh et al., 2002; Laridi et al., 2003; Were et al., 2003). Encapsulation or surface attachment of sensor materials as well as the subsequent release of liposome contents provide a simple and effective approach for signal amplification and transduction. Liposomes are compatible with current sensor technologies, including semiconductor QDs, nanoparticles, immunoassay, and electrochemical, fluorescence, and optical spectroscopy, etc. (Inbaraj and Chen, 2016). In addition, by incorporating different sensor materials and target molecules, liposome-based sensors can easily be modified for multi-target detection and sensing.

Previously, a number of different strategies of using GNP and SNP-based applications have been approved as potential alternatives in food industries for their use as detection biosensors, food fortifiers, enzyme and bacteriocin encapsulators and food stabilizers (Carbone et al., 2016; Fahim et al., 2016;

Kang et al., 2016; Üzer et al., 2016). In spite of this, there are numerous other possible uses of liposomes in the food industries, including protection and stability of hypersensitive ingredients, and to improve the adequacy of food preservatives (Carbone et al., 2016; Fahim et al., 2016; Üzer et al., 2016). Encapsulation of liposome entrapped material has shown a wide range of chemical and environmental stabilities against chemical and enzymatic modifications, including temperature and pH stability. Also, implementation of any newly developed liposome-based formulation and complications on their regulation authorities can be promptly overcome because of their natural chemical composition. However, their successful commercialization is limited by liposome-associated cytotoxic effects. Further, many liposomes leak their payloads almost immediately after administration. In addition, it has been reported that charged liposomes are toxic (Alhajlan et al., 2013), and it is possible that some liposome production methods result in trace amounts of organic solvents in the final preparations. In addition, the storage stability of liposomes is a limiting factor in their application due to their low thermodynamic stability. Liposomes aggregate, fuse, and eventually precipitate overtime. Moreover, degradation could induce the release of encapsulated materials and shorten the shelf-life of liposomes during storage. Particle composition, pH, ionic strength, physicochemical properties of encapsulated materials, light, and environmental temperature can all reduce or alter the stabilities of liposomal products.

Liposomes also have some manufacturing-associated issues such as batch-to-batch irreproducibility, lack of effective sterilization methods, stability problems, and most importantly, scale-up problems. Classical methods such as use of transmembrane gradients as well as novel methods such as microfluidization, freeze-thawing, supercritical reverse phase evaporation, and spray-drying have been used to improve encapsulation efficiency and size reproducibility, and production of lyophilized liposomal powders for reconstitution can be used to improve stabilities. Further, no universal method has been devised for liposome sterilization, as filtration through 0.22 μm filters remains the most commonly used technique. However, this raises serious concerns regarding batch sizes and is perhaps the greatest concern regarding the scalability of liposome technology. Multifunctional liposomes also have production issues, which in combination with raw material costs means that liposomes tend to be expensive. Nevertheless, given the technological advancements made during the past decade, it is evident that liposome-based research developments will promote the use of liposome nanoparticles as effective detection and other food safety aspects.

CONCLUSION

This review highlights the potential uses of liposomes and other nanomaterials for food analysis. Liposome-based sensors bind or react with the biological components of targets and generate detectable signals, which enable the rapid detection of food contaminants and enhance food safety by allowing prompt preventive action. In addition, they provide rapid, sensitive, and user-friendly detection assays that are portable and suitable for in-field applications. However, several issues such as interference during real-sample analysis, reproducibility, and toxicity remain to be resolved, and thus the need remains for more efficient liposome-based nanosensors for the food and medical sciences.

AUTHOR CONTRIBUTIONS

SS and VKB drafted and wrote the manuscript. SS, YSH, YH, SKH, and Y-KH contributed interpretation and analyzed data.

VKB, YSH, and Y-KH contributed to conception and provided technical support.

FUNDING

The authors gratefully acknowledge the financial support from the Basic Science Research Program through the National Research Foundation of Korea (NRF) funded by the Ministry of Science, ICT and Future Planning (2016R1A2B4013374 and 2017R1D1A1B03035373).

REFERENCES

Ahn-Yoon, S., DeCory, T. R., Baeumner, A. J., and Durst, R. A. (2003). Ganglioside-liposome immunoassay for the ultrasensitive detection of cholera toxin. *Anal. Chem.* 75, 2256–2261. doi: 10.1021/ac026428t

Ahn-Yoon, S., DeCory, T. R., and Durst, R. A. (2004). Ganglioside-liposome immunoassay for the detection of botulinum toxin. *Anal. Bioanal. Chem.* 378, 68–75.

Aishwarya, M., Subin, R. C. K. R., Quan Sophia, H., Laurent, B., and Chibuike, C. U. (2015). Encapsulation of food protein hydrolysates and peptides: a review. *RSC Adv.* 5, 79270–79278. doi: 10.1039/C5RA13419F

Alhajlan, M., Alhariri, M., and Omri, A. (2013). Efficacy and safety of liposomal clarithromycin and its effect on *Pseudomonas aeruginosa* virulence factors. *Antimicrob. Agents Chemother.* 57, 2694–2704. doi: 10.1128/AAC.00235-13

Baeumner, A. J., Cohen, R. N., Miksic, V., and Min, J. (2003). RNA biosensor for the rapid detection of viable *Escherichia coli* indrinking water. *Biosens. Bioelectron.* 18, 405–413. doi: 10.1016/S0956-5663(02)00162-8

Bangham, A. D., and Horne, R. W. (1964). Negative staining of phospholipids and their structural modification by-surface active agents as observed in electron microscope. *J. Mol. Biol.* 8, 660–668. doi: 10.1016/S0022-2836(64)80115-7

Banville, C., Vuillemard, J. C., and Lacroix, C. (2000). Comparison of different methods for fortifying Cheddar cheese with vitamin D. *Int. Dairy J.* 10, 375–382. doi: 10.1016/S0958-6946(00)00054-6

Baeumner, A. J., and Schmid, R. D. (1998). Development of a new immunosensor for pesticide detection: a disposable system with liposome-enhancement and amperometric detection. *Biosens. Bioelectron.* 13, 519–528. doi: 10.1016/S0956-5663(97)00131-0

Bender, N., Portmann, M., Heg, Z., Hofmann, K., Zwahlen, M., and Egger, M. (2014). Fish or n3-PUFA intake and body composition: a systematic review and meta-analysis. *Obes. Rev.* 15, 657–665. doi: 10.1111/obr.12189

Benech, R. O., Kheadr, E. E., Lacroix, C., and Fliss, I. (2003). Impact of nisin producing culture and liposome-encapsulated nisin on ripening of *Lactobacillus* added-cheddar cheese. *J. Dairy Sci.* 86, 1895–1909. doi: 10.3168/jds.S0022-0302(03)73776-X

Bui, M. P. N., Ahmed, S., and Abbas, A. (2015). Single-digit pathogen and attomolar detection with the naked eye using liposome-amplified plasmonic immunoassay. *Nano Lett.* 15, 6239–6246. doi: 10.1021/acs.nanolett.5b02837

Carbone, M., Donia, D. T., Sabbatella, G., and Antiochia, R. (2016). Silver nanoparticles in polymeric matrices for fresh food packaging. *J. King Saud. Univ. Sci.* 4, 273–279. doi: 10.1016/j.ijfoodmicro.2010.07.001

Chaize, B., Winterhalter, M., and Fournier, D. (2003). Encapsulation of acetylcholinesterase in preformed liposomes. *BioTechniques* 34, 1158–1162.

Charych, D., Cheng, Q., Reichert, A., Kuziemko, G., Stroh, M., Nagy, J. O., et al. (1996). A 'litmus' test for molecular recognition using artificial membranes. *Chem. Biol.* 3, 113–120. doi: 10.1016/S1074-5521(96)90287-2

Charych, D. H., Nagy, J. O., Spevak, W., and Bednarski, M. D. (1993). Direct colorimetric detection of a receptor-ligand interaction by a polymerized bilayer assembly. *Science* 261, 585–588. doi: 10.1126/science.8342021

Chau, C. F., Wu, S. H., and Yen, G. C. (2007). The development of regulations for food nanotechnology. *Trends Food Sci. Technol.* 18, 269–280. doi: 10.1016/j.tifs.2007.01.007

Chellaram, C., Murugaboopathi, G., John, A. A., Sivakumar, R., Ganesan, S., Krithika, S., et al. (2014). Significance of nanotechnology in food industry. *APCBEE Proced.* 8, 109–113. doi: 10.1016/j.apcbee.2014.03.010

Chen, C. S., Baeumner, A. J., and Durst, R. A. (2005). Protein G-liposomal nanovesicles as universalreagents for immunoassays. *Talanta* 67, 205–211. doi: 10.1016/j.talanta.2005.02.018

Chen, C. S., and Durst, R. A. (2006). Simultaneous detection of *Escherichia coli* O157:H7, *Salmonella* spp. and *Listeria monocytogenes* with an array-based immunosorbent assayusing universal protein G-liposomal nanovesicles. *Talanta* 69, 232–238. doi: 10.1016/j.talanta.2005.09.036

Chen, H., Weiss, J., and Shahidi, F. (2006). Nanotechnology in nutraceuticals and functional foods. *Food Technol.* 60, 30–32.

Chu, P. T., and Wen, H. W. (2013). Sensitive detection and quantification of gliadin contamination in gluten-free food with immunomagnetic beads based liposomal fluorescence immunoassay. *Anal. Chim. Acta* 17, 246–253. doi: 10.1016/j.aca.2013.05.014

Colletier, J. P., Chaize, B., Winterhalter, M., and Fournier, D. (2002). Protein encapsulation in liposomes: efficiency depends on interactions between protein and phospholipid bilayer. *BMC Biotechnol.* 2:9. doi: 10.1186/1472-6750-2-9

Connelly, J. T., Nugen, S. R., Wysocki, W. B., Durst, R. A., Montagana, R. A., and Baeumner, A. J. (2008). Human pathogenic *Cryptosporidium* species bioanalytical detection method with single oocyst detection capability. *Anal. Bioanal. Chem.* 391, 487–495. doi: 10.1007/s00216-008-1967-2

Critchley, P., and Dimmock, N. J. (2004). Binding of an influenza A virus to a neomembrane measured by surface plasmon resonance. *Biorg. Med. Chem.* 12, 2773–2780. doi: 10.1016/j.bmc.2004.02.042

da Silva Malheiros, P., Daroit, D. J., and Brandelli, A. (2010). Food applications of liposome-encapsulated antimicrobial peptides. *Trends Food Sci. Technol.* 21, 284–292. doi: 10.1016/j.colsurfb.2016.05.080

Damhorst, G. L., Smith, C. E., Salm, E. M., Sobieraj, M. M., Ni, H., Kong, H., et al. (2013). A liposome-based ion release impedance sensor for biological detection. *Biomed. Microdev.* 15, 895–905. doi: 10.1007/s10544-013-9778-4

Darwish, I. A. (2006). Immunoassay methods and their applications in pharmaceutical analysis: basic methodology and recent advances. *Int. J. Biomed. Sci.* 2, 217–235.

DeCory, T. R., Durst, R. A., Zimmerman, S. J., Garringer, L. A., Paluca, G., DeCory, H. H., et al. (2005). Development of an immunomagnetic bead immunoliposome fluorescence assay for rapid detection of *Escherichia coli* O157:H7 in aqueous samples and comparison of the assay with a standard microbiological method. *Appl. Environ. Microbiol.* 71, 1856–1864. doi: 10.1128/AEM.71.4.1856-1864.2005

Edwards, K. A., and Baeumner, A. J. (2013). Periplasmic binding protein-based detection of maltose using liposomes: a new class of biorecognition elements in competitive assays. *Anal. Chem.* 85, 2770–2778. doi: 10.1021/ac303258n

Egashira, N., Morita, S., Hifumi, E., Mitoma, Y., and Uda, T. (2008). Attomole detection of hemagglutinin molecule of influenza virus by combining an electrochemiluminescence sensor with an immunoliposome that encapsulates a Ru complex. *Anal. Chem.* 80, 4020–4025. doi: 10.1021/ac702625d

Emanuel, N., Kedar, E., Bolotin, E. M., Smorodinsky, N. I., and Barenholz, Y. (1996). Preparation and characterization of doxorubicin-loaded sterically stabilized immunoliposomes. *Pharm. Res.* 13, 352–359. doi: 10.1023/A:1016028106337

Fahim, H. A., Ahmed, S. K., and Ahmed, O. E. (2016). Nanotechnology: a valuable strategy to improve bacteriocin formulations. *Front. Microbiol.* 7:1385. doi: 10.3389/fmicb.2016.01385

Fan, Y., Ma, X., Hou, W., Guo, C., Zhang, J., Zhang, W., et al. (2016). The adjuvanticity of ophiopogon polysaccharide liposome against an inactivated porcine parvovirus vaccine in mice. *Int. J. Biol. Macromol.* 82, 264–272. doi: 10.1016/j.ijbiomac.2015.10.084

Findlay, J. W., Smith, W. C., Lee, J. W., and Nordblom, G. D. (2000). Validation of immunoassays for bioanalysis: a pharmaceutical industry perspective. *J. Pharm. Biomed. Anal.* 21, 1249–1273. doi: 10.1016/S0731-7085(99)00244-7

Food Safety Authority of Ireland [FSAI] (2008). *The Relevance for Food Safety of Applications of Nanotechnology in the Food and Feed Industries.* Available at: http://www.fsai.ie/assets/0/86/204/b81b142b-9ef7-414c-9614-3a969835b392.pdf [accessed December 17, 2013].

Ghaffar, K. A., Marasini, N., Giddam, A. K., Batzloff, M. R., Good, M. F., Skwarczynski, M., et al. (2016). Liposome-based intranasal delivery of lipopeptide vaccine candidates against group A streptococcus. *Acta Biomaterial.* 41, 161–168. doi: 10.1016/j.actbio.2016.04.012

Ghorbanzade, T., Jafari, S. M., Akhavan, S., and Hadavi, R. (2017). Nano-encapsulation of fish oil in nano-liposomes and its application in fortification of yogurt. *Food Chem.* 216, 146–152. doi: 10.1016/j.foodchem.2016.08.022

Gibbs, B. F., Kermasha, S., Alii, I., and Mulligan, C. N. (1999). Encapsulation in the food industry: a review. *Int. J. Food Sci. Nutr.* 50, 213–224. doi: 10.1080/096374899101256

Gutierrez, M., Ferrer, M. L., Tartaj, P., and Monte, F. (2009). "Biomedical applications of organic–inorganic hybrid nanoparticles," in *Hybrid Nanocomposites for Nanotechnology*, ed. L. Merhari (Boston, MA: Springer).

Hall, G. L., Mourer, C. R., and Shibamoto, T. (1997). Development and validation of an analytical method for naled and dichlorvos in Air. *J. Agric. Food Chem.* 45, 145–148. doi: 10.1021/jf9601224

He, X., and Hwang, H. M. (2016). Nanotechnology in food science: Functionality, applicability, and safety assessment. *J. Food Drug Anal.* 24, 671–681. doi: 10.1016/j.jfda.2016.06.001

Hidari, K., Shimada, S., Suzuki, Y., and Suzuki, T. (2007). Binding kinetics of influenza viruses to sialic acid-containing carbohydrates. *Glyco. J.* 24, 583–590. doi: 10.1007/s10719-007-9055-y

Hirsh, D. J., Schieler, B. M., Fomchenko, K. M., Jordan, E. T., and Bidle, K. D. (2016). A liposome-encapsulated spin trap for the detection of nitric oxide. *Free Radic. Biol. Med.* 96, 199–210. doi: 10.1016/j.freeradbiomed.2016.04.026

Ho, J. A., Hsu, H. W., and Huang, M. R. (2004). Liposome-based microcapillary immunosensor for detection of *Escherichia coli* O157:H7. *Anal. Biochem.* 330, 342–349. doi: 10.1016/j.ab.2004.03.038

Ho, J. A., and Wauchope, R. D. (2002). A strip liposome immunoassay for aflatoxin B1. *Anal. Chem.* 74, 1493–1496. doi: 10.1021/ac010903q

Hsieh, Y. F., Chen, T. L., Wang, Y. T., Chang, J. H., and Chang, H. M. (2002). Properties of liposomes prepared with various lipids. *J. Food Sci.* 67, 2808–2813. doi: 10.1111/j.1365-2621.2002.tb08820.x

Iafelice, G., Caboni, M. F., Cubadda, R., Criscio, T. D., Trivisonno, M. C., and Marconi, E. (2008). Development of functional spaghetti enriched with long chain omega-3 fatty acids. *Cereal Chem.* 85, 146–151. doi: 10.1094/CCHEM-85-2-0146

Inbaraj, B. S., and Chen, B. H. (2016). Nanomaterial-based sensors for detection of foodborne bacterial pathogens and toxins as well as pork adulteration in meat products. *J. Food Drug Anal.* 24, 15–28. doi: 10.1016/j.jfda.2015.05.001

Kaes, J., and Sackmann, E. (1991). Shape transitions and shape stability of giant phospholipid vesicles in pure water induced by area-to-volume changes. *Biophys. J.* 60, 825–844. doi: 10.1016/S0006-3495(91)82117-8

Kakhi, Z., Frisch, B., Bourel-Bonnet, L., Hemmerlé, J., Pons, F., and Heurtault, B. (2015). Airway administration of a highly versatile peptide-based liposomal construct for local and distant antitumoral vaccination. *Int. J. Pharm.* 496, 1047–1056. doi: 10.1016/j.ijpharm.2015.11.027

Kampers, F. (2016). *Nanotechnology can Make us Healthier.* Available at: https://www.theguardian.com/what-is-nano/nanotechnology-food-more-than-question-taste

Kang, H., Hwang, Y. G., Lee, T. G., Jin, C. R., Cho, C. H., Jeong, H. Y., et al. (2016). Use of gold nanoparticle fertilizer enhances the ginsenoside contents and anti-inflammatory effects of red ginseng. *J. Microbiol. Biotechnol.* 10, 1668–1674. doi: 10.4014/jmb.1604.04034

Kheadr, E. E., Vuillemard, J. C., and El-Deeb, S. A. (2002). Acceleration of Cheddar cheese lipolysis by using liposome-entrapped Upases. *J. Food Sci.* 67, 485–492. doi: 10.1111/j.1365-2621.2002.tb10624.x

Khosravi-Darani, K., Khoosfi, E. M., and Hossenini, H. (2016). Encapsulation of *Zataria multiflora* Boiss. Essential oil in liposome: antibacterial activity against *E. coli* O157:H7 in broth media and minced beef. *J. Food Safety* 36, 515–523. doi: 10.1111/jfs.12271

Kim, D., Kim, J., Kwak, C. H., Heo, N. S., Oh, S. Y., Lee, H., et al. (2016a). Rapid and label-free bioanalytical method of alpha fetoprotein detection using LSPR chip. *J. Cryst. Growth* 469, 131–135. doi: 10.1016/j.jcrysgro.2016.09.066

Kim, D., Kim, Y., Hong, S., Kim, J., Heo, N., Lee, M. K., et al. (2016b). Development of lateral flow assay based on size-controlled gold nanoparticles for detection of hepatitis B surface antigen. *Sensors* 16:2154. doi: 10.3390/s16122154

Kim, M., Oh, S. J., and Durst, R. A. (2003). Detection of *Escherichia coli* O157:H7 using combined procedure of immunomagnetic separation and test strip liposome immunoassay. *J. Microbiol. Biotechnol.* 13, 509–516.

Kirby, C. J. (1991). Microencapsulation and controlled delivery of food ingredients. *Food Sci. Technol. Today* 5, 74–78.

Kolanowski, W. (2010). Omega-3 LC PUFA contents and oxidative stability of encapsulated fish oil dietary supplements. *Int. J. Food Proper.* 13, 498–511. doi: 10.3382/ps.2009-00232

Laridi, R., Kheadr, E. E., Benech, R. O., Vuillemard, J. C., Lacroix, C., and Fliss, I. (2003). Liposome encapsulated nisin Z: optimization, stability, and release during milk fermentation. *Int. Dairy J.* 13, 325–336. doi: 10.1016/S0958-6946(02)00194-2

Law, B. A., and King, J. S. (1985). Use of liposomes for proteinase addition to Cheddar cheese. *J. Dairy Res.* 52, 183–188. doi: 10.1017/S0022029900024006

Lee, S. C., Yuk, H. G., Lee, D. H., Lee, K. E., Ludescher, Y. I., and Ludescher, R. D. (2002). Stabilization of retinol through incorporation into liposomes. *J. Biochem. Mol. Biol.* 35, 358–363.

Lee, S. K., Han, J. H., and Decker, E. A. (2002). Antioxidant activity of phosvitin in phosphatidylcholine liposomes and meat model systems. *J. Food Sci.* 67, 37–41. doi: 10.1111/j.1365-2621.2002.tb11355.x

Leonard, P., Hearty, S., Brennan, J., Dunne, L., Quinn, J., Chakraborty, T., et al. (2003). Advances in biosensors for detection of pathogens in food and water. *Enzyme Microb. Technol.* 32, 3–13. doi: 10.1016/S0141-0229(02)00232-6

Lopez, B. P., and Merkoci, A. (2011). Nanomaterials based biosensors for food analysis applications. *Trends Food Sci. Technol.* 2, 625–639. doi: 10.1016/j.tifs.2011.04.001

Ma, T., Liu, Y., Cheng, J., Liu, Y., Fan, W., Cheng, Z., et al. (2016). Liposomes containing recombinant E protein vaccine against duck Tembusu virus in ducks. *Vaccine* 34, 2157–2163. doi: 10.1016/j.vaccine.2016.03.030

Moghimipour, E., and Handali, S. (2013). Liposomes as drug delivery systems: properties and applications. *Res. J. Pharm. Biol. Chem. Sci.* 4, 163–185.

Mozafari, M. R., Johnson, C., Hatziantoniou, S., and Demetzos, C. J. (2008). Nanoliposomes and their applications in food nanotechnology. *J. Liposome Res.* 18, 309–327. doi: 10.1080/08982100802465941

Oberoi, H. S., Yorgensen, Y. M., Morasse, A., Evans, J. T., and Burkhart, D. J. (2016). PEG modified liposomes containing CRX-601 adjuvant in combination with methylglycol chitosan enhance the murine sublingual immune response to influenza vaccination. *J. Cont. Release* 223, 64–74. doi: 10.1016/j.jconrel.2015.11.006

Papahadjopoulos, D. (1978). Liposomes and their use in biology and medicine. *Ann. N. Y. Acad. Sci.* 308, 1–462. doi: 10.1111/j.1749-6632.1978.tb22009.x

Park, S., Oh, S., and Durst, R. A. (2004). Immunoliposomes sandwich fluorometric assay (ILSF) for detection of *Escherichia coli* O157:H7. *J. Food Sci.* 69, 151–156. doi: 10.1111/j.1365-2621.2004.tb11002.x

Patricia, G. P., and Durst, R. A. (2000). Determination of potato glycoalkaloids using a liposome immunomigration, liquid-phase competition immunoassay. *J. Agric. Food Chem.* 48, 1678–1683. doi: 10.1021/jf990349

Pattni, B. S., Chupin, V. V., and Torchilin, V. P. (2015). New developments in liposomal drug delivery. *Chem. Rev.* 115, 10938–10966. doi: 10.1021/acs. chemrev.5b00046

Peng, S., Zou, L., Liu, W., Li, Z., Liu, W., Hu, X., et al. (2017). Hybrid liposomes composed of amphiphilic chitosan and phospholipid: Preparation, stability and bioavailability as a carrier for curcumin. *Carbohyd. Polym.* 156, 322–332. doi: 10.1016/j.carbpol.2016.09.060

Pieniaszek, H. J., Davidson, A. F., Walton, H. L., and Pinto, D. J. (2003). A double antibody radioimmunoassay for the determination of XV459, the active hydrolysis metabolite of roxifiban, in human plasma. *J. Pharm. Biomed. Anal.* 30, 1441–1449. doi: 10.1016/S0731-7085(02) 00481-8

Priyadarshini, E., and Pradhan, N. (2017). Gold nanoparticles as efficient sensors in colorimetric detection of toxic metal ions: a review. *Sens. Actuators B* 238, 888–902. doi: 10.1016/j.snb.2016.06.081

Rashidi, L., and Khosravi-Darani, K. (2011). The applications of nanotechnology in food industry. *Crit. Rev. Food Sci. Nutr.* 51, 723–730. doi: 10.1080/ 10408391003785417

Ravichandran, R. (2010). Nanotechnology applications in food and food processing: Innovative green approaches, opportunities and uncertainties for global market. *Int. J. Green Nanotechnol.* 1, 72–96. doi: 10.1080/ 19430871003684440

Reichert, A., Nagy, J. O., Spevak, W., and Charych, D. (1995). Polydiacetylene liposomes functionalized with sialic acid bind and colorimetrically detect influenza virus. *J. Am. Chem. Soc.* 117, 829–830. doi: 10.1021/ja001 07a032

Reineccius, G. A. (1995). "Liposomes for controlled release in the food industry," in *Encapsulation and Controlled Release of Food Ingredients*, eds S. J. Risch and G. A. Reineccius (Washington, DC: ACS Symposium Series), 113–131. doi: 10.1021/bk-1995-0590.ch011

Rongen, H. A. H., Vannierop, T., Vanderhorst, H. M., Rombouts, R. F. M., Vandermeide, P. H., Bult, A., et al. (1995). Biotinylated and streptavidinylated liposomes as labels in cytokine immunoassays. *Anal. Chimi. Acta* 306, 333–341. doi: 10.1016/0003-2670(95)00014-Q

Rudolph, M. J. (2004). Cross-industry technology transfer. *Food Technol.* 58, 32–34.

Sebaaly, C., Jraij, A., Fessi, H., Charcosset, C., and Greige-Gerges, H. (2015). Preparation and characterization of clove essential oil-loaded liposomes. *Food Chem.* 178, 52–62. doi: 10.1016/j.foodchem.2015.01.067

Sharma, V. K., Mishra, D. N., Sharma, A. K., and Srivastava, B. (2010). Liposomes: present prospective and future challenges. *Int. J. Curr. Pharmaceut. Rev. Res.* 1, 7–16.

Sherry, M., Charcosset, C., Fessi, H., and Greige-Gerges, H. (2013). Essential oils encapsulated in liposomes: a review. *J. Liposome Res.* 23, 268–275. doi: 10.3109/ 08982104.2013.819888

Shin, J., and Kim, M. (2008). Development of liposome immunoassay for *Salmonella* spp. Using immunomagnetic separation and immunoliposome. *J. Microbiol. Biotechnol.* 18, 1689–1694.

Shinde, S. B., Fernandes, C. B., and Patravale, V. B. (2012). Recent trends in *in-vitro* nanodiagnostics for detection of pathogens. *J. Control Release* 159, 164–180. doi: 10.1016/j.jconrel.2011.11.033

Shukla, S., Bang, J., Heu, S., and Kim, M. H. (2012). Development of immunoliposome-based assay for the detection of *Salmonella* Typhimurium. *Eur. Food Res. Technol.* 234, 53–59. doi: 10.1007/s00217-011-1606-6

Shukla, S., Hong, S. Y., Chung, S. H., and Kim, M. (2016c). Rapid detection strategies for the global threat of Zika virus: current state, new hypotheses, and limitations. *Front. Microbiol.* 7:1685. doi: 10.3389/fmicb.2016. 01685

Shukla, S., Lee, G., Song, X., Park, J. H., Cho, H., Lee, E. J., et al. (2016b). Detection of *Cronobacter sakazakii* in powdered infant formula using an immunoliposome-based immunomagnetic concentration and separation assay. *Sci. Rep.* 6, e34721. doi: 10.1038/srep34721

Shukla, S., Lee, G., Song, X., Park, S., and Kim, M. (2016a). Imunoliposome-based immunomagnetic concentration and separation assay for rapid detection of *Cronobacter sakazakii*. *Biosens. Bioelectron.* 77, 986–994. doi: 10.1016/j.bios. 2015.10.077

Shukla, S., Leem, H., and Kim, M. H. (2011). Development of liposome based immunochromatographic strip assay for the detection of *Salmonella*. *Anal. Bioanal. Chem.* 401, 2581–2590. doi: 10.1007/s00216-011-5327-2

Singh, A. K., Harrison, S. H., and Schoeniger, J. S. (2000). Gangliosides as receptors for biological toxins: development of sensitive fluoroimmunoassays using ganglioside-bearing liposomes. *Anal. Chem.* 72, 6019–6024. doi: 10.1021/ ac000846l

Singh, A. K., Kilpatrick, P. K., and Carbonell, R. G. (1995). Non-competitive immunoassaysusing bifunctionalunilamellar vesicles (or liposomes). *Biotechnol. Prog.* 11, 333–341. doi: 10.1021/bp0003 3a014

Siro, I., Kapolna, E., Kapolna, B., and Lugasi, A. (2008). Functional food. Product development, marketing and consumer acceptance-A review. *Appetite* 51, 456–467. doi: 10.1016/j.appet.2008.05.060

Song, X., Shukla, S., Lee, G., and Kim, M. (2016). Immunochromatographic strip assay for detection of *Cronobacter sakazakii* in pure culture. *J. Microbiol. Biotechnol.* 28, 1855–1862. doi: 10.4014/jmb.1606.06004

Stat Nano. (2017). *Ranking of Countries in Nanotechnology Publications in 2016.* Available from: http://statnano.com/news/57105

Sung, T. C., Chen, W. Y., Shah, P., and Chen, C. S. (2016). A replaceable liposomal aptamer for the ultrasensitive and rapid detection of biotin. *Sci. Rep.* 6, e21369. doi: 10.1038/srep21369

Suzuki, Y., Arakawa, H., and Maeda, M. (2003). The immunoassay of methotrexate by capillary electrophoresis with laser-induced fluorescence detection. *Anal. Sci.* 19, 111–115. doi: 10.2116/analsci.19.111

Taylor, T. M., Weiss, J., Davidson, P. M., and Bruce, B. D. (2005). Liposomal nanocapsules in food science and agriculture. *Crit. Reve. Food Sci. Nutri.* 45, 587–605. doi: 10.1080/10408390591001135

Tiwari, D., Haque, S., Tiwari, R. P., Jawed, A., Govender, T., and Kruger, H. G. (2015). Fast and efficient detection of tuberculosis antigens using liposome encapsulated secretory proteins of *Mycobacterium tuberculosis*. *J. Microbiol. Immunol. Infect.* 50, 189–198. doi: 10.1016/j.jmii.2015.05.014

Umalkar, D. G., Rajesh, K. S., Bangale, G. S., Rathinaraj, B. S., Shinde, G. V., and Panicker, P. S. (2011). Applications of liposomes in medicine-A review. *Pharma Sci. Monit. Int. J. Pharma. Sci.* 2, 24–39.

Üzer, A., Sağlam, S., Can, Z., Erçağ, E., and Apak, R. (2016). Electrochemical determination of food preservative nitrite with gold nanoparticles/p-aminothiophenol-modified gold electrode. *Int. J. Mol. Sci.* 2, 17–18. doi: 10.3390/ijms17081253

Valdés, M. G., González, A. C. V., Calzón, J. A. G., and Díaz-García, M. E. (2009). Analytical nanotechnology for food analysis. *Microchim. Acta* 166, 1–19. doi: 10.3390/ijms17081253

Vamvakaki, V., and Chaniotakis, N. A. (2007). Pesticide detection with a liposome-based nano-biosensor. *Biosens. Bioelectron.* 22, 2848–2853. doi: 10.1007/ s00604-009-0165-z

Vamvakaki, V., Fournier, D., and Chaniotakis, N. A. (2005). Fluorescence detection of enzymatic activity within a liposome based nano-biosensor. *Biosens. Bioelectron.* 21, 384–388. doi: 10.1016/j.bios.2006.11.024

Wen, H. W., Borejsza, W. W., DeCory, T. R., Baeumner, A. J., and Durst, R. A. (2005). A novel extraction method for peanut allergenic proteinsin chocolate and their detection by a liposome-based lateral flow assay. *Eur. Food Res. Technol.* 221, 564–569. doi: 10.1016/j.bios.2004.10.028

Were, L. M., Bruce, B. D., Davidson, P. M., and Weiss, J. (2003). Size, stability, and entrapment efficiency of phospholipid nanocapsules containing polypeptide antimicrobials. *J. Agric. Food Chem.* 51, 8073–8079. doi: 10.1007/s00217-005-1202-8

Yao, S., Meyer, A., and Henze, G. (1991). Comparison of amperometric and UV-spectrophotometric monitoring in the HPLC analysis of pesticides. *Fresen. J. Anal. Chem.* 339, 207–211. doi: 10.1021/jf0348368

Yu, H. W., Wang, Y. S., Li, Y., Shen, G. L., Wu, H. L., and Yu, R. Q. (2011). One step highly sensitive piezoelectric agglutination method for cholera toxin detection using GM1 incorporated liposome. *Proc. Environ. Sci.* 8, 248–256. doi: 10.1016/j.proenv.2011.10.040

Zaytseva, N. V., Goral, V. N., Montagna, R. A., and Baeumner, A. J. (2005). Development of a microfluidic biosensor module for pathogen detection. *Lab Chip* 5, 805–811. doi: 10.1039/b503856a

Zhang, H., Shi, Y., Lan, F., Pan, Y., Lin, Y., Lv, J., et al. (2014). Detection of single-digit foodborne pathogens with the naked eye using carbon nanotube-based multiple cycle signal amplification. *Chem. Commun.* 50, 1848–1850. doi: 10.1039/c3cc48417c

Zhao, J., Jedlicka, S. S., Lannu, J. D., Bhunia, A. K., and Rickus, J. L. (2006). Liposome-doped nanocomposites as artificial-cell-based biosensors: detection of listeriolysin O. *Biotechnol. Prog.* 22, 32–37. doi: 10.1021/bp050154o

Zhen, Y., Wang, N., Gao, Z., Ma, X., Wei, B., Deng, Y., et al. (2015). Multifunctional liposomes constituting microneedles induced robust systemic and mucosal immunoresponses against the loaded antigens via oral mucosal vaccination. *Vaccine* 33, 4330–4340. doi: 10.1016/j.vaccine.2015.03.081

Conflict of Interest Statement: The authors declare that the research was conducted in the absence of any commercial or financial relationships that could be construed as a potential conflict of interest.

5

Anti-adhesion and Anti-biofilm Potential of Organosilane Nanoparticles against Foodborne Pathogens

Eleni N. Gkana[1], Agapi I. Doulgeraki[1,2]*, Nikos G. Chorianopoulos[2]* and George-John E. Nychas[1]

[1] Laboratory of Microbiology and Biotechnology of Foods, Department of Food Science and Human Nutrition, Faculty of Foods, Biotechnology and Development, Agricultural University of Athens, Athens, Greece, [2] Institute of Technology of Agricultural Products, Hellenic Agricultural Organization-DEMETER, Athens, Greece

*Correspondence:
Nikos G. Chorianopoulos
nchorian@nagref.gr
Agapi I. Doulgeraki
adoulgeraki@aua.gr

Nowadays, modification of surfaces by nanoparticulate coatings is a simple process that may have applications in reducing the prevalence of bacterial cells both on medical devices and food processing surfaces. To this direction, biofilm biological cycle of *Salmonella* Typhimurium, *Listeria monocytogenes*, *Escherichia coli* O157:H7, *Staphylococcus aureus,* and *Yersinia enterocolitica* on stainless steel and glass surfaces, with or without nanocoating was monitored. To achieve this, four different commercial nanoparticle compounds (two for each surface) based on organo-functionalized silanes were selected. In total 10 strains of above species (two for each species) were selected to form biofilms on modified or not, stainless steel or glass surfaces, incubated at 37°C for 72 h. Biofilm population was enumerated by bead vortexing-plate counting method at four time intervals (3, 24, 48, and 72 h). Organosilane based products seemed to affect bacterial attachment on the inert surfaces and/or subsequent biofilm formation, but it was highly dependent on the species and material of surfaces involved. Specifically, reduced bacterial adhesion (at 3 h) of *Salmonella* and *E. coli* was observed ($P < 0.05$) in nanocoating glass surfaces in comparison with the control ones. Moreover, fewer *Salmonella* and *Yersinia* biofilm cells were enumerated on stainless steel coupons coated with organosilanes, than on non-coated surfaces at 24 h ($P < 0.05$). This study gives an insight to the efficacy of organosilanes based coatings against biofilm formation of foodborne pathogens, however, further studies are needed to better understand the impact of surface modification and the underlying mechanisms which are involved in this phenomenon.

Keywords: organosilanes, nanoparticles, biofilms, foodborne pathogens, anti-adhesion

INTRODUCTION

During the last decades, it has become increasingly clear that biofilms are the predominant mode of bacterial growth in most of the natural environments (Lindsay and von Holy, 2006; Giaouris et al., 2013). Biofilm formation consists of at least two stages of development: the adherence of cells to an inert surface which may occur very rapidly and the formation of multilayered cell clusters surrounded by exopolysaccharides produced by bacteria (Götz, 2002). Initial adhesion process depends on bacterial species, interaction medium and inert surface (Pereni et al., 2006). Biofilm control or eradication occurs a considerable issue for food and medicine sector, since this complex bacterial community is resistant to antimicrobial and disinfectant agents (Hoyle and Costerton, 1991; Finlay and Falkow, 1997; Araújo et al., 2011; Bridier et al., 2011). Regarding the important medical and economic consequences of biofilm formation, the understanding of colonization process would be helpful in the design of surface modifications capable of preventing biofilm formation (Prigent-Combaret et al., 1999). Surface properties can be practically modified to reduce bacterial adhesion and further biofouling, which is a principal objective for food industries (Pereni et al., 2006). Surface modification refers to the alteration of physical and chemical properties of an inert substratum (roughness, hydrophobicity, etc.), leading to specific biochemical interactions that prevent bacterial attachment and thus biofilm formation (Kasimanickam et al., 2013).

Following this approach, nanomaterials were proposed as an interventional strategy for the management of biofilm formation due to their high surface area to volume ratio and unique chemical and physical properties (Morones et al., 2005). Nanomaterials were developed for a variety of food applications (food additives, food contact surfaces, food packaging, etc.) and for medical devices (catheter materials, dental acrylics, implants, etc.) (Harris and Graffagnini, 2007; Handford et al., 2014). Due to their small size (1–100 nm) and their ability to cover much larger surface to volume, they possessed altered physicochemical properties in comparison with larger sized material (Oberdörster et al., 2005; Bouwmeester et al., 2014). Nanoparticles such as ZnO (Heinlaan et al., 2008), TiO_2 (Kim et al., 2003; Adams et al., 2006; Chorianopoulos et al., 2011) CuO (Heinlaan et al., 2008), and Al_3O_2 (Ansari et al., 2013). Compared to the quantum of published reports on physical and chemical properties of nanofilms, only limited information is available on the antibacterial properties of these nanomaterials.

Organo-functional silanes could be potential candidates for surface modifications, as can be used to modify the surface energy or wettability of substrates through the interaction of boundary layers of solids with water, effecting variable degrees of hydrophobicity or hydrophilicity (Mittal, 2009). Monomeric silicon chemicals are known as silanes and when they contain at least one silicon carbon bond (e.g., Si-CH_3) are called organosilanes (Kregiel and Niedzielska, 2014). Organo-functional silanes are molecules carrying two different reactive groups on their silicon atom so that they can react with inorganic substrates such as glass and stainless steel and form stable covalent bonds and organic substitution (Thames and Panjnani, 1996; Sepeur, 2008). Several studies have examined the antimicrobial activity of nanoparticulate coatings constituted of silica and organosilanes; however, results retrieved are controversial.

Based on the above, the current study aimed to assess the potential anti-adhesion and anti-biofilm activity of commercial organosilane products applied on stainless steel and glass surfaces against common foodborne pathogens. To achieve this,

biofilm biological cycle of *Salmonella* Typhimurium, *Listeria monocytogenes*, *Escherichia coli* O157:H7, *Staphylococcus aureus*, and *Yersinia enterocolitica* on stainless steel and glass surfaces, with or without nanocoating was monitored.

MATERIALS AND METHODS

Bacterial Strains and Inocula Preparation

All the microorganisms used in this study are presented in **Table 1**. They consist of two strains of each species, specifically for *L. monocytogenes* (FMCC B-125, ScottA, serotype 4b, epidemic strain, human isolate; FMCC B-129, isolated from ready-to-eat frozen meal, minced meat based), *S.* Typhimurium (FMCC B-137, human isolate epidemic; FMCC B-193, isolated from calf bowel), *E. coli* O157:H7 (FMCC B-15 and FMCC B-16, both isolated from human feces), *S. aureus* [FMCC B-410, methicillin-resistant (MRSA) strain COL, isolated from hospital; FMCC B-135, isolated from human lesions], and *Y. enterocolitica* (FMCC B-89, CITY 650; FMCC B-90, CITY 844). Before each experiment the stock cultures (frozen at −80°C) were sub-cultured twice on 10 ml of Tryptic Soy Broth (TSB, LAB M Limited, Lancashire, United Kingdom) at 37°C for 24 and 16 h, respectively (pre-cultures). Cells from exponential phase (16 h) of cultures were collected by centrifugation (5000 × g for 10 min at 4°C), washed twice with 1/4 Ringer solution and re-suspended in 1/4 Ringer solution (working cultures) in order to be used as inoculum for biofilm assays.

Biofilm Formation and Quantification on Polystyrene Microplates

The ability of 10 bacterial strains to form biofilms on polystyrene (PS) microtiter plates was evaluated by using the method described by Jena et al. (2012) with some adaptations. Working culture of above bacteria was diluted 1:100 into fresh medium TSB. Diluted culture (20 μl) was added to the 96-well plates containing 180 μl of TSB. The strains were grown in defined medium (TSB) at 37°C for 24 and 48 h in 96-wells microtiter plates under static conditions.

Following incubation, planktonic bacteria were removed by violently turning upside down the plate to remove growth medium and each well was then washed twice with 200 μl 1/4 Ringer solution to remove the loosely attached cells. The remaining adherent bacteria (biofilms) were fixed for 15 min with 200 μl of methanol per well (Stepanović, 2000). The methanol was discarded and the plates were left to air dry in room temperature for 20 min. Biofilm cells were stained with 100 μl of 1% Crystal Violet solution which was added at each well. After washing with 200 μl 1/4 Ringer three times to remove excess stain, the crystal violet was solubilized with 100 μl ethanol (95%) for 15 min. Dye absorbance at 575 nm (A575) was measured using a microtiter plate reader (Sunrise, Tecan, Männedorf, Switzerland). For each strain eight replicates were performed. Regarding the obtained spectrometric measurement of optical densities the strains were classified into the four categories; non-biofilm producing (OD <= 0.2), weakly (0.2 < OD <= 0.4), moderately (0.4 < OD <= 0.8), and strongly (0.8 < OD)

TABLE 1 | Bacterial species used in this study*.

Microorganism	Strain number	Strain characteristics	Origin
Listeria monocytogenes	FMCC B-125	Scott A, Serotype 4b	Human isolated[a]
	FMCC B-129	21350	RTE frozen meal – minced meat based
Salmonella Typhimurium	FMCC B-137	DT 193 Multi-drug resistant	Human isolate epidemic[b]
	FMCC B-193	4/74	Isolated from calf bowel[c]
E. coli O157:H7	FMCC B-15	NCTC 13125, Verocytoxins negative	Human faeces[d]
	FMCC B-18	NCTC 13127, Verocytoxins negative	Human faeces[d]
Staphylococcus aureus	FMCC B-410	MRSA strain COL	English hospital[e]
	FMCC B-135	NCBF 1499	
Yersinia enterocolitica	FMCC B-89	CITY650	INCO[a]
	FMCC B-90	CITY844	INCO[a]

* From bacterial culture collection of Laboratory of Microbiology and Biotechnology of Foods (FMCC), Agricultural University of Athens.
[a] Kindly provided by Dr. E. Smid, ATO-DLO Netherlands.
[b] Food Microbiology Culture Collection of Agricultural University of Athens.
[c] Kindly provided by Dr. P. Skandamis.
[d] Kindly provided by Dr. E. Drosinos.
[e] Kindly provided by Dr. S. Kathariou, North Carolina State University, United States.

biofilm producing strains according to the method proposed by Stepanović et al. (2004).

Application of Commercial Organosilane Products for Modification of Stainless Steel and Glass Surfaces

Four organosilane based commercial products for coating of non-absorbing surfaces; two (2) for glass and two (2) for stainless steel, specific to each material surface according to manufacturers, were used. Specifically, three (3) commercial products that were obtained from Liquid Glass Nanotech[1] with EINECS (European Inventory of Existing Commercial Chemical Substances) registration were used. The active agent was silicon-free siloxane and consists of polymers made of silanes. One (1) organosilane product for glass (OSG1) (Liquid Glass Nanotech for glass and ceramic surfaces, LGN-600-1) and two organosilane products for stainless steel (OSS1, OSS2) (Universal antimicrobial for non-absorbent/hard surfaces, LGN-671-ANTI; Polish for Metals and Plastics for non-absorbent/hard surfaces, LGN-660-1) were used. Moreover one (1) commercial organosilane based product for glass (OSG2) (NANO-SKIN [HOME]) from BFP Hellas Company[2] was obtained, that is approved by General Chemical State Laboratory of Greece.

All the products were delivered as pump sprays for easy application and were applied following manufacturers' instructions. Briefly, for products OSG1, OSS1, and OSS2 the application consisted of cleaning the surfaces with isopropyl alcohol and then rinsing with deionized water, spraying the coating on surface and evenly distribute the coating with a lint free microfiber cloth across the surface, polish off residue after 30 min and let the coating seal for at least 12–24 h. Nano-Skin product consists of three liquid mixtures (an emulsion and two sprays) which are applied sequentially. Pretreatment with emulsion NANO-SKIN (1) based on a specific composition,

which restores the glass in its initial condition, was required. Then, NANO-SKIN (2) – an alcohol activating solution and NANO-SKIN (3) based on silicon oligomers, both sprayed subsequently to glass surface and spread with microfiber cloth, making gentle circular motions. All the aforementioned products sprayed onto a hard surface form a nano-film by self-organization during evaporation of the solvent (Sepeur, 2008). The film arises from the sol-gel process (Hench and West, 1990; Schmidt, 2006) that involves series of hydrolysis and condensation reactions between organo-functionalized silanes that result in a network of functionalized siloxanes (Nørgaard et al., 2014).

Biofilm Formation on Stainless Steel and Glass Coupons
Preparation of Stainless Steel and Glass Surfaces

Stainless steel is the surface used extensively throughout the food processing industry. On the other hand glass was selected due to its high hydrophilicity and excellent silane effectiveness on this material. In addition, it is well known that significant portion of food deposits is made of glass (e.g., doors and coverings of refrigerators in super markets). Stainless steel is the surface used extensively throughout the food processing industry. Stainless steel (SS) coupons (3 by 1 by 0.1 cm, type AISI-304; Halyvourgiki, Inc., Athens, Greece) and glass (G) coupons (3 by 1 by 0.1 cm cut from microscope slides) were initially soaked in acetone (overnight) to remove any manufacturing process debris and grease. Coupons were then washed by soaking overnight at room temperature in a 2% (vol/vol) solution of the commercial detergent RBS 35 (Fluka/Life Science Chemilab, S.A.) with shaking, rinsed thoroughly with tap water followed by distilled water and air dried. The coupons were coated by the procedure mentioned above with commercial nano-coatings. Glass and stainless steel coupon without coating were used as control. Finally, cleaned coupons were individually placed in empty glass test tubes (length, 10 cm; diameter, 1.5 cm) and autoclaved at 121°C for 15 min.

---

[1]https://www.liquidglassnanotech.com/
[2]http://bfphellas.gr/

Biofilm Formation and Enumeration

Ten strains (S. Typhimurium 137, 193, S. aureus 135, 410, Y. enterocolitica 89, 90, E. coli O157:H7 15, 18, and L. monocytogenes 125, 129) were selected to examine biofilm formation on stainless steel and glass surfaces, coated or not with organosilanes. Strains with different isolation origins (i.e., clinical, food, or environment) were selected in an attempt to pursue variability. The study was performed according to the protocol described by Kostaki et al. (2012) with minor modifications. The working cultures were diluted at 1:100 and 0.5 ml was added in 4.5 ml Ringer that contained a stainless steel or glass coupon. For the attachment step, 0.5 ml of each bacterial suspension in 4.5 ml quarter-strength Ringer solution, containing ca. 10^6 CFU/ml, was poured into each glass test tube containing a sterilized coupon and incubated at 15°C for 3 h under static conditions. This temperature, representative of food industry during non-production hours (15°C) was incorporated in this study to investigate the adherent properties of abovementioned foodborne pathogens.

Following the attachment step, each coupon was carefully removed from the glass test tube using sterile forceps and individually introduced into a new sterile glass test tube containing 5 ml of TSB and subsequently incubated at 37°C for 3 days (72 h), under static conditions, to allow biofilm development on the coupon, with no growth medium renewal. Each experiment included three replications and sampling was performed at 3, 24, 48, and 72 h. A higher temperature (37°C) of incubation to determine biofilm formation was selected because previous studies have shown that biofilm production is increased when bacteria allowed growing next to or at their optimal temperature (Morton et al., 1998; da Silva Meira et al., 2012; Kadam et al., 2013). Furthermore, it was evaluated that at 37°C, L. monocytogenes biofilm exhibited a complex system, in terms of cell number and EPS produced, due to advanced state of growth rate. Therefore, this temperature (37°C) represents the worst-case scenario of biofilm formation in order to determine if there is a potential anti-biofilm activity of organosilanes.

Briefly, each coupon was aseptically removed from the glass test tube and was then rinsed by pipetting twice with 10 ml of quarter-strength Ringer solution (each time). The coupon was transferred to a falcon centrifuge tube containing 6 ml of quarter-strength Ringer solution and 10 sterile glass beads (diameter, 3 mm) and then vortexed for 2 min at maximum speed to detach biofilm cells from the coupon. Detached cells obtained by bead vortexing method (Giaouris and Nychas, 2006) were subsequently enumerated on Tryptone Soy Agar (TSA; Lab M), after 10-fold serial dilutions. Stainless steel and glass surfaces were examined under conventional fluorescence microscope using acridine orange stain to determine the absence of residual biofilm remained on substrate (data not shown).

Data Analysis

Univariate analysis of Variance (n-way ANOVA) for each stainless steel and glass surfaces was performed to test the main interaction effects of independent factors: (a) three different materials of surfaces (one non-coated and two coated surfaces), (b) five pathogen species (S. Typhimurium. S. aureus,

L. monocytogenes, Y. enterocolitica, and E. coli O157:H7), and (c) four different time points (3, 24, 48, and 72 h) to bacterial attached cells as expressed by log CFU/cm^2 (dependent). Thus, a 3*5*4 factorial design was constructed and when probability of F-values were less than 0.05 for any independent or combinations of independents, it was concluded that the variable has an effect on the depended. Each experiment was conducted using three replicates for each. The Tukey post hoc test was used to compare the means at the 95% confidence level. The statistical analysis was conducted using the IBM® SPSS® Statistics for Windows software, Version 22.0 (IBM Corp., Armonk, NY, United States).

RESULTS

The biofilm forming capacity of five foodborne pathogens at strain level was initially examined in this study by crystal violet method. Briefly, two strains of each pathogen, i.e., S. Typhimurium, L. monocytogenes, E. coli, Y. enterocolitica, and S. aureus were left to form biofilm on microtiter plate at 37°C to check the strain variability on this phenomenon. In addition, the influence of incubation time, i.e., 24 and 48 h was estimated. The average optical density (OD575) values were calculated for all tested strains at 24 and 48 h (**Figure 1**).

Listeria monocytogenes FMCC-125 was classified as strongly biofilm producing strain, while E. coli O157:H7 FMCC-16 was evaluated as non-biofilm producer. In addition, both strains of S. aureus (FMCC-135, 410), both strains of Y. enterocolitica (FMCC 89, 90) and one strain of E. coli O157:H7 (FMCC-15) was classified as weak biofilm producers. The rest three strains, consisted of both strains of S. Typhimurium (FMCC-137, 193) and a strain of L. monocytogenes were classified as moderate biofilm producers.

The previous tested strains were left to form dual strain biofilm on stainless steel and glass surfaces. In accordance to the previous analyzed results, it was observed that biofilm formation was influenced by the bacterial species and incubation time; however, the effect of surface was also estimated. Briefly, a statistical difference was detected between biofilm formation on glass and stainless steel both at attachment step and formed biofilm (24 and 48 h). More specifically, in the case of glass surface the attached and biofilm cell population was found to be lower than on stainless steel surface.

Regarding the observations related to the attachment ability of the pathogens on non-coated glass surfaces (assessment of the population at 3 h), it seems that S. Typhimurium was attached in higher populations (about 4.32 log CFU/cm^2), while S. aureus, E. coli, L. monocytogenes, and Y. enterocolitica were attached at significant lower concentrations (1.6–2.7 log CFU/cm^2) (**Figure 2**; $P < 0.05$). However, L. monocytogenes and E. coli biofilm population was the highest and lowest, respectively ($P < 0.05$), while S. Typhimurium, S. aureus, and Y. enterocolitica biofilm populations were in similar levels at 24 h. Similar observations reported above regarding the data obtained from the microtiter plates assay. After 48 h, Y. enterocolitica biofilm population was significant lower than those of S. Typhimurium and L. monocytogenes while S. Typhimurium and

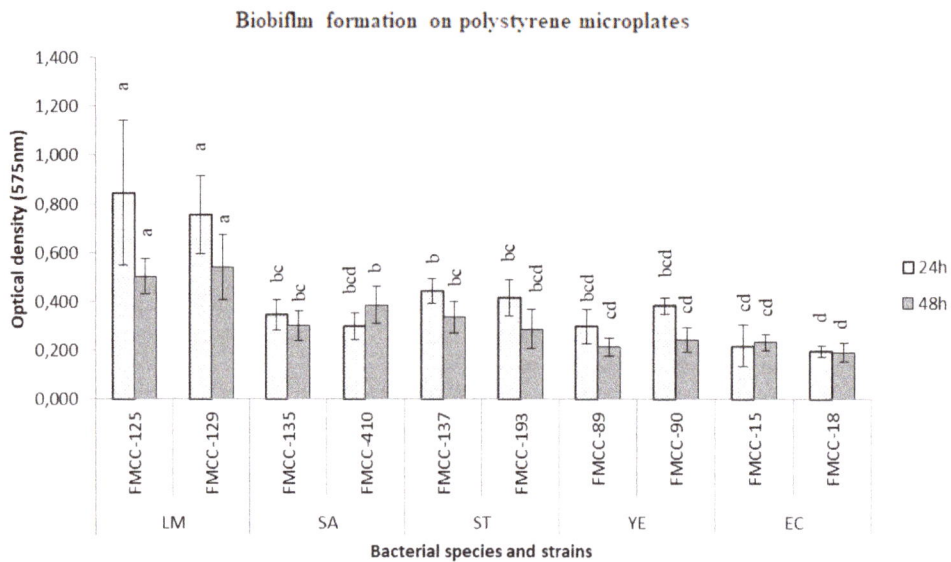

FIGURE 1 | Biofilm formation on polystyrene microtiter plates of different strains after 24 and 48 h of incubation at 37°C. Biofilm cells were indirectly quantified by crystal violet staining and absorbance measurements at 575 nm. Bars represent means ± standard deviations. Different letters at 24 or 48 h indicate significant differences between biofilm formation of strains ($P < 0.05$).

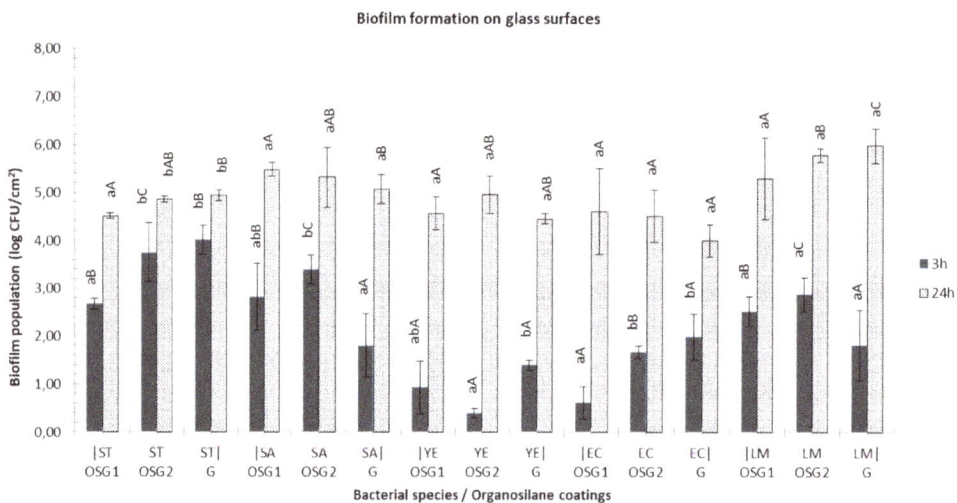

FIGURE 2 | Biofilm formation (log CFU/cm^2) on glass coupons with (OSG1/OSG2) or without (G) coating, using two strains of *Salmonella* Typhimurium (ST), *Staphylococcus aureus* (SA), *Yersinia enterocolitica* (YE), *Escherichia coli* (EC) or *Listeria monocytogenes* (LM) at 3 h (gray bars) and 24 h (white bars) of incubation at 37°C. Bars represent means ± standard deviations. Different lowercase letters indicate differences on cells attachment (3 h) or biofilm formation (24 h) according to coating for the same species. Similarly, different uppercase letters point out differences according to species adherence/biofilm formation for the same coated or non-surfaces.

E. coli were found to maintain higher level of sessile cells than *Y. enterocolitica*, at 72 h.

Biofilm cycles of *S.* Typhimurium and *S. aureus* had similar trend as they reached the higher biofilm formation at 24 h, while a significant reduction of sessile cells was observed at 72 h. *E. coli* had a different respond as remained throughout incubation period at approximately same numbers of 24 h biofilm population. Lower numbers of cells were retrieved after 48 h

of incubation as concern *L. monocytogenes* and *Y. enterocolitica* compared to biofilm formation of 24 h. At 72 h, *L. monocytogenes* sessile cells were remained at levels estimated at 48 h, while a further reduction was observed for *Y. enterocolitica*.

Staphylococcus aureus was found to be attached on stainless steel surfaces at a significant higher level compared to glass surfaces. Similar adhesion to glass and stainless steel surfaces and no correlation between materials surface hydrophobicity was obtained for all other species. Biofilm formation at 24 h was found to be significant lower on glass surfaces for *S.* Typhimurium, *Y. enterocolitica,* and *E. coli* ($P < 0.05$).

The application of organosilane products was found to affect the adhesion of the pathogens (estimation of population at 3 h) on glass surfaces, however, their effect influenced by bacterial species (**Figure 2**). More specific, product OSG1 reduced adhesion of *S.* Typhimurium and *E. coli* compared to bare glass surfaces at approximately 1.4 log CFU/cm^2. On the other hand, product OSG2 was found to induce the attachment of *S. aureus* at the level of 1.8 log CFU/cm^2 compared to the non-coated glass coupons. However, it seems that the application of both products affected only the first steps of biofilm formation as no significant differences were observed between coated or not glass surfaces after 24, 48, and 72 h.

On the other hand, significant differences were detected in the case of organosilanes application on stainless steel surfaces compared to bare ones, which highly depended on the bacterial species and time of incubation (**Figure 3**). Briefly, both OSS1 and OSS2 were able to reduce biofilm formation of *S.* Typhimurium at approximately 0.5 log CFU/cm^2, at 24 h ($P < 0.05$). Similarly, a reduction of *S. aureus* biofilm cells was observed at the level of 0.8–1.2 log CFU/cm^2 at 48 h. Biofilm formation of *Y. enterocolitica* was also affected after the application of OSS2 as an approximately 1.8 log CFU/cm^2 reduction of population was observed at 24 h.

DISCUSSION

Physicochemical properties of inert substratum and bacterium cell surface are known to have impact on bacterial attachment and biofilm formation, however, the exact correlation with discrete characteristics is difficult as the system is very complex. Hydrophobicity of surfaces has been reported as an important factor affecting the attachment of bacteria on surfaces. Specifically hydrophobicity seems to decrease the adhesion of microorganisms on inert surfaces (van Loosdrecht et al., 1987; Dickson and Daniels, 1991; Bonsaglia et al., 2014) and in the same time increase the detachment of sessile cells (Pereni et al., 2006). Stainless steel is considered a hydrophobic material (Lafuma and Quéré, 2003), while glass a hydrophilic material (Robert et al., 2001). Modification of surfaces with organosilanes usually increases the hydrophobic qualities and low surface free energy of native surfaces (Kregiel and Niedzielska, 2014).

Regarding the present results organosilanes found to eliminate adherence of *S.* Typhimurium and *E. coli* on modified glass surfaces, but this effect was not evident on stainless steel surfaces. A considerable alteration on physical properties of glass surfaces from hydrophilic to hydrophobic may be the reason of the anti-adherent properties observed. In addition, low surface energy chemistry and nano-textured morphology of the coating (homogeneity of the organosilane layer on glass surfaces) could also result in reduced protein adsorption and inhibition of bacterial attachment (Chen et al., 2013).

Significant reductions on biofilm formation (24 and 48 h) were pointed out for *S. aureus, S.* Typhimurium, and *Y. enterocolitica* on modified with organosilanes stainless steel surfaces as compared to their respective controls. A positive correlation between substratum hydrophobicity and the detachment of adherent biofilm was established by other studies. According to this approach, bacteria attached to hydrophobic materials were more easily removed from them (Harkes et al., 1992; Reid et al., 1993; Eginton et al., 1995; Bos et al., 2000; Gómez-Suárez et al., 2001). On the other hand, *S. aureus* found to attach more effectively on stainless steel surfaces in comparison

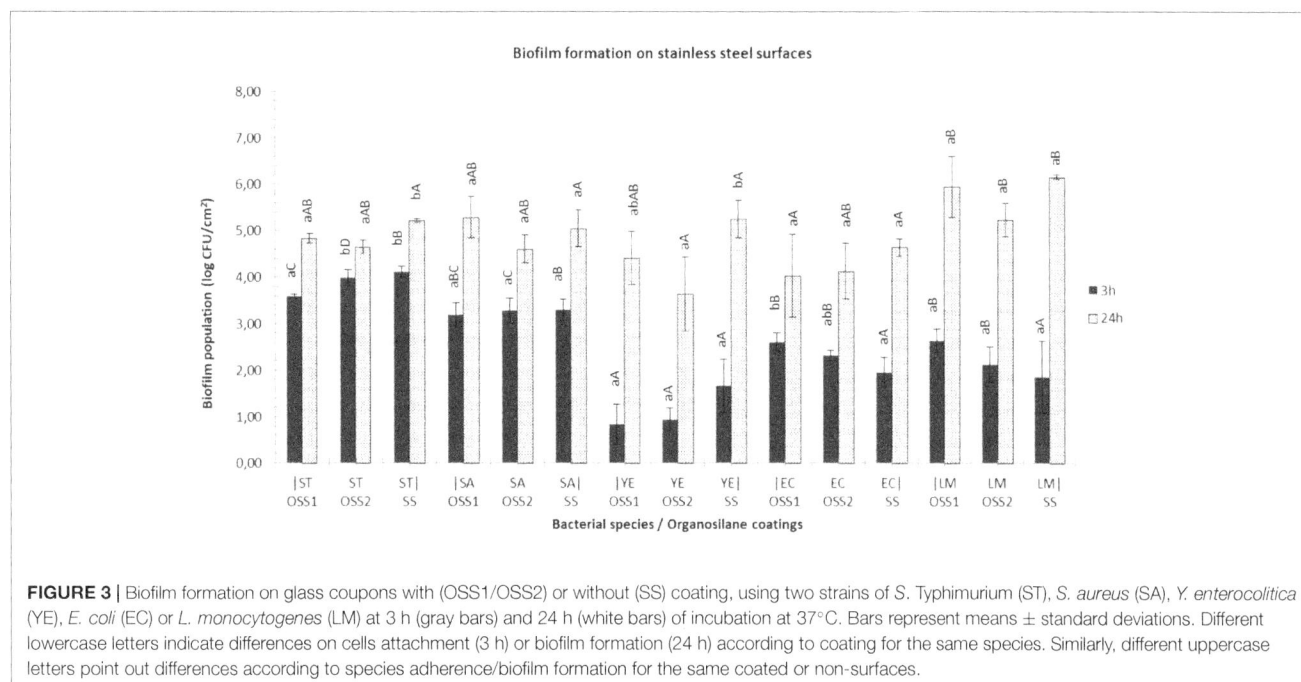

FIGURE 3 | Biofilm formation on glass coupons with (OSS1/OSS2) or without (SS) coating, using two strains of *S.* Typhimurium (ST), *S. aureus* (SA), *Y. enterocolitica* (YE), *E. coli* (EC) or *L. monocytogenes* (LM) at 3 h (gray bars) and 24 h (white bars) of incubation at 37°C. Bars represent means ± standard deviations. Different lowercase letters indicate differences on cells attachment (3 h) or biofilm formation (24 h) according to coating for the same species. Similarly, different uppercase letters point out differences according to species adherence/biofilm formation for the same coated or non-surfaces.

with glass ones, while organosilanes enhance the adherence of bacterium to modified glass surfaces. It seems that a correlation between hydrophobicity and the number of attached cells was resulted. Organosilanes had no effect on eliminating *L. monocytogenes* attached cells or biofilm formation. No differences were also observed regarding different non-modified glass or stainless steel surfaces. These results are in agreement with other studies, too. Teixeira et al. (2007) claimed that adhesion of *L. monocytogenes* to abiotic surfaces was not influenced by substratum hydrophobicity and roughness.

Silica nanoparticles have been found to eliminate *Candida albicans* adhesion and surface associated growth (Cousins et al., 2007). Another study found that concentration of silicon dioxide above 1000 ppm was required to achieve antibacterial activity against *Bacillus subtilis* and *E. coli* (Adams et al., 2006). Polyethylene surfaces, following activation by plasma processing and modification with active organosilanes, exhibit anti-adhesive and anti-biofilm properties against *Aeromonas hydrophila* (Kregiel and Niedzielska, 2014). Glass surfaces coated with hydrophobic silane (alkyl functionalized silane) modified silica nanoparticles exhibited inhibition performance against the growth of *E. coli, S. aureus,* and *Deinococcus geothermalis* compared to that of pristine silica nanoparticles (Song et al., 2011). Reduction of *S. aureus* and *P. aeruginosa* adherence on super-hydrophobic surfaces synthesized by fluorinated silica colloids was also demonstrated (Privett et al., 2011). On the other hand, silica nanoparticles against oral pathogenic species of *Streptococcus mutans* had limited antibacterial effects, using minimum inhibitory concentration assay for planktonic growth, in 96-well microplates (Besinis et al., 2014). Evaluation of two organosilane products applied on high-touch surfaces in patient rooms of a health care facility revealed that no significant residual antimicrobial activity was observed (Boyce et al., 2014).

Numerous previous studies have described the ability of aforementioned foodborne pathogens to attach to various surfaces and form biofilms (Joseph et al., 2001; Stepanović et al., 2004; Kim et al., 2008; Dourou et al., 2011; Oniciuc et al., 2016), with this ability to be depended on the interaction between intrinsic and extrinsic factors such as the bacterial cells, the attachment surface and the surrounding environmental conditions (Giaouris et al., 2014). However, most of these previous studies were performed by constructing single-strain biofilms, with obtaining results not to be necessarily representative of the bacterial species as whole. Undoubtedly, bacterial strains, even the ones belonging to the same species, may greatly differ in many phenotypic responses, including biofilm formation, and this variability should be

always taken into account (Lianou and Koutsoumanis, 2013). This is the reason why in the present study were selected two different strains for each species to form multi-strain biofilm communities. The observed phenotypic variability in biofilm formation which ranges from strong to non-biofilm formers even at strain level underlies the importance of strain level studies related to survival and spread of bacteria.

CONCLUSION

To the best of our knowledge, this is the first study evaluating modification of stainless steel and glass surfaces with organosilane based products in order to investigate anti-adhesion and anti-biofilm potential against foodborne pathogens. In conclusion, the current study was able to demonstrate anti-adhesion and anti-biofilm activity of specific organosilane based products, but this aspect highly depended on the species of pathogens used in this study and time of incubation (3, 24, 48, and 72 h). Further studies are needed to establish the underlying mechanisms regarding the role of organosilane based products modification on various surfaces types and bacterial species. On the other hand, nanomaterials could have a fundamental impact on the food and medicine sector, potentially offering benefits as concerning the battle against biofouling.

However, any potential risks for consumers are still required to be estimated and assessed in order to ensure public health. The risk of certain nanomaterial should be evaluated as concern the application, the use and final disposal (Contado, 2015). Furthermore, the risk of consumer exposure to nanoparticles directly from medical implants or indirectly through possible migration from surfaces to foodstuffs should be evaluated, since a knowledge gap exist with regards to absorbance, metabolism, and elimination of nanoparticles from the human body.

AUTHOR CONTRIBUTIONS

EG designed the studies, performed the experiments, and wrote the paper. AD designed the studies, performed the experiments, and wrote the paper. NC designed the studies and wrote the paper. G-JN wrote the paper.

ACKNOWLEDGMENT

EG and AD would like to thank the John S. Latsis Public Benefit Foundation for their financial support.

REFERENCES

Adams, L. K., Lyon, D. Y., and Alvarez, P. J. J. (2006). Comparative eco-toxicity of nanoscale TiO2, SiO2, and ZnO water suspensions. *Water Res.* 40, 3527–3532. doi: 10.1016/j.watres.2006.08.004

Ansari, M. A., Khan, H. M., Khan, A. A., Cameotra, S. S., Saquib, Q., and Musarrat, J. (2013). Interaction of Al2O3 nanoparticles with *Escherichia*

coli and their cell envelope biomolecules. *J. Appl. Microbiol.* 116, 772–783. doi: 10.1111/jam.12423

Araújo, P., Lemos, M., Mergulhão, F., Melo, L., and Simões, M. (2011). "Antimicrobial resistance to disinfectants in biofilms," in *Science against Microbial Pathogens: Communicating Current Research and Technological Advances*, eds P. Araújo, M. Lemos, F. Mergulhão, L. Melo, and M. Simões (Badajoz: Formatex), 826–834.

Besinis, A., De Peralta, T., and Handy, R. D. (2014). The antibacterial effects of silver, titanium dioxide and silica dioxide nanoparticles compared to the dental disinfectant chlorhexidine on *Streptococcus mutans* using a suite of bioassays. *Nanotoxicology* 8, 1–16. doi: 10.3109/17435390.2012.742935

Bonsaglia, E. C. R., Silva, N. C. C., Fernades, A. Jr., Araújo, J. P. Jr., Tsunemi, M. H., and Rall, V. L. M. (2014). Production of biofilm by *Listeria monocytogenes* in different materials and temperatures. *Food Control* 35, 386–391. doi: 10.1016/j.foodcont.2013.07.023

Bos, R., Mei, H. C., Gold, J., and Busscher, H. J. (2000). Retention of bacteria on a substratum surface with micro-patterned hydrophobicity. *FEMS Microbiol. Lett.* 189, 311–315. doi: 10.1111/j.1574-6968.2000.tb09249.x

Bouwmeester, H., Brandhoff, P., Marvin, H. J. P., Weigel, S., and Peters, R. J. B. (2014). State of the safety assessment and current use of nanomaterials in food and food production. *Trends Food Sci. Technol.* 40, 200–210. doi: 10.1016/j.tifs.2014.08.009

Boyce, J. M., Havill, N. L., Guercia, K. A., Schweon, S. J., and Moore, B. A. (2014). Evaluation of two organosilane products for sustained antimicrobial activity on high-touch surfaces in patient rooms. *Am. J. Infect. Control* 42, 326–328. doi: 10.1016/j.ajic.2013.09.009

Bridier, A., Briandet, R., Thomas, V., and Dubois-Brissonnet, F. (2011). Resistance of bacterial biofilms to disinfectants: a review. *Biofouling* 27, 1017–1032. doi: 10.1080/08927014.2011.626899

Chen, M., Yu, Q., and Sun, H. (2013). Novel strategies for the prevention and treatment of biofilm related infections. *Int. J. Mol. Sci.* 14, 18488–18501. doi: 10.3390/ijms140918488

Chorianopoulos, N. G., Tsoukleris, D. S., Panagou, E. Z., Falaras, P., and Nychas, G. E. (2011). Use of titanium dioxide (TiO_2) photocatalysts as alternative means for *Listeria monocytogenes* biofilm disinfection in food processing. *Food Microbiol.* 28, 164–170. doi: 10.1016/j.fm.2010.07.025

Contado, C. (2015). Nanomaterials in consumer products: a challenging analytical problem. *Front. Chem.* 3:48. doi: 10.3389/fchem.2015.00048

Cousins, B. G., Allison, H. E., Doherty, P. J., Edwards, C., Garvey, M. J., Martin, D. S., et al. (2007). Effects of a nanoparticulate silica substrate on cell attachment of *Candida albicans*. *J. Appl. Microbiol.* 102, 757–765. doi: 10.1111/j.1365-2672.2006.03124.x

da Silva Meira, Q. G., de Medeiros Barbosa, I., Alves Aguiar Athayde, A. J., de Siqueira-Júnior, J. P., and de Souza, E. L. (2012). Influence of temperature and surface kind on biofilm formation by *Staphylococcus aureus* from food-contact surfaces and sensitivity to sanitizers. *Food Control* 25, 469–475. doi: 10.1016/j.foodcont.2011.11.030

Dickson, J. S., and Daniels, E. K. (1991). Attachment of *Salmonella* typhimurium and *Listeria monocytogenes* to glass as affected by surface film thickness, cell density, and bacterial motility. *J. Ind. Microbiol.* 8, 281–283. doi: 10.1007/BF01576068

Dourou, D., Beauchamp, C. S., Yoon, Y., Geornaras, I., Belk, K. E., Smith, G. C., et al. (2011). Attachment and biofilm formation by *Escherichia coli* O157:H7 at different temperatures, on various food-contact surfaces encountered in beef processing. *Int. J. Food Microbiol.* 149, 262–268. doi: 10.1016/j.ijfoodmicro.2011.07.004

Eginton, P. J., Gibson, H., Holah, J., Handley, P. S., and Gilbert, P. (1995). Quantification of the ease of removal of bacteria from surfaces. *J. Ind. Microbiol.* 15, 305–310. doi: 10.1007/BF01569984

Finlay, B. B., and Falkow, S. (1997). Common themes in microbial pathogenicity revisited. *Microbiol. Mol. Biol. Rev.* 61, 136–169.

Giaouris, E., Heir, E., Hébraud, M., Chorianopoulos, N., Langsrud, S., Møretrø, T., et al. (2013). Attachment and biofilm formation by foodborne bacteria in meat processing environments: causes, implications, role of bacterial interactions and control by alternative novel methods. *Meat Sci.* 97, 298–309. doi: 10.1016/j.meatsci.2013.05.023

Giaouris, E., Heir, E., Hébraud, M., Chorianopoulos, N., Langsrud, S., Møretrø, T., et al. (2014). Attachment and biofilm formation by foodborne bacteria in meat processing environments: causes, implications, role of bacterial interactions and control by alternative novel methods. *Meat Sci.* 97, 289–309. doi: 10.1016/j.meatsci.2013.05.023

Giaouris, E. D., and Nychas, G. J. E. (2006). The adherence of *Salmonella* Enteritidis PT4 to stainless steel: the importance of the air-liquid interface and nutrient availability. *Food Microbiol.* 23, 747–752. doi: 10.1016/j.fm.2006.02.006

Gómez-Suárez, C., Busscher, H. J., and van der Mei, H. C. (2001). Analysis of bacterial detachment from substratum surfaces by the passage of air-liquid interfaces. *Appl. Environ. Microbiol.* 67, 2531–2537. doi: 10.1128/AEM.67.6.2531-2537.2001

Götz, F. (2002). *Staphylococcus* and biofilms. *Mol. Microbiol.* 43, 1367–1378. doi: 10.1046/j.1365-2958.2002.02827.x

Handford, C. E., Dean, M., Henchion, M., Spence, M., Elliott, C. T., and Campbell, K. (2014). Implications of nanotechnology for the agri-food industry: opportunities, benefits and risks. *Trends Food Sci. Technol.* 40, 226–241. doi: 10.1016/j.tifs.2014.09.007

Harkes, G., Dankert, J., and Feijen, J. (1992). Growth of uropathogenic *Escherichia coli* strains at solid surfaces. *J. Biomater. Sci. Polym. Ed.* 3, 403–418. doi: 10.1163/156856292X00213

Harris, D. L., and Graffagnini, M. J. (2007). Nanomaterials in medical devices: a snapshot of markets, technologies and companies. *Nanotechnol. Bus.* 255, 415–422.

Heinlaan, M., Ivask, A., Blinova, I., Dubourguier, H. C., and Kahru, A. (2008). Toxicity of nanosized and bulk ZnO, CuO and TiO_2 to bacteria *Vibrio fischeri* and crustaceans *Daphnia magna* and *Thamnocephalus platyurus*. *Chemosphere* 71, 1308–1316. doi: 10.1016/j.chemosphere.2007.11.047

Hench, L. L., and West, J. K. (1990). The sol-gel process. *Chem. Rev.* 90, 33–72. doi: 10.1021/cr00099a003

Hoyle, B. D., and Costerton, J. W. (1991). Bacterial resistance to antibiotics: the role of biofilms. *Prog. Drug Res.* 37, 91–105. doi: 10.1007/978-3-0348-7139-6_2

Jena, P., Mohanty, S., Mallick, R., Jacob, B., and Sonawane, A. (2012). Toxicity and antibacterial assessment of chitosan-coated silver nanoparticles on human pathogens and macrophage cells. *Int. J. Nanomed.* 7, 1805–1818. doi: 10.2147/IJN.S28077

Joseph, B., Otta, S. K., Karunasagar, I., and Karunasagar, I. (2001). Biofilm formation by *Salmonella* spp. on food contact surfaces and their sensitivity to sanitizers. *Int. J. Food Microbiol.* 64, 367–372. doi: 10.1016/S0168-1605(00)00466-9

Kadam, S. R., den Besten, H. M. W., van der Veen, S., Zwietering, M. H., Moezelaar, R., and Abee, T. (2013). Diversity assessment of *Listeria monocytogenes* biofilm formation: impact of growth condition, serotype and strain origin. *Int. J. Food Microbiol.* 165, 259–264. doi: 10.1016/j.ijfoodmicro.2013.05.025

Kasimanickam, R. K., Ranjan, A., Asokan, G. V., Kasimanickam, V. R., and Kastelic, J. P. (2013). Prevention and treatment of biofilms by hybrid- and nanotechnologies. *Int. J. Nanomed.* 8, 2809–2819. doi: 10.2147/IJN.S44100

Kim, B., Kim, D., Cho, D., and Cho, S. (2003). Bactericidal effect of TiO_2 photocatalyst on selected food-borne pathogenic bacteria. *Chemosphere* 52, 277–281. doi: 10.1016/S0045-6535(03)00051-1

Kim, T. J., Young, B. M., and Young, G. M. (2008). Effect of flagellar mutations on *Yersinia enterocolitica* biofilm formation. *Appl. Environ. Microbiol.* 74, 5466–5474. doi: 10.1128/AEM.00222-08

Kostaki, M., Chorianopoulos, N., Braxou, E., Nychas, G.-J., and Giaouris, E. (2012). Differential biofilm formation and chemical disinfection resistance of sessile cells of *Listeria monocytogenes* strains under monospecies and dual-species (with *Salmonella enterica*) conditions. *Appl. Environ. Microbiol.* 78, 2586–2595. doi: 10.1128/AEM.07099-11

Kregiel, D., and Niedzielska, K. (2014). Effect of plasma processing and organosilane modifications of polyethylene on *Aeromonas hydrophila* biofilm formation. *Biomed Res. Int.* 2014:232514. doi: 10.1155/2014/232514

Lafuma, A., and Quéré, D. (2003). Superhydrophobic states. *Nat. Mater.* 2, 457–460. doi: 10.1038/nmat924

Lianou, A., and Koutsoumanis, K. P. (2013). Strain variability of the behavior of foodborne bacterial pathogens: a review. *Int. J. Food Microbiol.* 167, 310–321. doi: 10.1016/j.ijfoodmicro.2013.09.016

Lindsay, D., and von Holy, A. (2006). Bacterial biofilms within the clinical setting: what healthcare professionals should know. *J. Hosp. Infect.* 64, 313–325. doi: 10.1016/j.jhin.2006.06.028

Mittal, K. (2009). *Silanes and Other Coupling Agents*, Vol. 5. Boca Raton, FL: CRC Press

Morones, J. R., Elechiguerra, J. L., and Camacho, A. (2005). The bactericidal effect of silver nanoparticles. *Nanotechnology* 16, 2346–2353. doi: 10.1088/0957-4484/16/10/059

Morton, L. H. G., Greenway, D. L. A., Gaylarde, C. C., and Surman, S. B. (1998). Consideration of some implications of the resistance of biofilms to biocides. *Int. Biodeterior. Biodegrad.* 41, 247–259. doi: 10.1016/S0964-8305(98)00026-2

Nørgaard, A. W., Hansen, J. S., Sørli, J. B., Levin, M., Wolkoff, P., Nielsen, G. D., et al. (2014). Pulmonary toxicity of perfluorinated silane-based nanofilm spray products: solvent dependency. *Toxicol. Sci.* 137, 179–188. doi: 10.1093/toxsci/kft225

Oberdörster, G., Oberdörster, E., and Oberdörster, J. (2005). Nanotoxicology: an emerging discipline evolving from studies of ultrafine particles. *Environ. Health Perspect.* 113, 823–839. doi: 10.1289/ehp.7339

Oniciuc, E.-A., Cerca, N., and Nicolau, A. I. (2016). Compositional analysis of biofilms formed by *Staphylococcus aureus* isolated from food sources. *Front. Microbiol.* 7:390. doi: 10.3389/fmicb.2016.00390

Pereni, C. I., Zhao, Q., Liu, Y., and Abel, E. (2006). Surface free energy effect on bacterial retention. *Colloids Surf. B Biointerfaces* 48, 143–147. doi: 10.1016/j.colsurfb.2006.02.004

Prigent-Combaret, C., Vidal, O., Dorel, C., and Lejeune, P. (1999). Abiotic surface sensing and biofilm-dependent regulation of gene expression in *Escherichia coli*. *J. Bacteriol.* 181, 5993–6002.

Privett, B. J., Youn, J., Hong, S. A., Lee, J., Han, J., Shin, J. H., et al. (2011). Antibacterial fluorinated silica colloid superhydrophobic surfaces. *Langmuir* 27, 9597–9601. doi: 10.1021/la201801e

Reid, G., Lam, D., Policova, Z., and Neumann, A. W. (1993). Adhesion of two uropathogens to silicone and lubricious catheters: influence of pH, urea and creatinine. *J. Mater. Sci. Mater. Med.* 4, 17–22. doi: 10.1007/BF0012 2972

Robert, J. M. I., Toguchi, A., and Harshey, R. M. (2001). *Salmonella enterica* serovar typhimurium swarming mutants with altered biofilm-forming abilities: surfactin inhibits biofilm formation *Salmonella enterica* serovar typhimurium swarming mutants with altered biofilm-forming abilities: surfactin inhibits bio. *J. Bacteriol.* 83, 5848–5854. doi: 10.1128/JB.183.20.5848

Schmidt, H. (2006). Considerations about the sol-gel process: from the classical sol-gel route to advanced chemical nanotechnologies. *J. Sol Gel Sci. Technol.* 40, 115–130. doi: 10.1007/s10971-006-9322-6

Sepeur, S. (2008). *Nanotechnology: Technical Basis and Applications.* Hanover: Vincentz Network.

Song, J., Kong, H., and Jang, J. (2011). Bacterial adhesion inhibition of the quaternary ammonium functionalized silica nanoparticles. *Colloids Surf. B Biointerfaces* 82, 651–656. doi: 10.1016/j.colsurfb.2010.10.027

Stepanović, S., Cirković, I., Ranin, L., and Svabić-Vlahović, M. (2004). Biofilm formation by *Salmonella* spp. and *Listeria monocytogenes* on plastic surface. *Lett. Appl. Microbiol.* 38, 428–432. doi: 10.1111/j.1472-765X.2004.01513.x

Stepanović, S., Vuković, D., Dakić, I., Savić, B., and Švabić-Vlahović, M. (2000). A modified microtiter-plate test for quantification of staphylococcal biofilm formation. *J. Microbiol. Methods* 40, 175–179. doi: 10.1016/S0167-7012(00) 00122-6

Teixeira, P., Silva, S. C., Araújo, F., Azeredo, J., and Oliveira, R. (2007). "Bacterial adhesion to food contacting surfaces," in *Communicating Current Research and Educational Topics and Trends in Applied Microbiology*, ed. A. Mendez-Vilas (Badajoz: Formatex Research Center), 13–20.

Thames, S. F., and Panjnani, K. G. (1996). Organosilane polymer chemistry: a review. *J. Inorg. Organomet. Polym.* 6, 59–94. doi: 10.1007/BF01098320

van Loosdrecht, M. C., Lyklema, J., Norde, W., Schraa, G., and Zehnder, A. J. (1987). Electrophoretic mobility and hydrophobicity as a measured to predict the initial steps of bacterial adhesion. *Appl. Environ. Microbiol.* 53, 1898–1901. doi: 10.1007/BF00878244

Conflict of Interest Statement: The authors declare that the research was conducted in the absence of any commercial or financial relationships that could be construed as a potential conflict of interest.

The reviewer Dr. AC and handling Editor declared their shared affiliation, and the handling Editor states that the process neverthless met the standards of a fair and objective review.

6

Nanotechnology in Sustainable Agriculture: Recent Developments, Challenges, and Perspectives

Ram Prasad[1]*, Atanu Bhattacharyya[2] and Quang D. Nguyen[3]

[1] *Amity Institute of Microbial Technology, Amity University, Noida, India,* [2] *Department of Entomology, University of Agricultural Sciences, Gandhi Krishi Vigyan Kendra, Bengaluru, India,* [3] *Research Centre of Bioengineering and Process Engineering, Faculty of Food Science, Szent István University, Budapest, Hungary*

Edited by:
Jayanta Kumar Patra,
Dongguk University Seoul,
South Korea

Reviewed by:
Karthik Loganathan,
East China University of Science
and Technology, China
Durgesh Kumar Tripathi,
Banaras Hindu University, India
Lucia Mendoza,
Consejo Nacional de Investigaciones
Científicas y Técnicas, Argentina

*Correspondence:
Ram Prasad
rprasad@amity.edu;
rpjnu2001@gmail.com

Nanotechnology monitors a leading agricultural controlling process, especially by its miniature dimension. Additionally, many potential benefits such as enhancement of food quality and safety, reduction of agricultural inputs, enrichment of absorbing nanoscale nutrients from the soil, etc. allow the application of nanotechnology to be resonant encumbrance. Agriculture, food, and natural resources are a part of those challenges like sustainability, susceptibility, human health, and healthy life. The ambition of nanomaterials in agriculture is to reduce the amount of spread chemicals, minimize nutrient losses in fertilization and increased yield through pest and nutrient management. Nanotechnology has the prospective to improve the agriculture and food industry with novel nanotools for the controlling of rapid disease diagnostic, enhancing the capacity of plants to absorb nutrients among others. The significant interests of using nanotechnology in agriculture includes specific applications like nanofertilizers and nanopesticides to trail products and nutrients levels to increase the productivity without decontamination of soils, waters, and protection against several insect pest and microbial diseases. Nanotechnology may act as sensors for monitoring soil quality of agricultural field and thus it maintain the health of agricultural plants. This review covers the current challenges of sustainability, food security and climate change that are exploring by the researchers in the area of nanotechnology in the improvement of agriculture.

Keywords: sustainable agriculture, nanotechnology, nanofertilizer, nanopesticides, nanoencapsulation, nanoemulsions

INTRODUCTION

Agriculture is always most important and stable sector because it produces and provides raw materials for food and feed industries. The limit of natural resources (production land, water, soil, etc.) and the growth of population in the world claim the agricultural development to be economically further, viable, environmentally and efficiently. This alteration will be the vital for achieving many factors in the recent year (Johnston and Mellor, 1961; Yunlong and Smit, 1994; Mukhopadhyay, 2014). Agricultural nutrient balances are differed noticeably with economic growth, and especially from this surmise, the development of the soil fertility is very much significant in developing countries (Campbell et al., 2014).

The development of agriculture is compulsory phenomena for the purge of poverty and hunger which must be getting rid of from the present situation. Therefore, we should have to take one bold step for agriculture development. In this world mainstream of peoples are below poverty level which are being scatted in the rural area where agriculture enlargement has not so being effective.

Nowadays, the most vital obsession is to create flanked by, agriculture poverty and nutritional process getting food. Therefore, new technology should have to adopt that decidedly focuses on getting better agricultural production (Yunlong and Smit, 1994). Recently, food and nutritional security are fully embedded in the novel knowledge. The agriculture development also depends on the social inclusion, health, climate changes, energy, ecosystem processes, natural resources, good supremacy, etc., must also be documented in specific target oriented goals. Therefore, sustainable agricultural strengthening the practical opportunity to get rid of poverty and hunger of the people. The agriculture on the road to recovery, thus the environmental performance is required and at the same time participation of food chain ecosystems are required in relation to agricultural food production (Thornhill et al., 2016).

No doubt that the sustainable growth of agriculture totally depends on the new and innovative techniques like nanotechnology. Naturally, it haunts us to know what is this important technology? If we like to go in the year 1959 Feynman's lecture on "Plenty of room at the bottom," from this very day, the nanoprocess is in underway (Feynman, 1996). Later on Professor Norio Tanaguchi (1974) proposed the actual term of nanotechnology (Bulovic et al., 2004). Afterward, nanotechnology develops more vivid way, as because, more recent instruments develops to consider or isolate nanomaterials in accurate way (Bonnell and Huey, 2001; Gibney, 2015). Additionally, the number of publications related to the term of "nano" was also grown exponentially. **Figure 1** demonstrates the number of documents on scopus.com (accessed date: March 15, 2017) with the search term of "Nano and (Food and Agriculture)." In 2016, about 14,000 documents with nanotechnology in food or agriculture were listed meaning high activities of this field. Also about 2707 patents matched this criteria are found in world patent database[1]. The world market size of nanotechnology in 2002 was about US\$ 110.6 billion and predicted to grow to US\$ 891.1 billion in 2015 according to analysis of Helmut Kaiser Consultancy[2]. The developments of nanotechnology in materials and electronics have higher dynamics than other applications (**Figure 2A**). Recently, food and agriculture also require high amount of nanomaterials especially in packaging. The NAFTA region shares the biggest slice from the market size (**Figure 2B**), but Europe and Asia especially China, Japan, and India also come up very dynamically.

It is well known that one billionth of a meter is one nanometer (nm). Why this nano will change its property? It is due to alteration in the atoms and develops a magnetic power. It can be projected that the smaller size of nanomaterial possesses larger surface area and exhibits more active. With this course of action nanotechnology is knocking the doors of perception. The magnetic property of polymer develops due to tellurium atoms; antimony-bismuth; and sulfur atoms. Moreover, it has been observed that when the atoms of do pant and atoms of europium interact together, and then the entire molecules carry out the magnetic property. Thus the alter property of nanomaterials is related with more reactive in the most sectors including in biological process (Pokropivny et al., 2007; Prasad, 2014). This ultimate technology possesses several unique electronic association, plasmonic and optical properties which are related with the quantum confinement effects, the alteration of the electronic energy levels may appear due to the surface area in relation to volume ratio (Sun, 2007; Aziz et al., 2015; Prasad et al., 2016). In the present century, there is a big demand for fast, reliable, and low-cost systems for the detection, monitoring, and diagnosis for biological host molecules in agricultural sectors (Vidotti et al., 2011; Sagadevan and Periasamy, 2014). The application of chemically synthesize nanomaterials now a days considered as toxic in the nature, in attract of this, nanomaterials may synthesis from plant system and it considered as green nanotechnology (Prasad, 2014). Green nanotechnology is a safe process, energy efficient, reduces waste and lessens greenhouse gas emissions. Use of renewable materials in production of such products is beneficial, thus these processes have low influence on the environment (Prasad et al., 2014, 2016). Nanomaterials are eco-environmentally sustainable and significant advances have been made in the field of green nanotechnology. In the present decade, it is more shift toward the green nano in a faster rate for implementation its functions. Still it is not clear how the environmental sustainability of green nanotechnology will be achieved in future? These risks must be mitigated in advancing green nanotechnology solutions (Kandasamy and Prema, 2015).

In modern agriculture, sustainable production and efficiency are unimaginable without the use of agrochemicals such as pesticides, fertilizers, etc. However, every agrochemical has some potential issues including contamination of water or residues on food products that threat the human being and environmental health, thus the precise management and control of inputs could allow to reduce these risks (Kah, 2015). The development of the high-tech agricultural system with use of engineered smart nanotools could be excellent strategy to make a revolution in agricultural practices, and thus reduce and/or eliminate the influence of modern agriculture on the environment as well as to enhance both the quality and quantity of yields (Sekhon, 2014; Liu and Lal, 2015).

The development of biosensors is also a good field for exploitation of many strengths of nanotechnology, thus nanotechnology is there and plays an essential role. Due to special properties of nanomaterials, on one hand, the sensitivity and performance of biosensors could be improved significantly in their applications (Fraceto et al., 2016); on another hand, many new signal transduction technologies are let to be introduced in biosensors (Sertova, 2015). Additionally, use of nanomaterials let to miniaturize many (bio)sensors to small and compact/smart devices such as nanosensors and other nanosystems that are very important in biochemical analysis (Viswanathan and Radecki,

---

[1] patentscope.wipo.int (accessed date: March 15, 2017).
[2] hkc22.com

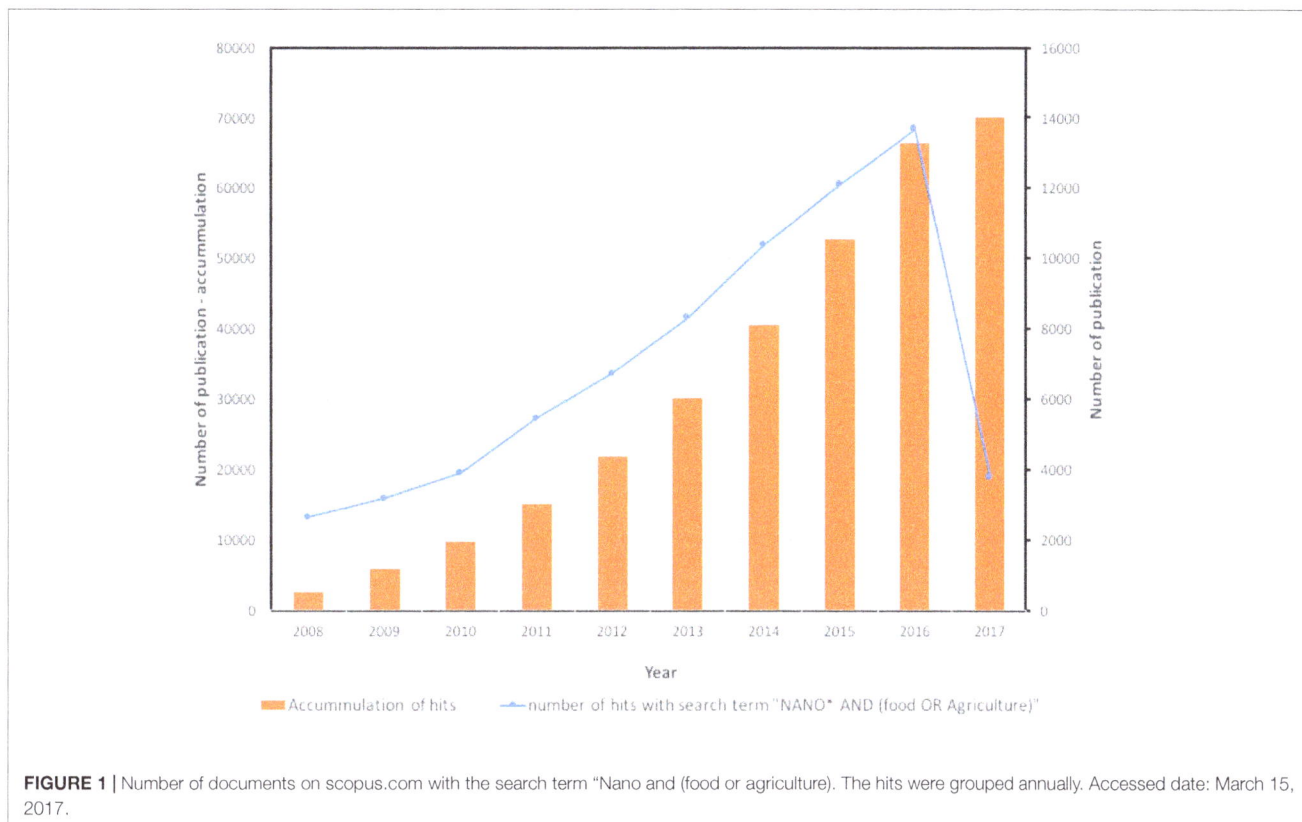

FIGURE 1 | Number of documents on scopus.com with the search term "Nano and (food or agriculture). The hits were grouped annually. Accessed date: March 15, 2017.

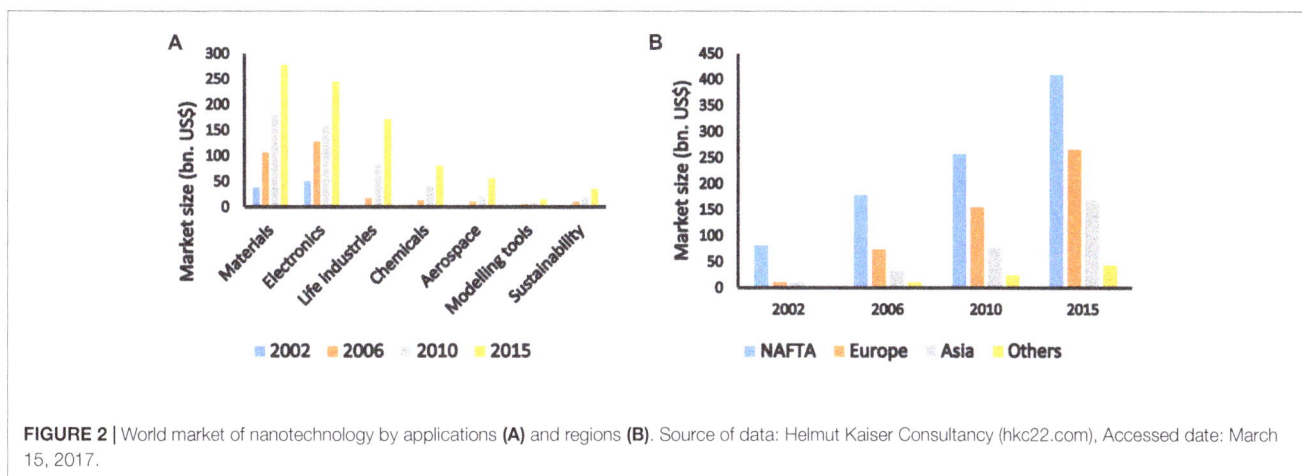

FIGURE 2 | World market of nanotechnology by applications **(A)** and regions **(B)**. Source of data: Helmut Kaiser Consultancy (hkc22.com), Accessed date: March 15, 2017.

2008; Sertova, 2015; Fraceto et al., 2016). It also helps to detect the mycotoxins present in several foods and their functions are very rapid (Sertova, 2015).

NANOPARTICLES AND THEIR FUNCTIONS

Carbon Nanotubes (CNTs)

It is a new form of carbon, equivalent to two dimensional graphene sheet rolled into a tube. Two main types of nanotubes are single-walled nanotubes (SWNTs) and multi-walled nanotubes (MWNTs). Its tensile strength ~200 GPa, thus ideal for reinforced composites and nanoelectro mechanical systems. Moreover, metallic or semiconducting and offers amazing possibilities to create electronic circuits, or even complete nanodevices. Structurally, the nanotube systems consist of graphitic layers seamlessly wrapped into cylinders. Recently, fluorescent nanoparticles (NPs) or quantum dots (QDs) have been developed for labeling the plant proteins (Pyrzynska, 2011; Chahine et al., 2014).

No doubt that properties (mechanical, electronic, thermal, optical, elastic, etc.), and thus applicability of CNTs were determined by geometrical dimensions especially by diameter.

Diameter of most SWNTs is about 1 nm and strongly correlated to synthesis techniques, mixing of σ and α bonds and electron orbital rehybridization. Exploitation of these properties of CNTs definitely will open new possibilities to develop many types of nanodevices which confers unique conductive, optical and thermal properties for applications in agri-field and in development of sustainable agricultural conditions (Raliya et al., 2013). Agrochemicals or other substances can be targeted to hosts by delivery systems based on CNTs, thus led to reduce the amount of chemicals released into the environment as well as the damage to other plant tissues (Raliya et al., 2013; Hajirostamlo et al., 2015).

Camilli et al. (2014) reported that the absorption of the toxic organic solvent dichlorobenzene from water increased about 3.5 times by some CNTs nano-sponges than CNT powder. Generally, the use of CNT nano-sponges containing sulfur and iron increases efficiency in soaking up water contaminants such as pesticides, fertilizers, oil, and pharmaceuticals. Unfortunately, under certain circumstances CNTs may cause vitality of human cells by penetrability and accumulability in the cytoplasm (Porter et al., 2007).

Quantum Dots

Generally, semiconductor QDs are high quantum yield and molar extinction coefficients, broad absorption spectra with narrow, symmetric fluorescence spectra spanning the ultraviolet to near-infrared, large effective excitation, high resistance to photobleaching and exceptional resistance to photochemical degradation. Thus these are excellent fluorescence, quantum confinement of charge carrier's materials and possess size tunable band energy (Bulovic et al., 2004; Androvitsaneas et al., 2016). QDs have unique spectral properties compared with traditional organic dyes, thus recently, they have been applied as a new generation of fluorophores in bioimaging and biosensing (Bakalova et al., 2004). QDs also function as photocatalysts for the light driven chemical conversion of water into hydrogen as a pathway to solar fuel (Konstantatos and Sargent, 2009). QDs at low concentration revealed no detectable cytotoxicity for seed germination and seedling growth. Therefore, based on this transport approach, QDs can be utilized for live imaging in plant root systems to verify known physiological processes (Hu et al., 2010; Das et al., 2015).

Nanorods

Multifunctional plasmonic materials which can couple sensing phenomenon well and size tunable energy regulation, can be coupled with MEMS, and induce specific field responses (Bulovic et al., 2004). The gold nanorods significantly physiological changes occurred of watermelon plant and confirmed phytotoxicity toward plant particularly at high concentration (Wan et al., 2014) and also ability to transport auxin growth regulator 2,4-D, which resulted in a significant influence on the regulation of tobacco cell culture growth (Nima et al., 2014).

Micro- and Nanoencapsulation

Encapsulation is defined as process in which the given object is surrounded by a coating or embedded in homogeneous or heterogeneous matrix, thus this process result capsules with many useful properties (Rodríguez et al., 2016). The benefits of encapsulation methods are for protection of substances/objects from adverse environments, for controlled release, and for precision targeting (Ezhilarasi et al., 2012; Ozdemir and Kemerli, 2016).

Depending on size and shape of capsules different encapsulation technologies are mentioned, while the (macro) encapsulation/coating results capsules in macroscale, whereas the micro- and nanoencapsulation will give particles in micro- and nanoscale size (Ozdemir and Kemerli, 2016). Nanocapsules are vesicular systems in which the substances are confined to a cavity consisting of an inner liquid core enclosed by a polymeric membrane (Couvreur et al., 1995). Recently, micro and NPs are getting significant attention for delivery of drugs, for protection and increase in bioavailability of food components or nutraceuticals, for food fortification and for the self-healing of several materials, and also it possesses big prospective phenomenon in plant science (Ozdemir and Kemerli, 2016). Some drugs such as peptides or anti-inflammatory compounds are successfully nanoencapsulated (Puglisi et al., 1995; Hildebrand and Tack, 2000; Haolong et al., 2011). The development of nanoencapsulated methods for ligation of targeted tissues to NPs which will make possible to deliver several biologically active compounds to the target tissues (Pohlmann et al., 2008). Furthermore, the development of this technology will build more possibility to create new drugs with precise therapeutic action on embattled tissues. Nanocapsules can potentially be used as MRI-guided nanorobots or nanobots (Vartholomeos et al., 2011).

Nanoemulsions

Nanoemulsions are formed by very small emulsion nanoscale droplets (oil/water system) exhibiting sizes lower than ~100 nm (Gutiérrez et al., 2008; Anton and Vandamme, 2011). Although fundamentally significant differences between nanoemulsions and microemulsions could not be exists, but in fact, the physical properties of nanoemulsions can be quite different from those of microscale emulsions (Mason et al., 2006; Gupta et al., 2016). Due to the size of droplets, the ratio of surface area to volume, Laplace pressure and elastic modulus of nanoemulsions are significantly larger than that of ordinary emulsions. Moreover unlike general emulsions, most of nanoemulsions appear optically transparent that, thus, technically have many advantages make such us incorporation into drinks. Unfortunately, the formulation of nanoemulsion needs very high energy, thus it requires some special devices that are able to generate extreme shear stress such as, high pressure homogenizer or ultrasonic generator (Asua, 2002; Gupta et al., 2016). Tadros et al. (2004) reported "low-energy" method for formation of nanoemulsions and in this process, two liquid phases (one is a homogeneous liquid consisted of lipophilic phase and hydrophilic surfactant plus potentially a solvent, polymer or drug, and the other is an aqueous phase, even pure water) are bought into contact of this phase. Then the hydrophilic species contained in the oily phase is rapidly solubilized into the aqueous one, inducing the demixation of the oil in the form of nano-droplets, instantly stabilized by the amphiphiles (Anton and Vandamme, 2009; Gupta et al., 2016). This method thus is seemed to be simplest and does not require any special devices with high energy.

NANOTECHNOLOGY AND AGRICULTURAL SUSTAINABLE DEVELOPMENT

The nanotechnology can takes an important part in the productivity through control of nutrients (Gruère, 2012; Mukhopadhyay, 2014) as well as it can also participate in the monitoring of water quality and pesticides for sustainable development of agriculture (Prasad et al., 2014). Nanomaterials have such diverse assets and activities that it is impossible to deliver a general assessment of their health and environmental risks (Prasad et al., 2014). Properties (other than size) of NPs have the influence on toxicity include chemical composition, shape, surface structure, surface charge, behavior, extent of particle aggregation (clumping) or disaggregation, etc. may associate with engineered NPs (Ion et al., 2010). For this reason even nanomaterials of the same chemical composition that have different sizes or shapes can exhibit their different toxicity. The implication of the nanotechnology research in the agricultural sector is become to be necessary even key factor for the sustainable developments. In the agri-food areas pertinent applications of nanotubes, fullerenes, biosensors, controlled delivery systems, nanofiltration, etc. were observed (Ion et al., 2010; Sabir et al., 2014). This technology was proved to be as good in resources management of agricultural field, drug delivery mechanisms in plants and helps to maintain the soils fertility. Moreover, it is being also evaluated steadily in the use of biomass

and agricultural waste as well as in food processing and food packaging system as well as risk assessment (Floros et al., 2010). Recently, nanosensors are widely applied in the agriculture due to their strengths and fast for environmental monitoring of contamination in the soils and in the water (Ion et al., 2010). Several sensors based on nano-detection technology such as viz. biosensors, electrochemical sensors, optical sensors, and devices will be the main instruments for detecting the heavy metals in trace range (Ion et al., 2010).

Nanomaterials not only directly catalyze degradation of waste and toxic materials but it also aids improve the efficiency of microorganisms in degradation of waste and toxic materials. Bioremediation uses living organisms to break down or remove toxins and harmful substances from agricultural soil and water. In particular, some other terms are also generally used such as bioremediation (beneficial microbes), phytoremediation (plants), and mycoremediation (fungi and mushrooms). Thus, with the bioremediation the heavy metals can be removed from soil and water environmentally and efficiently by microorganisms (Dixit et al., 2015). Therefore, the agricultural bioremediation helps in sustainable remediation technologies to resolve and restore the natural situation of the soil. It is an interesting phenomena in considering the nano–nano interaction to remove the toxic component of the agricultural soil and make it sustainable (Ion et al., 2010; Dixit et al., 2015).

Nanofertilizers

In the recent decade nanofertilizers are freely available in the market, but particularly the agricultural fertilizers are still not shaped by the major chemical companies (**Table 1**). Nanofertilizers may contain nano zinc, silica, iron and titanium dioxide, ZnCdSe/ZnS core shell QDs, InP/ZnS core shell QDs, Mn/ZnSe QDs, gold nanorods, core shell QDs, etc. as well as should endorse control release and improve the its quality. Studies of the uptake, biological fate and toxicity of several metal oxide NPs, viz. Al_2O_3, TiO_2, CeO_2, FeO, and ZnONPs were

TABLE 1 | Some commercial product of nanofertilizers.

Commercial product	Content	Company
Nano-Gro™	Plant growth regulator and immunity enhancer	Agro Nanotechnology Corp., FL, United States
Nano Green	Extracts of corn, grain, soybeans, potatoes, coconut, and palm	Nano Green Sciences, Inc., India
Nano-Ag Answer®	Microorganism, sea kelp, and mineral electrolyte	Urth Agriculture, CA, United States
Biozar Nano-Fertilizer	Combination of organic materials, micronutrients, and macromolecules	Fanavar Nano-Pazhoohesh Markazi Company, Iran
Nano Max NPK Fertilizer	Multiple organic acids chelated with major nutrients, amino acids, organic carbon, organic micro nutrients/trace elements, vitamins, and probiotic	JU Agri Sciences Pvt. Ltd, Janakpuri, New Delhi, India
Master Nano Chitosan Organic Fertilizer	Water soluble liquid chitosan, organic acid and salicylic acids, phenolic compounds	Pannaraj Intertrade, Thailand
TAG NANO (NPK, PhoS, Zinc, Cal, etc.) fertilizers	Proteino-lacto-gluconate chelated with micronutrients, vitamins, probiotics, seaweed extracts, humic acid	Tropical Agrosystem India (P) Ltd, India

carried out intensively in the present decade for agricultural production (Dimkpa, 2014; Zhang et al., 2016). The deficiency of zinc has been documented as one of the main problems in limiting agricultural productivity in the alkaline nature of soils (Sadeghzadeh, 2013).

Metal oxide NPs are radiolabeled by direct proton bombardment or enriched during synthesis with ^{18}O to generate ^{18}F (Llop et al., 2014). Size, degree of aggregation and zeta potential of the metal oxide NPs are studied in the presence of proteins and cell media (Llop et al., 2014; Marzbani et al., 2015). Moreover, NP uptake and intracellular fate are followed by ion beam microscopy, transmission electron microscopy, Raman chemical imaging spectroscopy, and confocal laser scanning microscopy (Marzbani et al., 2015). In the future, sustainable bio-based economy that uses eco-efficient bio-processes and renewable bio-resource, will continue decrease and substitute the harmful materials in established applications, and thus it will play a major role (the key strategic challenge) in the development of the technologies desired to address to 21st century (Prasad et al., 2014; Marzbani et al., 2015). Accumulation of knowledge in fields of ecology, biology, biodiversity, material science, biotechnology, and engineering opens possibilities to increase biomass productivity as well as to utilize biomass and organic wastes at a highly efficient.

In the present century, the smart agriculture is a way to achieve priority of short and long term development in the countenance of climate change and serves as a link to others (Helar and Chavan, 2015). It seeks to support countries and other functional aspects in securing the necessary agricultural functions (Kandasamy and Prema, 2015). In the past few years, researches related to the expansion of resources in a nanometric extent and their inherent properties are intensively conducted and focused. Practically, when the crystallite size of inorganic materials are reduced to nanoscale, two different phenomena can occur. In the first one (quantum size effect), radical changes of the physical–chemical properties of material are observed. In this case, the performance is totally dependent on the semiconductors-NPs. On the other hand, due to the huge ratio of surface area to volume, NPs exhibit very good transduction properties which are being more interesting for analytical purpose of agricultural products (Kandasamy and Prema, 2015). Nanostructures materials exposed several advantages in logical sciences when used as transducers or as a part of the appreciation in a macro-sized sensing device. In this facts the gold NPs (AuNPs) has its intrinsic properties, and may use as transducers for several improvements of agricultural products. The AuNPs have well-known surface plasmon band that is visible around 520 nm. Moreover, AuNPs have high surface areas and distinctive physicochemical assets that can be easily tuned and thus making them ideal candidates for developing biosensing devices. Additionally, these NPs possess attracted attention in biological studies owing to their low toxicity, biocompatibility and unique optical properties. Biological tests measuring the presence or activity of selected analytics become quicker, more sensitive and flexible when nanoscale particles are put together (Vidotti et al., 2011; Kandasamy and Prema, 2015). Thus, application of nanoscale particles results numerous advantages over traditional procedures.

Nanopesticides

The use of nanomaterials in plant protection and production of food is under-explored area in the future. It is well known that insect pests are the predominant ones in the agricultural fields and also in its products, thus NPs may have key role in the control of insect pests and host pathogens (Khota et al., 2012; **Table 2**). The recent development of a nanoencapsulated pesticide formulation has slow releasing properties with enhanced solubility, specificity, permeability and stability (Bhattacharyya et al., 2016). These assets are mainly achieved through either protecting the encapsulated active ingredients from premature degradation or increasing their pest control efficacy for a longer period. Formulation of nanoencapsulated pesticides led to reduce the dosage of pesticides and human beings exposure to them which is environmentally friendly for crop protection (Nuruzzaman et al., 2016). So, development of non-toxic and promising pesticide delivery systems for increasing global food production while reducing the negative environmental impacts to ecosystem (de Oliveira et al., 2014; Kah and Hofmann, 2014; Bhattacharyya et al., 2016; Grillo et al., 2016).

Microencapsulation-like nanoencapsulation is used to develop the quality of products of desired chemicals delivery to the target biological process. Recently, few chemical companies openly promote nanoscale pesticides for sale as "microencapsulated pesticides." Some products from Syngenta (Switzerland) such as Karate ZEON, Subdue MAXX, Ospray's Chyella, Penncap-M, and microencapsulated pesticides from BASF may fighting fit for nanoscale (Gouin, 2004). Syngenta also markets in the Australia some products such as the Primo MAXX, Banner MAXX, Subdue MAXX, etc. Despite they are known as microemulsions in the market, however, they are really nanoscale emulsions. It confirms very thin interface between the term of microemulsion and nanoemulsion. This technique is commonly used for formulations of organic NPs (Gouin, 2004) containing active agrochemicals or substances of interest.

Ecotoxicological Implications of the Nanoparticles

The advancement of nanotechnologies has presented significant extents of manufactured NPs into the environment. In order to protect human health and plant from the prospective antagonistic effects of a wide range of nanomaterials, an increasing number of research have focused on the assessment of the toxicity of the NPs normally used in industry (Yang and Watts, 2005; Rana and Kalaichelvan, 2013; Du et al., 2017; Tripathi et al., 2017a,b,c). The toxicity of a metal depends upon several factors like solubility, binding specificity to a biological site, and so forth. Metal NPs exhibit antibacterial, anticandidal, and antifungal activities (Aziz et al., 2016; Patra and Baek, 2017). Metal NPs exert cytotoxicity depending on the charge at membrane surface, of course, the efficiency of nanotoxic effects of NPs are definitely depending on structure of targeted cell-wall, thus the sensitive order should be

TABLE 2 | A list of studies on nanopesticides/herbicides and its application.

Carrier system	Agent	Purpose	Method	Reference
Chitosan	Imazapic and Imazapyr	Cytotoxicity assays	Encapsulation	Maruyama et al., 2016
Silica	Piracetam, pentoxifylline, and pyridoxine	Perfused brain tissue	Suspension	Jampilek et al., 2015
Alginate	Imidacloprid	Cytotoxicity, sucking pest (leafhoppers)	Emulsion	Kumar et al., 2014
Polyacetic acid-polyethylene glycol-polyacetic acid	Imidacloprid	Decrease the lethal concentration	Encapsulation	Memarizadeh et al., 2014
Carboxymethyl chitosan	Methomyl	Control release for longer time-period	Encapsulation	Sun et al., 2014
Chitosan/tripolyphosphate	Paraquat	Lower cyto- and genotoxicity	Encapsulation	Grillo et al., 2014
Chitosan/tripolyphosphate Chitosan-saponin Chitosan-Cu	Chitosan, saponin, $CuSO_4$	Antifungal activity	Cross-linking	Saharan et al., 2013
Xyloglucan/poloxamer	Tropicamide	Have significantly higher corneal permeation across excised goat cornea Less toxic and non-irritant	Encapsulation	Dilbaghi et al., 2013
Wheat gluten	Ethofumesate	Reduce its diffusivity	entrapment/extrusion	Chevillard et al., 2012
Alginate	Azadirachtin	Slower release	Encapsulation	Jerobin et al., 2012
Surfactants/oil/water	Glyphosate	Increase in bio-efficacy, alleviating the negative effect of pesticide formulations into environment	Emulsion	Jiang et al., 2012
Alginate/chitosan	Paraquat	Increased period of action of the chemical on precise targets, while reducing problems of ecological toxicity	Pre-gelation of alginate then complexation between alginate and chitosan	Silva Mdos et al., 2011
Polyhydroxybutyrate-co-hydroxyvalerate	Atrazine	Decreased genotoxicity and increased biodegradability	Encapsulation	Grillo et al., 2010
Organic-inorganic nanohybrid	2,4-Dichlorophenoxyacetate	Control release	Self-assembly	Hussein et al., 2005

mould > yeast > Gram-negative > Gram-positive. Nanotoxicity may be accredited to electrostatic interaction between NPs with membrane and their accumulation in cytoplasm (Rana and Kalaichelvan, 2011; Aziz et al., 2015, 2016).

Several NPs (TiO$_2$, ZnO, SiO$_2$, and Fullerenes) are photochemically active. When they are exposed to light, the excited electrons are generated that then form superoxide radicals in the presence of oxygen by direct electron transfer (Hoffmann et al., 2007). Thus, this ecotoxicity is surprised in the act when organisms are simultaneously exposed to NPs and UV light (particularly UV light has higher energy than visible light). In this case, the cells respond to oxidative stress by increasing a number of protective enzymatic or genetic constitutions that can easily be measured (Kovochich et al., 2005; Vannini et al., 2014), thus generation of reactive oxygen species (ROS) is oxidative stress parameter that can be exploited in determination of the context of toxicity and ecotoxicity. *In vitro* studies on the toxicity of NPs have confirmed the generation of ROS, for example, by

TiO$_2$ and fullerenes (Sayes et al., 2004), while on other hand, some authors revealed that NPs (fullerenes and silicon NPs) may protect against oxidative stress (Daroczi et al., 2006; Tripathi et al., 2016b, 2017d; Venkatachalam et al., 2017). Much more researches related to interactions between cells and NPs as well as mechanistic facets of NPs metabolism in organisms and specific cells are needed to clarify this dichotomy.

Ecotoxicological research would increasingly attention on the environmental consequence of the materials and complexity of natural systems. Extensive research would be necessary to determine delayed impacts of environmental exposure to NPs and to help determine possible adaptive mechanisms (Cox et al., 2017; Singh et al., 2017). More research on bioaccumulation in the food chain and interaction of NPs with other pollutants in the environment. NPs in plants enter cellular system, translocate them shoot and accumulate in various aerial parts, the possibility of their cycling in the ecosystem increases through various trophic levels. After accumulation of NPs effect rate of transpiration, respiration, altering the process of photosynthesis, and interfere with translocation of food material (Shweta et al., 2016; Tripathi et al., 2016b; Du et al., 2017). The degree of toxicity is linked to this surface and to the surface properties of the NPs. The ecotoxicity of NPs is thus very important as it creates a direct link between the adverse effects of NPs and the organisms including microorganisms, plants, and other organisms including humans at various trophic levels (Rana and Kalaichelvan, 2013; Tripathi et al., 2016a).

Growth of Cultivated Plants and Its Ecotoxicological Sustainability

The agriculture host plants take the main part in food chain. Recently, the plants do not only grow on agricultural lands, but they are also developed on aqueous medium too. Naturally, several NPs of iron oxide (magnetite), a magnetic form of iron ore can deposit in the plant host. It is interesting to propose that the iron (II, III) oxide NPs (Fe_3O_4 -NPs) have the ability to accumulate in *Lepidium sativum* and *Pisum sativum* plants. Therefore, this type of observation clearly proposes that the roles to mention NPs are present in the natural ecosystem (Bystrzejewska-Piotrowska et al., 2012; Abbas et al., 2016). Moreover, the uses of polymeric NPs in the agricultural field, especially loaded with insecticides of plant origin are unique and increasingly permeated (Chakravarthy et al., 2012; Perlatti et al., 2013). No doubt that microorganism plays crucial role in maintaining soil health, ecosystem, and crop productivity (Mishra and Kumar, 2009). Therefore, it is very essential to know the ecotoxicological aspects of the considered agricultural field. If nanomaterials containing agricultural plants are devoid of any toxic nanocomposite, then the unique possibility of more production of agricultural crops. Thus, the introduction of engineered (either chemical or green) NPs in the agricultural field should always be a routine check-up to sustain an eco-friendly in the agricultural field (**Figure 3**).

NANOBIOSENSORS

Many advantages of physical–chemical properties of nanoscale materials are also exploitable in field of biosensors development of biosensors. Sagadevan and Periasamy (2014) stated that the sensitivity and performance of biosensors can be improved by using nanomaterials through new signal transduction technologies. The tremendous advancement in the nanobiosensors are due to the great technological demand for rapid, sensitive and cost-effective nanobiosensor systems in vital areas of human activity such as health care, agriculture, genome analysis, food and drink, the process industries, environmental monitoring, defense, and security. At present, the nanotechnology-based biosensors are at the early stage of development (Fogel and Limson, 2016). The improvement of tools and procedures used to fabricate, measure and image nanoscale objects, has led to the development of sensors. The nanomaterials such as metal (gold, silver, cobalt, etc.) NPs, CNT, magnetic NPs, and QDs have been actively investigated for their applications in biosensors which have become a new

interdisciplinary frontier between biological detection and material science. Thus a biosensor is a device that combines a biological recognition element with physical or chemical principles. It integrates a biological one with an electronic component to yield a measurable signal component, and the biological recognition is through the transducer process and the signal processing through electronic achievement. The higher specificity and sensitivity of biosensor systems over the conventional methods are due to the presence of the bioreceptor (biological element) that is combined with a suitable transducer which produces a signal after interaction with the target molecule of interest. Recently, different natural and artificial bioreceptors are developed and applied such as enzymes, dendrimers, thin films, etc. Therefore, by biosensor, an analytical device, this converts a biological response into an electrical signal. It is concerned with these parts of biological elements like, an antibody, an enzyme, a protein, or a nucleic acid. The transducer and the associated electronics or signal processors that are primarily responsible for detection of the functions (Rai et al., 2012). The micro cantilever-based DNA biosensor that uses AuNPs have been developed and used widely to detect low level DNA concentration during a hybridization reaction (Brolo, 2012).

NANOTECHNOLOGIES IN FOOD INDUSTRY

Nanoscale biosensors can take part in pathogen detection and diagnosis. Nanotechnology has the ability to supply bioactive ingredients in foodstuffs to hosts while improvement of knowledge of food materials at the nanoscale (Martirosyan and Schneider, 2014). It also helps in nanoscale filtration systems for improved texture modification of food. Nano biosensors interact with food, attractive surface and thus maintain the glaziers and colors of food, magnetic nanocomposite for tag sensors. Nanoprinted, intelligent packaging, controlled release (Ghaani et al., 2016), nano-additives (Khond and Kriplani, 2016), nanocoding of plastics and paper materials (Bhushani and Anandharamakrishnan, 2014), for authentication and identification purposes (Khond and Kriplani, 2016; Savina et al., 2016). In food quality tests, some important aspects should be covered such as sensing ability of label and package, *in situ* sensors, food quality monitoring (e.g., color, smell, taste, texture), control and nutraceuticals delivery, portable DNA/protein chips, etc. Most of the time, nanomaterials produced by bottom–up methods (Siegrist et al., 2008).

Food Process

Recently nanotechnology is widely applied in food processing such as nanocarrier systems for delivery of nutrients and supplements, organic nano-sized additives for food, supplements, and animal feed. Many food products already contain NP naturally. Milk contains casein, a form of milk protein present at the nanoscale or meat is made up of protein filaments that are also be classified into nanomaterial group. The texture and

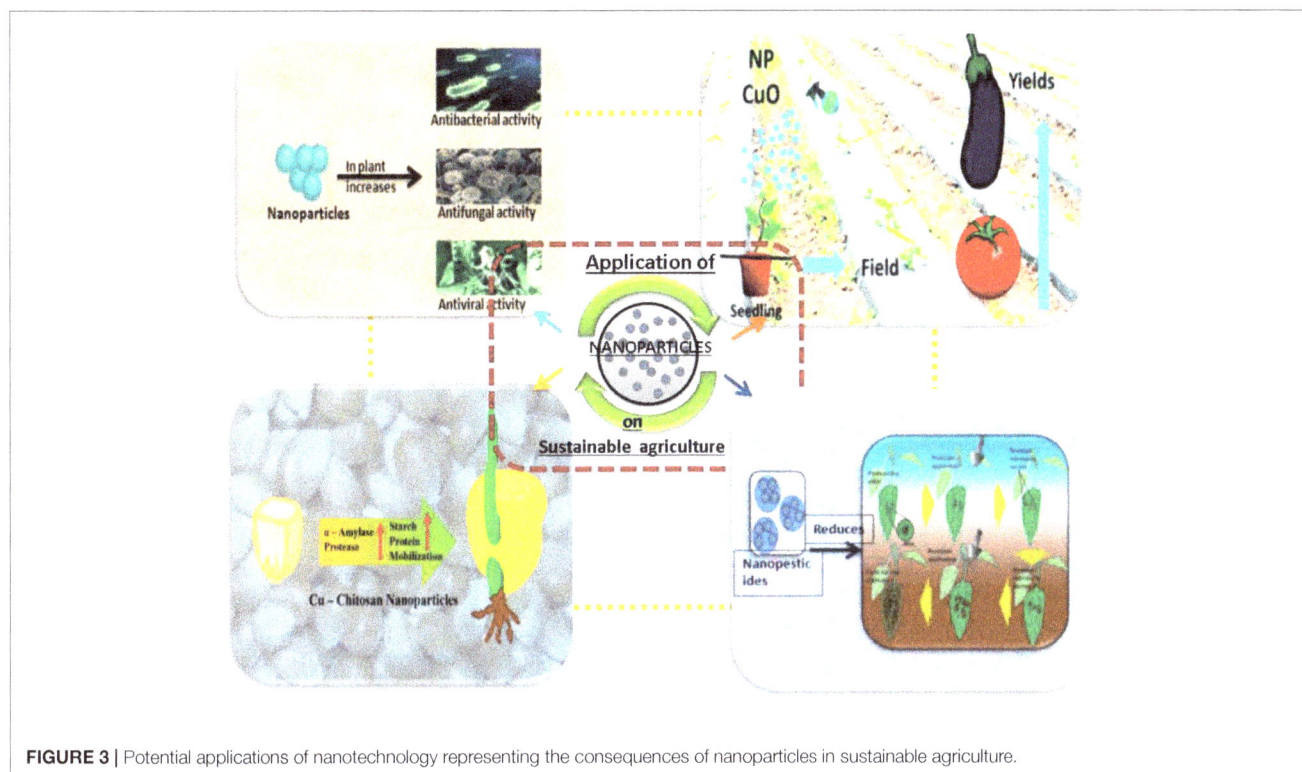

FIGURE 3 | Potential applications of nanotechnology representing the consequences of nanoparticles in sustainable agriculture.

properties of these products are determined by the organization and structures of proteins inside.

Recently, some nutrients mainly vitamins are encapsulated and delivered into the bloodstream through digestion system with very high efficiency. Some foods and drinks were fortified with these NPs without affecting the taste or appearance. NP emulsions are being used in ice cream and spreads of this nanoemulsion can improve the texture and uniformity of the ice cream (Berekaa, 2015). A real example: KD Pharma BEXBACH GMBH (Germany) provides encapsulated Omega-3 fatty acids in two different forms—suspension and powder. The capsulation technology used the resulted particles in nano- and microscale.

Food Packaging and Labeling

In the food industry, maintaining some important factors such as quality, safe, freshness, taste, etc. whole supply chain requires producers to packaging and labeling their products. Development of smart packages that can provide useful information is still big challenge for researchers and producers. Recently, some packaging materials incorporated with "nanosensors" to detect the oxidation process in food have been produced and used in food industry. Working scheme is quite simple: when the oxidation occurs in the food package, NP-based sensors indicate the color change and information about the nature of the packed foods can be observed. This technology have been successfully applied in package of milk and meat (Bumbudsanpharoke and Ko, 2015).

Due to nanostructure, NPs are good barriers for diffusion gases such as oxygen, carbon dioxide, thus it can be exploited in food packaging. Some drinks (beer, soda waters, etc.) naturally have to keep appreciate amount of carbon dioxide can be packaged in the bottles made with nanocomposites because of minimization of CO_2 lost, decrease in weight of packaging materials, increase in shelf life, etc. Other exploitation way is the incorporation of NPs in packaging and this technology will slow down some biochemical processes such as oxidation, degradation, etc. thus it help to extend the shelf-life of food products. In the food packaging industry, the most used materials are plastic polymers that can be incorporated or coated with nanomaterials for improved mechanical or functional properties (Berekaa, 2015). Moreover, nanocoatings on food contact surfaces act as barrier or antimicrobial properties. Silver NPs have been successfully embedded in the plastic for making food storage bins, and this acts like disinfection of bins, thus minimizing harmful bacterial growth. Therefore, the nanotechnology is a forward looking process, it acts as an agricultural biosecurity (Bumbudsanpharoke and Ko, 2015).

NPs in precise have revealed broad-spectrum antibacterial properties against both Gram-positive and Gram-negative bacteria. ZnO NPs were found to inhibit *Staphylococcus aureus* (Liu et al., 2009) and AgNPs exhibit concentration-dependent antimicrobial activity against *Escherichia coli, Aeromonas hydrophila*, and *Klebsiella pneumoniae* (Aziz et al., 2016). The antimicrobial mechanism of action of NPs is typically considered as of few prototypes such as oxidative stress and cell damage, metal ion release, or non-oxidative mechanisms (Wang et al., 2017). These mechanisms can happen concurrently. Firm studies have suggested that Ag NPs quick neutralization of the surface electric charge of the bacterial membrane and change its permeability, ultimately leading to apoptosis. Moreover, the generation of ROS prevents the antioxidant defense system and causes physiochemical damage to the intrinsic cell membrane. According to current investigation, the major processes causal the antibacterial effects of NPs are as follows: disruption of the bacterial cell membrane; generation of ROS; penetration of the bacterial cell membrane by passive or facilitated diffusion and induction of intracellular antibacterial effects, including interfaces with DNA replication, and inhibition of protein synthesis (Aziz et al., 2015; Wang et al., 2017).

Nanosensors help in food labeling and in combination with NP-based intelligent inks or reactive nanolayers may provide smart recognition of relevant food product. Printed labels in the food package that can indicate the following highlights: temperature, time, pathogens, freshness, humidity, etc. Nanobarcode particles with different patterns of gold which can form template and also with silver stripes are possible to synthesize itself. Lastly, we can suggest that the contaminant or nutrient sorption on NPs surfaces has attracted the attention of researchers for more studies on soil chemistry, showing that NPs have high sorption capacities for metal and anionic contaminants (Li et al., 2016). It was found that the contaminant sequestration was accomplished mainly by surface complication. It is likely that the sorbet surface species can be encapsulated within interior surfaces of NPs. A phenomenon with significant consequences for contaminant dispersion or remediation processes can exhibit. Moreover, metallic species as Ni can be linked to natural short-ordered aluminosilicates, TiO_2 surfaces, humic acids, and aromatic compounds by MWCNTs and these association may be considered as very potent bioremediation in nano agricultural system (Raliya et al., 2013; Hajirostamlo et al., 2015).

FUTURE PERSPECTIVES

Sustainable agriculture must be taken as an ecosystem method, where abiotic–biotic-living beings live in accord with a co-ordinated stability of food chains and their related energy balances. New technologies, modernization, increased in use of nano-chemicals, specialization and government policies are adapted to maximize the production in agriculture. To overcome the situation, it is mandatory to establish the recent technology in the food industry. Therefore, the new and future technology is nanotechnology that possesses very unique property in food supply chain (from the field to table: crop production, use of agro-chemicals such as nanofertilizers, nanopesticides, nanoherbicides, etc., precision farming techniques, intelligent feed, enhancement of food texture and quality, and bioavailability/nutrient values, packaging and labeling, etc.) round the world agricultural sector. Some focused areas may need more attention in near future researches in the field of agricultural nanotechnology or nanofoods:

- New environmental and safety delivery systems for carrying special food/feed compounds, plant nutrients, etc. These systems also can have pharmaceutical application potentials.
- The (bio)sensors related nanotechnology have effective role in insect pest control and food products of agriculture. Consumers always can get actual information of the state of certain food product via intelligent food packaging corporated with nanosensors.
- The properties of nanomaterials such as size, dose, exposure time, surface chemistry, structures, immune response, accumulation, retention time, etc., and other effects should be accessed carefully. New analytical methods are needed to develop to detect, validate and access the effects of each nanomaterials/nanofoods in whole ecosystems. Life-cycle analysis of nanomaterials/nanofoods should be done. Improvement of wide-ranging databank as well as international collaboration for policy, idea and regulation are needed for manipulation of this knowledge. Additionally, the authorities should provide clear guidelines and roadmaps for reducing risks of the use of nanotechnological products.
- New communication channels and debates should be opened with participation of different sides such as consumers, researchers, authorities, industrial sectors, etc. to discuss impacts of this technology in human life, economy, and science.

This technology in the long term may provide innovative and economical development routes for human nutrition worldwide.

CONCLUSION

Agriculture which is the only provider of human's food that should produce from transitional and final inputs with well-known technologies. Thus, it is necessary to take a modern knowledge in agriculture. In spite of being relative advantages in agriculture process, still developing countries are suffering from lack of high importance of food products. Despite a lot of information about individual nanomaterials are available, but toxicity level of many NPs is still indefinable, thus the application of these materials is limited due to the lack of knowledge of risk assessments and effects on human health. Development of comprehensive database and alarm system, as well as international cooperation for regulation and legislation are necessary for exploitation of this technology.

AUTHOR CONTRIBUTIONS

RP, AB, and QN developed the idea and wrote the manuscript. All authors proofread and approved the final manuscript.

REFERENCES

Abbas, S. S., Haneef, M., Lohani, M., Tabassum, H., and Khan, A. F. (2016). Nanomaterials used as a plants growth enhancer: an update. *Int. J. Pharm. Sci. Rev. Res.* 5, 17–23.

Androvitsaneas, P., Young, A. B., Schneider, C., Maier, S., Kamp, M., Höfling, S., et al. (2016). Charged quantum dot micropillar system for deterministic light-matter interactions. *Phys. Rev. B* 93:241409. doi: 10.1103/physrevb.93.241409

Anton, N., and Vandamme, T. F. (2009). The universality of low-energy nano-emulsification. *Int. J. Pharm.* 377, 142–147. doi: 10.1016/j.ijpharm.2009.05.014

Anton, N., and Vandamme, T. F. (2011). Nano-emulsions and micro-emulsions: clarifications of the critical differences. *Pharm. Res.* 28, 978–985. doi: 10.1007/s11095-010-0309-1

Asua, J. M. (2002). Miniemulsion polymerization. *Prog. Polym. Sci.* 27, 1283–1346. doi: 10.1016/S0079-6700(02)00010-2

Aziz, N., Faraz, M., Pandey, R., Sakir, M., Fatma, T., Varma, A., et al. (2015). Facile algae-derived route to biogenic silver nanoparticles: synthesis, antibacterial and photocatalytic properties. *Langmuir* 31, 11605–11612. doi: 10.1021/acs.langmuir.5b03081

Aziz, N., Pandey, R., Barman, I., and Prasad, R. (2016). Leveraging the attributes of Mucor hiemalis-derived silver nanoparticles for a synergistic broad-spectrum antimicrobial platform. *Front. Microbiol.* 7:1984. doi: 10.3389/fmicb.2016.01984

Bakalova, R., Zhelev, Z., Ohba, H., Ishikawa, M., and Baba, Y. (2004). Quantum dots as photosensitizers? *Nat. Biotechnol.* 22, 1360–1361. doi: 10.1038/nbt1104-1360

Berekaa, M. M. (2015). Nanotechnology in food industry; advances in food processing, packaging and food Safety. *Int. J. Curr. Microbiol. App. Sci.* 4, 345–357.

Bhattacharyya, A., Duraisamy, P., Govindarajan, M., Buhroo, A. A., and Prasad, R. (2016). "Nano-biofungicides: emerging trend in insect pest control," in *Advances and Applications through Fungal Nanobiotechnology*, ed. R. Prasad (Cham: Springer International Publishing), 307–319. doi: 10.1007/978-3-319-42990-8_15

Bhushani, J. A., and Anandharamakrishnan, C. (2014). Electrospinning and electrospraying techniques: potential food based applications. *Trends Food Sci. Technol.* 38, 21–33. doi: 10.1016/j.tifs.2014.03.004

Bonnell, D. A., and Huey, B. D. (2001). "Basic principles of scanning probe microscopy," in *Scanning Probe Microscopy and Spectroscopy: Theory, Techniques, and Applications*, ed. D. A. Bonnell (New York, NY: Wiley-VCH).

Brolo, A. G. (2012). Plasmonics for future biosensors. *Nat. Photonics* 6, 709–713. doi: 10.1038/nphoton.2012.266

Bulovic, V., Mandell, A., and Perlman, A. (2004). Molecular Memory Device. US 20050116256, A1.

Bumbudsanpharoke, N., and Ko, S. (2015). Nano-food packaging: an overview of market, migration research, and safety regulations. *J. Food Sci.* 80, R910–R923. doi: 10.1111/1750-3841.12861

Bystrzejewska-Piotrowska, G., Asztemborska, M., Steborowski, R., Polkowska-Motrenko, H., Danko, J., and Ryniewicz, B. (2012). Application of neutron activation for investigation of Fe3O4 nanoparticles accumulation by plants. *Nukleonika* 57, 427–430.

Camilli, L., Pisani, C., Gautron, E., Scarselli, M., Castrucci, P., D'Orazio, F., et al. (2014). A three-dimensional carbon nanotube network for water treatment. *Nanotechnology* 25:065701. doi: 10.1088/0957-4484/25/6/065701

Campbell, B. M., Thornton, P., Zougmoré, R., van Asten, P., and Lipper, L. (2014). Sustainable intensification: what is its role in climate smart agriculture? *Curr. Opin. Environ. Sustain.* 8, 39–43. doi: 10.1016/j.cosust.2014.07.002

Chahine, N. O., Collette, N. M., Thomas, B. C., Genetos, D. C., and Loots, G. G. (2014). Nanocomposite scaffold for chondrocyte growth and cartilage tissue engineering: effects of carbon nanotube surface functionalization. *Tissue Eng. Part A.* 20, 2305–2315. doi: 10.1089/ten.TEA.2013.0328

Chakravarthy, A. K., Bhattacharyya, A., Shashank, P. R., Epidi, T. T., Doddabasappa, B., and Mandal, S. K. (2012). DNA-tagged nano gold: a new tool for the control of the armyworm, *Spodoptera litura* Fab. (Lepidoptera: Noctuidae). *Afr. J. Biotechnol.* 11, 9295–9301. doi: 10.5897/AJB11.883

Chevillard, A., Angellier-Coussy, H., Guillard, V., Gontard, N., and Gastaldi, E. (2012). Controlling pesticide release via structuring agropolymer and nanoclays based materials. *J. Hazard. Mater.* 20, 32–39. doi: 10.1016/j.jhazmat.2011.11.093

Couvreur, P., Dubernet, C., and Puisieux, F. (1995). Controlled drug delivery with nanoparticles: current possibilities and future trends. *Eur. J. Pharm. Biopharm.* 41, 2–13.

Cox, A., Venkatachalam, P., Sahi, S., and Sharma, N. (2017). Reprint of: silver and titanium dioxide nanoparticle toxicity in plants: a review of current research. *Plant Physiol. Biochem.* 110, 33–49. doi: 10.1016/j.plaphy.2016.08.007

Daroczi, B., Kari, G., McAleer, M. F., Wolf, J. C., Rodeck, U., and Dicker, A. P. (2006). *In vivo* radioprotection by the fullerene nanoparticle DF-1 as assessed in a zebra fish model. *Clin. Cancer Res.* 12, 7086–7091. doi: 10.1158/1078-0432.CCR-06-0514

Das, S., Wolfson, B. P., Tetard, L., Tharkur, J., Bazata, J., and Santra, S. (2015). Effect of N-acetyl cysteine coated CdS:Mn/ZnS quantum dots on seed germination and seedling growth of snow pea (*Pisum sativum* L.): imaging and spectroscopic studies. *Environ. Sci.* 2, 203–212. doi: 10.1039/c4en00198b

de Oliveira, J. L., Campos, E. V., Bakshi, M., Abhilash, P. C., and Fraceto, L. F. (2014). Application of nanotechnology for the encapsulation of botanical insecticides for sustainable agriculture: prospects and promises. *Biotechnol. Adv.* 32, 1550–1561. doi: 10.1016/j.biotechadv.2014.10.010

Dilbaghi, N., Kaur, H., Ahuja, M., and Kumar, S. (2013). Evaluation of tropicamide-loaded tamarind seed xyloglucan nanoaggregates for ophthalmic delivery. *Carbohydr. Polym.* 94, 286–291. doi: 10.1016/j.carbpol.2013.01.054

Dimkpa, C. O. (2014). Can nanotechnology deliver the promised benefits without negatively impacting soil microbial life? *J. Basic Microbiol.* 54, 889–904. doi: 10.1002/jobm.201400298

Dixit, R., Wasiullah, Malaviya, D., Pandiyan, K., Singh, U. B., Sahu, A., et al. (2015). Bioremediation of heavy metals from soil and aquatic environment: An overview of principles and criteria of fundamental processes. *Sustainability* 7, 2189–2212. doi: 10.3390/su7022189

Du, W., Tan, W., Peralta-Videa, J. R., Gardea-Torresdey, J. L., Ji, R., Yin, Y., et al. (2017). Interaction of metal oxide nanoparticles with higher terrestrial plants: Physiological and biochemical aspects. *Plant Physiol. Biochem.* 110, 210–225. doi: 10.1016/j.plaphy.2016.04.024

Ezhilarasi, P. N., Karthik, P., Chhanwal, N., and Anandharamakrishnan, C. (2012). Nanoencapsulation techniques for food bioactive components: a Review. *Food Bioprocess Technol.* 6, 628–647. doi: 10.1007/s11947-012-0944-0

Feynman, R. P. (1996). *No Ordinary Genius: The Illustrated Richard Feynman.* New York, NY: W.W. Norton & Company.

Floros, J. D., Newsome, R., Fisher, W., Barbosa-Cánovas, G. V., Chen, H., Dunne, C. P., et al. (2010). Feeding the world today and tomorrow: the importance of food science and technology. *Compr. Rev. Food Sci. Food Saf.* 9, 572–599. doi: 10.1111/j.1541-4337.2010.00127.x

Fogel, R., and Limson, J. (2016). Developing biosensors in developing countries: South Africa as a case study. *Biosensors* 6:5. doi: 10.3390/bios6010005

Fraceto, L. F., Grillo, R., de Medeiros, G. A., Scognamiglio, V., Rea, G., and Bartolucci, C. (2016). Nanotechnology in agriculture: which innovation potential does it have? *Front. Environ. Sci.* 4:20. doi: 10.3389/fenvs.2016.00020

Ghaani, M., Cozzolino, C. A., Castelli, G., and Farris, S. (2016). An overview of the intelligent packaging technologies in the food sector. *Trends Food Sci. Tech.* 51, 1–11. doi: 10.1016/j.tifs.2016.02.008

Gibney, E. (2015). Buckyballs in space solve 100-year-old riddle. *Nature News* doi: 10.1038/nature.2015.17987

Gouin, S. (2004). Microencapsulation: industrial appraisal of existing technologies and trends. *Trends Food Sci. Technol.* 15, 330–347. doi: 10.1038/nature.2015.17987

Grillo, R., Abhilash, P. C., and Fraceto, L. F. (2016). Nanotechnology applied to bio-encapsulation of pesticides. *J. Nanosci. Nanotechnol.* 16, 1231–1234. doi: 10.1016/j.tifs.2003.10.005

Grillo, R., Melo, N. F. S., de Lima, R., Lourenço, R. W., Rosa, A. H., and Fraceto, L. F. (2010). Characterization of atrazine-loaded biodegradable poly(hydroxybutyrate-co-hydroxyvalerate) microspheres. *J. Polym. Environ.* 18, 26–32. doi: 10.1166/jnn.2016.12332

Grillo, R., Pereira, A. E., Nishisaka, C. S., de Lima, R., Oehlke, K., Greiner, R., et al. (2014). Chitosan/tripolyphosphate nanoparticles loaded with paraquat herbicide: an environmentally safer alternative for weed control. *J. Hazard. Mater.* 27, 163–171. doi: 10.1016/j.jhazmat.2014.05.079

Gruère, G. P. (2012). Implications of nanotechnology growth in food and agriculture in OECD countries. *Food Policy* 37, 191–198. doi: 10.1016/j.jhazmat.2014.05.079

Gupta, A., Eral, H. B., Hatton, T. A., and Doyle, P. S. (2016). Nanoemulsions: formation, properties and applications. *Soft Matter* 12, 2826–2841. doi: 10.1039/e5sm02958a

Gutiérrez, J. M., González, C., Maestro, A., Solè, I., Pey, C. M., and Nolla, J. (2008). Nano-emulsions: new applications and optimization of their preparation. *Curr. Opin. Colloid Interface Sci.* 13, 245–251. doi: 10.1039/C5SM02958A

Hajirostamlo, B., Mirsaeedghazi, N., Arefnia, M., Shariati, M. A., and Fard, E. A. (2015). The role of research and development in agriculture and its dependent concepts in agriculture [Short Review]. *Asian J. Appl. Sci. Eng.* 4, doi: 10.1016/j.cocis.2008.01.005

Haolong, L., Yang, Y., Yizhan, W., Chunyu, W., Wen, L., and Lixin, W. (2011). Self-assembly and ion-trapping properties of inorganic nanocapsule-surfactant hybrid spheres. *Soft Matter* 7, 2668–2673. doi: 10.1039/c0sm01044h

Helar, G., and Chavan, A. (2015). Synthesis, characterization and stability of gold nanoparticles using the fungus *Fusarium oxysporum* and its impact on seed. *Int. J. Recent Sci. Res.* 6, 3181–3318.

Hildebrand, G. E., and Tack, J. W. (2000). Microencapsulation of peptides and proteins. *Int. J. Pharm.* 196, 173–176. doi: 10.1016/S0378-5173(99)00415-9

Hoffmann, M., Holtze, E. M., and Wiesner, M. R. (2007). "Reactive oxygen species generation on nanoparticulate material," in *Environmental Nanotechnology. Applications and Impacts of Nanomaterials*, eds M. R. Wiesner and J. Y. Bottero (New York, NY: McGraw Hill), 155–203.

Hu, Y., Li, J., Ma, L., Peng, Q., Feng, W., Zhang, L., et al. (2010). High efficiency transport of quantum dots into plant roots with the aid of silwet L-77. *Plant Physiol. Biochem.* 48, 703–709. doi: 10.1016/j.plaphy.2010.04.001

Hussein, M. Z., Yahaya, A. H., Zainal, Z., and Kian, L. H. (2005). Nanocomposite-based controlled release formulation of an herbicide, 2,4-dichlorophenoxyacetate incapsulated in zinc-aluminium-layered double hydroxide. *Sci. Technol. Adv. Mater.* 6, 956–962. doi: 10.1016/j.stam.2005.09.004

Ion, A. C., Ion, I., and Culetu, A. (2010). Carbon-based nanomaterials: Environmental applications. *Univ. Politehn. Bucharest* 38, 129–132.

Jampilek, J., Zaruba, K., Oravec, M., Kunes, M., Babula, P., Ulbrich, P., et al. (2015). Preparation of silica nanoparticles loaded with nootropics and their *in vivo* permeation through blood-brain barrier. *BioMed Res. Int.* 2015:812673. doi: 10.1155/2015/812673

Jerobin, J., Sureshkumar, R. S., Anjali, C. H., Mukherjee, A., and Chandrasekaran, N. (2012). Biodegradable polymer based encapsulation of neem oil nanoemulsion for controlled release of Aza-A. *Carbohydr. Polym.* 90, 1750–1756. doi: 10.1016/j.carbpol.2012.07.064

Jiang, L. C., Basri, M., Omar, D., Rahman, M. B. A., Salleh, A. B., Zaliha, R. N., et al. (2012). Green nano-emulsion intervention for water-soluble glyphosate isopropylamine (IPA) formulations in controlling *Eleusine indica* (E. indica). *Pest. Biochem. Physiol.* 102, 19–29. doi: 10.1016/j.pestbp.2011.10.004

Johnston, B. F., and Mellor, J. W. (1961). The role of agriculture in economic development. *Am. Econ. Rev.* 51, 566–593.

Kah, M. (2015). Nanopesticides and nanofertilizers: emerging contaminants or opportunities for risk mitigation? *Front. Chem.* 3:64. doi: 10.3389/fchem.2015.00064

Kah, M., and Hofmann, T. (2014). Nanopesticides research: current trends and future priorities. *Environ. Int.* 63, 224–235. doi: 10.1016/j.envint.2013.11.015

Kandasamy, S., and Prema, R. S. (2015). Methods of synthesis of nano particles and its applications. *J. Chem. Pharm. Res.* 7, 278–285.

Khond, V. W., and Kriplani, V. M. (2016). Effect of nanofluid additives on performances and emissions of emulsified diesel and biodiesel fueled stationary CI engine: a comprehensive review. *Renew. Sustain. Energy Rev.* 59, 1338–1348. doi: 10.1016/j.rser.2016.01.051

Khota, L. R., Sankarana, S., Majaa, J. M., Ehsania, R., and Schuster, E. W. (2012). Applications of nanomaterials in agricultural production and crop protection: a review. *Crop Prot.* 35, 64–70. doi: 10.1016/j.cropro.2012.01.007

Konstantatos, G., and Sargent, E. H. (2009). Solution-processed quantum dot photodetectors. *Proc. IEEE* 97, 1666–1683. doi: 10.1109/JPROC.2009.2025612

Kovochich, M., Xia, T., Xu, J., Yeh, J. I., and Nel, A. E. (2005). Principles and procedures to assess nanoparticles. *Environ. Sci. Technol.* 39, 1250–1256.

Kumar, S., Bhanjana, G., Sharma, A., Sidhu, M. C., and Dilbaghi, N. (2014). Synthesis, characterization and on field evaluation of pesticide loaded sodium alginate nanoparticles. *Carbohydr. Polym.* 101, 1061–1067. doi: 10.1016/j.carbpol.2013.10.025

Li, H., Shan, C., Zhang, Y., Cai, J., Zhang, W., and Pan, B. (2016). Arsenate adsorption by hydrous ferric oxide nanoparticles embedded in cross-linked anion exchanger: effect of the host pore structure. *ACS Appl. Mater. Interfaces* 8, 3012–3020. doi: 10.1021/acsami.5b09832

Liu, R., and Lal, R. (2015). Potentials of engineered nanoparticles as fertilizers for increasing agronomic productions. *Sci. Total Environ.* 514, 131–139. doi: 10.1016/j.scitotenv.2015.01.104

Liu, Y., He, L., Mustapha, A., Li, H., Hu, Z. Q., and Lin, M. (2009). Antibacterial activities of zinc oxide nanoparticles against *Escherichia coli* O157:H7. *J. Appl. Microbiol.* 107, 1193–1201. doi: 10.1111/j.1365-2672.2009.04303.x

Llop, J., Estrela-Lopis, I., Ziolo, R. F., González, A., Fleddermann, J., Dorn, M., et al. (2014). Uptake, biological fate, and toxicity of metal oxide nanoparticles. *Part. Part. Syst. Charact.* 31, 24–35. doi: 10.1002/ppsc.201300323

Maruyama, C. R., Guilger, M., Pascoli, M., Bileshy-José, N., Abhilash, P. C., Fraceto, L. F., et al. (2016). Nanoparticles based on chitosan as carriers for the combined herbicides imazapic and imazapyr. *Sci. Rep.* 6:23854. doi: 10.1038/srep23854.

Martirosyan, A., and Schneider, Y. J. (2014).Engineered nanomaterials in food: implications for food safety and consumer health. *Int. J. Environ. Res. Public Health* 11, 5720–5750. doi: 10.3390/ijerph110605720

Marzbani, P., Afrouzi, Y. M., and Omidvar, A. (2015). The effect of nano-zinc oxide on particleboard decay resistance. *Maderas Cienc. Tecnol.* 17, 63–68. doi: 10.4067/s0718-221x2015005000007

Mason, T. G., Wilking, J. N., Meleson, K., Chang, C. B., and Graves, S. M. (2006). Nanoemulsions: formation, structure, and physical properties. *J. Phys. Condens. Matt.* 18, R635-R666. doi: 10.1088/0953-8984/18/41/r01

Memarizadeh, N., Ghadamyari, M., Adeli, M., and Talebi, K. (2014). Preparation, characterization and efficiency of nanoencapsulated imidacloprid under laboratory conditions. *Ecotoxicol. Environ. Saf.* 107, 77–83. doi: 10.1016/j.ecoenv.2014.05.009

Mishra, V. K., and Kumar, A. (2009). Impact of metal nanoparticles on the plant growth promoting rhizobacteria. *Dig. J. Nanomater. Biostruct.* 4, 587–592.

Mukhopadhyay, S. S. (2014). Nanotechnology in agriculture: prospects and constraints. *Nanotechnol. Sci. Appl.* 7, 63–71. doi: 10.2147/NSA.S39409

Nima, A. Z., Lahiani, M. H., Watanabe, F., Xu, Y., Khodakovskaya, M. V., and Biris, A. S. (2014). Plasmonically active nanorods for delivery of bio-active agents and high-sensitivity SERS detection in planta. *RSC Adv.* 4, 64985–64993. doi: 10.1039/C4RA10358K

Nuruzzaman, M., Rahman, M. M., Liu, Y., and Naidu, R. (2016). Nanoencapsulation, nano-guard for pesticides: a new window for safe application. *J. Agric. Food Chem.* 64, 1447–1483. doi: 10.1021/acs.jafc.5b05214

Ozdemir, M., and Kemerli, T. (2016). "Innovative applications of micro and nanoencapsulation in food packaging," in *Encapsulation and Controlled Release Technologies in Food Systems*, ed. J. M. Lakkis (Chichester: John Wiley & Sons, Ltd).

Patra, J. K., and Baek, K.-H. (2017). Antibacterial activity and synergistic antibacterial potential of biosynthesized silver nanoparticles against foodborne pathogenic bacteria along with its anticandidal and antioxidant effects. *Front. Microbiol.* 8:167. doi: 10.3389/fmicb.2017.00167

Perlatti, B., Bergo, P. L. S., Silva, M. F. G., Fernandes, J. B., and Forim, M. R. (2013). "Polymeric nanoparticle-based insecticides: a controlled release purpose for agrochemicals," in: *Insecticides-Development of Safer and More Effective Technologies*, ed. S. Trdan (Rijeka: InTech), 523–550. doi: 10.5772/53355

Pohlmann, R., Beck, R. C. R., Lionzo, M. I. Z., Coasta, T. M. H., Benvenutti, E. V., Re, M. I., et al. (2008). Surface morphology of spray-dried nanoparticle-coated microparticles designed as an oral drug delivery system. *Braz. J. Chem. Eng.*, 25, 389–398

Pokropivny, V., Lohmus, R., Hussainova, I., Pokropivny, A., and Vlassov, S. (2007). *Introduction to Nanomaterials and Nanotechnology*. Tartu: University of Tartu, 225.

Porter, A. E., Gass, M., Muller, K., Skepper, J. N., Midgley, P. A., and Welland, M. (2007). Direct imaging of single-walled carbon nanotubes in cells. *Nat. Nanotechnol.* 2, 713–717. doi: 10.1038/nnano.2007.347

Prasad, R. (2014). Synthesis of silver nanoparticles in photosynthetic plants. *J. Nanopart.* 2014:963961. doi: 10.1155/2014/963961

Prasad, R., Kumar, V., and Prasad, K. S. (2014). Nanotechnology in sustainable agriculture: present concerns and future aspects. *Afr. J. Biotechnol.*13, 705–713. doi: 10.5897/AJBX2013.13554

Prasad, R., Pandey, R., and Barman, I. (2016). Engineering tailored nanoparticles with microbes: quo vadis. *WIREs Nanomed. Nanobiotechnol.* 8, 316–330. doi: 10.1002/wnan.1363

Puglisi, G., Fresta, M., Giammona, G., and Ventura, C. A. (1995). Influence of the preparation conditions on poly (ethylcyanoacrylate) nanocapsule formation. *Int. J. Pharm.* 125, 283–287. doi: 10.1016/0378-5173(95)00142-6

Pyrzynska, K. (2011). Carbon nanotubes as sorbents in the analysis of pesticides. *Chemosphere* 83, 1407–1413. doi: 10.1016/j.chemosphere.2011.01.057

Rai, V., Acharya, S., and Dey, N. (2012). Implications of nanobiosensors in agriculture. *J. Biomater. Nanobiotechnol.* 3, 315–324. doi: 10.4236/jbnb.2012.322039

Raliya, R., Tarafdar, J. C., Gulecha, K., Choudhary, K., Ram, R., Mal, P., et al. (2013). Review article; scope of nanoscience and nanotechnology in agriculture. *J. Appl. Biol. Biotechnol.* 1, 041–044.

Rana, S., and Kalaichelvan, P. T. (2013). Ecotoxicity of nanoparticles. *ISRN Toxicol.* 2013:574648. doi: 10.1155/2013/574648

Rana, S., and Kalaichelvan, P. T. (2011). Antibacterial effects of metal nanoparticles. *Adv. Biotech.* 2, 21–23.

Rodríguez, J., Martín, M. J., Ruiz, A. M., and Clares, B. (2016). Current encapsulation strategies for bioactive oils: from alimentary to pharmaceutical perspectives. *Food Res. Int.* 83, 41–59. doi: 10.1016/j.foodres.2016.01.032

Sabir, S., Arshad, M., and Chaudhari, S. K. (2014). Zinc oxide nanoparticles for revolutionizing agriculture: synthesis and applications. *Sci. World J.* 2014:8. doi: 10.1155/2014/925494

Sadeghzadeh, B. (2013). A review of zinc nutrition and plant breeding. *J. Soil Sci. Plant Nutr.* 13, 905–927. doi: 10.4067/S0718-95162013005000072

Sagadevan, S., and Periasamy, M. (2014). Recent trends in nanobiosensors and their applications - a review. *Rev. Adv. Mater. Sci.* 36, 62–69.

Saharan, V., Mehrotra, A., Khatik, R., Rawal, P., Sharma, S. S., and Pal, A. (2013). Synthesis of chitosan based nanoparticles and their in vitro evaluation against phytopathogenic fungi. *Int. J. Biol. Macromol.* 62, 677–683. doi: 10.1016/j.ijbiomac.2013.10.012

Savina, E., Karlsen, J. D., Frandsen, R. P., Krag, L. A., Kristensen, K., and Madsen, N. (2016). Testing the effect of soak time on catch damage in a coastal gillnetter and the consequences on processed fish quality. *Food Control* 70, 310–317. doi: 10.1016/j.foodcont.2016.05.044

Sayes, C. M., Fortner, J. D., Guo, W., Lyon, D., Boyd, A. M., Ausman, K. D., Tao, Y. J., et al. (2004). The differential cytotoxicity of water-soluble fullerenes. *Nano Lett.* 4, 1881–1887. doi: 10.1002/btpr.707

Sekhon, B. S. (2014). Nanotechnology in agri-food production: an overview. *Nanotechnol. Sci. Appl.* 7, 31–53. doi: 10.2147/NSA.S39406

Sertova, N. M. (2015). Application of nanotechnology in detection of mycotoxins and in agricultural sector. *J. Cent. Eur. Agric.* 16, 117–130. doi: 10.5513/JCEA01/16.2.1597

Shweta, Tripathi, D. K., Singh, S., Singh, S., Dubey, N. K., and Chauhan, D. K. (2016). Impact of nanoparticles on photosynthesis: challenges and opportunities. *Mater Focus* 5, 405–411. doi: 10.1166/mat.2016.1327

Siegrist, M., Stampfli, N., Kastenholz, H., and Keller, C. (2008). Perceived risks and perceived benefits of different nanotechnology foods and nanotechnology food packaging. *Appetite* 51, 283–290. doi: 10.1016/j.appet.2008.02.020

Silva Mdos, S., Cocenza, D. S., Grillo, R., de Melo, N. F., Tonello, P. S., de Oliveira, L. C., et al. (2011). Paraquat-loaded alginate/chitosan nanoparticles: preparation, characterization and soil sorption studies. *J. Hazard. Mater.* 190, 366–374. doi: 10.1016/j.jhazmat.2011.03.057

Singh, S., Vishwakarma, K., Singh, S., Sharma, S., Dubey, N.K., Singh, V.K., et al. (2017). Understanding the plant and nanoparticle interface at transcriptomic and proteomic level: a concentric overview. *Plant Gene* (in press). doi: 10.1016/j.plgene.2017.03.006

Sun, C., Shu, K., Wang, W., Ye, Z., Liu, T., Gao, Y., et al. (2014). Encapsulation and controlled release of hydrophilic pesticide in shell cross-linked nanocapsules containing aqueous core. *Int. J. Pharm.* 463, 108–114. doi: 10.1016/j.ijpharm.2013.12.050

Sun, C. Q. (2007). Size dependence of nanostructures: impact of bond order deficiency. *Prog. Solid State Chem.* 35, 1–159. doi: 10.1016/j.progsolidstchem.2006.03.001

Tadros, T. F., Izquierdo, P., Esquena, J., and Solans, C. (2004). Formation and stability of nanoemulsions. *Adv. Colloid Interface Sci.* 108-109, 303–318. doi: 10.1016/j.cis.2003.10.023

Thornhill, S., Vargyas, E., Fitzgerald, T., and Chisholm, N. (2016). Household food security and biofuel feedstock production in rural Mozambique and Tanzania. *Food Sec.* 8, 953–971. doi: 10.1007/s12571-016-0603-9

Tripathi, D. K., Mishra, R. K., Singh, S., Singh, S., Vishwakarma, K., Sharma, S., et al. (2017b). Nitric oxide ameliorates zinc oxide nanoparticles phytotoxicity in wheat seedlings: implication of the ascorbate-glutathione cycle. *Front. Plant Sci,* 8:1. doi: 10.3389/fpls.2017.00001

Tripathi, D. K., Shweta, Singh, S., Singh, S., Pandey, R., Singh, V. P., et al. (2016a). An overview on manufactured nanoparticles in plants: uptake, translocation, accumulation and phytotoxicity. *Plant Physiol. Biochem.* 110, 2–12. doi: 10.1016/j.plaphy.2016.07.030.

Tripathi, D. K., Singh, S., Singh, V. P., Prasad, S. M., Chauhan, D. K., and Dubey, N. K. (2016b). Silicon nanoparticles more efficiently alleviate arsenate toxicity than silicon in maize cultiver and hybrid differing in arsenic tolerance. *Front. Environ. Sci.* 4:46. doi: 10.3389/fenvs.2016.00046

Tripathi, D. K., Singh, S., Singh, S., Srivastava, P. K., Singh, V. P., Singh, S., et al. (2017a). Nitric oxide alleviates silver nanoparticles (AgNps)-induced phytotoxicity in Pisum sativum seedlings. *Plant Physiol. Biochem.* 110, 167–177. doi: 10.1016/j.plaphy.2016.06.015

Tripathi, D. K., Tripathi, A., Shweta, S. S., Singh, Y., Vishwakarma, K., Yadav, G., et al. (2017c). Uptake, accumulation and toxicity of silver nanoparticle in autotrophic plants, and heterotrophic microbes: a concentric review. *Front. Microbiol.* 8:07. doi: 10.3389/fmicb.2017.00007

Tripathi, D. K., Singh, S., Singh, V. P., Prasad, S. M., Dubey, N. K., and Chauhan, D. K. (2017d). Silicon nanoparticles more effectively alleviated UV-B stress than silicon in wheat (*Triticum aestivum*) seedlings. *Plant Physiol. Biochem.* 110, 70–81. doi: 10.1016/j.plaphy.2016.06.026

Vannini, C., Domingo, G., Onelli, E., De Mattia, F., Bruni, I., Marsoni, M., et al. (2014). Phytotoxic and genotoxic effects of silver nanoparticles exposure on germinating wheat seedlings. *J. Plant Physiol.* 171, 1142–1148. doi: 10.1016/j.jplph.2014.05.002

Vartholomeos, P., Fruchard, M., Ferreira, A., Mavroidis, C. (2011). MRI-guided nanorobotic systems for therapeutic and diagnostic applications. *Annu. Rev. Biomed. Eng.*13, 157–184. doi: 10.1146/annurev-bioeng-071910-124724

Venkatachalam, P., Jayaraj, M., Manikandan, R., Geetha, N., Rene, E. R., Sharma, N. C., et al. (2017). Zinc oxide nanoparticles (ZnONPs) alleviate heavy metal-induced toxicity in *Leucaena leucocephala* seedlings: a physiochemical analysis. *Plant Physiol. Biochem.* 110, 59–69. doi: 10.1016/j.plaphy.2016.08.022

Vidotti, M., Carvalhal, R. F., Mendes, R. K., Ferreira, D. C. M., and Kubota, L. T. (2011). Biosensors based on gold nanostructures. *J. Braz. Chem. Soc.* 22, 3–20. doi: 10.1590/S0103-50532011000100002

Viswanathan, S., and Radecki, J. (2008). Nanomaterials in electrochemical biosensors for food analysis- a review. *Pol. J. Food Nutr. Sci.* 58, 157–164.

Wan, Y., Li, J., Ren, H., Huang, J., and Yuan, H. (2014). Physiological investigation of gold nanorods toward watermelon. *J. Nanosci. Nanotechnol.* 14, 6089–6094. doi: 10.1166/jnn.2014.8853

Wang, L., Hu, C., and Shao, L. (2017). The antimicrobial activity of nanoparticles: present situation and prospects for the future. *Int. J. Nanomed.* 12, 1227–1249. doi: 10.2147/IJN.S121956

Yang, L., and Watts, D. J. (2005). Particle surface characteristics may play an important role in phytotoxicity of alumina nanoparticles. *Toxicol. Lett.* 158, 122–132. doi: 10.1016/j.toxlet.2005.03.003

Yunlong, C., and Smit, B. (1994). Sustainability in agriculture: a general review. *Agric. Ecosyst. Environ.* 49, 299–307. doi: 10.1016/0167-8809(94)90059-0

Zhang, Q., Han, L., Jing, H., Blom, D. A., Lin, Y., Xin, H. L., et al. (2016). Facet control of gold nanorods. *ACS Nano* 10, 2960–2974. doi: 10.1021/acsnano.6b00258

Conflict of Interest Statement: The authors declare that the research was conducted in the absence of any commercial or financial relationships that could be construed as a potential conflict of interest.

7

Catalytic Interactions and Molecular Docking of Bile Salt Hydrolase (BSH) from *L. plantarum* RYPR1 and its Prebiotic Utilization

Ruby Yadav[1†], Puneet K. Singh[1†], Anil K. Puniya[2,3] and Pratyoosh Shukla[1]*

[1] Enzyme Technology and Protein Bioinformatics Laboratory, Department of Microbiology, Maharshi Dayanand University, Rohtak, India, [2] Division of Dairy Microbiology, Indian Council of Agricultural Research (ICAR) – National Dairy Research Institute (NDRI), Karnal, India, [3] College of Dairy Science and Technology, Guru Angad Dev Veterinary and Animal Sciences University, Ludhiana, India

Edited by:
Jayanta Kumar Patra,
Dongguk University, South Korea

Reviewed by:
Kiiyukia Matthews Ciira,
Mount Kenya University, Kenya
Xiaoxia Zhu,
University of Miami, USA
Renu Agrawal,
Central Food Technological Research
Institute (CFTRI), India

*Correspondence:
Pratyoosh Shukla
pratyoosh.shukla@gmail.com

[†] These authors have contributed
equally to this work.

Prebiotics are the non-digestible carbohydrate, which passes through the small intestine into unmetabolized form, reaches the large intestine and undergoes fermentation by the colonic bacteria thus; prebiotics stimulate the growth of probiotic bacteria. Further, bile salt hydrolase (BSH) is an enzyme that catalyses the deconjugation of bile salt, so it has enormous potential toward utilizing such capability of *Lactobacillus plantarum* RYPR1 toward detoxifying through BSH enzyme activity. In the present study, six isolates of *Lactobacillus* were evaluated for the co-aggregation assay and the isolate *Lactobacillus plantarum* RYPR1 was further selected for studies of prebiotic utilization, catalytic interactions and molecular docking. The prebiotic utilization ability was assessed by using commercially available prebiotics lactulose, inulin, xylitol, raffinose, and oligofructose P95. The results obtained revealed that RYPR1 is able to utilize these probiotics, maximum with lactulose by showing an increase in viable cell count (7.33 $\pm$ 0.02 to 8.18 $\pm$ 0.08). In addition, the molecular docking of BSH from *Lactobacillus plantarum* RYPR1 was performed which revealed the binding energy −4.42 and 7.03 KJ/mol. This proves a considerably good interactions among BSH and its substrates like Taurocholic acid (−4.42 KJ/mol) and Glycocholic acid (−7.03 KJ/mol). These results from this study establishes that *Lactobacillus plantarum* RYPR1 possesses good probiotic effects so it could be used for such applications. Further, molecular dynamics simulations were used to analyze the dynamic stability of the of modeled protein to stabilize it for further protein ligand docking and it was observed that residues Asn12, Ile8, and Leu6 were interacting among BSH and its substrates, i.e., Taurocholic acid and Lys88 and Asp126 were interacting with Glycocholic acid. These residues were interacting when the docking was carried out with stabilized BSH protein structure, thus, these residues may have a vital role in stabilizing the binding of the ligands with the protein.

Keywords: probiotics, prebiotic utilization, molecular docking, Glycocholic acid, bile salt hydrolase (BSH), *Lactobacillus plantarum*

INTRODUCTION

Probiotics are live microbial food supplements which, when administrated in adequate amounts, exerts various health benefits to consumers (Vinderola et al., 2008). Probiotics is a promising field in dairy and food industry with tremendous growth potential (Mitropoulou et al., 2013). These bacteria exert various health benefits to the host, such as immunomodulation, lipid and cholesterol reduction, anticancer, antimicrobial, antiallergic, antioxidative properties, prevention of gastrointestinal infections, improvement of lactose metabolism, etc. (Lee et al., 2014). Probiotics produce diverse inhibitory substances (organic acids, antimicrobial substances, exoploysaccharides, bacteriocins etc.) which depress growth of pathogenic microorganisms in the gut (Pessione, 2012; Maldonado et al., 2015; Yadav and Shukla, 2015). There are various *in vitro* tests for the selection and study of functional properties of a probiotic strain. The co-aggregation study of probiotic bacteria with pathogens helps in evaluating the pathogen interaction with bacteria which prevents pathogen colonization in the gut (Gupta and Malik, 2007). The interactions of probiotics with prebiotics have a beneficial role in improving the growth of normal microflora, resulting in immune system modulation of the host. A number of systems biology tools have been studied to comprehend the interactions between microorganisms and plant or human cell (Kumar et al., 2016). Furthermore, various genetic modifications which involve the introduction of desired genes may also have a constructive impact in the probiotic field (Gupta and Shukla, 2015). There are many reports on molecular docking of enzymes, which gives good insights of various protein interactions and their effective binding patterns (Singh and Shukla, 2011, 2014; Singh et al., 2011, 2016; Karthik et al., 2012; Baweja et al., 2015, 2016).

In the present study probiotic properties, prebiotic utilization and the molecular docking of *Lactobacillus plantarum* RYPR1, isolated from indigenous fermented beverage raabadi was performed. The development of nano-encapsulated probiotics is an emerging field and showing new possibilities of probiotics in food industry. The viability of probiotic bacteria in the human body could extend by using nanoencapsulated bacteria, so that it could show better interaction with receptors of the gastrointestinal tract. The results reported from our previous studies showed that *L. plantarum* RYPR1 possess good antimicrobial activity so its probiotic effect could be further improved by the development of nano-encapsulated probiotics by using nanotechnology applications.

MATERIALS AND METHODS

Isolation and Probiotic Properties of *Lactobacillus* Isolates

A total of 11 curd (6) and raabadi (5) samples were collected from different regions of Haryana, India following the standard microbiological protocols. Moreover, the Kanji (fermented beverage made up of carrot) samples were prepared in laboratory under aseptic conditions for isolating lactic acid bacteria. The isolation and purification of lactic acid bacteria, was done using De Man Rogose Sharpe (MRS) medium (Goyal et al., 2013). The purified cultures isolated from these samples were tested for grams staining, endospore staining, catalase test and further tested for various probiotic properties as reported in our previous studies (Yadav et al., 2016).

Co-aggregation Assay

Co-aggregation involves the process of aggregation of bacterial cells of more than one type (Kumar et al., 2012). Co-aggregation ability provides a close interaction of probiotic bacteria with pathogenic bacteria (Singh et al., 2012). In this experiment, *E. coli* was taken as indicator organism which can co-aggregate with selected isolates. Overnight grown Lactobacilli (16–18 h) and *E. coli* cultures were centrifuged (10000 rpm, 15 min) and the pellets obtained were washed twice with phosphate buffer saline (PBS) solution (pH 6.0). The pellets were resuspended in PBS, vortexed and the absorbance was set 0.5 at 600 nm. After this, 500 µl of culture and 500 µl of the pathogen were mixed and optical density (OD) was measured at 600 nm and incubated at 37°C for 2 h. Upper phase was carefully removed and absorbance was measured at 600 nm. Decrease in absorbance was taken as a measure of cell co-aggregation. The co-aggregation percentage was calculated by using the following formula:

$$\text{Percent co-aggregation- } [(OD_1 + OD_2) - 2(OD_3)/(OD_1 + OD_2) \times 100]$$

OD_1: optical density of *Lactobacillus* isolates, OD_2: optical density of *E. coli*, OD_3: optical density of mixture.

Prebiotic Utilization

Commercially available five prebiotics Lactulose, Xylitol, D+ raffinose, Inulin and Oligofructose P95 were used for the test. Prebiotics were solubilized in distilled water and filter sterilized. Isolates were inoculated with 3 ml of modified MRS medium (2% of each probiotic) and incubated at 37°C for 24 h under anaerobic conditions. OD of each culture was measured at 560 nm and the cell growth rate was calculated by using the formula:

$$\text{Prebiotic utilization- } (MRSp - MRSb) \times 100/MRSg - MRSb$$

Molecular Dynamic Simulation

Molecular dynamics was performed with Gromacs 4.5.5 using the Gromos96 force field. All the water molecules were deleted and polar hydrogen atoms were added. The hydrogen atoms were minimized with 500 steps of Steepest Descent (SD) optimization; spc water was added in a sphere with a radius of 18 Å around the reaction center (the Cl and NA ions). Before the unconstrained MD simulation, the solvent was subjected to 1000 steps of SD minimization, and equilibrated for 2.5 ps at 300 K with solute fixed. The production simulation was carried out for 5,000 ps (5 ns). The average conformation was calculated for the desired represented frame of MD simulations. This was achieved by averaging the snapshots of the last 500 ps, then choosing a typical structure with the lowest RMSD to the average conformation, and using this in the binding mode analysis.

Molecular Docking and Analysis of Bile Salt Hydrolase (BSH)

The Bile Salt Hydrolase (BSH) activity of the selected isolate as reported previously was further taken as standard for the catalytic interaction. The study of enzyme modeling was performed with SWISS-MODEL, it is a fully automated protein structure homology-modeling server, accessible via the ExPASy web server, or from the program DeepView (Swiss Pdb-Viewer). SWISS-MODEL provides graphical representation as well as numerical calculations for the alignment of structures. In SWISS-MODEL we have to submit sequence of protein of which we have to model structure and it will provide a structure after few hours (Arnold et al., 2006; Guex et al., 2009; Kiefer et al., 2009). The Stereochemical quality of a protein was checked by PROCHECK it analyses the structure by analyzing residue-by-residue geometry and overall structural geometry of the modeled structure (Laskowski et al., 1993, 1996). ERRAT Analyzes the statistics of non-bonded interactions between different atom types and compare with the highly refined structures.

RESULTS

Isolation and Probiotic Properties of *Lactobacillus* Isolates

A total of 119 isolates were isolated from curd, kanji, and raabadi samples and 90 were purified. On the basis of colony morphology, gram staining, endospore staining and catalase test 54 isolates were identified and selected as *Lactobacillus* isolates. These isolates were further tested for probiotic properties. It has been shown that among the tested *Lactobacillus* isolates, isolate *Lactobacillus plantarum* RYPR1 as identified by 16S rRNA sequencing and phylogenetic analysis (GenBank accession number KX620369) showed the maximum probiotic potential and it was selected for further studies. Moreover, cell co-aggregation, prebiotic utilization, BSH activity and *in silico* studies are presented in this paper.

Cell Co-aggregation

Cell co-aggregation involves interaction of probiotic microorganism with surface components of pathogenic bacteria. The co-aggregation activity involves biofilm formation which helps the host by prevention of pathogen colonization in the gut. The co-aggregating cell clumps together and settled at the bottom of the tube, resulting in decreasing absorbance of suspension. Percent co-aggregation of selected isolates ranged from 17 to 40% (**Figure 1**). Strain RYPR1 showed highest co-aggregation potential followed by RYPR9.

Prebiotic Utilization

The viable cell counts of *L. plantarum* with prebiotics after 24 h incubation are presented in **Table 1**. Based on viable cell count, it

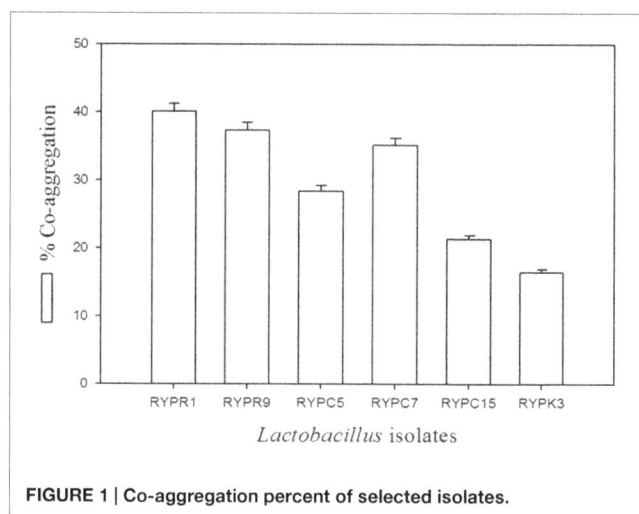

FIGURE 1 | Co-aggregation percent of selected isolates.

was observed that isolate RYPR1 showed the maximum survival with lactulose followed by raffinose and inulin. So, with this study it was concluded that RYPR1 growth could be stimulated by tested probiotics.

Molecular Docking Analysis of BSH

Isolate RYPRI was grown in the presence of bile salts (sodium tauroglycocholate, sodium taurocholate, sodium taurodeoxycholate) to evaluate its ability to hydrolyze high concentration of bile salts. The results obtained from this study concluded that RYPR1 is not only able to survive the toxicity of bile salts, but also carries out bile salt deconjugation which helps in the colonization of bacteria to intestinal epithelial cells. The results from *in vitro* studies were further confirmed by *in silico* studies.

Homology Modeling and Structure Validation

The sequence for modeling was submitted to Swiss Model[1] for structural modeling. During the study, the template chosen for modeling demonstrated similarity of 70.01% with 4wl3 chain B, Crystal structure determination of BSH from *Enterococcus feacalis* having resolution of 2.01 Å. Structure validation was performed using SAVES server.

---

[1]http://swissmodel.expasy.org

TABLE 1 | Prebiotic substrate utilization by *Lactobacillus plantarum* RYPR1 after 24 h incubation.

Prebiotic substrate	log cfu/ml (after incubation)	
	0 h	24 h
Lactulose	7.33 ± 0.02	8.18 ± 0.08
Xylitol	7.1 ± 0.01	6.44 ± 0.07
D+ raffinose	7.98 ± 0.17	8.05 ± 0.10
Inulin	7.91 ± 0.09	7.90 ± 0.04
Oligofructose P95	7.62 ± 0.10	6.88 ± 0.09

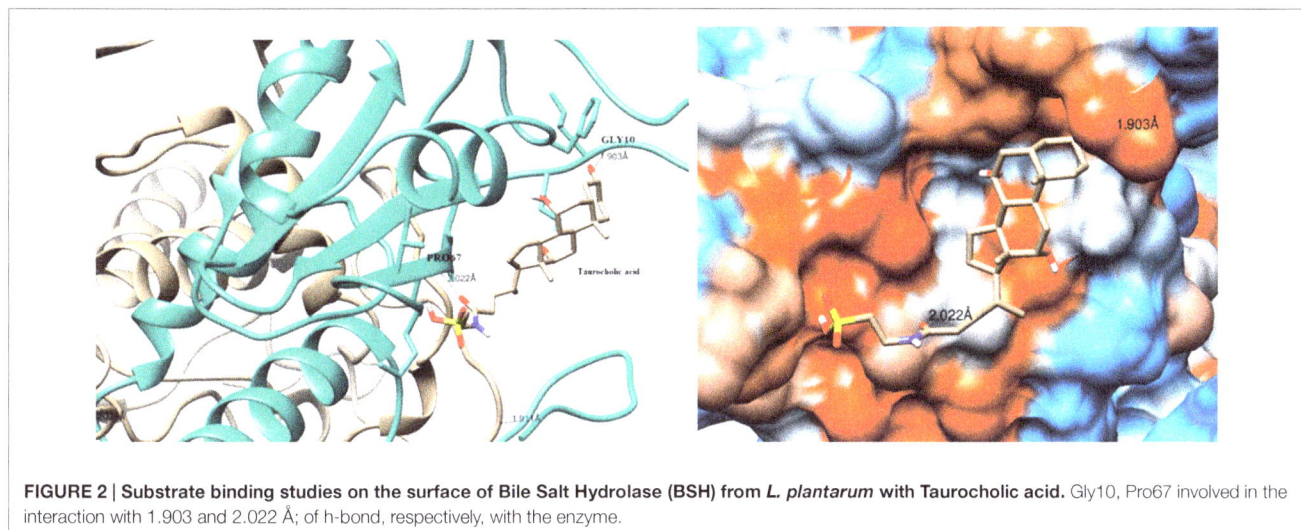

FIGURE 2 | Substrate binding studies on the surface of Bile Salt Hydrolase (BSH) from _L. plantarum_ with Taurocholic acid. Gly10, Pro67 involved in the interaction with 1.903 and 2.022 Å; of h-bond, respectively, with the enzyme.

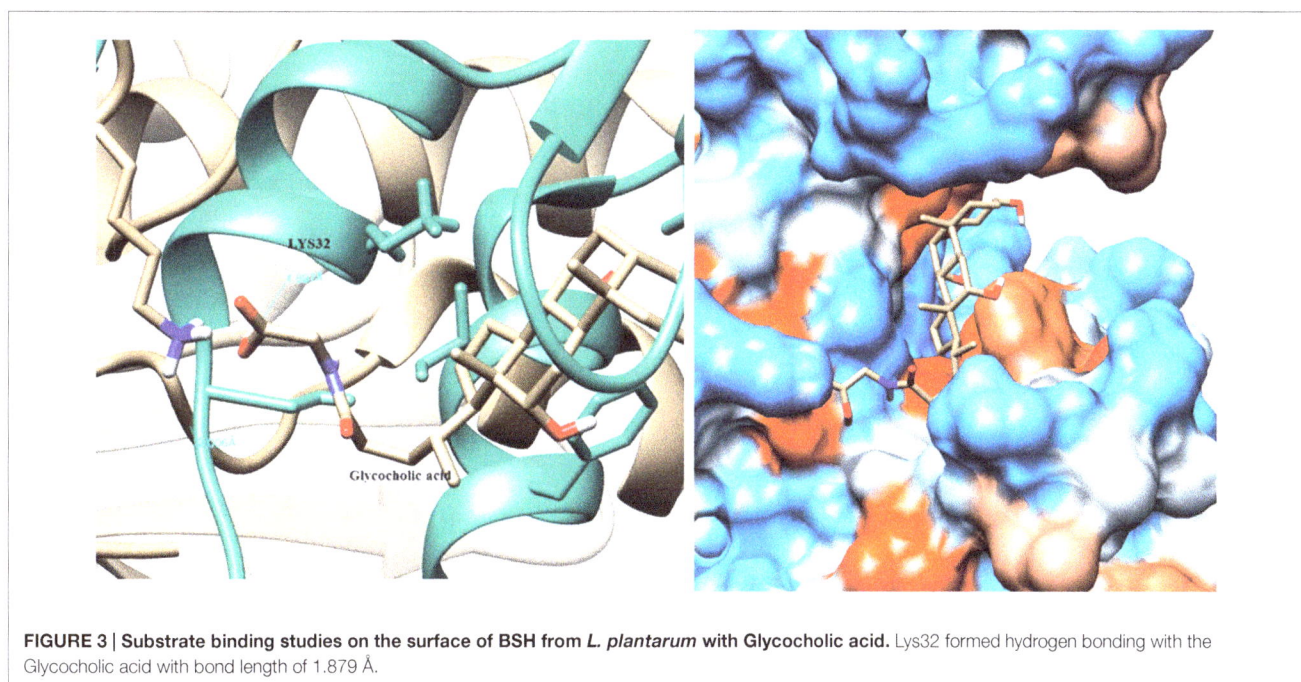

FIGURE 3 | Substrate binding studies on the surface of BSH from _L. plantarum_ with Glycocholic acid. Lys32 formed hydrogen bonding with the Glycocholic acid with bond length of 1.879 Å.

TABLE 2 | Docking studies of Bile Salt Hydrolase (BSH) from _L. plantarum_ with Taurocholic acid and Glycocholic acid.

Ligand	Protein	Binding energy (kc/mol)	Inhibition constant	Hydrogen bonds	Hydrogen bond length
Taurocholic acid	BSH	−4.42	580.35 mM	Gly10, Pro67	1.903 and 2.022 Å
Taurocholic acid (after stabilizing protein)	BSH (after stabilizing protein)	−3.01	6.26 mM	Asn12, Ile8 and Leu6	2.570, 2.037, and 2.940 Å
Glycocholic acid	BSH	−4.91	252.24 mM	Lys32	1.879 Å
Glycocholic acid (after stabilizing protein)	BSH (after stabilizing protein)	−3.45	7.66 mM	Asp136 and Lys88	2.042 and 2.155 Å

ERRAT Overall quality factor was obtained was 96.349 which represent a stable structure. PROCHECK also exhibited favorable result for protein model to proceed for molecular docking.

Molecular Dynamic Simulation and Docking Analysis

The modeled BSH from _L. plantarum_ was stabilized by molecular dynamic simulation of 5,000 ps (5 ns) (**Figure 4**). Further, the stabilized protein was docked with Taurocholic acid and

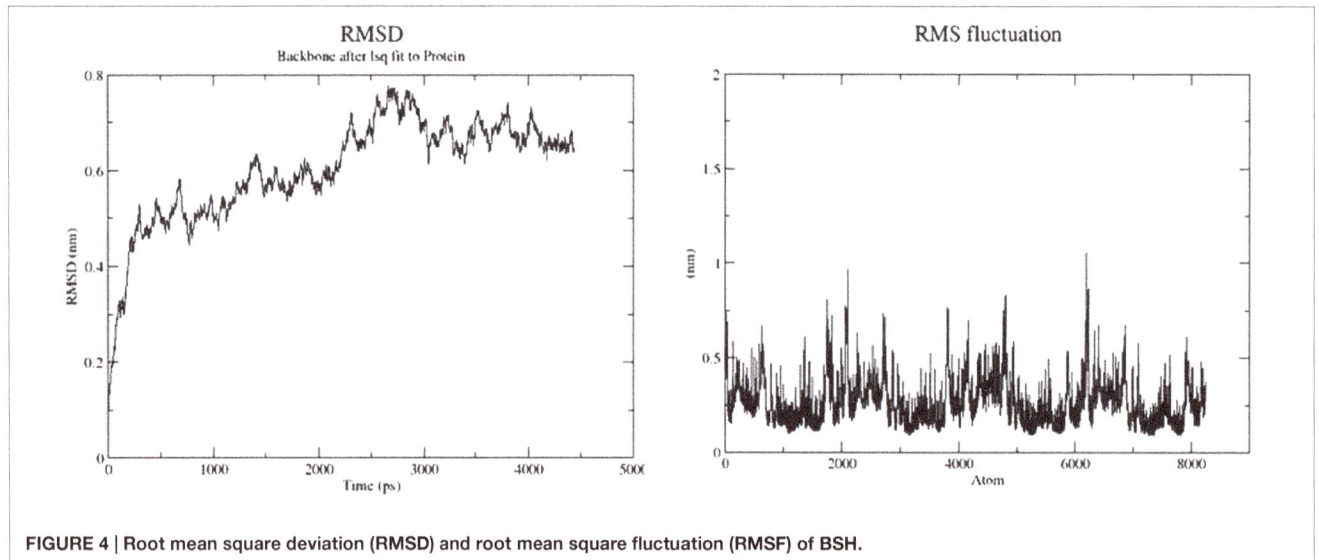

FIGURE 4 | Root mean square deviation (RMSD) and root mean square fluctuation (RMSF) of BSH.

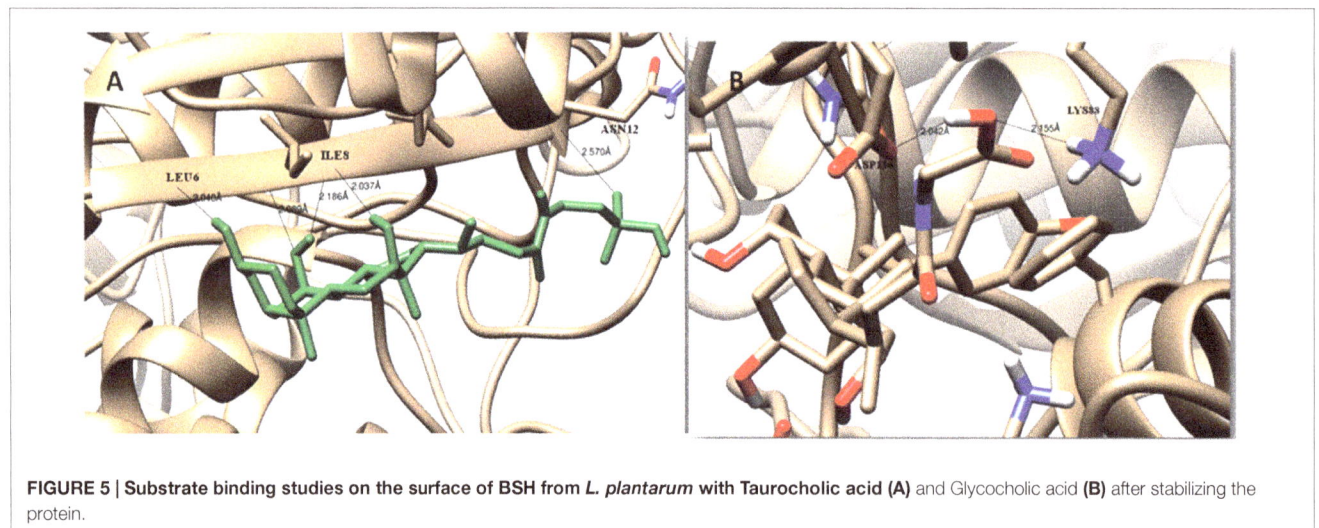

FIGURE 5 | Substrate binding studies on the surface of BSH from *L. plantarum* with Taurocholic acid (A) and Glycocholic acid (B) after stabilizing the protein.

Glycocholic acid and the result was compared with the docking result of unstabilized protein. The BSH activity of the selected isolate as reported previously was further taken as standard for the catalytic interaction of BSH with Taurocholic acid and Glycocholic acid. The docking was carried out with AutoDock4[2]. In AutoDock4, enzyme BSH from *L. plantarum* docked with Taurocholic acid and Glycocholic acid The result was recorded as the least binding energy with Glycocholic acid as –7.03 KJ/mol, followed by docking with Taurocholic acid with –4.42 KJ/mol, least binding energy signifies the strong binding between substrate and enzyme. The minimum inhibition constant of Glycocholic acid also came to be minimum, i.e., 252.24 μl. Lys32 formed hydrogen bonding with the Glycocholic acid with bond length of 1.879 Å. Gly10, Pro67 involved in the interaction with 1.903 and 2.022 Å of h-bond, respectively, with the enzyme (**Figures 2** and **3**). The comparison of the minimum binding energy of both the substrate observed with AutoDock **Table 2**.

[2]http://autodock.scripps.edu

The docking result of BSH from *L. plantarum* docked with Taurocholic acid and Glycocholic acid after stabilizing the protein gave different results. The least binding energy raised, however, Lys88 was involved in the interaction with the Glycocholic acid, which implies that Lys plays an important role in the active site of BSH (**Figure 5**).

DISCUSSION

The objective of the present study was to assess the probiotic potential of *Lactobacillus* isolates from food samples. Among tested isolates, isolate *L. plantarum* RYPR1 showed good probiotic potential and therefore it was selected for further studies. It is a commonly used and well studied probiotic strain and it is used for development of various probiotic based food products (Pisano et al., 2008). *In vitro* assessment of co-aggregation ability of isolate with *E. coli* was also studied as it is also an important selection criterion. The co-aggregation rate was observed for 2 h and it was observed that RYPR1 showed the maximum co-aggregation (40%). Similar studies were conducted by Ramos et al. (2013) and reported that *L. plantarum* CH41 showed highest co-aggregation ability with *E. coli*. Furthermore, another study reported that *L. plantarum* S1 showed the maximum co-aggregation ability (37–41%) with common enteric pathogens (Jankovic et al., 2012). Based on viable cell count after 24 h incubation with probiotics it was analyzed that RYPR1 is able to utilize probiotics. The prebiotic study of RYPR1 with commercially available prebiotics is important as prebiotics stimulates their growth in GIT (Macfarlane et al., 2007). A few studies have been reported with *L. plantarum* which confirms a correlation of prebiotics and β-galactosidase enzyme (Pennacchia et al., 2006). Moreover, few other researchers conducted studies related to probiotics functionality, safety, γ-amino butyric acid production; genomics and metabolomics etc., however, the catalytic binding and interaction studies are included in the present work which provides further lead to carry out such work in prebiotic utilizations. (Devi et al., 2016; Shekh et al., 2016; Stefanovic et al., 2017). In the present study, the BSH activity of RYPR1 was further taken as standard for studying the catalytic interaction of BSH with Taurocholic acid and Glycocholic acid. The study was done using SWISS-MODEL which provides a model structure of tested protein. The BSH enzyme from *L. plantarum* was docked with Taurocholic acid and Glycocholic acid and the results revealed that Glycocholic acid showed the least binding energy (–7.03 KJ/mol) followed by Taurocholic acid (–4.42 KJ/mol). Minimum the binding energy more will be the interaction resulting in good BSH activity. The study showed that *L. plantarum* RYPR1 is able to hydrolyse these salts, which assume that it can survive the toxicity of bile salts and also carry out deconjugation of these salts which may help in their colonization in the intestine.

CONCLUSION

In our previous study, we have reported the probiotic potential of *L. plantarum* RYPR1. Consequently, concluded that it can be used as a starter culture for the preparation of probiotic food products. The results of co-aggregation studies with pathogenic bacteria indicate that *L. plantarum* RYPR1 could be used to prevent pathogen colonization in the gut. The present study concludes that *L. plantarum* RYPR1 is able to utilize most of the prebiotics. However, the best growth was observed among lactulose followed by raffinose. Thus, we could use these prebiotics along with our selected strain to develop an effective synbiotic, which can stimulate the overall human gut microflora. Due to its good antimicrobial activity and other aspects, this indigenous isolate could be used in other relevant applications. Furthermore, the catalytic interaction of BSH with Taurocholic acid and Glycocholic acid proves further that it can act as an excellent source for various probiotic applications.

AUTHOR CONTRIBUTIONS

All authors listed, have made substantial, direct and intellectual contribution to the work, and approved it for publication.

ACKNOWLEDGMENTS

The authors duly acknowledge the University Research Scholarship (URS) to RY (Ref. 10-RUR-8058) by Maharshi Dayanand University, Rohtak, India. The authors acknowledge Dr. Abhijeet Ganguli, Thapar University, Patiala for providing basic training to RY.

REFERENCES

Arnold, K., Bordoli, L., Kopp, J., and Schwede, T. (2006). The SWISS-MODEL Workspace: a web-based environment for protein structure homology modelling. *Bioinformatics* 22, 195–201. doi: 10.1093/bioinformatics/bti770

Baweja, M., Singh, P. K., and Shukla, P. (2015). "Enzyme technology, functional proteomics and systems biology towards unraveling molecular basis for functionality and interactions in biotechnological processes," in *Frontier Discoveries and Innovations in Interdisciplinary Microbiology*, ed. P. Shukla (Berlin: Springer-Verlag), 207–212.

Baweja, M., Tiwari, R., Singh, P. K., Nain, L., and Shukla, P. (2016). An alkaline protease from *Bacillus pumilus* MP 27: functional analysis of its binding model toward its applications as detergent additive. *Front. Microbiol.* 7:1195. doi: 10.3389/fmicb.2016.01195

Devi, S. M., Aishwarya, S., and Halami, P. M. (2016). Discrimination and divergence among *Lactobacillus plantarum*-group (LPG) isolates with reference to their probiotic functionalities from vegetable origin. *Syst. Appl. Microbiol.* 39, 562–570. doi: 10.1016/j.syapm.2016.09.005

Goyal, S., Raj, T., Banerjee, C., Imam, J., and Shukla, P. (2013). Isolation and ecological screening of indigenous probiotic microorganisms from curd and chili sauce samples. *Int. J. Probiotics Prebiotics.* 8:91.

Guex, N., Peitsch, M. C., and Schwede, T. (2009). Automated comparative protein structure modeling with SWISS-MODEL and Swiss-PdbViewer: a historical perspective. *Electrophoresis* 30, 162–173. doi: 10.1002/elps.200900140

Gupta, H., and Malik, R. K. (2007). Incidence of virulence in bacteriocin-producing enterococcal isolates. *Le Lait* 87, 587–601. doi: 10.1051/lait:2007031

Gupta, S. K., and Shukla, P. (2015). Advanced technologies for improved expression of recombinant proteins in bacteria: perspectives and applications. *Crit. Rev. Biotechnol.* 18, 1–10.

Jankovic, T., Frece, J., Abram, M., and Gobin, I. (2012). Aggregation ability of potential probiotic *Lactobacillus plantarum* strains. *Int. J. Sanit. Eng. Res.* 6, 19–24.

Karthik, M. V. K., Deepak, N. S., and Shukla, P. (2012). Explication of interactions between HMGCR isoform 2 and various statins through In silico modeling and docking. *Comput. Biol. Med.* 42, 156–163. doi: 10.1016/j.compbiomed.2011.11.003

Kiefer, F., Arnold, K., Kunzli, M., Bordoli, L., and Schwede, T. (2009). The SWISS-MODEL repository and associated resources. *Nucleic Acids Res.* 37, 387–392. doi: 10.1093/nar/gkn750

Kumar, M., Ghosh, M., and Ganguli, A. (2012). Mitogenic response and probiotic characteristics of lactic acid bacteria isolated from indigenously pickled vegetables and fermented beverages. *World J. Microbiol. Biotechnol.* 28, 703–711. doi: 10.1007/s11274-011-0866-4

Kumar, V., Baweja, M., Singh, P. K., and Shukla, P. (2016). Recent developments in systems biology and metabolic engineering of plant–microbe interactions. *Front. Plant Sci.* 7:1421. doi: 10.3389/fpls.2016.01421

Laskowski, R. A., MacArthur, M. W., Moss, D. S., and Thornton, J. M. (1993). PROCHECK - a program to check the stereochemical quality of protein structures. *J. App. Cryst.* 26, 283–291. doi: 10.1107/S0021889892009944

Laskowski, R. A., Rullmannn, J. A., MacArthur, M. W., Kaptein, R., and Thornton, J. M. (1996). AQUA and PROCHECK-NMR: programs for checking the quality of protein structures solved by NMR. *J. Biomol. NMR* 8, 477–486. doi: 10.1007/BF00228148

Lee, N.-K., Kim, S.-Y., Han, K. J., Eom, S. J., and Paik, H.-D. (2014). Probiotic potential of *Lactobacillus* strains with anti-allergic effects from kimchi for yogurt starters. *LWT Food Sci. Technol.* 58, 130–134. doi: 10.1016/j.lwt.2014.02.028

Macfarlane, G. T., Steed, H., and Macfarlane, S. (2007). Bacterial metabolism and health related effects of galacto-oligosaccharides and other prebiotics. *J. Appl. Microbiol.* 104, 305–344. doi: 10.1111/j.1365-2672.2007.03520.x

Maldonado, G. C., Lemme-Dumit, J., Thieblemont, N., Carmuega, E., Weill, R., and Perdigón, G. (2015). Stimulation of innate immune cells induced by probiotics: participation of toll-like receptors. *J. Clin. Cell. Immunol.* 6:283. doi: 10.4172/2155-9899.1000283

Mitropoulou, G., Nedovic, V., Goyal, A., and Kourkoutas, Y. (2013). Immobilization technologies in probiotic food production. *J. Nutr. Metab.* 2013:716861. doi: 10.1155/2013/716861

Pennacchia, C., Vaughan, E. E., and Villani, F. (2006). Potential probiotic Lactobacillus strains from fermented sausages: further investigations on their probiotic properties. *Meat Sci.* 73, 90–101. doi: 10.1016/j.meatsci.2005.10.019

Pessione, E. (2012). Lactic acid bacteria contribution to gut microbiota complexity: lights and shadows. *Front. Cell. Infect. Microbiol.* 2:86. doi: 10.3389/fcimb.2012.00086

Pisano, M., Casula, M., Corda, A., Fadda, M., Deplano, M., and Cosentino, S. (2008). In vitro probiotic characteristics of lactobacillus strains isolated from fiore sardo cheese. *Ital. J. Food Sci.* 20, 505–516.

Ramos, C. L., Thorsen, L., Schwan, R. F., and Jespersen, L. (2013). Strain-specific probiotics properties of Lactobacillus fermentum, *Lactobacillus plantarum* and *Lactobacillus brevis* isolates from Brazilian food products. *Food Microbiol.* 36, 22–29. doi: 10.1016/j.fm.2013.03.010

Shekh, S. L., Dave, J. M., and Vyas, B. R. M. (2016). Characterization of *Lactobacillus plantarum* strains for functionality, safety and γ-amino butyric acid production. *LWT Food Sci. Technol.* 74, 234–241. doi: 10.1016/j.lwt.2016.07.052

Singh, K., Kallali, B., Kumar, A., and Thaker, V. (2011). Probiotics: a review. *Asian Pac. J. Trop. Biomed.* 1, 287–290. doi: 10.1016/S2221-1691(11)60174-3

Singh, P. K., Joseph, J., Goyal, S., Grover, A., and Shukla, P. (2016). Functional analysis of the binding model of microbial inulinases using docking and molecular dynamics simulation. *J. Mol. Model.* 22:69. doi: 10.1007/s00894-016-2935-y

Singh, P. K., and Shukla, P. (2011). Molecular modeling and docking of microbial inulinases towards perceptive enzyme-substrate interactions. *Ind. J. Microbiol.* 52, 373–380. doi: 10.1007/s12088-012-0248-0

Singh, P. K., and Shukla, P. (2014). Systems biology as an approach for deciphering microbial interactions. *Brief. Funct. Genomics* 14, 166–168. doi: 10.1093/bfgp/elu023

Singh, T. P., Kaur, G., Malik, R. K., Schillinger, U., Guigas, C., and Kapila, S. (2012). Characterization of Intestinal *Lactobacillus reuteri* strains as potential probiotics. *Probiotics Antimicrob. Proteins* 4, 47–58. doi: 10.1007/s12602-012-9090-2

Stefanovic, E., Fitzgerald, G., and McAuliffe, O. (2017). Advances in the genomics and metabolomics of dairy lactobacilli: a review. *Food Microbiol.* 61, 33–49. doi: 10.1016/j.fm.2016.08.009

Vinderola, G., Capellini, B., Villareal, F., Suarez, V., Quiberoni, A., and Reinheimer, J. (2008). Usefulness of a set of simple in vitro tests for the screening and identification of probiotic candidate strains for dairy use. *LWT Food Sci. Technol.* 41, 1678–1688. doi: 10.1016/j.lwt.2007.10.008

Yadav, R., Puniya, A. K., and Shukla, P. (2016). Probiotic properties of *Lactobacillus plantarum* RYPR1 from an indigenous fermented beverage raabadi. *Front. Microbiol.* 7:1683. doi: 10.3389/fmicb.2016.01683

Yadav, R., and Shukla, P. (2015). An overview of advanced technologies for selection of probiotics and their expediency: a review. *Crit. Rev. Food Sci. Nutr.* doi: 10.1080/10408398.2015.1108957 [Epub ahead of print].

Conflict of Interest Statement: The authors declare that the research was conducted in the absence of any commercial or financial relationships that could be construed as a potential conflict of interest.

8

Detecting *Lactococcus lactis* Prophages by Mitomycin C-Mediated Induction Coupled to Flow Cytometry Analysis

Joana Oliveira[1], Jennifer Mahony[1,2], Laurens Hanemaaijer[3], Thijs R. H. M. Kouwen[3], Horst Neve[4], John MacSharry[2] and Douwe van Sinderen[1,2]*

[1] *School of Microbiology, University College Cork, Cork, Ireland,* [2] *APC Microbiome Institute, University College Cork, Cork, Ireland,* [3] *DSM Biotechnology Center, Delft, Netherlands,* [4] *Max Rubner-Institut, Kiel, Germany*

Most analyzed *Lactococcus lactis* strains are predicted to harbor one or more prophage genomes within their chromosome; however, the true extent of the inducibility and functionality of such prophages cannot easily be deduced from sequence analysis alone. Chemical treatment of lysogenic strains with Mitomycin C is known to cause induction of temperate phages, though it is not always easy to clearly identify a lysogenic strain or to measure the number of released phage particles. Here, we report the application of flow cytometry as a reliable tool for the detection and enumeration of released lactococcal prophages using the green dye SYTO-9.

Keywords: temperate phages, lysogeny, chemical inductions, prophage, flow cytometry

Edited by:
Spiros Paramithiotis,
Agricultural University of Athens,
Greece

Reviewed by:
Xingmin Sun,
University of South Florida,
United States
Daniel M. Linares,
Teagasc – The Irish Agriculture
and Food Development Authority,
Ireland

*Correspondence:
Douwe van Sinderen
d.vansinderen@ucc.ie

INTRODUCTION

Lactococcus lactis is a non-pathogenic Gram-positive lactic acid bacterium (LAB), which is used as a starter culture for the manufacture of a variety of fermented dairy products (Rousseau and Moineau, 2009; Mahony and van Sinderen, 2014). Members of the two recognized *L. lactis* subspecies, subsp. *lactis* and subsp. *cremoris*, each impart organoleptic properties that contribute to flavor and textural characteristics of the final fermented product (Fernandez et al., 2011). Inconsistencies in the dairy production process are frequently due to (bacterio)phage infection of the lactococcal starter culture(s), which has prompted many detailed scientific studies on such bacterial viruses (Chopin et al., 2001; Marco et al., 2012). Currently, ten genetically distinct groups of lactococcal phages are known to exist, all being members of the *Caudovirales* order. The majority of these lactococcal phage groups exhibit typical characteristics of *Siphoviridae* (representing phages that possess a long, non-contractile tail), while two groups are classified as *Podoviridae* (i.e., phages possessing a short tail) (Deveau et al., 2006). Despite the apparent diversity of lactococcal phages, three groups are most commonly isolated in the environment of commercial dairy fermentations: the virulent 936 and c2 groups, and phages belonging to the P335 group whose members may be virulent or temperate (Chopin et al., 2001; Deveau et al., 2006). Lysogenic lactococcal phages that have been identified and/or characterized to date have, by and large, been assigned to the P335 group and their presence as a prophage in a lactococcal chromosome may either be considered beneficial or undesirable. In the integrated state, prophages may provide superinfection exclusion and homo-immunity against (super)infecting phages, thereby protecting their lysogenic host (McGrath et al., 2002; Labrie et al., 2010). In contrast, the threat of prophage induction and consequent cell lysis, or prophage conversion to a strictly lytic derivative constitutes a realistic risk factor to the fermentation industry (Marco et al., 2012).

The plaque assay is the standard technique used for the detection and quantification of infectious phage particles, although it is a labor-intensive and time-consuming technique, detecting infectious and virulent particles within the overall phage population (Anderson et al., 2011). Genetic techniques, in particular PCR, in combination with mitomycin C (MmC)-mediated induction growth profiles, have been described as a useful approach to identify lysogenic strains (Martín et al., 2006), where integrase-specific primers (O'Sullivan et al., 2000), and oligonucleotide primers previously described for the lactococcal P335 group detection (Labrie and Moineau, 2000) have been applied. This approach has had mixed success, as PCR-based detection only indicates the presence of a target phage sequence, but not necessarily a functional prophage, while false positive results were also reported (Martín et al., 2006). Furthermore, MmC-mediated induction growth curves do not appear to represent a very reliable procedure to detect lysogenic strains (Martín et al., 2006). In recent years, alternative techniques have been described for the purpose of virus particle detection and enumeration, including real-time PCR, the so-called nanoparticle tracking analysis (NTA)-based approach using NanoSight (NS) technology, transmission electron microscopy (TEM), fluorescence-staining methods, e.g., epifluorescence microscopy (EFM), and flow cytometry (FCM) (Marie et al., 1999; Chen et al., 2001; Anderson et al., 2011). The two former techniques provide reproducible data which are consistent with those obtained by plaque assays, though with a reduced turnaround time. Furthermore, TEM, EFM, and FCM have been assessed in detail to determine their usefulness in detecting and enumerating phage particles, revealing that EFM- and TEM-based counts often underestimate the actual virion number, whereas FCM was demonstrated to be a sensitive, (relatively) rapid and reproducible detection technique (Marie et al., 1999; Chen et al., 2001). As mentioned above, plaque assays represent a traditional methodology to detect the ability of a single phage particle to infect a permissive bacterial host, thus resulting in progeny formation (Anderson et al., 2011). However, in situations where a sensitive host may not be available for a phage, such as in the case of an induced (pro)phage, it is important to know if and how many viral particles are present (Marco et al., 2012). In the current study we report on the use of FCM as a reliable detection and enumeration method for phage particles that are released from lysogenic L. lactis strains following MmC treatment.

MATERIALS AND METHODS

Bacterial Growth and Prophage Induction Conditions

Lactococcus lactis strains (L. lactis strain details are described in **Table 1** and Supplementary Table S1) were grown at 30°C in M17 broth (Oxoid) supplemented with 0.5% glucose (GM17) and prophages were chemically induced using 1.3 or 3 μg.ml^{-1} of MmC (Sigma). Four well studied laboratory L. lactis strains were employed in this study, i.e., L. lactis 3107 (Braun et al., 1989)

and SMQ-86 (Emond et al., 1997), as potential sensitive hosts for lytic propagation of (some of the) temperate phages identified in this study (phage-host survey described in Supplementary Table S2), and L. lactis NZ9000-TP901-1*erm* (Koch et al., 1997; Stockdale et al., 2013) and UC509.9 (Costello, 1988), which were used as a positive and negative controls for flow cytometric analysis, respectively (**Table 1**). For phage propagations, 0.01 M CaCl$_2$ was added to the growth medium. Phage particles were resuspended in TBT buffer (0.05 M Tris HCl [pH 7.0]; 0.1 M NaCl; 0.01 M MgCl$_2$.6H$_2$0) for phage DNA extraction, phage grouping by multiplex PCR, plaque assays or FCM analysis.

Small and Large Scale Prophage Inductions

Small-scale prophage induction trials were performed in 96-well microtitre plates by (in each well) inoculating 200 μl of GM17 broth with 2% fresh overnight culture of a particular L. lactis strain. Chemical induction was performed in early exponential phase (OD$_{600\,nm}$~0.2) by the addition of MmC at a final concentration of 0 (i.e., without MmC; negative control), 1.3 or 3 μg.ml^{-1}. Incubation was continued at 30°C and bacterial growth was followed for 8 h. A final OD$_{600\,nm}$ reading was obtained 24 h after the lactococcal strains were inoculated in the 96-well microplate, and growth profiles were then generated. In order to obtain cell lysates, MmC-mediated inductions were performed as described before, but at a large scale. Briefly, this involved the addition of 3 μg.ml^{-1} of MmC (final concentration) to early exponential phase cultures grown in a 50 ml volume of GM17 broth, followed by overnight incubation at room temperature. Cell debris was removed by centrifugation at 7560 $\times$ g for 20 min and lysates were then passaged through a 0.45 μm filter (Sarstedt, Nümbrecht Germany).

Phage Propagation and Isolation

To assess if prophage induction resulted in the production of infective phage particles, lysates (see previous section) were employed for a phage susceptibility analysis, involving two strains previously described as a sensitive host to several P335 phages, i.e., L. lactis 3107 and SMQ-86. This phage-host range analysis was performed by plaque assays as previously described (Lillehaug, 1997). Where a host was identified for a (MmC-induced) phage, propagations using single plaques were performed to ensure that a pure phage with a single genotype was propagated, while serial propagations in GM17 broth were performed in order to increase the phage titer. A final propagation of temperate phage(s) was performed using a 1% inoculum of the bacterial host in 50 ml of broth supplemented with 0.01 M CaCl$_2$ (final concentration) and to which 1% (v/v) of phage lysate with a titer between 10^8 and 10^9 plaque-forming units per ml (pfu.ml^{-1}) was added. The phage-host mixture was incubated at 30°C until visible lysis was observed, after which remaining cells were removed from the lysate by passage through a 0.45 μm filter, followed by storage at 4°C.

TABLE 1 | Shortlist of lactococcal strains and phages used in this study (for a description of all lactococcal strains used see Supplementary Table S1).

Bacterial strains and phages	Relevant features	Reference
***L. lactis* strains**		
UC509.9	Prophage-free lactococcal strain (used as negative control)	Costello, 1988
NZ9000-TP901-1*erm*	Strain carrying the inducible prophage TP901-1*erm* (used as positive control)	Koch et al., 1997; Stockdale et al., 2013
SMQ-86	Sensitive host strain for phage-host survey and propagation of induced (pro)phage	Emond et al., 1997
3107	Sensitive host strain for phage-host survey and propagation of induced (pro)phage	Braun et al., 1989
(Pro)phages		
TP901-1*erm*	Inducible temperate phage, used as a positive control	Koch et al., 1997; Stockdale et al., 2013
56701	Temperate phage induced from *L. lactis* DS68567	Oliveira et al., 2016
98201	Temperate phage induced from *L. lactis* DS64982	Oliveira et al., 2016
28201	Temperate phage induced from *L. lactis* DS70282	Oliveira et al., 2016
50901	Temperate phage induced from *L. lactis* DS68509	Oliveira et al., 2016
62501	Temperate phage induced from *L. lactis* DS63625	Oliveira et al., 2016
18301	Temperate phage induced from *L. lactis* DS72183	This study
58501	Temperate phage induced from *L. lactis* DS68585	This study
38501	Temperate phage induced from *L. lactis* DS70385	This study
16001	Temperate phage induced from *L. lactis* DS72160	This study
15901	Temperate phage induced from *L. lactis* DS72159	This study
07501	Temperate phage induced from *L. lactis* DS69075	This study
63301	Temperate phage induced from *L. lactis* DS63633	Oliveira et al., 2016
50101	Temperate phage induced from *L. lactis* DS68501	Oliveira et al., 2016
58601	Temperate phage induced from *L. lactis* DS68586	This study
86501	Temperate phage induced from *L. lactis* DS71865	Oliveira et al., 2016
24801	Temperate phage induced from *L. lactis* DS70248	This study
06701	Temperate phage induced from *L. lactis* DS69067	This study
49801	Temperate phage induced from *L. lactis* DS68498	This study
51801	Temperate phage induced from *L. lactis* DS68518	This study
49501	Temperate phage induced from *L. lactis* DS68495	This study

Phage DNA Extraction and Multiplex PCR

Filtered phage lysates were treated with DNase and RNase to remove residual host chromosomal DNA and RNA, and incubated at 37°C for 40 min before adding polyethylene glycol (PEG$_{8000}$) to a final concentration of 10% w/v, followed by incubation at 4°C for 16 h. Phage DNA extraction was performed as described previously (Mahony et al., 2013), with minor modifications (Moineau et al., 1994). Phage genotyping was performed by a previously established multiplex PCR methodology (Labrie and Moineau, 2000), in which three different primer pairs, each based on group-specific regions of the three dominant lactococcal phage groups (i.e., the 936, P335 and c2 groups) were employed. Phage DNA or phage lysate was used as a template, and the PCR products were generated as previously described (Labrie and Moineau, 2000). PCR products were separated by gel electrophoresis on a 1% agarose gel and visualized by UV illumination.

Detection of Prophages by Flow Cytometry

Following large scale induction, as described above, released phage particles were detected by FCM using the LIVE/DEAD BacLight bacterial viability and counting kit (Life Technologies).

Briefly, bacterial debris were removed by centrifugation (9148 × *g* for 20 min) following addition of 0.5 M NaCl (final concentration) and subsequent incubation for 2 h at 4°C with agitation. Phage particles were precipitated by the addition of 10% PEG$_{8000}$ to filtered cell-free supernatant followed by overnight incubation at 4°C and subsequent recovery by centrifugation at 17,620 × *g* for 15 min. The resulting (virion-containing) pellet was resuspended in 1 ml TBT buffer followed by two washes (10,000 × *g*) in ¼ strength Ringer's solution as described in the LIVE/DEAD BacLight bacterial viability and counting kit (Thermo Fisher Scientific, Leiden, The Netherlands). Following incubation for 30–60 min at room temperature, a final wash step was performed and the pellet was then resuspended in 1 ml of ¼ strength Ringer's solution. A 1:10 dilution of this phage suspension was prepared in ¼ strength Ringer's solution, stained with 0.15% of the SYTO-9 nucleic acid dye (light protected) and analyzed by FCM in triplicate. Of note, the same phage suspension prepared in ¼ strength Ringer's solution was additionally used for plaque assays analysis, in order to see the reliability of the FCM in phage enumeration. FCM analysis was performed using the BD Accuri C6 flow cytometer by detection of excitation/emission wavelengths from SYTO-9-stained DNA (485/498 nm, respectively). Briefly, measurements were performed in logarithmic scale based on the following parameters: run limits for 5000 events.ml^{-1};

medium flow rate (33 µl.min^{-1}) and a threshold set on forward scatter (FSC-A) to allow for the discrimination of phages (from background noise; see Supplementary Figure S1). Phage particle quantification was obtained by applying the following formula: [(no. events detected/sample volume analyzed)*10] (pfu.ml^{-1}) for all lactococcal strains used. In order to obtain an accurate phage particle quantitation by FCM, each analysis was conducted to acquire an equal amount of events per sample (5000 events.ml^{-1}) and/or in instances where samples presented with low particle numbers, the same sampling time was maintained to ensure equal sampling. Positive (phage TP901-1 induced from NZ9000-TP901-1) (Koch et al., 1997; Stockdale et al., 2013) and negative (L. lactis strain UC509.9, which is prophage-free) (Costello, 1988) controls were also included in the FCM analysis. SYTO-9 emissions were detected in the FL-1 channel (BP Filter 530/30), and data analysis was performed using FCS 5 Express plus software (described in the Supplementary Figure S1).

Temperate Phage Particle Detection by Transmission Electron Microscopy (TEM)

For TEM analysis phage particles produced from MmC-treated cultures (in G-M17 growth broth) were adsorbed to a carbon-coated 400-mesh copper grid (Agar Scientific, Essex, United Kingdom) and negative staining with 2% (w/v) uranyl acetate was performed as described previously (Deasy et al., 2011). Specimens were examined with a Tecnai 10 transmission electron microscope (FEI Thermo Fisher Scientific, Eindhoven, The Netherlands) operated at an acceleration voltage of 80 kV. Micrographs were taken with a MegaView G2 charge-coupled-device camera (EMSIS, Münster, Germany). All measurements of the phage particle dimensions were performed using iTEM imaging software (EMSIS).

RESULTS

Small Scale Prophage Induction Profile Analysis

113 L. lactis strains were assessed for the presence of inducible prophages following treatment with two different MmC concentrations. Optical density of treated cultures was monitored over a period of 24 h at 30°C and compared to a corresponding untreated control culture (i.e., same strain but without MmC addition; **Figure 1**). Growth profiles were obtained and compared with growth profiles from two lactococcal strains used as a negative or positive control: L. lactis UC509.9, which is free of inducible prophages (Costello, 1988), and L. lactis NZ9000-TP901-1erm, which contains an inducible prophage (Koch et al., 1997; Stockdale et al., 2013), respectively. Following addition of 1.3 or 3 µg.ml^{-1} of MmC, the observed growth profiles for the positive control revealed a drastic growth reduction of the strain when exposed to MmC (as compared to untreated control), with an equal impact observed for either of the two MmC concentrations (**Figure 1B**; this growth behavior is referred to as growth profile B). In contrast, the negative control {L. lactis UC509.9 [a prophage-free strain] (Costello, 1988)} was

shown to exhibit a gradual impact on growth when exposed to MmC, with a more pronounced diminishment of growth at the higher MmC concentration, possibly due to the toxic effect of MmC (**Figure 1A**; this growth behavior is referred to as growth profile A). To try to correlate cell lysis with prophage induction, growth profile comparisons between 113 tested lactococcal strains and control strains was performed (**Figure 1** and Supplementary Table S1). Interestingly, in addition to the two growth profiles A and B as identified above, two additional and distinct MmC-mediated growth profiles were observed.

Detailed assessment of the growth profiles obtained from this lactococcal strain collection showed that the growth behavior of 33.63% of the lactococcal strains screened was reminiscent of growth profile B as observed for the positive control (NZ9000-TP901-1erm; similar profiles were observed in **Figures 1B,D**), suggesting that these strains may harbor one or more inducible prophages. Furthermore, the growth behavior of 36.28% of the lactococcal strain screened was similar to growth profile A, i.e., the negative control used in this study (UC509.9; similar profiles were observed in **Figures 1A,E**). Interestingly, 15.93% of L. lactis strains tested in this study displayed similar growth profiles to the corresponding untreated strain (**Figure 1C**), i.e., no significant growth arrest was observed following MmC addition (compared to growth in the absence of MmC), and this was taken as an indication that no prophage induction had taken place (this behavior is referred to as growth profile C). Finally, the remaining 14.16% of the tested lactococcal strains exhibited a more severe growth arrest (and substantial cell lysis) in the presence of 1.3 µg.ml^{-1} MmC compared to that observed for the same culture following exposure to 3 µg.ml^{-1} MmC (**Figure 1F** and referred here as growth profile D). The substantial drop in bacterial growth observed following addition of 1.3 µg.ml^{-1} of MmC (and less so at the higher concentration of MmC) may be due to phage induction (and consequent cell lysis) rather than MmC toxicity (as one may expect this to cause similar or incremental growth cessation profiles).

Characterization of Induced Temperate Phages

In order to validate the identification of putative lysogens among the 113 tested strains, chemical induction was performed on a larger scale (50 ml and addition of 3 µg.ml^{-1} MmC; see Materials and Methods). The resulting lysates were tested for the presence of infectious phages in a phage-host survey against two highly phage-sensitive L. lactis strains, 3107 (Braun et al., 1989) and SMQ-86 (Emond et al., 1997), in an effort to find a suitable propagation host. The results of this phage-host survey are summarized in Supplementary Table S2 with the identification of MmC-inducible phages from fifteen L. lactis strains capable of infecting either L. lactis SMQ-86 or L. lactis 3107. Interestingly, a third of these strains were shown to exhibit a growth profile that was similar to that of the negative control UC509.9 following MmC exposure (i.e., growth profile A), indicating that comparative growth profile analysis (following MmC treatment)

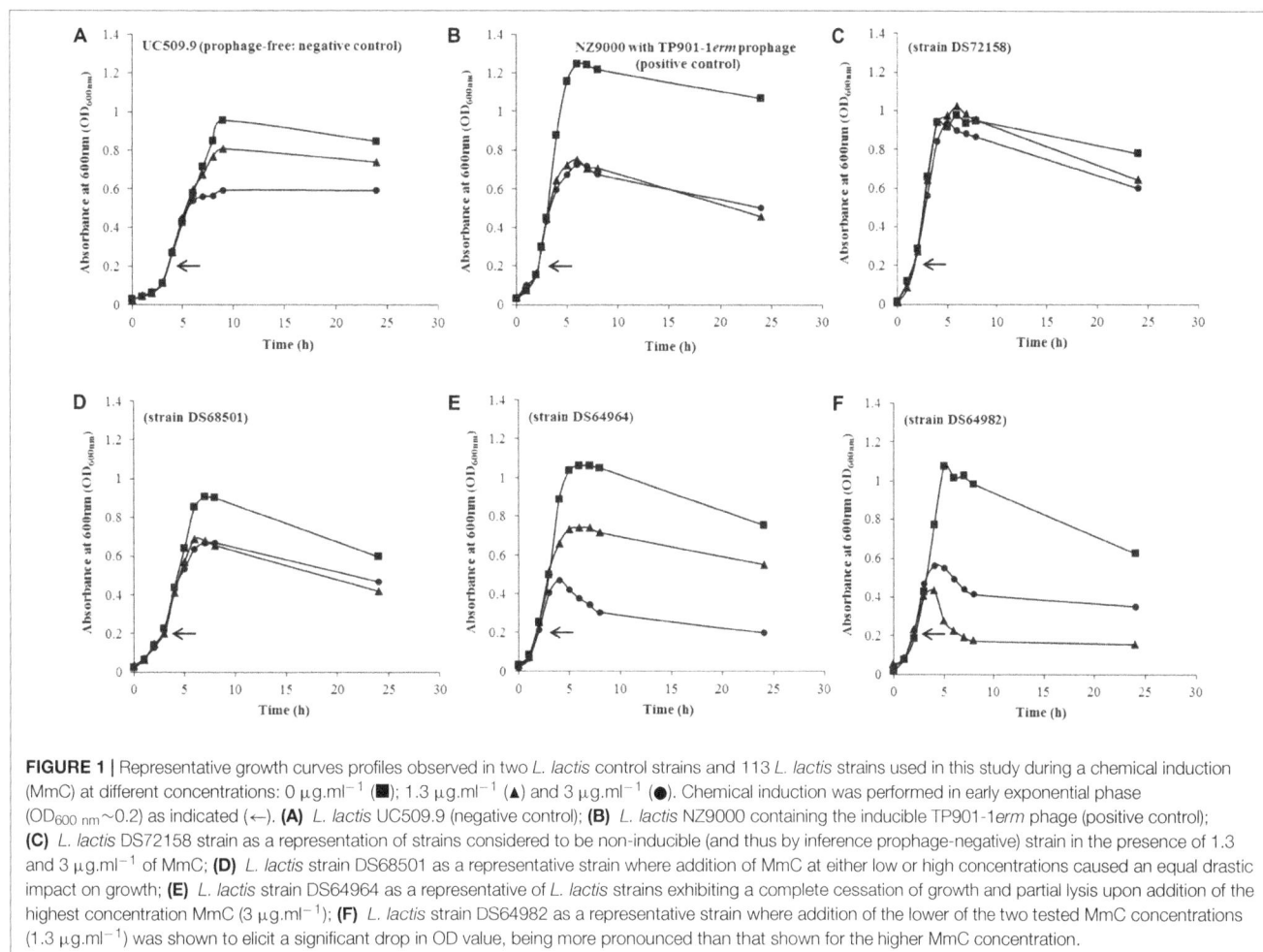

FIGURE 1 | Representative growth curves profiles observed in two *L. lactis* control strains and 113 *L. lactis* strains used in this study during a chemical induction (MmC) at different concentrations: 0 μg.ml^{-1} (■); 1.3 μg.ml^{-1} (▲) and 3 μg.ml^{-1} (●). Chemical induction was performed in early exponential phase ($OD_{600\ nm}$~0.2) as indicated (←). **(A)** *L. lactis* UC509.9 (negative control); **(B)** *L. lactis* NZ9000 containing the inducible TP901-1*erm* phage (positive control); **(C)** *L. lactis* DS72158 strain as a representation of strains considered to be non-inducible (and thus by inference prophage-negative) strain in the presence of 1.3 and 3 μg.ml^{-1} of MmC; **(D)** *L. lactis* strain DS68501 as a representative strain where addition of MmC at either low or high concentrations caused an equal drastic impact on growth; **(E)** *L. lactis* strain DS64964 as a representative of *L. lactis* strains exhibiting a complete cessation of growth and partial lysis upon addition of the highest concentration MmC (3 μg.ml^{-1}); **(F)** *L. lactis* strain DS64982 as a representative strain where addition of the lower of the two tested MmC concentrations (1.3 μg.ml^{-1}) was shown to elicit a significant drop in OD value, being more pronounced than that shown for the higher MmC concentration.

is not a very dependable method to determine if a strain harbors (an) inducible prophage(s).

Direct Detection of Temperate Phages Induced from Lysogenic Strains by Flow Cytometry Analysis

In order to assess if released phage particles (following induction from a lysogenic host) can be detected and quantified by FCM, we tested this technology by employing MmC-mediated lysates in which the DNA-binding dye SYTO-9 had been incorporated together with a calibrated suspension of microspheres (6.0 μm) to accurately estimate the volume analyzed (see Materials and Methods; Supplementary Figure S1). FCM optimization and subsequent lactococcal lysate analysis is discussed below.

Establishment of Flow Cytometry Procedure for Phage Particle Detection

Phage particle detection by means of FCM was first explored through the analysis of two different controls (**Figure 2**); one positive control (phage TP901-1 induced from the lysogenic strain NZ9000-TP901-1*erm*) and one negative control (UC509.9 [prophage-negative]). All data were acquired from three independent experiments, and using the FCS Express 5 Plus

software[1]. The obtained MmC-mediated lysates from these two controls were analyzed by FCM through the measurement of scattered light in the forward and side scatter directions (FSC and SSC, respectively) (**Figures 2A,B** and Supplementary Figure S1). Both FSC and SSC are unique for every particle, and the combination of these two scatter data sets allows the distinction between two different particles based on size and complexity: lysate particles (virions and other particles produced by cell lysis) and microspheres, represented by Gates 1 and 2, respectively. In order to identify phage particles (potentially) present in our lysate (Gate 1), the fluorescence emitted from stained phage DNA was measured using the FL1 channel (**Figures 2C,D** and Supplementary Figure S1) and detected in Gate 3. The fluorescence levels detected in Gate 3 for the positive and negative controls show that they emit clearly distinguishable fluorescent signals (**Figure 2E** and Supplementary Figure S1). Briefly, for the employed positive control (phage TP901-1 induced from NZ9000), a high level of fluorescence was detected in the virus particle gate region (Gate 3; Supplementary Figure S1). In detail, 80.60% fluorescence was detected for phage lysates of TP901.1, thus revealing the detection of SYTO-9 stained phage particles.

---

[1]www.denovosoftware.com

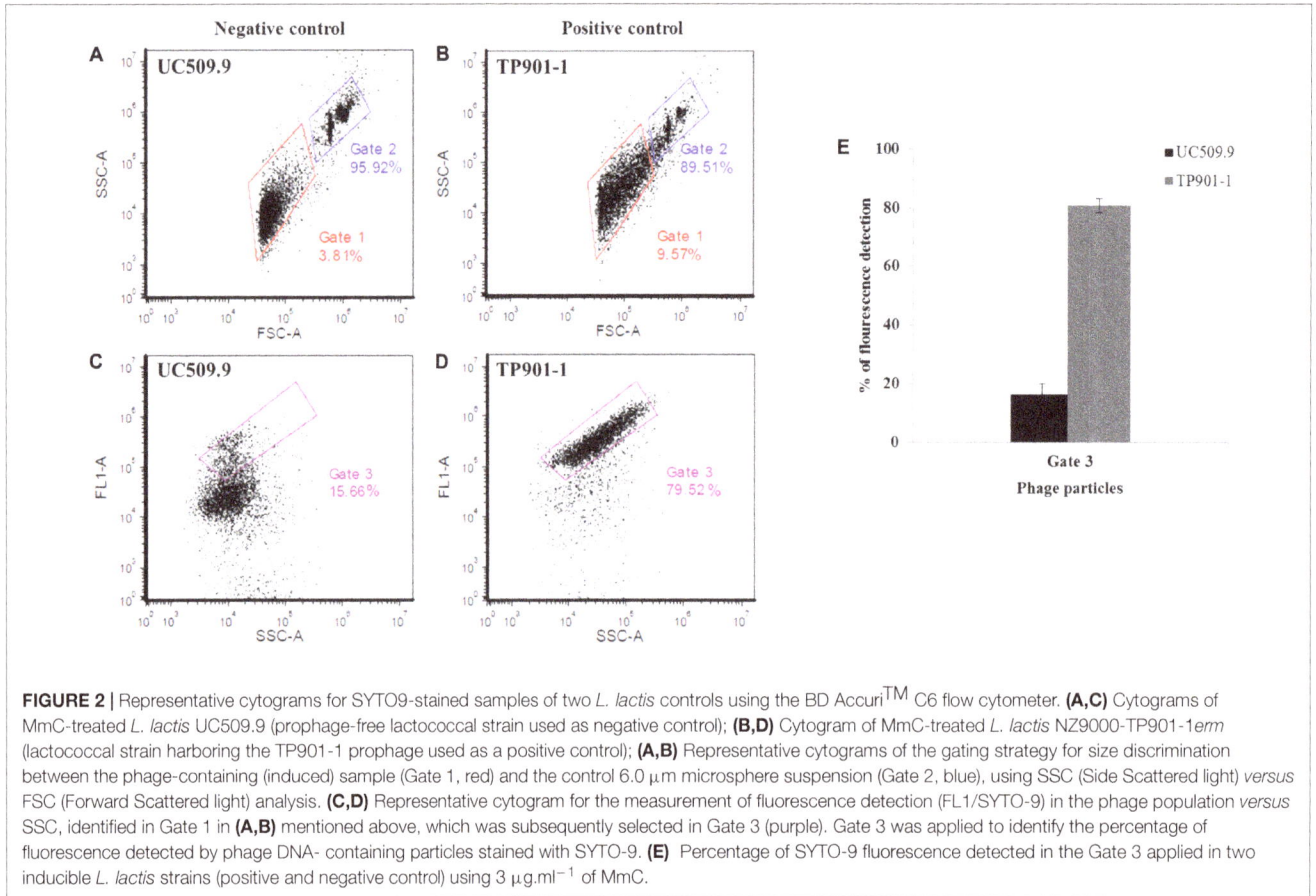

FIGURE 2 | Representative cytograms for SYTO9-stained samples of two *L. lactis* controls using the BD Accuri™ C6 flow cytometer. **(A,C)** Cytograms of MmC-treated *L. lactis* UC509.9 (prophage-free lactococcal strain used as negative control); **(B,D)** Cytogram of MmC-treated *L. lactis* NZ9000-TP901-1*erm* (lactococcal strain harboring the TP901-1 prophage used as a positive control); **(A,B)** Representative cytograms of the gating strategy for size discrimination between the phage-containing (induced) sample (Gate 1, red) and the control 6.0 μm microsphere suspension (Gate 2, blue), using SSC (Side Scattered light) *versus* FSC (Forward Scattered light) analysis. **(C,D)** Representative cytogram for the measurement of fluorescence detection (FL1/SYTO-9) in the phage population *versus* SSC, identified in Gate 1 in **(A,B)** mentioned above, which was subsequently selected in Gate 3 (purple). Gate 3 was applied to identify the percentage of fluorescence detected by phage DNA- containing particles stained with SYTO-9. **(E)** Percentage of SYTO-9 fluorescence detected in the Gate 3 applied in two inducible *L. lactis* strains (positive and negative control) using 3 μg.ml^{-1} of MmC.

For the negative control strain (UC509.9) employed in this study, a low and presumed background level of fluorescence (16.33%; **Figure 2** and Supplementary Figure S1) was emitted in the viral particle gate region (Gate 3), being consistent with the absence of phage particles in this MmC-treated strain. Based on the results obtained from these *L. lactis* controls it thus appeared feasible to establish a reliable FCM analysis for phage particle detection.

Flow Cytometry Analysis of MmC-Induced *L. lactis* Strains

The protocol developed to distinguish a lysogenic from a non-lysogenic strain (as described above) was subsequently applied to detect (temperate) phages released from MmC-treated lactococcal strains (**Figure 3** and **Table 2**). FCM analysis was performed using a selection of eighteen lactococcal strains: seven strains which upon MmC treatment produce lysates containing phages for which a suitable host had been identified (**Table 3** and Supplementary Table S2); and eleven strains, which represent each of the four previoulsy described MmC-mediated growth profiles (see above; Supplementary Table S1) and for which we could not detect phages in the resulting lysate as based on a plaque assay (**Figure 3** and **Tables 2**, **3**). From **Figure 3**, a pattern of SYTO-9-stained lysates can be observed that allows the identification of bacterial strains that either do or do not release temperate phages following MmC treatment (**Table 2**). As

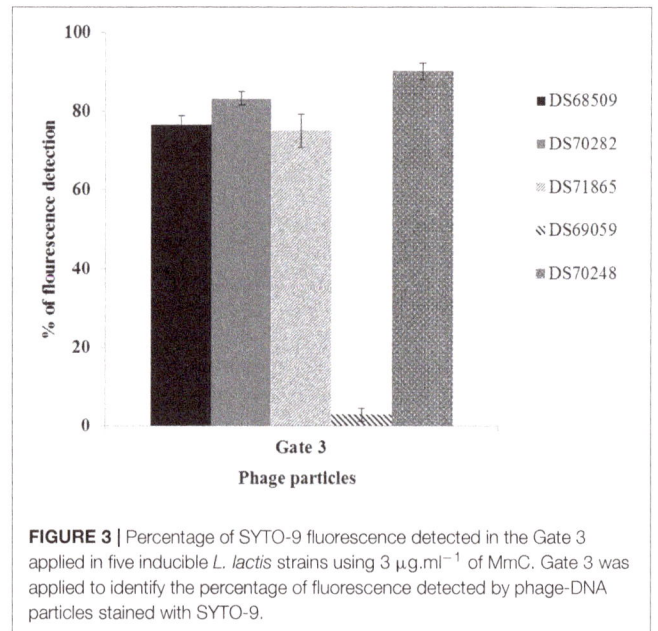

FIGURE 3 | Percentage of SYTO-9 fluorescence detected in the Gate 3 applied in five inducible *L. lactis* strains using 3 μg.ml^{-1} of MmC. Gate 3 was applied to identify the percentage of fluorescence detected by phage-DNA particles stained with SYTO-9.

expected, the seven *L. lactis* strains (**Table 3**), which upon MmC treatment released phages that lytically infect 3107 or SMQ-86 (see above), were all shown to emit high fluorescence signals due to the presence of DNA-stained phage particles (**Figure 3** and

TABLE 2 | Percentage of fluorescence detected in five inducible lysates stained with SYTO-9 dye.

L. lactis strains tested	Features		Flow cytometry analysis (% of fluorescence) associated with Gate 3	
	Induced prophage	L. lactis host	Virion-associated	Background
DS68509	50901	SMQ-86	76.49 ± 5.18	23.51 ± 5.18
DS70282	28201	SMQ-86	83.19 ± 4.37	16.81 ± 4.37
DS71865	86501	3107	75.03 ± 8.20	24.97 ± 8.20
DS69059	–	–	2.86 ± 0.43	97.14 ± 0.43
DS70248	24801	–	90.03 ± 5.21	9.97 ± 5.21

TABLE 3 | Phage particles detection in inducible L. lactis lysates and correlation between flow cytometry and MmC growth profiles.

L. lactis strains	MmC growth profile	Inducible prophage	L. lactis host	Plaque assays (pfu.ml^{-1})	Phage-group[1]	Flow cytometry analysis (pfu.ml^{-1})*	Gate 3 (% fluorescence)	Electron microscopy[#]
Controls								
NZ9000 (TP901-1 erm)	B	TP901-1	3107	2.43e05	P335	6.85e05	80.6	nd
UC509.9	A	Prophage-free	–	–	–	<1.0e04	16.3	nd
Strains tested								
DS68567	A	56701	SMQ-86	5.90e04	P335	1.74e06	48.6	nd
DS71865	A	86501	3107	8.53e03	P335	9.12e05	75.0	+
DS63633	B	63301	3107	2.07e05	P335	1.79e05	29.7	+
DS70282	B	28201	SMQ-86	8.30e03	P335	1.59e06	83.2	nd
DS63625	D	62501	SMQ-86	2.07e04	P335	5.47e05	72.7	nd
DS68509	D	50901	SMQ-86	8.88e04	P335	1.24e06	76.5	+
DS64982	D	98201	SMQ-86	6.55e04	P335	1.12e06	55.9	+
DS64964	A	No prophage detected	–	–	nd	<1.0e04	0.37	nd
DS601	A	No prophage detected	–	–	nd	<1.0e04	2.92	–
DS69067	A	06701	–	–	nd	2.07e07	99.79	+
DS68498	B	49801	–	–	nd	5.22e06	99.13	+
DS69059	B	No prophage detected	–	–	nd	<1.0e04	2.86	nd
DS68518	B	51801	–	–	nd	2.02e06	97.48	nd
DS66563	B	No prophage detected	–	–	nd	<1.0e04	5.28	–
DS68569	C	No prophage detected	–	–	nd	<1.0e04	11.14	nd
DS72158	C	No prophage detected	–	–	nd	<1.0e04	12.05	–
DS70248	D	24801	–	–	nd	3.80e06	90.03	+
DS68495	D	49501	–	–	nd	4.04e06	98.37	nd

Profile A: UC509.9 profile with a growth cessation at 3 $\mu g.ml^{-1}$; **Profile B:** TP901-1 profile with an equal growth cessation (1.3 and 3 $\mu g.ml^{-1}$); **Profile C:** No chemical effect; **Profile D:** Growth cessation at 1.3 $\mu g.ml^{-1}$; nd, Not determined; [1] Phage-group was determined by Multiplex PCR; *Phage particles quantification was performed using the following equation: [(no. events detected/sample volume analyzed)* 10] (pfu.ml^{-1}). [#]+: Presence of phage particles; [#]–: phage particles not detected.

Tables 2, **3**). As mentioned above, despite the identification of a sensitive host for the phages released by these seven lysogenic strains, only some of these strains were shown to exhibit MmC-mediated growth profile similar to that of TP901-1 (*L. lactis* strains DS63633 and DS70282; **Table 3** and Supplementary Table S1). The same approach was applied to test for the presence of particles in eleven lysates obtained upon MmC treatment for which no suitable host had been identified (**Tables 2**, **3**). It was observed that the lysates obtained (following MmC exposure)

from *L. lactis* strains DS64964, DS601, DS69059, DS66563, DS68569, and DS72158 were shown to be associated with a low level SYTO-9 fluorescence signal (<12.05%; **Tables 2**, **3**). This level of fluorescence is somewhat lower than that observed for the negative control strain (UC509.9; **Table 3** and Supplementary Figure S1) and it was therefore concluded that these strains do not release (DNA-containing) phage particles upon MmC treatment. Interestingly, despite their apparent inability to release phage particles, it was observed that two of these six strains exhibit a

MmC-mediated growth profile similar to that observed for the UC509.9 (growth profile A; negative for phage particles release; **Table 3**) and another two exhibited a growth profile C (above described).

The lysates of the remaining five strains (i.e., five out of the eleven with no suitable sensitive host identified), i.e., DS69067, DS68498, DS68518, DS70248, and DS68495, were shown to exhibit high fluorescence signals (>90%), clearly indicative of phage particle release. Notably, only two of these strains (DS68498 and DS68518) were shown to exhibit a MmC-mediated growth profile related to that observed for the positive control (growth profile B), while two other strains were shown to exhibit growth profile D (described above).

TEM Analysis of Temperate Lactococcal Phages

In order to substantiate the notion that FCM can reliably detect phage particles induced from lysogenic strains (in particular when no suitable sensitive host is available for such phages), TEM analysis was performed on MmC-mediated lysates of six strains, which were selected based on their distinct MmC-mediated growth profile and FCM results (DS601, DS69067, DS68498, DS66563, DS72158, and DS70248; **Table 3**). This TEM analysis revealed that the MmC-mediated lysates of three strains (i.e., DS69067, DS68498, and DS70248) harbored phage particles (**Figure 4**). Of these three strains, only DS68498 was shown to exhibit a MmC-mediated growth profile similar to that of the positive control. Analysis of the images revealed that all phages possess an isometric capsid and a non-contractile tail measuring with approximately 57–60 nm and 117–190 nm (**Figure 4**), in some cases complex baseplates were also observed (**Figure 4B**), similar to that noted for other members of the P335 phage group (Labrie et al., 2008; Mahony et al., 2017). In contrast, no observable phage particles were detected in the lysates of strains DS601, DS66563, and DS72158 by TEM with a limit of detection of 10^5 to 10^6 particles per ml (Zhang et al., 2013). Based on our

data it is clear that MmC-mediated growth profile analysis is not a reliable method to identify lysogenic strains; in contrast, the results obtained from FCM and TEM analyses are fully consistent with each other, suggesting that the former is a reliable and relatively rapid method to establish if a strain harbors inducible prophages.

Phage Detection Using Flow Cytometry: Plaque Assays versus FCM Total Virus Validation

We also wanted to validate FCM as a reliable method for the quantitation of viral particles (in particular when a suitable host is unavailable to perform plaque assays). For this purpose, using the ¼ strength Ringer's phage suspension following PEG precipitation of MmC-mediated lysates, phage particle quantification values obtained from FCM (see Material and Methods) and plaque assays were compared (**Figure 5**) for seven strains (out of fifteen lysates, for which a sensitive host strain had been identified by phage-host survey; Supplementary Table S2). Plaque assays on the inducible prophages were performed using two lactococcal hosts: *L. lactis* subsp. *lactis* SMQ-86 or *L. lactis* subsp. *cremoris* 3107 (**Figure 5** and Supplementary Table S2). The phage numbers obtained by plaque assays and FCM were compared, and *L. lactis* UC509.9 (phage-negative sample) was used to normalize the total phage count from FCM analysis and additionally, to estimate the limit of phage particle detection by FCM. Based on our comparisons, lactococcal lysates (following MmC treatment) that by FCM analysis exhibit fluorescence signals below 16% and a phage particle number of less than 10^4 pfu.ml^{-1} are considered to be negative for phage particles (**Table 3**). While FCM is unable to distinguish between infectious and non-infectious viral particles (Bosch et al., 2004), the number of phage particles detected by FCM was in most cases found to be in the same order of magnitude as the titre obtained from the plaque assay method (where this was possible). In cases where a more pronounced difference was observed [i.e., where FCM data indicates a more than 10-fold higher number of phage particles; e.g., *L. lactis* strains harboring prophages DS64982, DS71865, DS70282, and DS63625 (**Figure 5**)], this difference may be explained by the possibility that not all phage particles present are infectious (confirmed by detection of disintegrated particles and empty capsids by TEM, data not shown) or that the host encodes a phage-resistance system that reduces the efficacy of plaquing.

DISCUSSION

The current study assessed if FCM is a suitable method to detect virus particles induced from lysogenic *L. lactis* strains. Several studies have indicated that the majority of lactococcal strains harbor one or more prophage-like sequences in their genome (Chopin et al., 2001; Ventura et al., 2007; Mahony et al., 2008). MmC, UV and temperature treatments are among the most commonly reported methods to achieve prophage induction (Chopin et al., 2001; Nanda et al., 2015; Ho et al.,

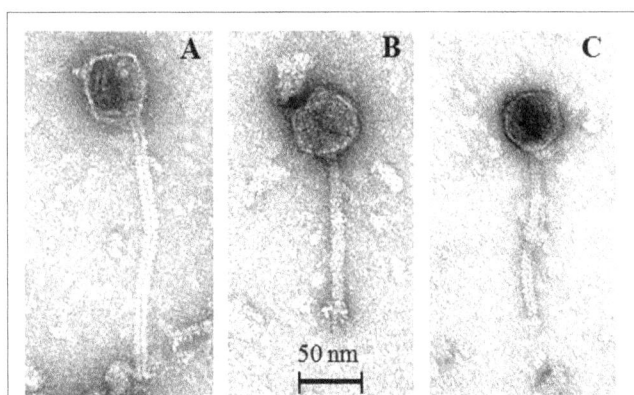

FIGURE 4 | Representative transmission electron micrographs of P335-like phages released from MmC-treated lactococcal strains. **(A)** Phage 06701 released from *L. lactis* strain DS69067; **(B)** Phage 24801 released from *L. lactis* strain DS70248; **(C)** Phage 49801 released from *L. lactis* strain DS68498.

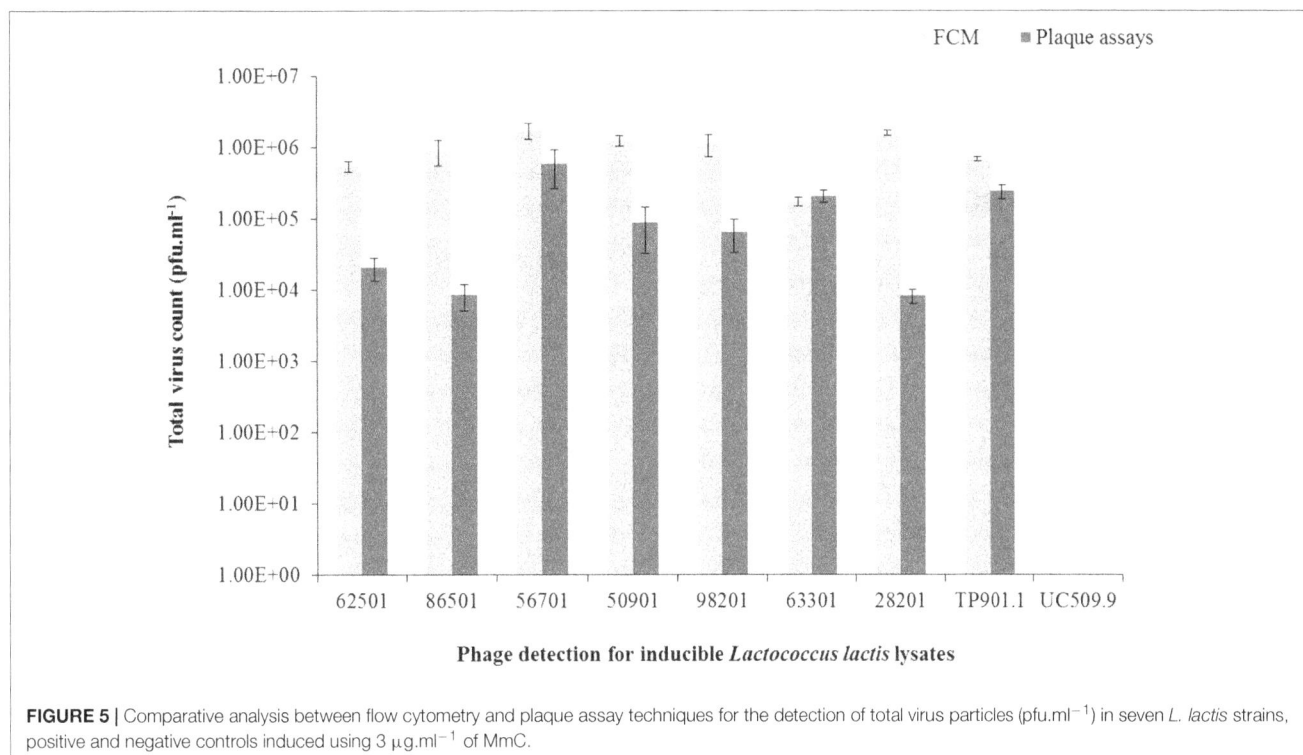

FIGURE 5 | Comparative analysis between flow cytometry and plaque assay techniques for the detection of total virus particles (pfu.ml^{-1}) in seven *L. lactis* strains, positive and negative controls induced using 3 μg.ml^{-1} of MmC.

2016). As reported here, MmC-mediated growth profile analysis is of limited value due to many false positive/negative results thereby highlighting the need for an accurate phage particle detection and enumeration method (Martín et al., 2006). We show here that FCM can be employed as a reliable method for the identification of lysogenic bacteria and detection of released phage particles. Marie and co-workers first described an approach in 1999 (Marie et al., 1999) to detect marine viruses by FCM based on viral infection of *Phaeocystis pouchetii*. Through analysis of uninfected host bacteria these authors showed that it was possible to differentiate between background signal and viral particles (Marie et al., 1999). In our study, using FCM it was possible to reliably detect and enumerate phage particles, although it is not possible to determine if such particles represent one or multiple distinct phages. Validation of FCM as a reliable method for the detection and quantitation of viral particles was achieved using (i) plaque assays (in cases where a suitable host was available to perform plaque assays), and (ii) TEM (where no suitable host was available to perform plaque assays). Despite the fact that the double-agar plaque assay is probably the most efficient method for intact (i.e., infectious) phage detection and quantification, this approach cannot be employed in situations where a suitable host cannot be identified for a released (pro)phage. Lambeth et al. (2005) demonstrated the application of FCM-based assays for dengue virus particle quantifications as a useful technique to titrate clinical isolates of dengue that frequently do not form clear plaques (Lambeth et al., 2005). Several techniques have been applied to detect (induced) temperate phages, such as real time PCR (Lunde et al., 2003), bacterial genome sequencing analysis (Chopin et al., 2001),

TEM and EFM analysis (Marie et al., 1999; Chen et al., 2001). FCM has previously been applied for real-time detection of *L. lactis* infected with c2-type phages in order to detect the early stages of infection, aimed at improving the management of dairy fermentation processes (Michelsen et al., 2007). In contrast to our approach for phage particle detection, the latter study was intended to differentiate between phage-infected and uninfected lactococcal populations.

Interestingly, our results obtained from FCM and TEM analyses are fully consistent with each other, where it was possible to identify intact phage particles with morphological features typical of the P335 phages (Mc Grath et al., 2006; Veesler et al., 2012; Mahony et al., 2017). Despite the fact that TEM analysis is a very useful means by which to visualize virus particles, this technique comes with certain limitations such as lengthy sample preparation procedures, expensive equipment and specialized expertise (Goldsmith and Miller, 2009; Vale et al., 2010; Zhang et al., 2013; Brown et al., 2015). Thus, only a relatively small number of strains were subjected to TEM analysis. FCM has shown to be a relatively fast and accurate tool for the identification of inducible phage particles when a sensitive host cannot be applied, reducing the need for TEM access. FCM is a rapid technique to perform an analysis of multiple parameters of individual cells and with important applications in food microbiology (Comas-Riu and Rius, 2009; Paparella et al., 2012). Prevention of phage infection has received a lot of attention in recent years, and several strategies have been employed such as the application of rotating cultures and phage-resistant starter strains (Michelsen et al., 2007). The FCM technique described here will assist in reliable identification of potential problematic starter cultures, and will thus ultimately reduce the risk of phage particles being released in the dairy environment.

AUTHOR CONTRIBUTIONS

JO wrote the manuscript. JO, JM, and DvS were involved in the idea conception and manuscript editing. JO, LH, HN, and JMS were involved in the experimental design and/or data analysis. JM, TK, and DvS were involved in manuscript editing.

FUNDING

JO is supported by DSM Food Specialties, the Netherlands. JM is the recipient of a Starting Investigator Research Grant (SIRG) (Ref. No. 15/SIRG/3430) funded by Science Foundation Ireland (SFI). DvS is supported by a Principal Investigator Award (ref. no. 13/IA/1953) through SFI.

ACKNOWLEDGMENT

This work was supported by the Science Foundation Ireland (SFI).

REFERENCES

Anderson, B., Rashid, M. H., Carter, C., Pasternack, G., Rajanna, C., Revazishvili, T., et al. (2011). Enumeration of bacteriophage particles: comparative analysis of the traditional plaque assay and real-time QPCR- and nanosight-based assays. Bacteriophage 1, 86–93. doi: 10.4161/bact.1.2.15456

Bosch, A., Pinto, R. M., Comas, J., and Abad, F. X. (2004). Detection of infectious rotaviruses by flow cytometry. Methods Mol. Biol. 268, 61–68. doi: 10.1385/1-59259-766-1:061

Braun, V., Hertwig, S., Neve, H., Geis, A., and Teuber, M. (1989). Taxonomic differentiation of bacteriophages of Lactococcus lactis by electron-microscopy, DNA-DNA hybridization, and protein profiles. J. Gen. Microbiol. 135, 2551–2560.

Brown, M. R., Camezuli, S., Davenport, R. J., Petelenz-Kurdziel, E., Ovreas, L., and Curtis, T. P. (2015). Flow cytometric quantification of viruses in activated sludge. Water Res. 68, 414–422. doi: 10.1016/j.watres.2014.10.018

Chen, F., Lu, J. R., Binder, B. J., Liu, Y. C., and Hodson, R. E. (2001). Application of digital image analysis and flow cytometry to enumerate marine viruses stained with SYBR gold. Appl. Environ. Microbiol. 67, 539–545. doi: 10.1128/AEM.67.2.539-545.2001

Chopin, A., Bolotin, A., Sorokin, A., Ehrlich, S. D., and Chopin, M. (2001). Analysis of six prophages in Lactococcus lactis IL1403: different genetic structure of temperate and virulent phage populations. Nucleic Acids Res. 29, 644–651. doi: 10.1093/nar/29.3.644

Comas-Riu, J., and Rius, N. (2009). Flow cytometry applications in the food industry. J. Ind. Microbiol. Biotechnol. 36, 999–1011. doi: 10.1007/s10295-009-0608-x

Costello, V. A. (1988). Characterization of Bacteriophage-Host Interactions in Streptococcus cremoris UC503 and Related Lactic Streptococci. Ph.D. thesis, National University of Ireland, Cork.

Deasy, T., Mahony, J., Neve, H., Heller, K. J., and van Sinderen, D. (2011). Isolation of a virulent Lactobacillus brevis phage and its application in the control of beer spoilage. J. Food Prot. 74, 2157–2161. doi: 10.4315/0362-028X.JFP-11-262

Deveau, H., Labrie, S. J., Chopin, M. C., and Moineau, S. (2006). Biodiversity and classification of lactococcal phages. Appl. Environ. Microbiol. 72, 4338–4346. doi: 10.1128/AEM.02517-05

Emond, E., Holler, B. J., Boucher, I., Vandenbergh, P. A., Vedamuthu, E. R., Kondo, J. K., et al. (1997). Phenotypic and genetic characterization of the bacteriophage abortive infection mechanism AbiK from Lactococcus lactis. Appl. Environ. Microbiol. 63, 1274–1283.

Fernandez, E., Alegria, A., Delgado, S., Martin, M. C., and Mayo, B. (2011). Comparative phenotypic and molecular genetic profiling of wild Lactococcus lactis subsp. lactis strains of the L. lactis subsp. lactis and L. lactis subsp. cremoris genotypes, isolated from starter-free cheeses made of raw milk. Appl. Environ. Microbiol. 77, 5324–5335. doi: 10.1128/AEM.02991-10

Goldsmith, C. S., and Miller, S. E. (2009). Modern uses of electron microscopy for detection of viruses. Clin. Microbiol. Rev. 22, 552–563. doi: 10.1128/Cmr.00027-09

Ho, C. H., Stanton-Cook, M., Beatson, S. A., Bansal, N., and Turner, M. S. (2016). Stability of active prophages in industrial Lactococcus lactis strains in the presence of heat, acid, osmotic, oxidative and antibiotic stressors. Int. J. Food Microbiol. 220, 26–32. doi: 10.1016/j.ijfoodmicro.2015.12.012

Koch, B., Christiansen, B., Evison, T., Vogensen, F. K., and Hammer, K. (1997). Construction of specific erythromycin resistance mutations in the temperate lactococcal bacteriophage TP901-1 and their use in studies of phage biology. Appl. Environ. Microbiol. 63, 2439–2441.

Labrie, S., and Moineau, S. (2000). Multiplex PCR for detection and identification of lactococcal bacteriophages. Appl. Environ. Microbiol. 66, 987–994. doi: 10.1128/Aem.66.3.987-994.2000

Labrie, S. J., Josephsen, J., Neve, H., Vogensen, F. K., and Moineau, S. (2008). Morphology, genome sequence, and structural proteome of type phage P335 from Lactococcus lactis. Appl. Environ. Microbiol. 74, 4636–4644. doi: 10.1128/AEM.00118-08

Labrie, S. J., Samson, J. E., and Moineau, S. (2010). Bacteriophage resistance mechanisms. Nat. Rev. Microbiol. 8, 317–327. doi: 10.1038/nrmicro2315

Lambeth, C. R., White, L. J., Johnston, R. E., and de Silva, A. M. (2005). Flow cytometry-based assay for titrating dengue virus. J. Clin. Microbiol. 43, 3267–3272. doi: 10.1128/JCM.43.7.3267-3272.2005

Lillehaug, D. (1997). An improved plaque assay for poor plaque-producing temperate lactococcal bacteriophages. J. Appl. Microbiol. 83, 85–90. doi: 10.1046/j.1365-2672.1997.00193.x

Lunde, M., Blatny, J. M., Lillehaug, D., Aastveit, A. H., and Nes, I. F. (2003). Use of real-time quantitative PCR for the analysis of phi LC3 prophage stability in lactococci. Appl. Environ. Microbiol. 69, 41–48. doi: 10.1128/Aem.69.1.41-48.2003

Mahony, J., Kot, W., Murphy, J., Ainsworth, S., Neve, H., Hansen, L. H., et al. (2013). Investigation of the relationship between lactococcal host cell wall polysaccharide genotype and 936 phage receptor binding protein phylogeny. Appl. Environ. Microbiol. 79, 4385–4392. doi: 10.1128/AEM.00653-13

Mahony, J., McGrath, S., Fitzgerald, G. F., and van Sinderen, D. (2008). Identification and characterization of lactococcal-prophage-carried superinfection exclusion genes. Appl. Environ. Microbiol. 74, 6206–6215. doi: 10.1128/AEM.01053-08

Mahony, J., Oliveira, J., Collins, B., Hanemaaijer, L., Lugli, G. A., Neve, H., et al. (2017). Genetic and functional characterisation of the lactococcal P335 phage-host interactions. BMC Genomics 18:146. doi: 10.1186/s12864-017-3537-5

Mahony, J., and van Sinderen, D. (2014). Current taxonomy of phages infecting lactic acid bacteria. Front. Microbiol. 5:7. doi: 10.3389/fmicb.2014.00007

Marco, M. B., Moineau, S., and Quiberoni, A. (2012). Bacteriophages and dairy fermentations. Bacteriophage 2, 149–158. doi: 10.4161/bact.21868

Marie, D., Brussaard, C. P. D., Thyrhaug, R., Bratbak, G., and Vaulot, D. (1999). Enumeration of marine viruses in culture and natural samples by flow cytometry. *Appl. Environ. Microbiol.* 65, 45–52.

Martín, M. C., Ladero, V., and Alvarez, M. A. (2006). PCR identification of lysogenic *Lactococcus lactis* strains. *J. Verbrauch. Lebensm.* 1, 121–124. doi: 10.1007/s00003-006-0020-7

Mc Grath, S., Neve, H., Seegers, J. F., Eijlander, R., Vegge, C. S., Brondsted, L., et al. (2006). Anatomy of a lactococcal phage tail. *J. Bacteriol.* 188, 3972–3982. doi: 10.1128/JB.00024-06

McGrath, S., Fitzgerald, G. F., and van Sinderen, D. (2002). Identification and characterization of phage-resistance genes in temperate lactococcal bacteriophages. *Mol. Microbiol.* 43, 509–520.

Michelsen, O., Cuesta-Dominguez, A., Albrechtsen, B., and Jensen, P. R. (2007). Detection of bacteriophage-infected cells of *Lactococcus lactis* by using flow cytometry. *Appl. Environ. Microbiol.* 73, 7575–7581. doi: 10.1128/AEM. 01219-07

Moineau, S., Pandian, S., and Klaenhammer, T. R. (1994). Evolution of a lytic bacteriophage via DNA acquisition from the *Lactococcus lactis* chromosome. *Appl. Environ. Microbiol.* 60, 1832–1841.

Nanda, A. M., Thormann, K., and Frunzke, J. (2015). Impact of spontaneous prophage induction on the fitness of bacterial populations and host-microbe interactions. *J. Bacteriol.* 197, 410–419. doi: 10.1128/JB.02230-14

Oliveira, J., Mahony, J., Lugli, G. A., Hanemaaijer, L., Kouwen, T., Ventura, M., et al. (2016). Genome sequences of eight prophages isolated from *Lactococcus lactis* dairy strains. *Genome Announc.* 4:e00906-16. doi: 10.1128/genomeA. 00906-16

O'Sullivan, D., Ross, R. P., Fitzgerald, G. F., and Coffey, A. (2000). Investigation of the relationship between lysogeny and lysis of *Lactococcus lactis* in cheese using prophage-targeted PCR. *Appl. Environ. Microbiol.* 66, 2192–2198. doi: 10.1128/Aem.66.5.2192-2198.2000

Paparella, A., Serio, A., and López, C. C. (2012). *Flow Cytometry Applications in Food Safety Studies.* Rijeka: INTECH Open Access Publisher.

Rousseau, G. M., and Moineau, S. (2009). Evolution of *Lactococcus lactis* phages within a cheese factory. *Appl. Environ. Microbiol.* 75, 5336–5344. doi: 10.1128/ AEM.00761-09

Stockdale, S. R., Mahony, J., Courtin, P., Chapot-Chartier, M. P., van Pijkeren, J. P., Britton, R. A., et al. (2013). The lactococcal phages Tuc2009 and TP901-1 incorporate two alternate forms of their tail fiber into their virions for infection specialization. *J. Biol. Chem.* 288, 5581–5590. doi: 10.1074/jbc.M112.444901

Vale, F., Correia, A., Matos, B., Moura Nunes, J., and Alves de Matos, A. (2010). "Applications of transmission electron microscopy to virus detection and identification," in *Microscopy: Science, Technology, Applications and Education*, eds A. Mendez-Vilas and J. Diaz (Badajoz: Formatex Research Center), 128–136.

Veesler, D., Spinelli, S., Mahony, J., Lichiere, J., Blangy, S., Bricogne, G., et al. (2012). Structure of the phage TP901-1 1.8 MDa baseplate suggests an alternative host adhesion mechanism. *Proc. Natl. Acad. Sci. U.S.A.* 109, 8954–8958. doi: 10.1073/pnas.1200966109

Ventura, M., Zomer, A., Canchaya, C., O'Connell-Motherway, M., Kuipers, O., Turroni, F., et al. (2007). Comparative analyses of prophage-like elements present in two *Lactococcus lactis* strains. *Appl. Environ. Microbiol.* 73, 7771–7780. doi: 10.1128/AEM.01273-07

Zhang, Y., Hung, T., Song, J. D., and He, J. S. (2013). Electron microscopy: essentials for viral structure, morphogenesis and rapid diagnosis. *Sci. China Life Sci.* 56, 421–430. doi: 10.1007/s11427-013-4476-2

Conflict of Interest Statement: JO is funded by DSM Food Specialties and LH and TK are employees of DSM Food Specialties.

The other authors declare that the research was conducted in the absence of any commercial or financial relationships that could be construed as a potential conflict of interest.

Interactions between Food Additive Silica Nanoparticles and Food Matrices

Mi-Ran Go, Song-Hwa Bae, Hyeon-Jin Kim, Jin Yu and Soo-Jin Choi *

Department of Applied Food System, Major of Food Science and Technology, Seoul Women's University, Seoul, South Korea

Nanoparticles (NPs) have been widely utilized in the food industry as additives with their beneficial characteristics, such as improving sensory property and processing suitability, enhancing functional and nutritional values, and extending shelf-life of foods. Silica is used as an anti-caking agent to improve flow property of powered ingredients and as a carrier for flavors or active compounds in food. Along with the rapid development of nanotechnology, the sizes of silica fall into nanoscale, thereby raising concerns about the potential toxicity of nano-sized silica materials. There have been a number of studies carried out to investigate possible adverse effects of NPs on the gastrointestinal tract. The interactions between NPs and surrounding food matrices should be also taken into account since the interactions can affect their bioavailability, efficacy, and toxicity. In the present study, we investigated the interactions between food additive silica NPs and food matrices, such as saccharides, proteins, lipids, and minerals. Quantitative analysis was performed to determine food component-NP corona using HPLC, fluorescence quenching, GC-MS, and ICP-AES. The results demonstrate that zeta potential and hydrodynamic radius of silica NPs changed in the presence of all food matrices, but their solubility was not affected. However, quantitative analysis on the interactions revealed that a small portion of food matrices interacted with silica NPs and the interactions were highly dependent on the type of food component. Moreover, minor nutrients could also affect the interactions, as evidenced by higher NP interaction with honey rather than with a simple sugar mixture containing an equivalent amount of fructose, glucose, sucrose, and maltose. These findings provide fundamental information to extend our understanding about the interactions between silica NPs and food components and to predict the interaction effect on the safety aspects of food-grade NPs.

Keywords: silica, interaction, quantitative analysis, sugar, protein, lipid, mineral

Edited by:
Jayanta Kumar Patra,
Dongguk University Seoul,
South Korea

Reviewed by:
Young-Rok Kim,
Kyung Hee University, South Korea
Qazi Mohd Rizwanul Haq,
Jamia Millia Islamia, India

*Correspondence:
Soo-Jin Choi
sjchoi@swu.ac.kr

INTRODUCTION

Nanotechnology and engineered nanoparticles (NPs) have attracted much attention for a wide range of applications in the field of chemical engineering, materials science, cosmetics, pharmaceutics, and medicine. NPs have been also widely applied to food products as additives, nutrient supplements, antimicrobial agents in food packaging, and delivery systems. According to Nanotechnology Consumer Products Inventory released in October 2013, the number of NPs-based commercial products in food and beverage have increased recently, of which gold, titanium dioxide, zinc oxide, and silica (SiO_2) are among the most widely applied NPs (Vance et al., 2015). In particular, SiO_2 NP is registered as a food additive E551 in the European Union (EU) and used as an anti-caking or anti-clumping agent (Dekkers et al., 2011; Wang et al., 2013). Humans are exposed to NPs through oral intake of NPs-containing food or beverage, which counts for 16% of total exposure (Vance et al., 2015). Oral ingestion being the third route of NP exposure, next to skin (58%) and inhalation (25%) (Vance et al., 2015; Cao et al., 2016), safety aspects of NPs to food sector duly need a critical consideration.

Toxicological effects of food-grade SiO_2 NPs on biological systems have not been extensively explored compared to other commercial types. Moreover, most *in vitro* and *in vivo* studies have focused on biological responses upon exposure to SiO_2 NPs, without considering potential interactions between NPs and biological or food components (Borak et al., 2012; Lee et al., 2014; Kim et al., 2016). As NPs in food are present in mixtures with food matrices containing carbohydrates, proteins, lipids, minerals, and other trace elements, interactions between NPs and food components can be critical factors affecting potential toxicity, oral absorption, biodistribution, and efficacy of NPs. It was reported that NPs affect nutrient absorption (Mahler et al., 2012; Dorier et al., 2015), but the interactions between food-grade NPs and food components have not been well explored. On the other hand, food components could also influence the absorption and toxicity of NPs as well (Wang et al., 2014; Bohmert et al., 2015; Docter et al., 2015; Lichtenstein et al., 2015; Jiang et al., 2016). Indeed, our previous research suggested that oral absorption of food-grade SiO_2 NPs is highly affected by food components, showing 2.4- and 2.5-fold enhanced absorption efficiencies in the presence of albumin and glucose, respectively (Lee et al., 2017).

The aim of the present study was to determine the interactions between SiO_2 NPs and food matrices. NP interactions with honey, skim milk, olive oil, or phosphate buffered saline (PBS), which represent saccharide, protein, lipid, and mineral matrices, respectively, were analyzed quantitatively. In addition, the role of trace nutrients on the NP interaction with saccharides and casein (major components in honey and skim milk, respectively) was also investigated.

MATERIALS AND METHODS
Materials and Characterization
Food-grade amorphous SiO_2 NPs were purchased from Evonik Industries AG (Essen, Germany). Materials used were as follow: acacia honey (Dongsuh Food Co., Ltd., Seoul, Republic of Korea), D-(+)-glucose (Sigma-Aldrich, St. Louis, MO, USA), D-(−)-fructose (Sigma-Aldrich), sucrose (Sigma-Aldrich), D-(+)-maltose monohydrate (Sigma-Aldrich), skim milk powder (Seoul Milk, Seoul, Republic of Korea), casein sodium salt from bovine milk (Sigma-Aldrich), extra virgin olive oil (imported from Spain, Beksul, CJ CheilJedang, Seoul, Republic of Korea), 37-component fatty acid methyl esters (FAMEs; Sigma-Aldrich), hexane (Sigma-Aldrich), sodium hydroxide (Sigma-Aldrich), biphenyl (Sigma-Aldrich), methanol (Samchun Chemical Co., Ltd., Gyeonggi-do, Republic of Korea), acetonitrile (HPLC grade, Samchun Chemical Co., Ltd.), water (HPLC grade, Samchun Chemical Co., Ltd.), and PBS buffer (NaCl 137 mM, KCl 2.7 mM, Na_2HPO_4 10 mM, KH_2PO_4 1.8 mM; Dongin Biotech. Co., Republic of Korea).

Particle size and morphology were determined by scanning electron microscopy (SEM; FEIQUANTA 250 FEG, Hillsboro, OR, USA). Zeta potentials and hydrodynamic radii of NPs in aqueous suspension and food matrices were measured with Zetasizer Nano System (Malvern Instruments, Worcestershire, UK). Stock solution of SiO_2 NPs [50 mg/ml in distilled water (DW)] was prepared, stirred for 30 min, and diluted to designated concentrations just prior to experiments.

Solubility
Particles (5 mg/ml) were dispersed in 10% honey, 1 mg/ml of skim milk solution, olive oil, and PBS buffer, respectively. After different incubation times at 25°C, supernatants were collected by ultracentrifugation (16,000 × g) for 15 min, and analysis of Si in the supernatants was performed as described previously (Paek et al., 2014).

Interaction between NPs and Saccharides
Different concentrations (1, 2, 5, and 10%) of acacia honey were prepared in DW and incubated with 5 mg/ml NPs with shaking at different temperatures (4, 25, and 40°C). 10% concentration of honey, containing ~42.4% fructose, ~29.6% glucose, ~0.2% sucrose, and ~0.1% maltose, was used as the highest concentration due to high viscosity over this level. Fructose, glucose, sucrose, and maltose at each concentration of 1, 2, and 5% were mixed and used as sugar mixtures to mimic honey matrix containing only equivalent amounts of each sugar, without trace nutrients. After designated incubation times (1, 24, 48 h, and 7 d), the samples were centrifuged at 23,000 × g for 1 h. The supernatants were analyzed after washing with distilled and deionized water (DDW) or without washing, and then, filtered through syringe filter (Agela Technologies, Wilmington, DE, USA). Saccharide concentrations were quantified by high performance liquid chromatography (HPLC) using a Shimadzu HPLC system (Kyoto, Japan), equipped with RID-10A refractive index detector, on a Hypersil APS-2 column (250 × 4.6 mm i.d., 5 μm, 120 Å, Thermo Fisher Scientific, MA, USA). The mobile phase was acetonitrile:water (80:20, v/v) and flow rate was set at 1 ml/min. Column temperature was maintained at a constant 40°C and injection volume of sample was 20 μl. Each experiment was repeated three times on separate days.

Interaction between NPs and Proteins
Different concentrations of SiO_2 NPs (1, 2.5, and 5 mg/ml) were suspended in 1 mg/ml of skim milk solution (in DW) or 0.35 mg/ml of casein solution (in DW), and incubated with shaking at different temperatures (4, 25, and 40°C). After designated incubation times (1, 24, 48 h, and 7 d), the suspensions were subjected to protein fluorescence quenching analysis using a luminescence spectrometer (SpectraMax® M3, Molecular Devices, CA, USA). Excitation wavelength was set at 280 nm and fluorescence emission intensity was measured at wavelength

from 300 to 420 nm. Quenching ratio was calculated as (I0-I)/I0, where I0 and I stand for basal fluorescence emission intensity in controls (NPs-untreated proteins) and experimental groups (NPs-treated proteins), respectively. Each experiment was repeated three times on separate days.

Interaction between NPs and Lipids

SiO_2 NPs (50 mg/ml) were dispersed in olive oil for 30 min and diluted to 20 mg/ml prior to experiments. NPs suspended in olive oil were incubated with shaking for 1, 24, and 48 h at different temperatures (4, 25, and 40°C). After centrifugation at 23,000 × g for 1 h, fatty acid methyl esters were prepared by alkaline transmethylation (Liang et al., 2011). The supernatants (1 μl) were spiked with internal standard solution (final concentration of 100 μg/ml), and 1 ml of 0.4 M $NaOH-CH_3OH$ was added and reacted for 5 min with ultrasonification. The methyl esters were extracted with 5 ml of hexane, diluted to 3-folds, and analyzed by gas chromatography-mass spectrometry (GC-MS). A standard mixture (FAMEs, 1 ml of 10 mg/ml) was dissolved in 9 ml of hexane and the standard solution was diluted to 10, 20, 50, 100, 200, and 500 μg/ml concentrations, and then, spiked with internal standard biphenyl (100 μg/ml). The adsorbed fatty acids on NPs were calculated after subtraction of reduced fatty acids in the supernatants from those in olive oil control. Each experiment was repeated three times on separate days. All GC-MS analyses were performed with an Agilent 5977E GC-MSD system (Agilent Technologies, CA, USA), including 7820A GC instrument coupled with a 5977E MS detector. A StabilWax® capillary column (30 m × 0.25 mm, 0.25 μm thickness, Restek, PA, USA) was used, and column temperature was hold for 1 min at 50°C, programmed from 50 to 200°C at the rate of 25°C/min, from 200 to 230°C at the rate of 3°C/min, and then hold for 23 min at 230°C. The injection temperature was kept at 250°C and the carrier gas was helium. The column flow was 1 ml/min and a sample of 3 μl was injected with a split ratio of 5:1. The ion source temperature was 230°C and the samples were ionized by electron impact ionization at 70 eV. Selected-ion monitoring was performed at m/z 55, 67, 74, 79, and 87.

Interaction between NPS and Minerals

Two different concentrations (5 and 10 mg/ml) of SiO_2 NPs were dispersed in 1 ml PBS and reacted with shaking for various times (1, 24, 48 h, and 7 d) at different temperatures (4, 25, and 40°C). To remove unbound minerals from NPs, the samples were centrifuged at 23,000 × g for 1 h and washed with DDW. This procedure was repeated three times. The aliquots were digested with 10 ml of ultrapure nitric acid at ~180°C and diluted with 3 ml of DDW. After filtering through syringe filter (0.22 μm, Agela Technolgies), total Na concentrations were determined by inductively coupled plasma-atomic emission spectroscopy (ICP-AES, JY2000 Ultrace, HORIBA Jobin Yvon, longjumeau, France). Each experiment was repeated three times on separate days.

Statistical Analysis

Results were expressed as means ± standard deviations. Experimental values were compared with corresponding untreated control values. Statistical analysis was performed using the Student's test for unpaired data and one-way analysis of variance (ANOVA) with Tukey's Test in SAS Version 9.4 (SAS Institute Inc., Cary, NC, USA) was carried out to determine the significances of intergroup differences. Statistical significance was accepted for p-values of <0.05.

RESULTS

Characterization

Particle morphology, primary particle size, and size distribution were examined by SEM. **Figure 1** shows that food-grade SiO_2 NPs dispersed in DW had irregular particle morphology with an average particle size of 24.1 ± 3.5 nm. Zeta potential and hydrodynamic size of SiO_2 NPs were determined to be −28.2 ± 1.0 mV and 287.6 ± 1.7 nm, respectively. Solubility of SiO_2 NPs in DW was not detected and did not increase in the presence of food matrices.

Interaction between NPS and Saccharides

Zeta potential values of SiO_2 NPs in honey changed to less negative charges as honey concentration and incubation time increased (**Table 1**), regardless of temperature (**Table 2**). Hydrodynamic radii of SiO_2 NPs in honey also significantly increased as honey concentration increased (**Table 1**), without effect of incubation time and temperature (**Tables 1, 2**). However, zeta potentials and hydrodynamic radii of NPs could not be detected at concentrations of more than 2% honey after incubation for 7 d due to the formation of high aggregation (Supplementary Figure 1).

Since honey contains different amounts of saccharides and other trace nutrients, the interactions between SiO_2 NPs and saccharides were further evaluated using sugar mixtures composed of equivalent amounts of fructose, glucose, sucrose, and maltose, in order to investigate the effects of only saccharides on interactions. It is worth noting that these are major four saccharide components in acacia honey. The highest concentration of each saccharide was set at 5%, based on maximum solubility of both fructose and maltose in DW. Zeta potentials of SiO_2 NPs in sugar mixtures did not significantly change, except at 5%, and were not affected by incubation time and temperature (**Tables 2, 3**). Interestingly, hydrodynamic sizes of NPs decreased in the presence of sugar mixtures in a time-dependent manner (**Table 3**). No aggregation tendency was observed in sugar mixtures (Supplementary Figure 1).

When the interactions between SiO_2 NPs and saccharides in honey were quantified by HPLC, **Figure 2A** shows that fructose and glucose, the most abundant two saccharides in honey, interacted with SiO_2 NPs in a concentration-dependent manner at 25°C, while no interactions between SiO_2 NPs and sucrose or maltose were found. NP interaction with fructose was slightly higher than that with glucose at more than 5% honey ($p < 0.05$). However, the interactions between SiO_2 NPs and glucose or fructose in honey were not affected by incubation time (**Figure 2B**) for 48 h. Meanwhile, the interactions between SiO_2 NPs and glucose or fructose increased as temperature increased, especially, low interaction levels were detected at 4°C, compared to 25 and 40°C (**Figure 2C**).

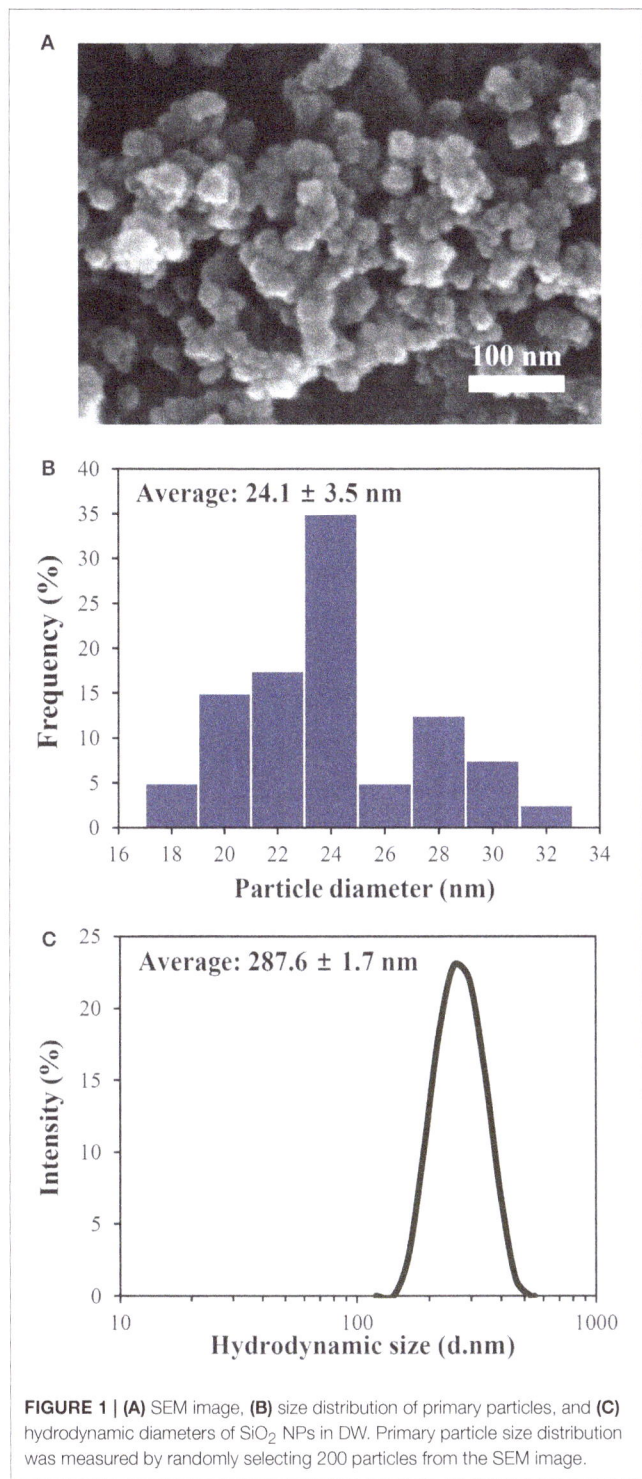

FIGURE 1 | (A) SEM image, (B) size distribution of primary particles, and (C) hydrodynamic diameters of SiO$_2$ NPs in DW. Primary particle size distribution was measured by randomly selecting 200 particles from the SEM image.

TABLE 1 | Zeta potentials and hydrodynamic diameters of SiO$_2$ NPs in honey at 25°C.

| | | Zeta potential (mV) | | | | Hydrodynamic size (nm) | | | |
| | | Concentration (%) | | | | Concentration (%) | | | |
DW	Time	1%	2%	5%	10%	1%	2%	5%	10%
−28.2 ± 1.0[A,a]	1 min	−27.8 ± 0.8[A,a]	−28.4 ± 0.4[A,a]	−25.0 ± 0.9[B,b]	−20.6 ± 1.0[BC,c]	311.5 ± 39.2[A,a]	346.1 ± 52.7[B,a]	537.4 ± 65.2[B,b]	871.0 ± 36.1[B,c]
	1 h	−27.0 ± 0.9[AB,a]	−24.4 ± 0.5[B,b]	−25.0 ± 0.5[B,b]	−21.6 ± 0.4[B,c]	310.9 ± 4.5[A,a]	355.2 ± 7.3[B,a]	533.7 ± 75.4[B,b]	882.7 ± 18.4[B,c]
	24 h	−26.7 ± 0.4[AB,ab]	−25.3 ± 0.8[B,b]	−22.7 ± 0.9[C,c]	−19.1 ± 0.8[C,d]	302.2 ± 14.0[A,a]	367.1 ± 0.7[B,a]	533.2 ± 29.2[B,b]	840.5 ± 111.3[B,c]
	48 h	−25.6 ± 1.0[B,b]	−24.9 ± 1.0[B,b]	−22.1 ± 0.4[C,c]	−18.7 ± 0.6[C,d]	298.9 ± 9.7[A,a]	382.9 ± 15.7[B,a]	565.4 ± 43.5[B,b]	799.5 ± 137.2[B,c]
287.6 ± 1.7[A,a]	7 d	−19.1 ± 0.8[C]	ND	ND	ND	304.5 ± 17.5[A]	ND	ND	ND

Different letters (A–C) in the same column indicate significant differences between incubation times (p < 0.05). Different letters (a–d) in the same row indicate significant differences between honey concentrations (p < 0.05). Zeta potential and hydrodynamic radius of NPs in DW were measured in the absence of honey as a control. Zeta potentials and hydrodynamic diameters could not be detected at 2–5% after incubation for 7 d due to aggregation (See Supplementary Figure 1). ND, not determined.

Similar tendency was found in sugar mixtures containing equivalent amounts of fructose, glucose, sucrose, and maltose. Concentration-, but not time-dependent interactions between SiO$_2$ NPs and saccharides were found (Figures 3A,B), and these interactions significantly increased when incubation temperature increased at 40°C (Figure 3C). In particular, all different types of

TABLE 2 | Zeta potentials and hydrodynamic diameters of SiO_2 NPs in honey, sugar mixtures, skim milk, casein, and PBS buffer after incubation for 24 h at different temperatures.

Matrix	Zeta potential (mV)			Hydrodynamic size (nm)		
	Temperature (°C)			Temperature (°C)		
	4°C	25°C	40°C	4°C	25°C	40°C
Honey (10%)	−18.7 ± 2.1	−19.1 ± 0.8	−17.2 ± 0.7	823.9 ± 13.8	840.5 ± 111.3	846.1 ± 126.2
Sugar mixture (5%)	−27.8 ± 1.6	−26.5 ± 0.4	−27.0 ± 0.6	272.7 ± 10.5	271.8 ± 0.4	273.0 ± 18.8
Skim milk (1 mg/ml)	−25.5 ± 1.8	−25.6 ± 1.3	−25.3 ± 1.3	259.6 ± 6.8	261.7 ± 1.6	268.0 ± 0.5
Casein (0.35 mg/ml)	−34.8 ± 0.6	−33.7 ± 0.7	−35.1 ± 1.8	299.9 ± 7.0	290.3 ± 8.2	290.4 ± 8.8
PBS	−36.4 ± 0.4	−36.7 ± 1.0	−35.5 ± 0.9	251.0 ± 4.2	249.6 ± 4.3	251.7 ± 1.7

No significant differences between different temperatures were found ($p > 0.05$). Sugar mixtures consist of equivalent amounts of fructose, glucose, sucrose, and maltose. Concentration of casein solution was adjusted to have an equivalent amount of protein content in 1 mg/ml of skim milk solution.

TABLE 3 | Zeta potentials and hydrodynamic diameters of SiO_2 NPs in sugar mixtures at 25°C.

		Zeta potential (mV)					Hydrodynamic size (nm)		
DW	Time	Concentration (%)			DW	Time	Concentration (%)		
		1%	2%	5%			1%	2%	5%
−28.2 ± 1.0[A,a]	1 min	−28.6 ± 0.1[A,a]	−27.8 ± 0.7[A,ab]	−26.2 ± 0.2[A,b]	287.6 ± 1.7[A,a]	1 min	270.4 ± 3.7[BC,b]	285.2 ± 8.7[A,a]	288.7 ± 6.4[A,a]
	1 h	−27.8 ± 1.2[A,a]	−27.6 ± 0.6[A,a]	−26.7 ± 0.6[A,a]		1 h	275.6 ± 3.9[B,b]	285.6 ± 5.2[A,a]	289.7 ± 2.1[A,a]
	24 h	−27.1 ± 0.5[A,a]	−27.3 ± 0.5[A,ab]	−26.5 ± 0.4[A,b]		24 h	264.0 ± 4.3[C,b]	284.6 ± 7.4[AB,a]	271.8 ± 0.4[B,b]
	48 h	−28.6 ± 1.0[A,a]	−27.9 ± 1.1[A,a]	−26.6 ± 2.0[A,a]		48 h	263.9 ± 3.3[C,b]	270.4 ± 1.8[AB,b]	270.7 ± 5.9[B,b]
	7 d	−28.1 ± 0.8[A,a]	−27.9 ± 0.7[A,a]	−26.2 ± 0.4[A,b]		7 d	262.3 ± 2.0[C,b]	252.0 ± 18.6[B,b]	257.8 ± 8.6[B,b]

Different letters (A–C) in the same column indicate significant differences between incubation times ($p < 0.05$). Different letters (a,b) in the same row indicate significant differences between sugar mixture concentrations ($p < 0.05$). Sugar mixtures consist of equivalent amounts of fructose, glucose, sucrose, and maltose. Zeta potential and hydrodynamic radius of NPs in DW were measured in the absence of sugar mixtures as a control.

saccharide interacted with NPs in a similar manner. Significant differences in the interactions between saccharide types were not remarkably found ($p > 0.05$). Overall, <4.6 and 3.1% of saccharides in honey and sugar mixtures, respectively, were determined to interact with SiO_2 NPs. It should be noted that quantitative analysis was performed after ultracentrifugation of reacted NPs with saccharides, collecting precipitated NPs, and detaching adsorbed saccharides on NPs with HPLC eluent. Whereas, no adsorbed saccharide was detected after washing the centrifuged and precipitated NPs with DDW.

Interaction between NPs and Proteins

Skim milk and its main protein component, casein were used to determine the interactions between SiO_2 NPs and proteins. Skim milk solution at 1 mg/ml (in DW) was applied for protein fluorescence measurement, because this concentration exhibited the highest fluorescence intensity without precipitation (data not shown). Based on casein content in skim milk, 0.35 mg/ml of casein solution was used for comparative study. Zeta potential values of SiO_2 NPs significantly changed to less negative direction in skim milk solution, whereas more negative zeta potentials were found in casein solution (**Table 4**). Effect of temperature on the interactions was not found (**Table 2**; $p > 0.05$). Aggregation was observed after incubation for 7 d at 25 and 40°C in both skim milk and casein solutions treated with NPs (Supplementary Figure 1). Meanwhile, hydrodynamic radii of NPs were not

remarkably influenced by incubation time, temperature, and the presence of skim milk or casein (**Tables 2, 4**).

Interactions between SiO_2 NPs and protein were further estimated by measuring protein fluorescence quenching ratio in the presence of NPs. **Figure 4** shows that fluorescence quenching ratio immediately increased just after adding NPs in skim milk solution at all temperatures tested in a NP concentration-dependent manner. Remarkably high fluorescence quenching ratio (more than 60% fluorescence quenching) was induced at 25°C after incubation for 7 d (**Figure 4B**), where aggregation was observed (Supplementary Figure 1). Meanwhile, fluorescence quenching ratio significantly increased at 25 and 40°C compared to 4°C. However, overall fluorescence quenching ratios were <40% and blue or red shift was not observed.

When the interactions between SiO_2 NPs and casein, a major protein in skim milk, was assessed, similar tendency was obtained (**Figure 5**). NP concentration-dependent increase in fluorescence quenching ratio was found, and 60–70% quenching was induced by NPs at 25°C after 7 d of incubation. Interactions between NPs and casein at 25°C were comparable to those at 40°C, except at 7 d post-incubation, while significantly reduced fluorescence quenching was observed at 4°C. Blue shift was only observed at 40°C after 48 h of incubation (Supplementary Figure 2). On the other hand, fluorescence quenching ratios of NPs in 0.35 mg/ml of casein solution (**Figure 5**) were similar to those in 1 mg/ml of skim milk solution (**Figure 4**).

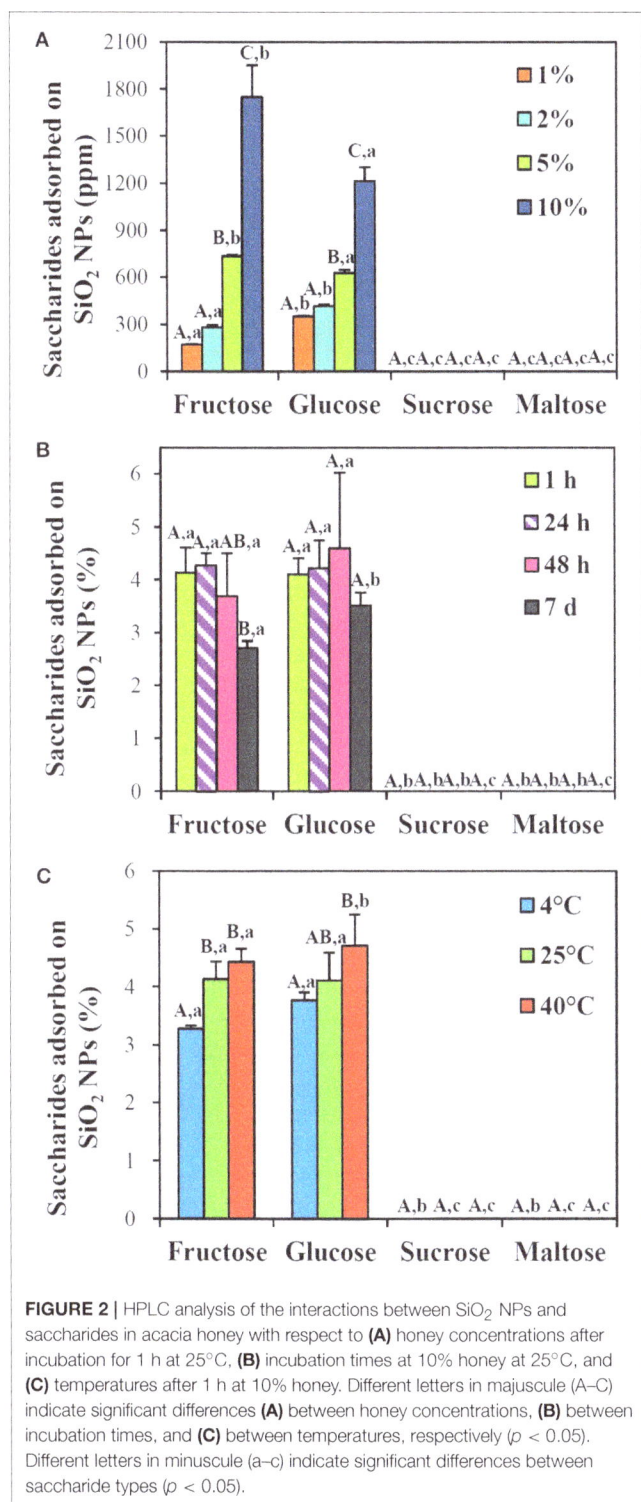

FIGURE 2 | HPLC analysis of the interactions between SiO_2 NPs and saccharides in acacia honey with respect to (A) honey concentrations after incubation for 1 h at 25°C, (B) incubation times at 10% honey at 25°C, and (C) temperatures after 1 h at 10% honey. Different letters in majuscule (A–C) indicate significant differences (A) between honey concentrations, (B) between incubation times, and (C) between temperatures, respectively ($p < 0.05$). Different letters in minuscule (a–c) indicate significant differences between saccharide types ($p < 0.05$).

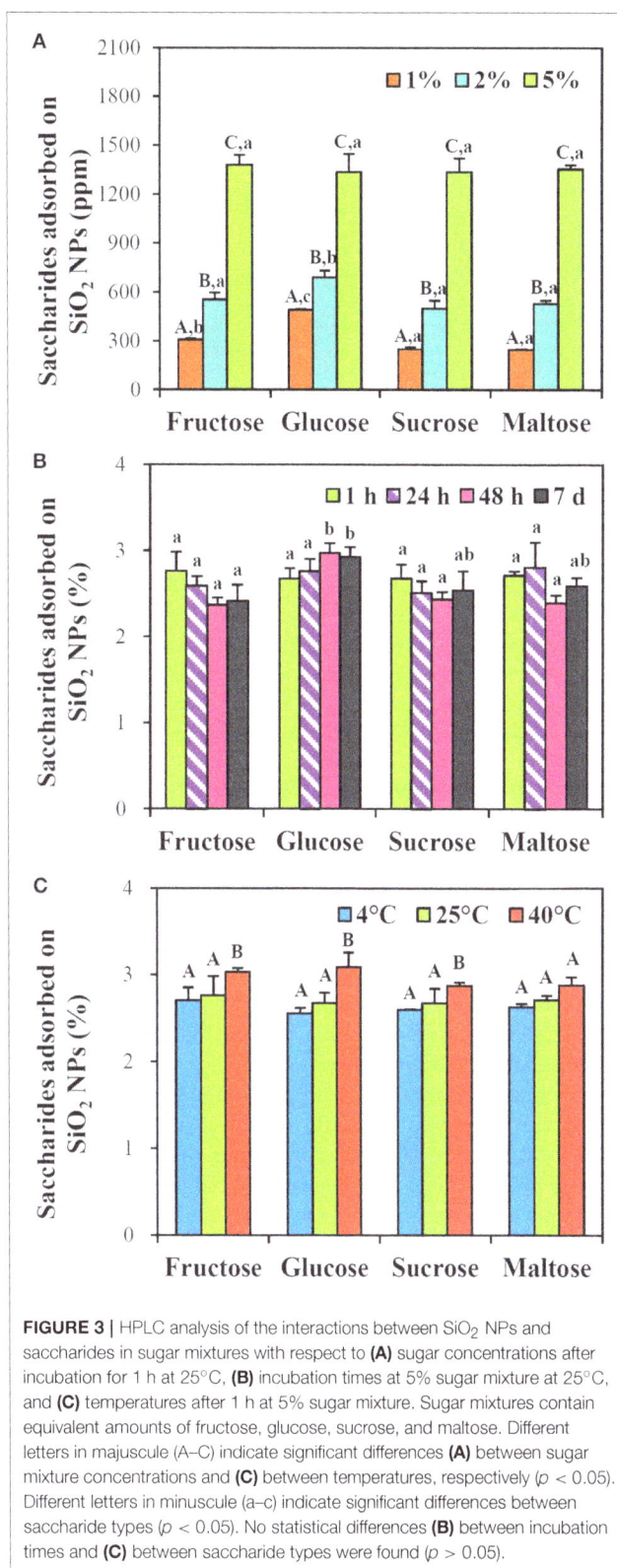

FIGURE 3 | HPLC analysis of the interactions between SiO_2 NPs and saccharides in sugar mixtures with respect to (A) sugar concentrations after incubation for 1 h at 25°C, (B) incubation times at 5% sugar mixture at 25°C, and (C) temperatures after 1 h at 5% sugar mixture. Sugar mixtures contain equivalent amounts of fructose, glucose, sucrose, and maltose. Different letters in majuscule (A–C) indicate significant differences (A) between sugar mixture concentrations and (C) between temperatures, respectively ($p < 0.05$). Different letters in minuscule (a–c) indicate significant differences between saccharide types ($p < 0.05$). No statistical differences (B) between incubation times and (C) between saccharide types were found ($p > 0.05$).

Interaction between NPs and Lipids

Zeta potentials and hydrodynamic radii of SiO_2 NPs in the presence of olive oil could not be detected because of intense yellow color and high viscosity of olive oil. GC-MS analysis reveals that olive oil used in the presence study contains ~59.3% oleic acid, ~11.4% palmitic acid, and ~5.7% linoleic acid. Because SiO_2 NPs have hydrophilic surface characteristics (Jesionowski and Krysztafkiewicz, 2002) and slightly bound fatty acids on the surface of NPs are likely to be easily detached during

TABLE 4 | Zeta potentials and hydrodynamic diameters SiO$_2$ NPs in skim milk and casein solutions at 25°C.

		Zeta potential (mV)				Hydrodynamic size (nm)	
DW	Time	Skim milk (1 mg/ml)	Casein (0.35 mg/ml)	DW	Time	Skim milk (1 mg/ml)	Casein (0.35 mg/ml)
−28.2 ± 1.0[A,a]	1 min	−27.3 ± 0.8[AB,a]	−43.9 ± 3.6[C,b]	287.6 ± 1.7[A,a]	1 min	284.5 ± 4.7[A,a]	317.87 ± 1.42[B,b]
	1 h	−27.1 ± 0.8[AB,a]	−36.0 ± 1.4[B,b]		1 h	322.0 ± 11.6[C,b]	344.3 ± 18.5[C,b]
	24 h	−25.6 ± 1.34[B,a]	−33.7 ± 0.7[B,b]		24 h	268.0 ± 9.2[B,b]	292.0 ± 4.3[A,a]
	48 h	−25.4 ± 1.2[B,b]	−32.8 ± 2.4[AB,c]		48 h	270.6 ± 4.8[B,b]	274.2 ± 1.8[A,b]
	7 d	−24.0 ± 1.9[C,b]	−32.4 ± 1.6[A,c]		7 d	296.6 ± 0.7[A,b]	287.9 ± 1.6[A,a]

Different letters (A–C) in the same column indicate significant differences between incubation times (p < 0.05). Different letters (a–c) in the same row indicate significant differences between skim milk and casein solutions (p < 0.05). Concentration of casein solution was adjusted to have an equivalent amount of protein content in 1 mg/ml of skim milk solution. Zeta potential and hydrodynamic radius of NPs in DW were measured in the absence of skim milk and casein solutions as a control.

washing with organic solvents, the interactions between SiO$_2$ NPs and lipids were estimated by measuring fatty acid composition in the supernatant after reaction with NPs, followed by subtraction of reduced fatty acids from those in olive oil control. The results show that <1.8% oleic acid and 0.4% palmitic and linoleic acids interacted with SiO$_2$ NPs, and the effects of incubation time and temperature on the interactions were not found (**Figure 6**; $p > 0.05$). However, interactions at 4°C after 48 h and at all temperatures tested after 7 d could not be detected, resulted from gelation of SiO$_2$ NPs in olive oil (Supplementary Figure 1).

Interaction between NPs and Minerals

PBS buffer was used to investigate the interaction effects of minerals on the surface of SiO$_2$ NPs. Zeta potentials of NPs significantly and rapidly changed to more negative charge in PBS without effects of incubation time, NP concentration, and temperature (**Tables 2, 5**). DLS data reveal that significantly reduced size distribution of NPs was found just after adding NPs in PBS and their reduced hydrodynamic radii were maintained for 7 d. Remarkable aggregation was not observed even after incubation for 7 d (Supplementary Figure 1).

Quantitative analysis was also performed by measuring the amount of bound Na$^+$ to NPs, the most abundant mineral in PBS, by ICP-AES. As shown in **Figure 7A**, the interactions between SiO$_2$ NPs and Na$^+$ increased as NP concentration and incubation time increased. However, no effect of temperature on the interactions was found (**Figure 7B**; $p > 0.05$). Overall, <1.7% of Na$^+$ was determine to interact with SiO$_2$ NPs (**Figure 7B**).

DISCUSSION

In the present study, acacia honey, skim milk, olive oil, and PBS buffer were used as representative foods to investigate the interactions between food additive SiO$_2$ NPs and food matrices, and then, NP interactions with main components in each representative food were further explored quantitatively. Physicochemical characterization results demonstrate that the primary particle size (24.1 ± 3.5 nm) of SiO$_2$ NPs increased in DW, as evidenced by increased hydrodynamic radius (287.6 ± 1.7 nm; **Figure 1**), indicating their high agglomeration or aggregation tendency. Solubility of SiO$_2$ NPs was not detected under all experimental conditions including DW and food

matrices, suggesting their particulate fate in food matrices. This result is highly consistent with our previous report (Kim et al., 2016).

SiO$_2$ NPs in honey had less negative surface charges and increased hydrodynamic radii as honey concentration increased compared to NPs in DW (**Table 1**), suggesting NP interactions with honey. Positively charged minor components in honey, such as amino acids and minerals (Cotte et al., 2003, 2004; Conti et al., 2007), seem to play a role in zeta potential changes. When the interactions were further quantified by HPLC analysis, fructose and glucose were found to be adsorbed on the surface of NPs in a honey concentration- and temperature-dependent manner (**Figure 2**), while no disaccharide, such as sucrose and maltose, did interact with NPs. This result seems to be closely related to the composition of saccharide in acacia honey (fructose: ~42%, glucose: ~30%, sucrose: ~ 0.2%, maltose: ~0.1%). Hence, elevated levels of fructose and glucose in honey contribute to their active interaction with NPs. Indeed, acacia honey contains high sugar concentration composed of various saccharides as well as small amount of amino acids and minerals (Cotte et al., 2004; Conti et al., 2007), and thus, the interactions between SiO$_2$ NPs and saccharides are surely influenced by trace nutrients. This hypothesis was assumed by investigating the interactions between NPs and sugar mixtures containing equivalent amounts (1, 2, and 5%) of fructose, glucose, sucrose, and maltose, respectively. Interestingly, zeta potential values of NPs did not statistically change in sugar mixtures as incubation time increased, but slightly decreased values were found at 5% compared to those at 1 or 2% (**Table 3**). Meanwhile, their hydrodynamic radii decreased as incubation time increased (**Table 3**), contrary to the results obtained in honey (**Table 1**). This result supports the role of trace nutrients in the interactions between NPs and saccharides in honey. Moreover, reduced hydrodynamic radii of SiO$_2$ NPs in sugar mixtures imply that saccharides can be used as dispersing agents, as reported by other researches (Montero et al., 2012; Maldiney et al., 2014; Strobl et al., 2014).

On the other hand, saccharide concentration- and temperature-dependent interactions between NPs and each saccharide in sugar mixtures were found, regardless of incubation time (**Figure 3**). However, statistical differences

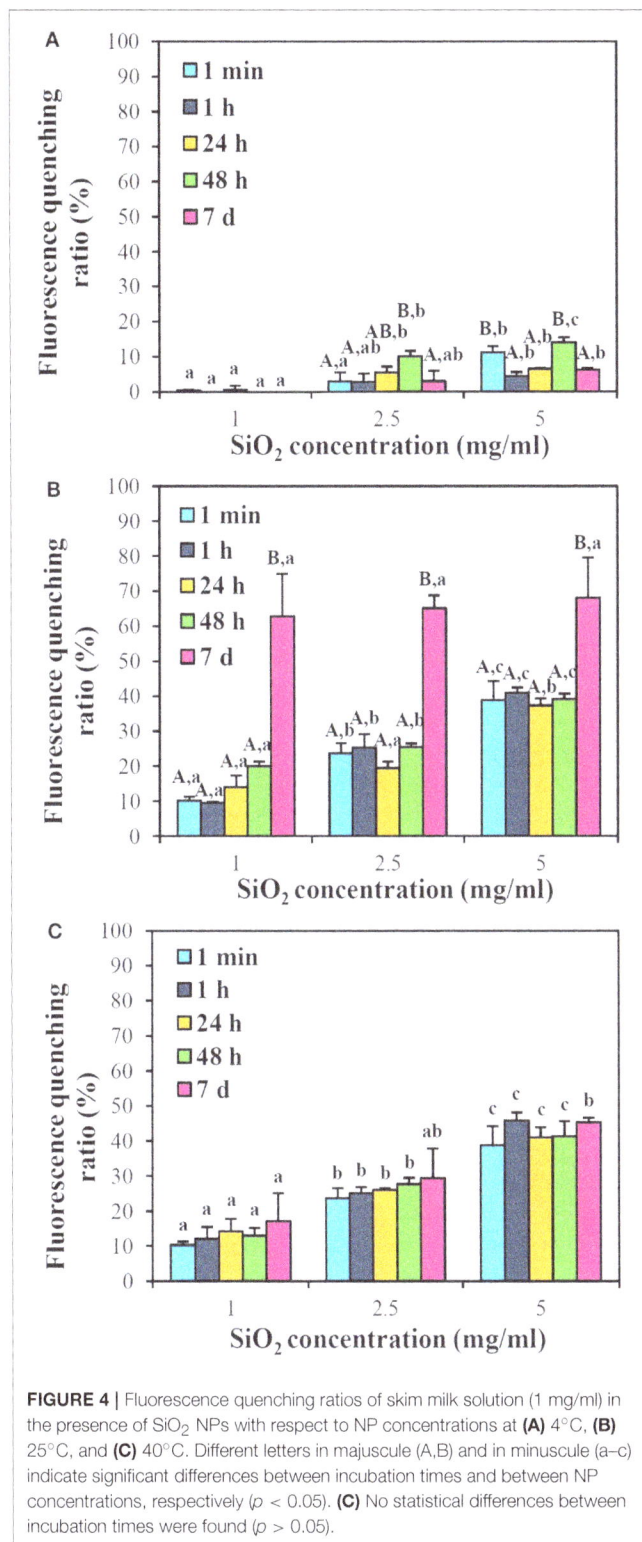

FIGURE 4 | Fluorescence quenching ratios of skim milk solution (1 mg/ml) in the presence of SiO$_2$ NPs with respect to NP concentrations at **(A)** 4°C, **(B)** 25°C, and **(C)** 40°C. Different letters in majuscule (A,B) and in minuscule (a–c) indicate significant differences between incubation times and between NP concentrations, respectively ($p < 0.05$). **(C)** No statistical differences between incubation times were found ($p > 0.05$).

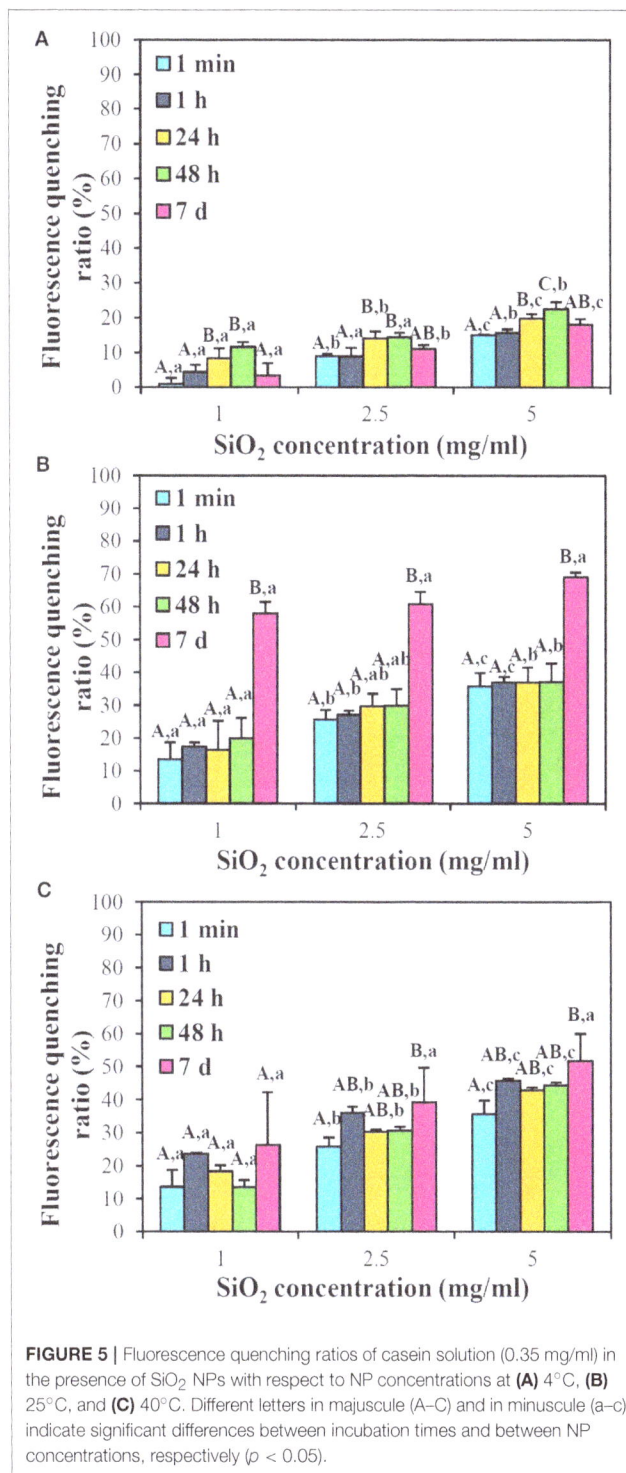

FIGURE 5 | Fluorescence quenching ratios of casein solution (0.35 mg/ml) in the presence of SiO$_2$ NPs with respect to NP concentrations at **(A)** 4°C, **(B)** 25°C, and **(C)** 40°C. Different letters in majuscule (A–C) and in minuscule (a–c) indicate significant differences between incubation times and between NP concentrations, respectively ($p < 0.05$).

in the interactions between saccharide types were not remarkably observed, suggesting that all saccharide types can be adsorbed on SiO$_2$ NPs in a similar manner. Overall interaction amounts of saccharides in honey and sugar mixtures were <4.6 and 3.1%,

respectively, suggesting that a small portion of saccharides can interact with SiO$_2$ NPs. It is worth noting that the maximum concentration of honey solution tested was 10%, which contains about 4.2% fructose, 3.0% glucose, 0.02% sucrose, and 0.01% maltose. Whereas, the highest sugar mixture was composed of 5% of each saccharide. It is, therefore, clear that NP interactions with saccharides are facilitated in the

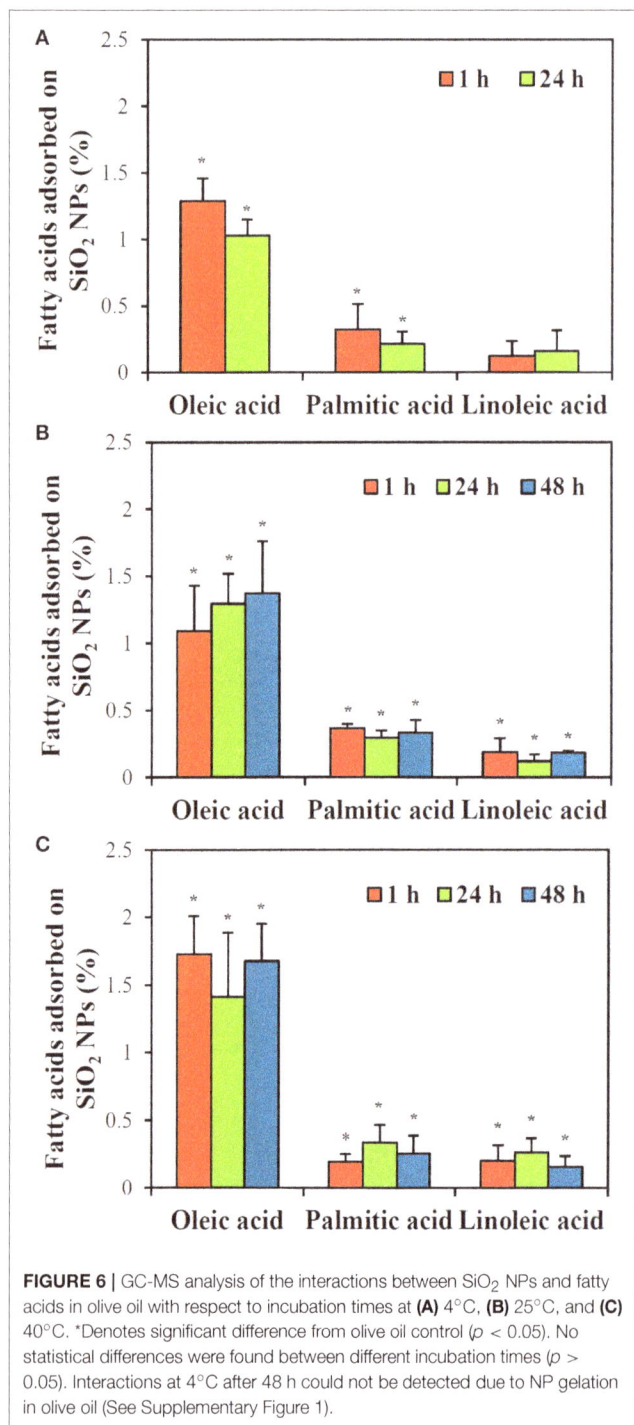

FIGURE 6 | GC-MS analysis of the interactions between SiO_2 NPs and fatty acids in olive oil with respect to incubation times at (A) 4°C, (B) 25°C, and (C) 40°C. *Denotes significant difference from olive oil control ($p < 0.05$). No statistical differences were found between different incubation times ($p > 0.05$). Interactions at 4°C after 48 h could not be detected due to NP gelation in olive oil (See Supplementary Figure 1).

This is also evidenced by increased hydrodynamic sizes in honey solution (**Table 1**), contrary to decreased DLS values in sugar mixtures (**Table 3**). This result also indicates that trace nutrients in honey play a role in the interactions between SiO_2 NPs or between SiO_2 NPs and saccharides. Reduced NP interaction with fructose in honey at 7 d post-incubation may be resulted from NP aggregation (**Figure 2B**, Supplementary Figure 1). On the other hand, no adsorbed saccharides on NPs were detected after washing with DDW, suggesting week interaction force between two materials. It is also probable that high solubility of saccharides in DW also facilitates to detach adsorbed saccharides from NPs during washing process (Alves et al., 2007).

Interactions between SiO_2 NPs and skim milk or casein were evidenced by changes in zeta potential in a time-dependent manner (**Table 4**). Interestingly, zeta potentials of NPs in skim milk solution changed to less negative direction, contrary to more negative charges found in casein solution. Skim milk powder is composed of ~34–37% protein (80% casein and 20% whey protein), 50–52% lactose, 8% minerals, and small amount of lipids as well as amino acids (Lagrange, 2005). The difference in composition between skim milk and casein is strongly likely to differently affect zeta potential change. Moreover, the isoelectric point of casein is 4.6, inducing negative charge under physiological condition. This can also induce zeta potential changes of NPs to more negative direction in casein solution. Interestingly, hydrodynamic size of SiO_2 NPs in both skim milk and casein solutions did not remarkably increase, in particular, as incubation time increased (**Table 4**). It is known that proteins can play a role as dispersants for NPs (Bihari et al., 2008; Ji et al., 2010; Jo et al., 2016; Vranic et al., 2017), which may explain this result.

Fluorescence quenching ratios of NPs in 1 mg/ml of skim milk solution (**Figure 4**) were similar to those in 0.35 mg/ml of casein solution (**Figure 5**), implying that other nutrients in skim milk did not affect NP interaction with proteins. It is notable that aggregation was observed at 25°C after incubation for 7 d (Supplementary Figure 1), probably leading to high fluorescence quenching ratio (**Figures 4B, 5B**). Less fluorescence quenching at 40°C (**Figures 4C, 5C**) than 25°C (**Figures 4B, 5B**) after 7 d might be attributed to protein denaturation, as observed by remarkably high protein precipitation (Supplementary Figure 1). Meanwhile, blue shift indicating protein denaturation or deformation was only observed in casein solution at 5 mg/ml NP concentration at 40°C after 48 h. Moreover, overall fluorescence quenching ratios were <40%, except incubation for 7 d at 25°C. These results suggest that the interactions between SiO_2 NPs and proteins are not strong compared to other NP interactions which induce more than 80% quenching (Iosin et al., 2009; Chatterjee et al., 2010; Lee et al., 2015).

On the other hand, small amounts of lipids and minerals were found to interact with SiO_2 NPs, as evaluated with olive oil and PBS. Oleic acid, palmitic acid, and linoleic acid, main components of olive oil, interacted with NPs without incubation time and temperature effects (**Figure 6**). However, total adsorbed

presence of small amount of trace nutrients, as indicated by high NP interactions in honey solution (**Figure 2**) compared to sugar mixtures (**Figure 3**). In both honey and sugar mixture solutions, incubation time was not a critical factor affecting interactions.

It is interesting to note that incubation of SiO_2 NPs with more than 2% of honey for 7 d induced strong aggregation, which was less evident in sugar mixtures (Supplementary Figure 1).

TABLE 5 | Zeta potentials and hydrodynamic diameters SiO_2 NPs in PBS buffer at $25°C$.

		Zeta potential (mV)				Hydrodynamic size (nm)		
DW	Time	SiO_2 concentration		DW	Time	SiO_2 concentration		
		5 mg/ml	10 mg/ml			5 mg/ml	10 mg/ml	
$-28.2 \pm 1.0^{A,a}$	1 min	$-36.3 \pm 1.0^{B,b}$	$-35.8 \pm 0.4^{B,b}$	$287.6 \pm 1.7^{A,a}$	1 min	$269.4 \pm 5.5^{B,b}$	$269.7 \pm 4.7^{B,b}$	
	1 h	$-37.3 \pm 1.2^{B,b}$	$-36.5 \pm 1.0^{B,b}$		1 h	$250.4 \pm 2.1^{C,b}$	$250.0 \pm 4.7^{C,b}$	
	24 h	$-36.4 \pm 1.8^{B,b}$	$-36.7 \pm 1.0^{B,b}$		24 h	$247.9 \pm 1.7^{C,b}$	$249.6 \pm 4.3^{C,b}$	
	48 h	$-36.2 \pm 1.1^{B,b}$	$-36.3 \pm 0.5^{B,b}$		48 h	$247.4 \pm 2.3^{C,b}$	$246.7 \pm 1.4^{C,b}$	
	7 d	$-37.1 \pm 1.1^{B,b}$	$-37.0 \pm 1.5^{B,b}$		7 d	$270.4 \pm 3.3^{B,b}$	$269.6 \pm 1.6^{B,b}$	

Different letters (A–C) in the same column indicate significant differences between incubation times ($p < 0.05$). Different letters (a,b) in the same row indicate significant differences between NP concentrations ($p < 0.05$). Zeta potential and hydrodynamic radius of NPs in DW were measured in the absence of PBS as a control.

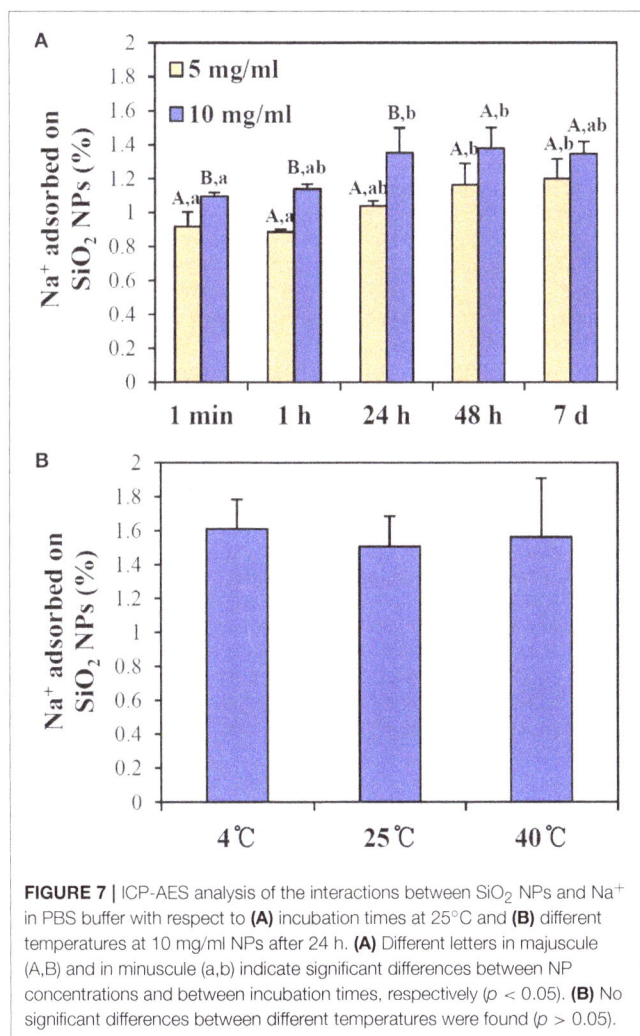

FIGURE 7 | ICP-AES analysis of the interactions between SiO_2 NPs and Na^+ in PBS buffer with respect to **(A)** incubation times at $25°C$ and **(B)** different temperatures at 10 mg/ml NPs after 24 h. **(A)** Different letters in majuscule (A,B) and in minuscule (a,b) indicate significant differences between NP concentrations and between incubation times, respectively ($p < 0.05$). **(B)** No significant differences between different temperatures were found ($p > 0.05$).

result is not surprising because SiO_2 NPs have hydrophilic surface characteristics (Jesionowski and Krysztafkiewicz, 2002), and thus, repulsion force between SiO_2 NP surfaces and fatty acids is surely predominant. Considering the fact that adsorbed fatty acids on NPs were quantified after subtracting reduced amounts in the supernatant from those in olive oil control, actual adsorbed fatty acids on NP surface seem to be less than the present result obtained. However, the interactions between NPs and lipids could strongly occur for polymer-based NPs or NPs with more hydrophobic surface, such as carbon-based materials. Meanwhile, PBS buffer was used to investigate NP interactions with minerals because mineral composition of PBS is well-known. When the interactions between SiO_2 NPs and minerals were evaluated by measuring the most abundant mineral component, Na^+ in PBS, the interactions were dependent on NP concentration and incubation time, but total adsorbed amount was <1.7% (**Figure 7**). It should be noted that the interactions were quantified after washing three times with DDW. Hence, it is clear that minerals could be strongly adsorbed on SiO_2 NPs, although the interaction amount was not that much. This interaction was also supported by significant changes in zeta potential values of NPs to more negative direction and decreased hydrodynamic size in PBS (**Table 5**). It is strongly likely that small amount of minerals can also act as dispersing agents, by reducing repulsion force between NPs (Dishon et al., 2009).

CONCLUSION

In conclusion, food additive SiO_2 NPs were found to interact with saccharides, proteins, fatty acids, and minerals. Maximum 4.6% of saccharides, 1.7% of fatty acids, and 1.6% of minerals were found to be adsorbed on NPs. The interaction between SiO_2 NPs and proteins was determined to be rather weak, which did not induce high fluorescence quenching ratio and protein deformation. Trace nutrients were found to play a role in the interactions, as observed by the increased interactions in honey compared to simple sugar mixtures. On the other hand, the degree of NP interaction with skim milk was not much different from that with casein solutions. Saccharides, proteins, and minerals are likely to act as NP

amounts of oleic acid and palmitic/linoleic acids on NPs were <1.8 and 0.4%, respectively, indicating that a small portion of fatty acids is adsorbed on NPs. High NP interaction with oleic acid was quantitatively determined because oleic acid is the most abundant fatty acid (~59%) in olive oil. This dispersants as well. Taken together, SiO_2 NPs interact with a small portion of food matrices and the interactions are not strong. Further study is required to assume the effects of interactions between NPs and food components on the potential toxicity of NPs as well as the absorption of nutrients and NPs.

AUTHOR CONTRIBUTIONS

Conceived and designed the experiments and wrote the manuscript: SC. Performed the experiments: MG, SB, HK, and JY. Performed statistical and data analysis: MG and JY.

ACKNOWLEDGMENTS

This research was supported by the Basic Science Research Program through the National Research Foundation of Korea (NRF) funded by the Ministry of Education (2015R1D1A1A01057150) and by a research grant from Seoul Women's University (2017).

REFERENCES

Alves, L. A., Silva, J. B. A., and Giulietti, M. (2007). Solubility of D-glucose in water and ethanol/water mixtures. *J. Chem. Eng. Data* 52, 2166–2170. doi: 10.1021/je700177n

Bihari, P., Vippola, M., Schultes, S., Praetner, M., Khandoga, A. G., Reichel, C. A., et al. (2008). Optimized dispersion of nanoparticles for biological *in vitro* and *in vivo* studies. *Part. Fibre Toxicol.* 5:14. doi: 10.1186/1743-8977-5-14

Bohmert, L., Niemann, B., Lichtenstein, D., Juling, S., and Lampen, A. (2015). Molecular mechanism of silver nanoparticles in human intestinal cells. *Nanotoxicology* 9, 852–860. doi: 10.3109/17435390.2014.980760

Borak, B., Biernat, P., Prescha, A., Baszczuk, A., and Pluta, J. (2012). *In vivo* study on the biodistribution of silica particles in the bodies of rats. *Adv. Clin. Exp. Med.* 21, 13–18. Available online at: http://www.advances.umed.wroc.pl/en/article/2012/21/1/13/

Cao, Y., Li, J., Liu, F., Li, X., Jiang, Q., Cheng, S., et al. (2016). Consideration of interaction between nanoparticles and food components for the safety assessment of nanoparticles following oral exposure: a review. *Environ. Toxicol. Pharmacol.* 46, 206–210. doi: 10.1016/j.etap.2016.07.023

Chatterjee, T., Chakraborti, S., Joshi, P., Singh, S., Gupta, V., and Chakrabarti, P. (2010). The effect of zinc oxide nanoparticles on the structure of the periplasmic domain of the *Vibrio cholerae* ToxR protein. *FEBS J.* 277, 4184–4194. doi: 10.1111/j.1742-4658.2010.07807.x

Conti, M. E., Stripeikis, J., Campanella, L., Cucina, D., and Tudino, M. B. (2007). Characterization of Italian honeys (Marche Region) on the basis of their mineral content and some typical quality parameters. *Chem. Cent. J.* 1:14. doi: 10.1186/1752-153X-1-14

Cotte, J. F., Casabianca, H., Giroud, B., Albert, M., and Grenier-Loustalot, M. F. (2004). Characterization of honey amino acid profiles using high-pressure liquid chromatography to control authenticity. *Anal. Bioanal. Chem.* 378, 1342–1350. doi: 10.1007/s00216-003-2430-z

Cotte, J. F., Casabianca, H., Chardon, S., Lheritier, J., and Grenier-Loustalot, M. F. (2003). Application of carbohydrate analysis to verify honey authenticity. *J. Chromatogr. A* 1021, 145–155. doi: 10.1016/j.chroma.2003.09.005

Dekkers, S., Krystek, P., Peters, R. J. B., Lankveld, D. P. K., Bokkers, B. G. H., Hoeven-Arentzen, P. H., et al. (2011). Presence and risks of nanosilica in food products. *Nanotoxicology* 5, 393–405. doi: 10.3109/17435390.2010.519836

Dishon, M., Zohar, O., and Sivan, U. (2009). From repulsion to attraction and back to repulsion: the effect of NaCl, KCl, and CsCl on the force between silica surfaces in aqueous solution. *Langmuir* 25, 2831–2836. doi: 10.1021/la803022b

Docter, D., Westmeier, D., Markiewicz, M., Stolte, S., Knauer, S. K., and Stauber, R. H. (2015). The nanoparticle biomolecule corona: lessons learned - challenge accepted? *Chem. Soc. Rev.* 44, 6094–6121. doi: 10.1039/C5CS00217F

Dorier, M., Brun, E., Veronesi, G., Barreau, F., Pernet-Gallay, K., Desvergne, C., et al. (2015). Impact of anatase and rutile titanium dioxide nanoparticles on uptake carriers and efflux pumps in Caco-2 gut epithelial cells. *Nanoscale* 7, 7352–7360. doi: 10.1039/C5NR00505A

Iosin, M., Toderas, F., Baldeck, P. L., and Astilean, S. (2009). Study of protein-gold nanoparticle conjugates by fluorescence and surface-enhanced Raman scattering. *J. Mol. Struct.* 924–926, 196–200. doi: 10.1016/j.molstruc.2009.02.004

Jesionowski, T., and Krysztafkiewicz, A. (2002). Preparation of the hydrophilic/hydrophobic silica particles. *Colloid. Surf. A. Eng. Asp.* 207, 49–58. doi: 10.1016/S0927-7757(02)00137-1

Ji, Z., Jin, X., George, S., Xia, T., Meng, H., Wang, X., et al. (2010). Dispersion and stability optimization of TiO_2 nanoparticles in cell culture media. *Sci. Technol.* 44, 7309–7314. doi: 10.1021/es100417s

Jiang, Q., Li, X., Cheng, S., Gu, Y., Chen, G., Shen, Y., et al. (2016). Combined effects of low levels of palmitate on toxicity of ZnO nanoparticles to THP-1 macrophages. *Environ. Toxicol. Pharmacol.* 48, 103–109. doi: 10.1016/j.etap.2016.10.014

Jo, M. R., Chung, H. E., Kim, H. J., Bae, S. H., Go, M. R., Yu, J., et al. (2016). Effects of zinc oxide nanoparticle dispersants on cytotoxicity and cellular uptake. *Mol. Cell. Toxicol.* 12, 281–288. doi: 10.1007/s13273-016-0033-y

Kim, M. K., Lee, J. A., Jo, M. R., and Choi, S. J. (2016). Bioavailability of silica, titanium dioxide, and zinc oxide nanoparticles in rats. *J. Nanosci. Nanotechnol.* 16, 6580–6586. doi: 10.1166/jnn.2016.12350

Lagrange, V. (2005). *Reference Manual for US and Milk Powders 2005 Revised Edn.* Arlington, VA: US Dairy Export Council.

Lee, J. A., Kim, M. K., Kim, H. M., Lee, J. K., Jeong, J., Kim, Y. R., et al. (2015). The fate of calcium carbonate nanoparticles administered by oral route: absorption and their interaction with biological matrices. *Int. J. Nanomed.* 10, 2273–2293. doi: 10.2147/IJN.S79403

Lee, J. A., Kim, M. K., Paek, H. J., Kim, Y. R., Lee, J. K., Jeong, J., et al. (2014). Tissue distribution and excretion kinetics of orally administered silica nanoparticles in rats. *Int. J. Nanomed.* 9, 251–260. doi: 10.2147/IJN.S57939

Lee, J. A., Kim, M. K., Song, J. H., Jo, M. R., Yu, J., Kim, K. M., et al. (2017). Biokinetics of food additive silica nanoparticles and their interactions with food components. *Colloid. Surf. B.* 150, 384–392. doi: 10.1016/j.colsurfb.2016.11.001

Liang, N. N., Zhang, L. X., Wang, X. L., Tan, B. B., and Liang, Y. Z. (2011). Identification of fatty acids in vegetable oils by mass spectrometry and equivalent chain length. *Chin. J. Anal. Chem.* 39, 1166–1170. doi: 10.3724/SP.J.1096.2011.01166

Lichtenstein, D., Ebmeyer, J., Knappe, P., Juling, S., Böhmert, L., Selve, S., et al. (2015). Impact of food components during *in vitro* digestion of silver nanoparticles on cellular uptake and cytotoxicity in intestinal cells. *Biol. Chem.* 396, 1255–1264. doi: 10.1515/hsz-2015-0145

Mahler, G. J., Esch, M. B., Tako, E., Southard, T. L., Archer, S. D., Glahn, R. P., et al. (2012). Oral exposure to polystyrene nanoparticles affects iron absorption. *Nat. Nanotechnol.* 7, 264–271. doi: 10.1038/nnano.2012.3

Maldiney, T., Bessière, A., Seguin, J., Teston, E., Sharma, S. K., Viana, B., et al. (2014). The *in vivo* activation of persistent nanophosphors for optical imaging of vascularization, tumours and grafted cells. *Nat. Mater.* 13, 418–426. doi: 10.1038/nmat3908

Montero, M., Molina, T., Szafran, M., Moreno, R., and Nieto, M. I. (2012). Alumina porous nanomaterials obtained by colloidal processing using D-fructose as dispersant and porosity promoter. *Ceram. Int.* 38, 2779–2784. doi: 10.1016/j.ceramint.2011.11.048

Paek, H. J., Chung, H. E., Lee, J. A., Kim, M. K., Lee, Y. J., Kim, M. S., et al. (2014). Quantitative determination of silica nanoparticles in biological matrices and their pharmacokinetics and toxicokinetics in rats. *Sci. Adv. Mater.* 6, 1605–1610. doi: 10.1166/sam.2014.1817

Strobl, F. G., Seitz, F., Westerhausen, C., Reller, A., Torrano, A. A., Bräuchle, C., et al. (2014). Intake of silica nanoparticles by giant lipid vesicles: influence of particle size and thermodynamic membrane state. *Beilstein J. Nanotechnol.* 5, 2468–2478. doi: 10.3762/bjnano.5.256

Vance, M. E., Kuiken, T., Vejerano, E. P., McGinnis, S. P., Hochella, M. F. Jr., Rejeski, D., et al. (2015). Nanotechnology in the real world: redeveloping the nanomaterial consumer products inventory. *Beilstein J. Nanotechnol.* 6, 1769–1780. doi: 10.3762/bjnano.6.181

Vranic, S., Gosens, I., Jacobsen, N. R., Jensen, K. A., Bokkers, B., Kermanizadeh, A., et al. (2017). Impact of serum as a dispersion agent for *in vitro* and *in vivo* toxicological assessments of TiO_2 nanoparticles. *Arch. Toxicol.* 91, 353–363. doi: 10.1007/s00204-016-1673-3

Wang, H., Du, L. J., Song, Z. M., and Chen, X. X. (2013). Progress in the characterization and safety evaluation of engineered inorganic nanomaterials in food. *Nanomedicine* 8, 2007–2025. doi: 10.2217/nnm.13.176

Wang, Y., Yuan, L., Yao, C., Ding, L., Li, C., Fang, J., et al. (2014). A combined toxicity study of zinc oxide nanoparticles and vitamin C in food additives. *Nanoscale* 6, 15333–15342. doi: 10.1039/C4NR05480F

Conflict of Interest Statement: The authors declare that the research was conducted in the absence of any commercial or financial relationships that could be construed as a potential conflict of interest.

10

Impact of Synergistic Association of ZnO-nanorods and Symbiotic Fungus *Piriformospora indica* DSM 11827 on *Brassica oleracea* var. botrytis (Broccoli)

Uma Singhal[1], Manika Khanuja[2] [*], Ram Prasad[1] [*] and Ajit Varma[1] [*]

[1] Amity Institute of Microbial Technology, Amity University, Noida, India, [2] Centre for Nanoscience and Nanotechnology, Jamia Millia Islamia, New Delhi, India

Edited by:
*Jayanta Kumar Patra,
Dongguk University Seoul,
South Korea*

Reviewed by:
*Mihir Kumar Mandal,
Oak Ridge Institute for Science
and Education (ORISE), United States
Durgesh Kumar Tripathi,
Banaras Hindu University, India
Mostafa Abdelwahed Abdelrahman,
Tohoku University, Japan*

*Correspondence:
*Ram Prasad
rprasad@amity.edu
Manika Khanuja
manikakhanuja@gmail.com
Ajit Varma
ajitvarma@amity.edu*

In the present work, novel nanotool called 'nano-embedded fungus' formed by impact of synergistic association of ZnO-nanorods and fungus *Piriformospora indica* DSM 11827, for growth of *Brassica oleracea* var. botrytis (Broccoli) is reported. ZnO-nanorods were synthesized by mechanical assisted thermal decomposition process and characterized by scanning electron microscopy (SEM) for morphology, X-ray diffraction for structural studies and UV-vis absorption spectroscopy for band gap determination. Nanoembedded fungus is prepared by optimizing ZnO-nanorods concentration (500 ppm) which resulted in the increased biomass of *P. indica*, as confirmed by dry weight method, spore count, spread plate and microscopy techniques viz. SEM and confocal microscopy. Enhancement in *B. oleracea* var. botrytis is reported on treatment with nanoembedded fungus. According to the authors, this is the first holistic study focusing on the impact of ZnO-nanorods in the enhancement of fungal symbiont for enhanced biomass productivity of *B. oleracea* plant.

Keywords: zinc oxide nanorods, *Piriformospora indica*, UV-vis, XRD, SEM, confocal microscopy

INTRODUCTION

Agricultural nanotechnology has the potential to overcome the challenges associated with undeveloped farming practices including unbalanced ecosystem and low productivity through nano-formulation of fertilizers (or pesticides, herbicides), effective management of soil and water resources through porous nanostructures. This leads to enrichment in nutritional quantity as well as quality, simultaneously rejuvenating soil fertility and stabilization of erosion-prone surfaces (Tebebu et al., 2015; Prasad et al., 2017).

Zinc oxide (ZnO) is considered to be one of the best exploited materials at nano dimensions because of its large excitonic binding energy and wide band gap which is important for both scientific and industrial applications (Wang et al., 2004; Bhuyan et al., 2015a; Kotzybik et al., 2016; Baral et al., 2017). ZnO nanostructures exhibit gigantic area of applications and potential to boost the yield, development of food crops and their use as food additive (Sawai et al., 1996; Huang et al., 2001; Song et al., 2006; Bhuyan et al., 2015a,b; Wang et al., 2015; Sharma et al., 2017). ZnO is currently listed as "generally recognized as safe (GRAS)" material by the Food and

Drug Administration (Rajiv et al., 2013). In previous reports, the colloidal solution of zinc oxide is used as 'nano-fertilizer' a plant nutrient which is more than a fertilizer because it not only supplies nutrients for the plant but also revives the soil to an organic state without the harmful factors of chemical fertilizer (Sabir et al., 2014; Taheri et al., 2016). *Piriformospora indica* DSM 11827 is a multifunctional fungus, recently named as 'Serendipita indica' acts as plant growth promoter, biofertilizer, metabolic regulator, bio-herbicide, immunomodulator, phytoremediator, bio-insecticide and bio-pesticide, antioxidant enhancer, etc. (Prasad et al., 2013; Gill et al., 2016). It has proven attributes for enhanced plant productivity and confers resistance against biotic (Andrade-Linares et al., 2013; Ansari et al., 2013) and abiotic stresses (Franken, 2012; Gill et al., 2016; Weiß et al., 2016).

This study targets to develop a nanotechnology-assisted fungal symbiont with an objective to enhance crop productivity and medicinal value of human food crops viz. *Brassica oleracea* to overcome the challenges associated with the conventional farming (Singh et al., 2016). In the present study ZnO-nanorods have been synthesized, and the impact on fungal symbiont was studied by optimizing ZnO-nanorod concentration on interaction with *P. indica* which results in enhanced biomass. The optimized nanorods interacted fungal symbiont is called "Nanoembedded fungus." Study was further performed on Broccoli (*Brassica olearacea* var. botrytis) plants. In this study, effect of ZnO-nanorod embedded *P. indica* was analyzed on *B. oleracea* plants. Two treatments were given to *B. oleracea* plants in triplets that are *B. oleracea* treated with (i) *P. indica*, (ii) ZnO-nanorod embedded *P. indica* and the plants without any treatment were taken as control.

MATERIALS AND METHODS

Experimental
Chemicals of analytical grade were used in all the experiments, directly without any further purification, procured from Sigma–Aldrich (India), Merck (India), and HiMedia (India). In all the conducted experiments, Millli-Q water or double distilled water (ddH$_2$O) was used. Glassware was rinsed with Milli-Q water and air-dried before use in experiments.

Pure ZnO-nanorods were prepared by mechanical-assisted thermal decomposition process (Bhuyan et al., 2015a,b). In ZnO-nanorods synthesis process, 2 gm of zinc acetate dihydrate [Zn (CH$_3$COO)$_2$.2H$_2$O] was grinded in mortar pestle for 45 min. The grinded powder was placed in an alumina crucible and heated in programmable furnace (ramp rate 4°C/min) at 300°C for 4 h. Therefore, the synthesis process is termed as mechanical assisted thermal decomposition process. Double distilled water was used to wash the resultant powder twice followed by drying in hot air oven at 100°C for 8 h.

Characterization
Zinc oxide-nanorod structure and surface morphology of the samples were observed using scanning electron microscopy (SEM) (Model: JEOL-JSM-6010LA) at an accelerating voltage

of 20 kV. The absorption spectrum was measured by Perkin-Elmer Lambda 35 UV-vis spectrometer. Band gap of the sample were calculated using Tauc's plot. X-ray diffractometer (Model: Bruker; D2-Phaser) was used to investigate crystalline structures of ZnO-nanorod. The diffractogram was recorded in the scan range of 5 to 80° using CuK$_\alpha$ ($\lambda = 1.5403$ Å) X-ray operated at10 kV, 30 mA.

Fungal Strain and Culture Conditions
Aspergillus medium (Hill and Kaefer, 2001; Prasad et al., 2013) was found to be the best among different synthetic media to grow the axenically grown fungus *P. indica* DSM 11827. Circular solidified disks (4 mm dia.) consisting of actively grown hypha and chlamydospores of *P. indica* were placed on solidified aspergillus medium (pH 6.8–7.0, 28 ± 2°C in dark) as well as in broth. After 7 days, the Petri plates were found to be completely filled up with the fungal biomass.

Scanning Electron Microscopy (SEM) and Energy Dispersive X-ray Spectroscopy (EDX)
Scanning electron microscopy (EVO 18 Special edition, ZEISS) analyses were carried on *P. indica* before and after treatment with ZnO-nanorods. The *P. indica* culture without ZnO-nanorods treatment was taken as control. The chemical fixation of *P. indica* was done in order to stabilize and preserve its chemical structure. *P. indica* disks were washed with 0.1 M sodium phosphate (pH 7.4) buffer for 30 min at room temperature, then put the sample in fixative, i.e., 2.5% glutaraldehyde for overnight. In order to remove the glutaraldehyde deposits, the suspension was sequentially washed with 0.1 M sodium phosphate buffer solution (pH-7.4) and distilled water, followed by centrifugation for further isolation. Sample was dehydrated with ascending series from 50 to 100% ethanol (EtOH), in 10% increments for 20 min each and finally kept for drying. The elemental analysis of the sample was carried with energy dispersive X-ray (EDX) spectroscopy facility (Oxford instruments, 51-ADD0048) using SMARTSEM software to confirm the presence to zinc oxide nanorods in the treated sample.

Dry Weight Method
The growth of fungal biomass on interaction with ZnO-nanorods, was signified in terms of increase in dry weight. *P. indica* was inoculated in 100 ml of Hill & Kaefer medium. The culture was incubated in dark (28 ± 2°C, 80 rpm). ZnO-nanorods in different concentrations (ppm) viz. 300, 400, 500, 600, 1000, and 2000 were added to the culture after 3 days of incubation. The culture was incubated for 4 more days. *P. indica* without any treatment was grown separately for maintaining control against treated sample. Fungal culture was filtered after 7 days of incubation. The dry cell weight was calculated using:

$$W = \left[\frac{C - C_o}{C_o} \right] \times 100$$

where, W is increase in dry weight of fungal biomass on treatment with ZnO-nanorods, C_o is dry weight of fungal biomass without any treatment called 'control' and C is the dry weight

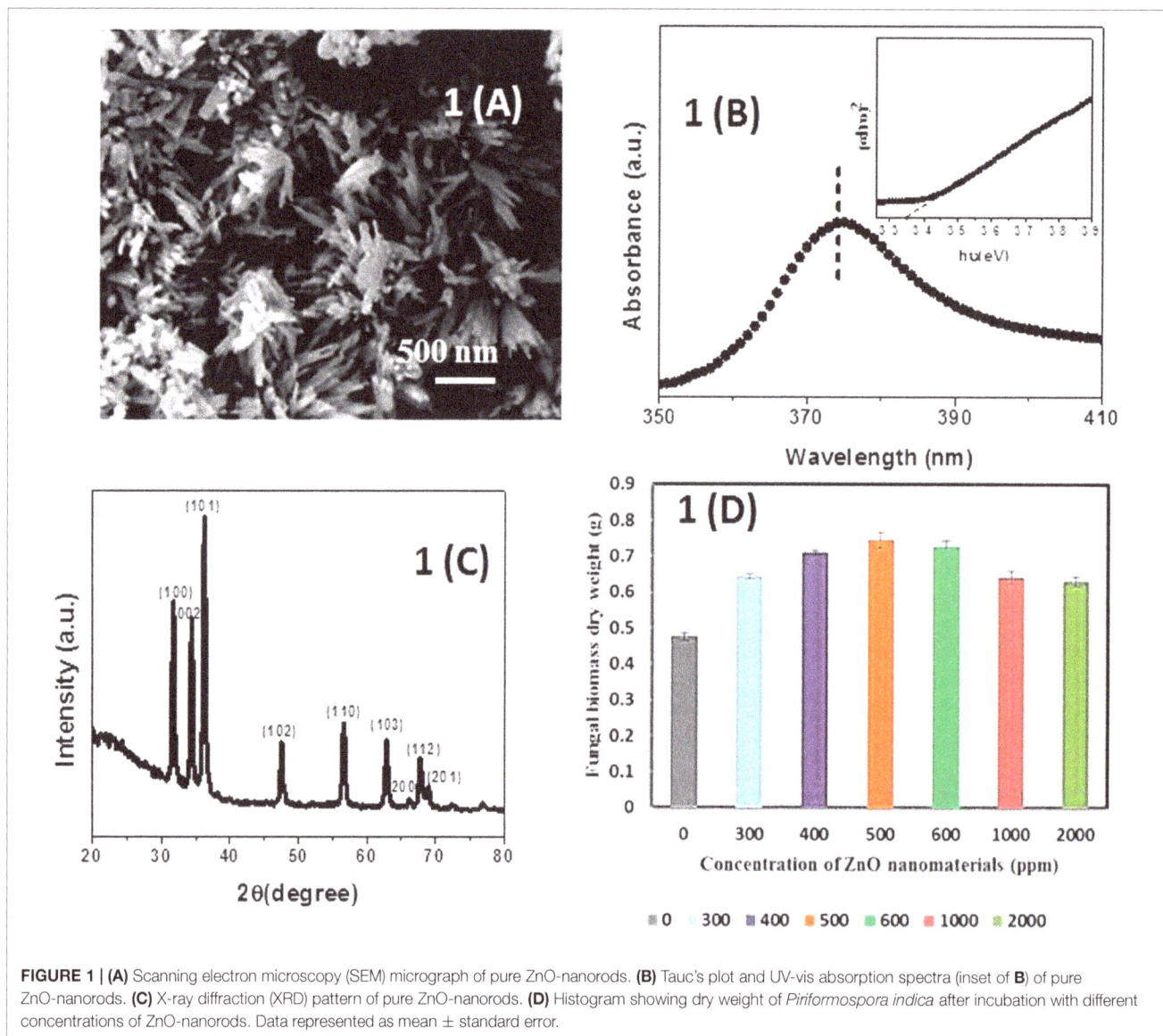

FIGURE 1 | (A) Scanning electron microscopy (SEM) micrograph of pure ZnO-nanorods. (B) Tauc's plot and UV-vis absorption spectra (inset of B) of pure ZnO-nanorods. (C) X-ray diffraction (XRD) pattern of pure ZnO-nanorods. (D) Histogram showing dry weight of *Piriformospora indica* after incubation with different concentrations of ZnO-nanorods. Data represented as mean ± standard error.

of ZnO-nanorods treated fungal biomass. The concentration of ZnO-nanorods for which maximum fungal biomass was obtained is termed as 'optimized ZnO.'

Quantification of Spore Using Hemocytometer

Spores were harvested from *P. indica* cultured on agar plate by flooding the culture with 5 ml of 0.05% (v/v) 'Tween 80' solution. The spores were carefully scraped off from the hyphae using sterile glass spreader. Spores were collected in 15 ml centrifuge tube and centrifuged for 5 min at 800 rpm to remove left over hypha fragments. Supernatant was discarded and pelleted spores were counted using hemocytometer.

Quantifying Colonies and to Study Their Morphology by Spread Plate Technique

Piriformospora indica culture was suspended in the test tubes containing distilled water with the dilution factor of (10^{-1}, 10^{-2},

10^{-3} upto 10^{-7} respectively). 1 μL of suspension (10^{-7}) and 1 mL of optimized ZnO-nanorods was poured onto the agar plates (triplicates). The prepared suspension was spread and incubated for 7 days at 28 ± 2°C. Petri plates without nanorods were taken as control.

Confocal Microscopy

Piriformospora indica culture with and without ZnO-nanorods were observed under a confocal laser scanning microscope LSM-780 (Carl-Zeiss, Inc., Jena, Germany). For culture staining, wheat germ agglutinin Alexa Fluor- 488 (WGA-AF488, Molecular Probes, Eugene, OR, United States) was used. Ethanol/chloroform/trichloroacetic acid in the ratio 1/4/0.15% v/v/w were used for the fixation of fungal biomass. The culture was washed three times with distilled H_2O, boiled in 10% KOH for 1 min, washed with phosphate buffered saline (PBS). Afterward, biomass was stained with PBS solution containing

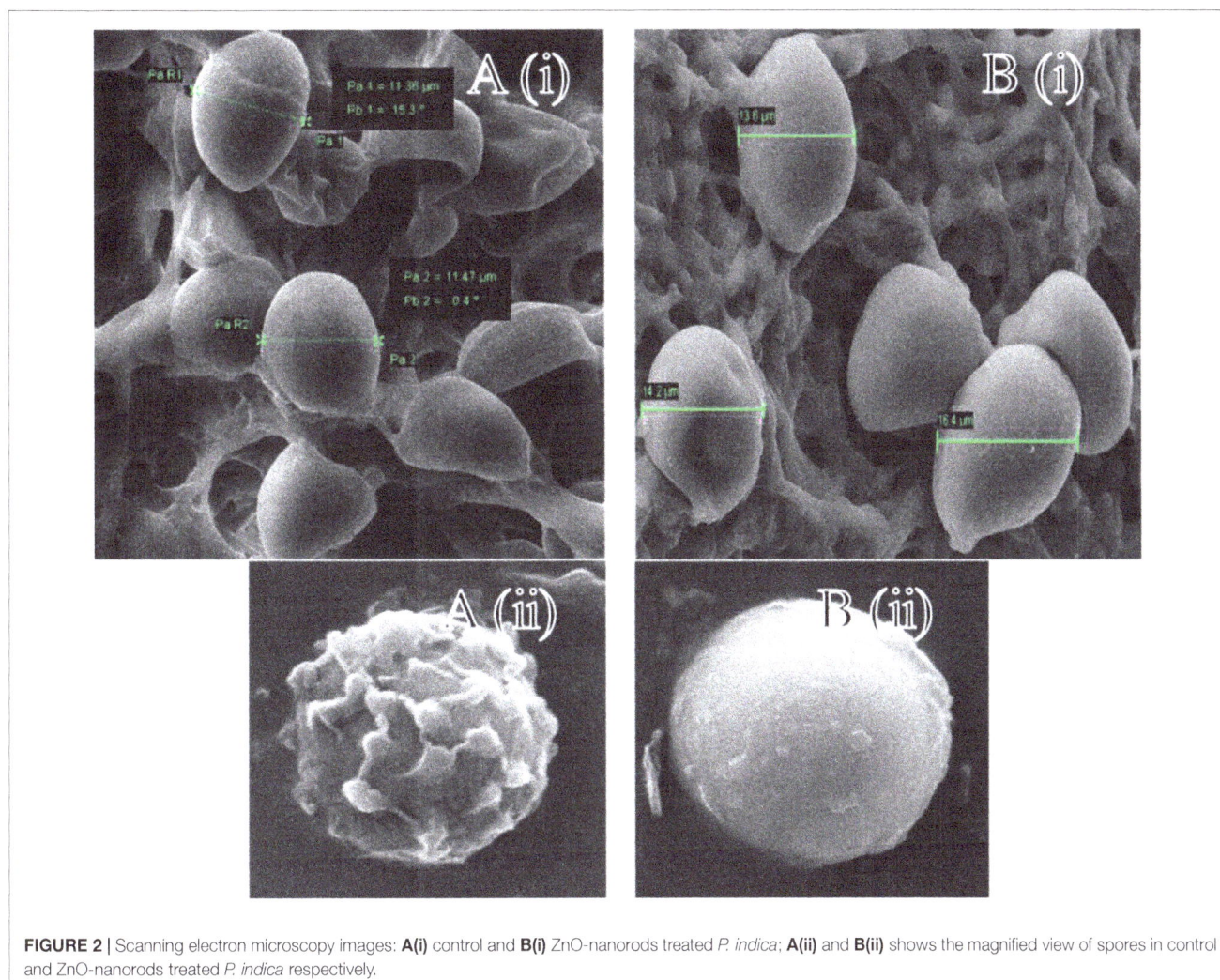

FIGURE 2 | Scanning electron microscopy images: **A(i)** control and **B(i)** ZnO-nanorods treated *P. indica*; **A(ii)** and **B(ii)** shows the magnified view of spores in control and ZnO-nanorods treated *P. indica* respectively.

0.2% Silwet L-77 and 50 μg/mL WGA- AF488. Vacuum infiltration of treated and control biomass in staining solution was done three times under 50 mm Hg vacuum. The cultures were transferred to PBS followed by removal of staining solution. The cultures were analyzed under confocal microscope, fungal hyphae was strained with WGA-AFA 488.

TABLE 1 | Dry weight of *Piriformospora indica* after addition of different concentrations of ZnO-nanorods.

ZnO-nanorods {concentration (ppm)}	Fungal biomass {dry weight (gm)}	Increase in biomass {dry weight (%)}
0	0.48	
300	0.65 ± 0.0065	35.5
400	0.71 ± 0.0062	49.16
500	0.75 ± 0.0226	56.51
600	0.73 ± 0.0156	52.94
1000	0.64 ± 0.0180	34.66
2000	0.63 ± 0.0121	32.56

Each value is mean ± standard error of three replicates.

Additionally, trypan blue was used to stain *P. indica* spores and hyphae, on binding to cells strong red fluorescence was radiated. Trypan blue staining facilitated quantification of cell size and cell wall volume under confocal microscope, thus enabling the quantification of morphological changes viz. spore size, hyphae thickness.

Treatment of *B. oleracea* with Nanoembedded Fungus

Brassica oleracea seeds brought from Indian Agriculture Research Institute (IARI), New Delhi, India of same sizes were planted in pots. The average germination rate of the seeds was 75% as shown on MS medium (Murashige and Skoog, 1962). To minimize errors in seed germination and seedling vigor, the seeds of uniform size were selected. The plants were incubated under humidity (60%), temperature (24 ± 2°C) and light (1000 lux, 16 h light and 8 h dark). After 15 days of exposure, roots and shoots were separated and washed with water to remove the growth medium and dried with wipes to remove the surface water. The growth parameters like stem and root length, fresh and dry weight per plant were recorded.

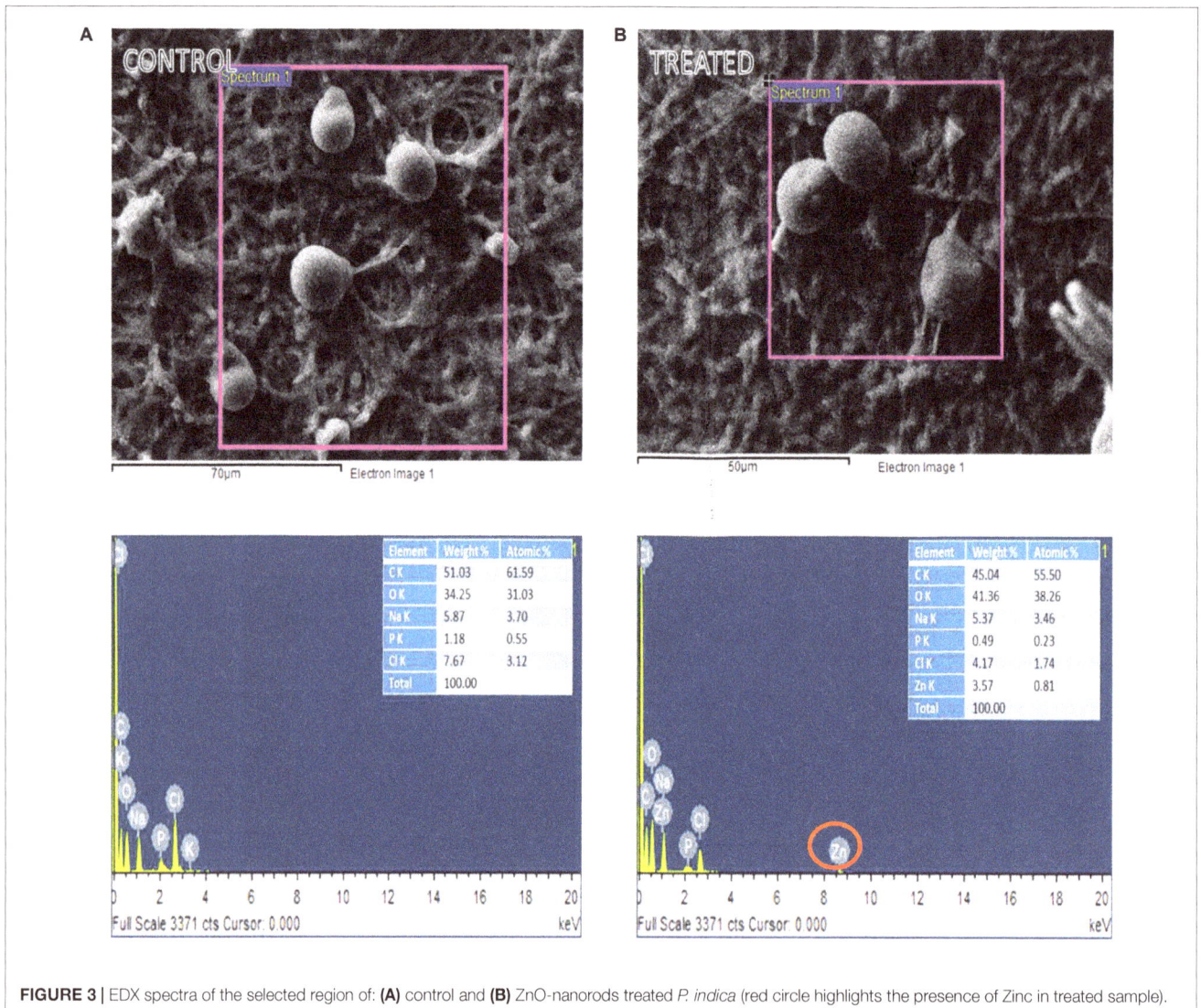

FIGURE 3 | EDX spectra of the selected region of: **(A)** control and **(B)** ZnO-nanorods treated *P. indica* (red circle highlights the presence of Zinc in treated sample).

Statistical Analysis

Each treatment was conducted in triplicates and the results were presented as mean ± standard error (SE) and analyzed by using one-way ANOVA.

RESULTS

Figure 1A showed the SEM of ZnO-nanorods. The ZnO-nanorods were of an average diameter 50 nm and length of 500 nm. **Figure 1B** showed the UV-vis absorption spectra of pure ZnO-nanorods. Band gap is calculated by Tauc's relation: $a = A(h\upsilon - E_g)^n/h\upsilon$, where a is the absorption coefficient, A is constant, E_g is the absorption band gap, n is subjected to the nature of the transitions, n may have values 1/2, 2, 3/2, and 3 corresponding to allow (direct and indirect), forbidden (direct and indirect) transitions, respectively. In this case, $n = 1/2$ for direct allowed transition. The band gap of 3.35 eV was obtained from the Tauc's plot (inset of **Figure 1B**).

Figure 1C shows the X-ray diffraction (XRD) pattern of pure ZnO-nanorods. After comparing with JCPDS File number (06-82151), all the peaks were labeled with (hkl) planes. Wurtzite structure for ZnO-nanorods was confirmed by XRD pattern. The lattice constants 'a' and 'c' can be calculated using the relations (a) and (b) given below:

$$a = \sqrt{\frac{1}{3}}\frac{\lambda}{\sin\theta} \quad (a)$$

$$c = \sqrt{\frac{\lambda}{\sin\theta}} \quad (b)$$

where, λ is the wavelength of incident X-ray beam and theta (θ) is angle of incidence.

The lattice constant 'a' and 'c' of the wurtzite structure of ZnO-nanorod were found to be 3.247 and 5.203 Å, respectively.

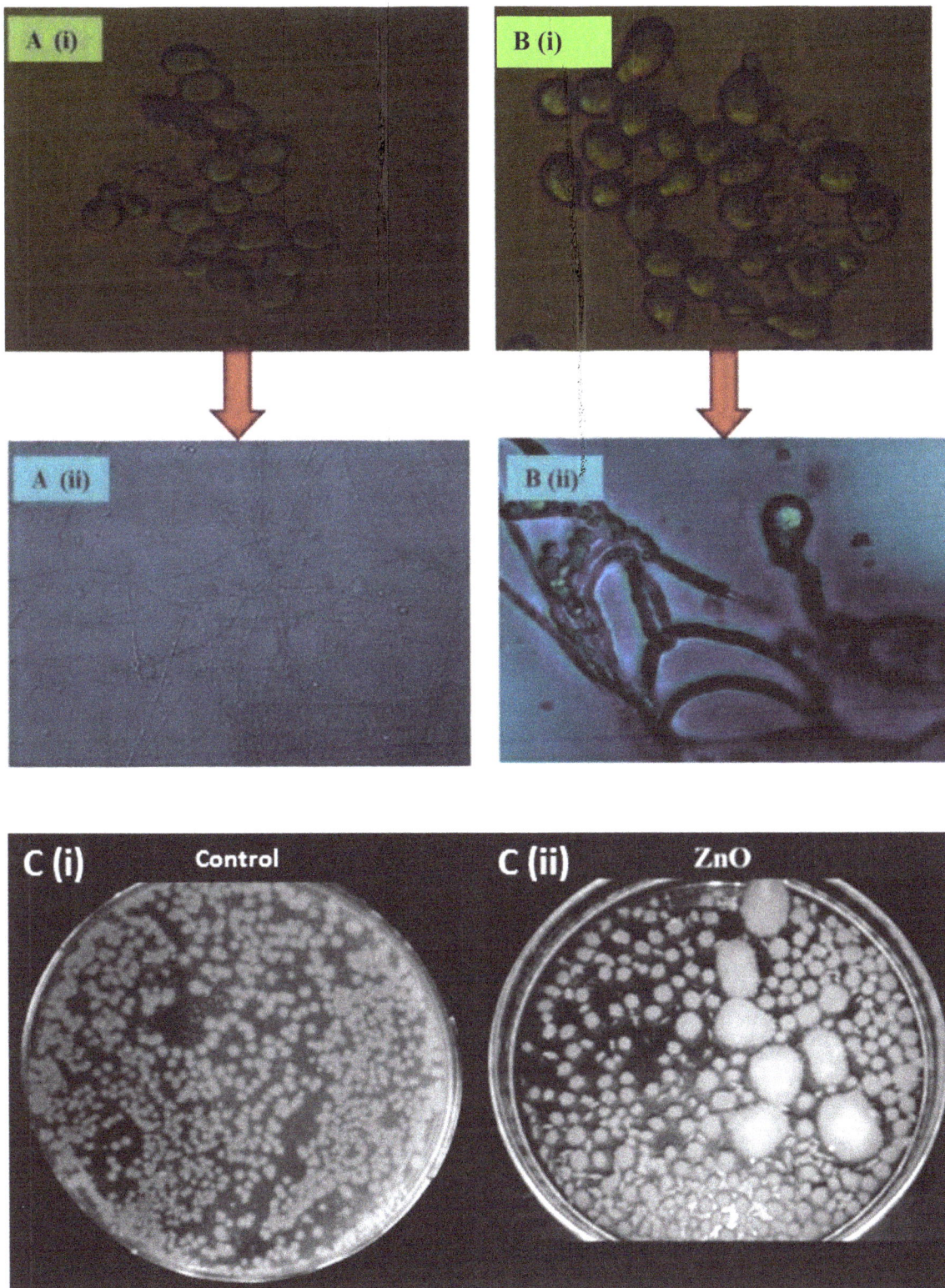

FIGURE 4 | Comparative analysis of *P. indica* spores using light microscopy: **(A)** Without staining **(i)** Control and **(ii)** ZnO-nanorods treated *P. indica,* showing more number of bigger spores. **(B)** With staining using trypan blue **(i)** control and **(ii)** ZnO-nanorods treated *P. indica,* shows the morphogenesis of hyphae and spores with early sporulation as compared to control. **(C)(i)** *P. indica* 'control' and **(ii)** ZnO-nanorods treated *P. indica.*

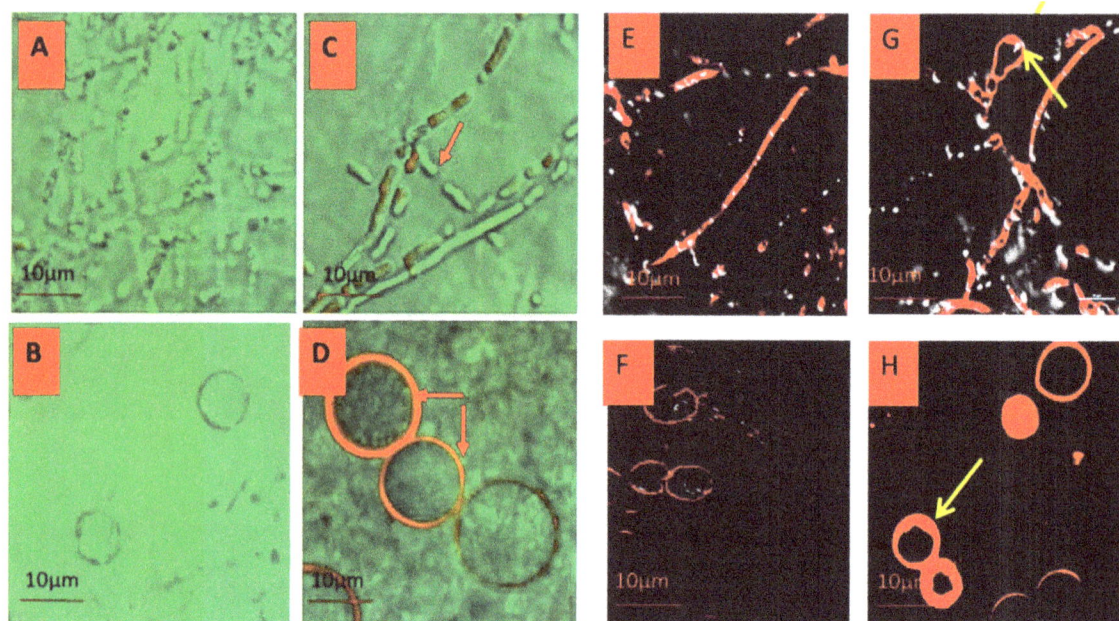

FIGURE 5 | Comparative morphology analysis of *P. indica* before **(A,B)** and after treating it with ZnO-nanorods **(C,D)** by Confocal microscopy using Alexafluor 488 dye and similar analysis was done by using Trypan Blue dye before **(E,F)** and after treatment **(G,H)** with ZnO-nanorods [red arrow shown the thick hyphae as compared to control and early sporulation **(G)**].

Dry Weight

Fresh culture of *P. indica* was treated with ZnO-nanorods of various concentrations of 300, 400, 500, 600, 1000, and 2000 ppm, respectively. The untreated culture was taken as control. The study was conducted in triplicates. The cultures were incubated at 27°C for 8 days. After that the samples were filtered by using Whatman filter paper, and followed by drying of filtered biomass at 70°C in hot air oven for 24 h. The dry weight of the samples was taken by using the metal balances (Balance AE240 Metler). The fungal biomass

observed to be 35.5, 49.2, 56.5, 52.9, 34.7, and 32.6% for ZnO-nanorod concentrations of 300, 400, 500, 600, 1000, and 2000 ppm, respectively. The best growth, i.e., 56.4% was observed when the fungus was incubated with 500 ppm of ZnO-nanorods concentration (**Figure 1D** and **Table 1**). The 500 ppm concentration of ZnO-nanorods is regarded as 'optimized ZnO-nanorods' for interaction with *P. indica*. In all followed studies, *P. indica* is treated with optimized ZnO-nanorods.

Morphological Changes in Fungal Spores When Treated with ZnO-Nanorods

The changes in the size and spores count of the primed fungal samples were analyzed by SEM as shown in **Figure 2**. The SEM images of the fungal culture clearly show pear-shaped chlamydospores. In control, spores were small (diameter $\sim$ 11.4 μm) and less in number whereas in ZnO-nanorods treated *Piriformospora indica* spores were large (diameter $\sim$ 16.4 μm) and more in number as shown in **Figures 2A(i),B(i)**, respectively. **Figures 2A(ii),B(ii)** showed the magnified view of spores in control and ZnO-nanorods treated *P. indica*, respectively. As evident, in control sample spores were rough whereas in ZnO-nanorods treated *P. indica*, spores were smooth.

Figures 3A,B shows the EDX of control and ZnO-nanorods treated *P. indica* samples, respectively. Elemental analysis confirmed the presence of Zinc in ZnO-nanorods treated *P. indica* sample as indicated by peak at 8.6 keV. Inset in the figures shows the atomic and weight percent of all the elements

TABLE 2 | Summarization of effect on *Brassica oleracea* plants when treated with (i) *P. indica* and ZnO-nanorods + *P. indica* in terms of various parameters.

Parameters	Control	*P. indica*	*P. indica* + ZnO-nanords
Seed germination rate ($\pm$ SE)	18.66 ± 1.15^a	20.33 ± 1.52^b	23 ± 1^c
Shoot length (cm) ($\pm$ SE)	5.23 ± 0.02^a	6.87 ± 0.02^b	10.46 ± 0.1528^c
Root length (cm) ($\pm$ SE)	1.51 ± 0.015^a	1.68 ± 0.02^a	2.52 ± 0.02^b
Fresh weight (g) ($\pm$ SE)	5.5 ± 0.2^a	6.13 ± 0.3^b	11.73 ± 0.32^c
Dry weight (g) ($\pm$ SE)	0.933 ± 0.15^a	1.2 ± 0.1^b	3.36 ± 0.25^c

The results of treatments were compared with control. Data are mean $\pm$ standard error of three replicates. Values with different letters within same column show significant differences at P < 0.05 level between treatments according to the Duncan's multiple range test.

FIGURE 6 | *Brassica oleracea* plants treated with fungus and nano embedded fungus: **(A)** Control, i.e., plants without any treatment, **(B)** *P. indica,* and **(C)** ZnO-nanorods + *P. indica.*

like C, Cl, P, Na, and O present in both the samples and Zn in the treated sample.

Spore Count Using Hemocytometer

Spore counts studies were carried out using hemocytometer. In control, the fungus yield was 5.34 ($\pm$ 0.28) $\times$ 10^9 spores/ml and in ZnO-nanorod treated *P. indica* sample; significant increase in sporulation viz. 7.18 ($\pm$ 0.32) $\times$ 10^9 spores per ml ($P < 0.0001$) were observed. **Figures 4A(i,ii),B(i,ii)** showed control (*P. indica* grown without any treatment) and ZnO-nanorod treated *P. indica*, respectively under light microscope. In control, spores were small in size and less in number, and hyphae is thin walled as shown in **Figure 4A(i,ii)**. On the other hand, the large and more number of bigger spores and early sporulation were observed in ZnO-nanorods treated *P. indica* as shown in **Figure 4B(i,ii)**. Based on the dry weight results, ZnO-nanorods (500 ppm) termed as 'optimized ZnO' is selected as a stimulatory agent for growth enhancement of *P. indica* and to study the effect of ZnO-nanorods treated *P. indica* on the growth of *B. oleracea*.

Spread Plate Technique

Effect of optimized ZnO-nanorods on *P. indica* is also studied by spread plate technique. **Figure 4C(i)(ii)** shows 7 days cultured *P. indica* (control) and ZnO-nanorods treated *P. indica* in Petri plates, respectively. In control, small with only 235 ($\pm$ 0.26) distinct colonies were observed whereas large number viz. 270 ($\pm$ 0.21) of bigger distinct colonies with an overall 50% enhancement of fungal biomass in ZnO-nanorods treated *P. indica* were observed.

Confocal Microscopy

Morphology of *P. indica* was viewed under confocal laser scanning microscopy (CLSM) using Alexa fluor 488 (**Figures 5A–D**) and trypan blue (**Figures 5E–H**). Laser excitation at 488 nm resulted in emission in the visible range. In the control specimen, the hyphae (**Figures 5A,E**) were thin walled; spore (**Figures 5B,F**) count was low with morphological deformities and disaggregation. In ZnO-nanorods treated *P. indica*, the hyphal walls were thick and hyaline (**Figures 5C,G**);

spores (**Figures 5D,H**) were large in size, count was more with smooth surface topology.

Interaction with *Brassica oleracea, P. indica* and ZnO-Nanorods

Two treatments were given to *B. oleracea* plants in triplets, *P. indica*, and ZnO-nanorods treated *P. indica* called 'nanoembedded fungus' and the plants without any treatment were taken as control. *B. oleracea* responded variably toward the treatments, results are summarized in **Table 2**. Seeds treated with nanoembedded fungus recorded significant germination rate, significant increase in dry and fresh weight as well as prominent increase in root and shoot length as compared to other treatments as mentioned in **Table 2** and **Figure 6**.

DISCUSSION

The dry weight results indicated that the fungal biomass was heightened approximately two times on interaction with optimized ZnO-nanorods (500 ppm) as compared to the control. The spores were more in number, large sized, smooth and round in ZnO-nanorod treated *P. indica* whereas in control sample, spores were less in number, small sized and rough as evident from SEM and confocal microscopy studies. SEM clearly indicated the stimulating effect on the size of the chlamydospores. The increase in size of the spores was almost 50% in comparison to the control. The results also exhibited the increased spore density of the test fungus on interaction with ZnO-nanorods.

Availability of different nutrients causing diauxic growth attributes to the observed results, as alternative pathways are activated beyond a certain physiological threshold value. The infused nanomaterial of different concentrations leads to the varied stress compensation pathway stimulation (Kotzybik et al., 2016). The interaction between the nanoparticle surface and cell wall impacts tremendously on dispersion of the nutrients and hence affects the growth rate (Suman et al., 2010; Ren et al., 2011; Feng et al., 2013). Researchers reported that cells when subjected to oxidative stress, leads to stimulation of stress compensation mechanisms, making them competent at sub-optimal growth conditions, as compared to those cells which were not pre-stressed (Fillinger et al., 2001; Alvarez-Peral et al., 2002; González-Párraga et al., 2003; Tripathi et al., 2016a,b, 2017).

It was observed that the stage of nanomaterials inclusion is essential. The antimicrobial property of nanomaterials was recorded when added at a late growth phase. However, the inclusion before media sterilization showed a positive and stimulating effect on the fungus, as the case with our present studies. Researchers (Aguilar-Uscanga and Francois, 2003; Siddhanta et al., 2016) reported thick walled hyphae and large number of spores of bigger size on interaction of nanomaterials with *P. indica* as evident by SEM and CLSM studies (Schermelleh et al., 2010). Roncero and Durán (1985) had observed that ZnO-nanorods treated cells were resulted in added multi-cellular aggregates with early germination of spores and abnormally thick septa.

The conflicting effect of nanomaterials at a specific stage of addition can be best explained by considering the mechanism of antimicrobial action. Antimicrobial behavior of the nanoparticles is reported to be due to the presence of electronic effects brought about as a result of changes in the local electronic structure of the surfaces due to small sizes (<100 nm) (Sharma et al., 2016). Nanomaterials, especially silver nanoparticles, strongly interact with the thiol groups of the vital enzymes and inactivate them (Aziz et al., 2015, 2016; Patra and Baek, 2017; Prasad et al., 2017). As a result, the DNA loses its stability to replicate (Morones et al., 2005). It also destabilizes the plasma membrane potential and results in the depletion of intracellular energy bond of ATP, thus resulting in cell death (Lok et al., 2006; Ramalingam et al., 2016). In the present set of experiments, incorporation of the ZnO-nanorods led to the growth promotion of *P. indica* which is supposed to act as media ingredients and carrier for the fast uptake of nutrients and gasses due to their small size, large surface area, and absorption capacity by the test fungus (Gleiter, 2000). Although, the various studies have been performed to understand the interaction between different nanoparticles and mycorrhizal fungi, but due to huge contradiction in the

findings reported for each study. A lot of research has to be done to find out the exact role of different nanoparticles and their actual interaction with mycorrhizal fungi (Kotzybik et al., 2016).

CONCLUSION

Zinc oxide nanorods have been successfully synthesized *via* mechanically assisted thermal decomposition method. In particular, ZnO-nanorods for the first time demonstrated the property of fungal symbiont productivity. Dry weight method has shown the maximum biomass of *P. indica* (about 60%) after interacting with optimized ZnO-nanorods (500 ppm) in comparison to control. Further, the interaction of *P. indica* with ZnO-nanorods significantly increases the number of fungal pellets, spore size, early sporulation, thick hyphae as confirmed by spore count method, scanning electron and confocal microscopic studies. Therefore, it is anticipated that the ZnO-nanorods on interaction with *P. indica* created a novel nanotool "nanoembedded fungus" which has the potential to significantly enhance the crop (*B. oleracea*) productivity as demonstrated in the present studies.

AUTHOR CONTRIBUTIONS

MK, RP, and AV: perceived and designed the experiments; US: conducted the experiments, MK, RP, and AV: analyzed the data; US: prepared the draft; RP, MK, and AV: proofread the final draft. All authors approved the final manuscript.

ACKNOWLEDGMENTS

The author will acknowledge with thanks the partial financial support by BIRAC, DBT FT/12/02/(96)/14/1228, DST SR/NM/NB-1039/2016, and DST-FIST for providing Confocal Microscopy facility. Authors are thankful to Mr. Anil Chandra, AIMT, Amity University for statistical analysis.

REFERENCES

Aguilar-Uscanga, B., and Francois, J. M. (2003). A study of the yeast cell wall composition and structure in response to growth conditions and mode of cultivation. *Lett. Appl. Microbiol* 37, 268–274. doi: 10.1046/j.1472-765X.2003.01394.x

Alvarez-Peral, F. J., Zaragoza, O., Pedreno, Y., and Argüelles, J. C. (2002). Protective role of trehalose during severe oxidative stress caused by hydrogen peroxide and the adaptive oxidative stress response in *Candida albicans*. *Microbiology* 148, 2599–2606. doi: 10.1099/00221287-148-8-2599

Andrade-Linares, D. R., Müller, A., Fakhro, A., Schwarz, D., and Franken, P. (2013). Impact of *Piriformospora indica* on tomato. *Soil Biol.* 33, 107–117.

Ansari, M. W., Bains, G., Shukla, A., Pant, R. C., and Tuteja, N. (2013). A critical review on fungi mediated plant responses with special emphasis to *Piriformospora indica* on improved production and protection of crops. *Plant Physiol. Biochem.* 70, 403–441. doi: 10.1016/j.plaphy.2013.06.005

Aziz, N., Faraz, M., Pandey, R., Shakir, M., Fatma, T., Varma, A., et al. (2015). Facile algae-derived route to biogenic silver nanoparticles: synthesis, antibacterial

and photocatalytic properties. *Langmuir* 31, 11605–11612. doi: 10.1021/acs.langmuir.5b03081

Aziz, N., Pandey, R., Barman, I., and Prasad, R. (2016). Leveraging the attributes of *Mucor hiemalis*-derived silver nanoparticles for a synergistic broad-spectrum antimicrobial platform. *Front. Microbiol.* 7:1984. doi: 10.3389/fmicb.2016.01984

Baral, A., Khanuja, M., Islam, S. S., Sharma, R., and Mehta, B. R. (2017). Identification and origin of visible transitions in one dimensional (1D) ZnO nanostructures: excitation wavelength and morphology dependence study. *J. Lumin.* 183, 383–390. doi: 10.1016/j.jlumin.2016.11.060

Bhuyan, T., Khanuja, M., Sharma, R., Patel, S., Reddy, M. R., Anand, S., et al. (2015a). A comparative study of pure and copper (Cu) doped ZnOnanorods for antibacterial and photocatalytic applications with their mechanism of action. *J. Nanopart. Res.* 17, 288. doi: 10.1007/s11051-015-3093-3

Bhuyan, T., Mishra, K., Khanuja, M., Prasad, R., and Varma, A. (2015b). Biosynthesis of zinc oxide nanoparticles from *Azadirachta indica* for antibacterial and photocatalytic applications. *Mater. Sci. Semicond. Process.* 32, 55–61. doi: 10.1016/j.mssp.2014.12.053

Feng, Y., Cui, X., He, S., Dong, G., Chen, M., and Wang, J. (2013). The role of metal Nanoparticles in influencing arbuscular mycorrhizal fungi effects on plant growth. *Environ. Sci. Technol.* 47, 9496–9504. doi: 10.1021/es402109n

Fillinger, S., Chaveroche, M. K., van Dijck, P., de Vries, R., Ruijter, G., and Thevelein, J. (2001). Trehalose is required for the acquisition of tolerance to a variety of stresses in the filamentous fungus *Aspergillus nidulans*. *Microbiology* 147, 1851–1862. doi: 10.1099/00221287-147-7-1851

Franken, P. (2012). The plant strengthening root endophyte *Piriformospora indica*: potential application and the biology behind. *Appl. Microbiol. Biotechnol.* 96, 1455–1464. doi: 10.1007/s00253-012-4506-1

Gill, S. S., Gill, R., Trivedi, D. K., Anjum, N. A., Sharma, K. K., Ansari, M. W., et al. (2016). *Piriformospora indica*: potential and significance in plant stress tolerance. *Front. Microbiol.* 7:332. doi: 10.3389/fmicb.2016.00332

Gleiter, H. (2000). Nanostructured materials: basic concepts and microstructure. *Acta Mater.* 48, 1–29. doi: 10.1016/S1359-6454(99)00285-2

González-Párraga, P., Hernández, J. A., and Argüelles, J. C. (2003). Role of antioxidant enzymatic defenses against oxidative stress (H2O2) and the acquisition of oxidative tolerance in *Candida albicans*. *Yeast* 20, 1161–1169. doi: 10.1002/yea.1029

Hill, T. W., and Kaefer, E. (2001). Improved protocols for *Aspergillus* medium: trace elements and minimum medium salt stock solutions. *Fungal Genet. News Lett.* 48, 20–21. doi: 10.4148/1941-4765.1173

Huang, M. H., Mao, S., and Feick, H. (2001). Room-temperature ultraviolet nanowire nanolasers. *Science* 292, 1897–1899. doi: 10.1126/science.1060367

Kotzybik, K., Gräf, V., Kugler, L., Stoll, D. A., Greiner, R., Geisen, R., et al. (2016). Influence of different nanomaterials on growth and mycotoxin production of *Penicillium verrucosum*. *PLOS ONE* 11:e0150855. doi: 10.1371/journal.pone.0150855

Lok, C. N., Ho, C. M., Chen, R., He, Q. Y., Yu, W. Y., Sun, H., et al. (2006). Proteomic analysis of the mode of antibacterial action of silver nanoparticles. *J. Proteome Res.* 5, 916–924. doi: 10.1021/pr0504079

Morones, J. R., Elechiguerra, J. L., Camacho, A., Holt, K., Kouri, J. B., Ramirez, J. T., et al. (2005). The bactericidal effect of silver nanoparticles. *Nanotechnology* 16, 2346–2353. doi: 10.1088/0957-4484/16/10/059

Murashige, T., and Skoog, F. (1962). A revised medium for rapid growth and bio assays with tobacco tissue cultures. *Physiol. Plant.* 15, 473–497. doi: 10.1111/j.1399-3054.1962.tb08052.x

Patra, J. K., and Baek, K.-H. (2017). Antibacterial activity and synergistic antibacterial potential of biosynthesized silver nanoparticles against foodborne pathogenic bacteria along with its anticandidal and antioxidant effects. *Front. Microbiol.* 8:167. doi: 10.3389/fmicb.2017.00167

Prasad, R., Bhattacharyya, A., and Nguyen, Q. D. (2017). Nanotechnology in sustainable agriculture: recent developments, challenges, and perspectives. *Front. Microbiol.* 8:1014. doi: 10.3389/fmicb.2017.01014

Prasad, R., Kamal, S., Sharma, P. K., Oelmueller, R., and Varma, A. (2013). Root endophyte *Piriformospora indica* DSM 11827 alters plant morphology, enhances biomass and antioxidant activity of medicinal plant *Bacopa monniera*. *J. Basic Microbiol.* 53, 1016–1024. doi: 10.1002/jobm.201200367

Rajiv, P., Sivaraj, R., and Rajendran, V. (2013). Bio-Fabrication of zinc oxide nanoparticles using leaf extract of *Parthenium hysterophorus* L. and its size-dependent antifungal activity against plant fungal pathogens. *Spectrochim. Acta A* 112, 384–387. doi: 10.1016/j.saa.2013.04.072

Ramalingam, B., Parandhaman, T., and Das, S. K. (2016). Antibacterial effects of biosynthesized silver nanoparticles on surface ultrastructure and nanomechanical properties of Gram-negative bacteria viz. *Escherichia coli* and *Pseudomonas aeruginosa*. *ACS Appl. Mater. Interfaces* 8, 4963–4976. doi: 10.1021/acsami.6b00161

Ren, H. X., Liu, L., Liu, C., He, S. Y., Huang, J., and Li, J. L. (2011). Physiological investigation of magnetic iron oxide nanoparticles towards *Chinese Mung Bean*. *J. Biomed. Nanotechnol.* 7, 677–684. doi: 10.1166/jbn.2011.1338

Roncero, C., and Durán, A. (1985). Effect of Calcofluor white and Congo red on fungal cell wall morphogenesis: *in vivo* activation of chitin polymerization. *J. Bacteriol.* 163, 1180–1185.

Sabir, S., Arshad, M., and Chaudhari, S. K. (2014). Zinc oxide nanoparticles for revolutionizing agriculture: synthesis and applications. *Sci. World J.* 2014:925494. doi: 10.1155/2014/925494

Sawai, J., Igarashi, H., Hashimoto, A., Kokugan, T., and Shimizu, M. (1996). Effect of particle size and heating temperature of ceramic powders on antibacterial activity of their slurries. *J. Chem. Eng. Data Jpn.* 29, 251–256. doi: 10.1252/jcej.29.251

Schermelleh, L., Heintzmann, R., and Leonhardt, H. (2010). A guide to super-resolution fluorescence microscopy. *J. Cell Biol.* 190, 165–175. doi: 10.1083/jcb.201002018

Sharma, R., Khanuja, M., Islam, S. S., Uma, and Varma, A. (2017). Aspect ratio dependent photoinduced antimicrobial and photocatalytic organic pollutant degradation efficiency of ZnOnanorods. *Res. Chem. Intermediates* 43, 5345–5364. doi: 10.1007/s11164-017-2930-7

Sharma, R., Uma, Singh, S., Varma, A., and Khanuja, M. (2016). Visible light induced bactericidal and photocatalytic activity of hydrothermally synthesized BiVO4 nano-octahedrals. *J. Photochem. Photobiol. B* 162, 266–272. doi: 10.1016/j.jphotobiol.2016.06.035

Siddhanta, S., Paidi, S. K., Bushley, K., Prasad, R., and Barman, I. (2016). Exploring morphological and biochemical linkages in fungal growth with label-free light sheet microscopy and Raman spectroscopy. *Chemphyschem* 18, 72–78. doi: 10.1002/cphc.201601062

Singh, S., Tripathi, D. K., Dubey, N. K., and Chauhan, D. K. (2016). Effects of nano-materials on seed germination and seedling growth: striking the slight balance between the concepts and controversies. *Mater. Focus* 5, 195–201. doi: 10.1166/mat.2016.1329

Song, J., Zhou, J., and Wang, Z. L. (2006). Piezoelectric and semiconducting coupled power generating process of a single ZnO belt/wire. *Nano Lett.* 6, 1656–1662. doi: 10.1021/nl060820v

Suman, V., Prasad, R., Jain, V. K., and Varma, A. (2010). Role of nanomaterials in symbiotic fungus growth enhancement. *Curr. Sci.* 99, 1189–1191.

Taheri, M., Qarache, H. A., Qarache, A. A., and Yoosefi, M. (2016). The effects of zinc-oxide nanoparticles on growth parameters of corn (SC704). *STEM Fellowship J.* 1, 17–20. doi: 10.17975/sfj-2015-011

Tebebu, T. Y., Steenhuis, T. S., Dagnew, D. C., Guzman, C. D., Bayabil, H. K., Zegeye, A. D., et al. (2015). Improving efficacy of landscape interventions in the (sub) humid Ethiopian highlands by improved understanding of runoff processes. *Front. Earth Sci.* 3:49. doi: 10.3389/feart.2015.00049

Tripathi, D. K., Mishra, R. K., Singh, S., Singh, S., Vishwakarma, K., Sharma, S., et al. (2017). Nitric oxide ameliorates zinc oxide nanoparticles phytotoxicity in wheat seedlings: implication of the ascorbate-glutathione cycle. *Front. Plant Sci.* 8:1. doi: 10.3389/fpls.2017.00001

Tripathi, D. K., Shweta, S. S., Singh, S., Pandey, R., Singh, V. P., Sharma, N. C., et al. (2016a). An overview on manufactured nanoparticles in plants: uptake, translocation, accumulation and phytotoxicity. *Plant Physiol. Biochem.* 110, 2–12. doi: 10.1016/j.plaphy.2016.07.030

Tripathi, D. K., Singh, S., Singh, S., Dubey, N. K., and Chauhan, D. K. (2016b). Impact of nanoparticles on photosynthesis: challenges and opportunities. *Mater. Focus* 5, 405–411. doi: 10.1166/mat.2016.1327

Wang, X., Ding, Y., Summers, C. J., and Wang, Z. L. (2004). Large-scale synthesis of six-nanometer-wide ZnOnanobelts. *J. Phys. Chem. B* 108, 8773–8777. doi: 10.1021/jp048482e

Wang, X., Yang, X., Chen, S., Li, Q., Wang, W., Hou, C., et al. (2015). Zinc oxide nanoparticles affect biomass accumulation and photosynthesis in *Arabidopsis*. *Front. Plant Sci.* 6:1243. doi: 10.3389/fpls.2015.01243

Weiß, M., Waller, F., Zuccaro, A., and Selosse, M. A. (2016). Sebacinales-one thousand and one interactions with land plants. *New Phytol.* 211, 20–40. doi: 10.1111/nph.13977

Conflict of Interest Statement: The authors declare that the research was conducted in the absence of any commercial or financial relationships that could be construed as a potential conflict of interest.

11

Differential Phytotoxic Impact of Plant Mediated Silver Nanoparticles (AgNPs) and Silver Nitrate (AgNO$_3$) on *Brassica* sp.

Kanchan Vishwakarma[1], Shweta[2], Neha Upadhyay[1], Jaspreet Singh[1], Shiliang Liu[3,4], Vijay P. Singh[5], Sheo M. Prasad[6], Devendra K. Chauhan[2], Durgesh K. Tripathi[7] and Shivesh Sharma[1,7]*

[1] Department of Biotechnology, Motilal Nehru National Institute of Technology Allahabad, Allahabad, India, [2] D D Pant Interdisciplinary Research Lab, Department of Botany, University of Allahabad, Allahabad, India, [3] College of Landscape Architecture, Sichuan Agricultural University, Chengdu, China, [4] College of Agriculture, Food and Natural Resources, University of Missouri, Columbia, MO, United States, [5] Government Ramanuj Pratap Singhdev Post Graduate College, Baikunthpur, India, [6] Ranjan Plant Physiology and Biochemistry Laboratory, Department of Botany, University of Allahabad, Allahabad, India, [7] Centre for Medical Diagnostic and Research, Motilal Nehru National Institute of Technology Allahabad, Allahabad, India

Edited by:
Jayanta Kumar Patra,
Dongguk University Seoul,
South Korea

Reviewed by:
Rupesh Kailasrao Deshmukh,
Laval University, Canada
Sowbiya Muneer,
Gyeongsang National University,
South Korea
Santosh Kumar,
University of Coimbra, Portugal

*Correspondence:
Shivesh Sharma
shiveshsharmamnnit@gmail.com;
ssnvsharma@gmail.com;
shiveshs@mnnit.ac.in

Continuous formation and utilization of nanoparticles (NPs) have resulted into significant discharge of nanosized particles into the environment. NPs find applications in numerous products and agriculture sector, and gaining importance in recent years. In the present study, silver nanoparticles (AgNPs) were biosynthesized from silver nitrate (AgNO$_3$) by green synthesis approach using *Aloe vera* extract. Mustard (*Brassica* sp.) seedlings were grown hydroponically and toxicity of both AgNP and AgNO$_3$ (as ionic Ag$^+$) was assessed at various concentrations (1 and 3 mM) by analyzing shoot and root length, fresh mass, protein content, photosynthetic pigments and performance, cell viability, oxidative damage, DNA degradation and enzyme activities. The results revealed that both AgNPs and AgNO$_3$ declined growth of *Brassica* seedlings due to enhanced accumulation of AgNPs and AgNO$_3$ that subsequently caused severe inhibition in photosynthesis. Further, the results showed that both AgNPs and AgNO$_3$ induced oxidative stress as indicated by histochemical staining of superoxide radical and hydrogen peroxide that was manifested in terms of DNA degradation and cell death. Activities of antioxidants, i.e., ascorbate peroxidase (APX) and catalase (CAT) were inhibited by AgNPs and AgNO$_3$. Interestingly, damaging impact of AgNPs was lesser than AgNO$_3$ on *Brassica* seedlings which was due to lesser accumulation of AgNPs and better activities of APX and CAT, which resulted in lesser oxidative stress, DNA degradation and cell death. The results of the present study showed differential impact of AgNPs and AgNO$_3$ on *Brassica* seedlings, their mode of action, and reasons for their differential impact. The results of the present study could be implied in toxicological research for designing strategies to reduce adverse impact of AgNPs and AgNO$_3$ on crop plants.

Keywords: AgNPs, AgNO$_3$, plant growth, *Brassica*, photosynthetic parameters

INTRODUCTION

Nanotechnology deals with the manipulation in material(s) at atomic and molecular level upto the dimension of 100 nanometers. Extensive research on effect of nanoparticles (NPs) on living systems including plants and other organisms has been carried out in last few decades. NPs, due to specific physicochemical properties, are widely being used in day to day life like in medical science, agriculture, environment and other areas under science and technology (Nigam et al., 2009; Chaloupka et al., 2010; Tripathi et al., 2012; Austin et al., 2014; Majdalawieh et al., 2014; Siripattanakul-Ratpukdi and Fürhacker, 2014). NPs show significant changes in physical and chemical characteristics in comparison to their bulk counterparts, which are due to the presence of free dangling bonds and large surface area (Daniel and Astruc, 2004). Although the potential of nanotechnology is widely recognized, however, its optimized use in agriculture to increase crop yield is still debated. Ions from the industrial discharge form cluster after reduction to form NPs, which might be absorbed by plants (Almutairi and Alharbi, 2015). Regulation of level of industrial impurity in water and soil to prevent their adverse effect on vegetation is still a great challenge (Mueller and Nowack, 2008).

The interaction between plants and nanomaterials is still not studied well (Mehta et al., 2016). Reported studies are contradicting in terms of biotransformation, translocation, toxicity, accumulation and absorption of nanomaterials in plants (Husen and Siddiqi, 2014). Investigation on the effect of silver nanoparticles (AgNPs) on plants is still going on (Kaegi et al., 2010; Nowack, 2010). The effect of AgNPs on plants at higher level seems to be dependent on the age and plants species, NP size and its concentration, various experimental conditions and the method and duration of experimental exposure. Studies have shown that AgNPs shed ionic silver into surrounding and cause oxidative stress and inhibition of respiratory enzymes by generating reactive oxygen species (ROS) (Cheng et al., 2010; Tripathi A. et al., 2017). By reviewing various studies, it was perceived that the reason of toxic effect of AgNPs is still not clear, whether it is caused by ionic sliver or it is caused by its intrinsic property. Current literature on the effect of AgNPs on plants is very limited and thus, it becomes important to check their effect, particularly when their constructive and damaging impacts on environment are unknown.

However, it has been shown that as the amount of AgNPs rises in plants, there is a decrement in root length and biomass, which indicates intensification in toxicity (Yin et al., 2011). Lin and Xing (2007) also showed that nanomaterial interaction with plant had variable effect on seed germination, root initiation and growth and it depends upon the properties and concentration of NPs, and species of the plant. AgNP bioaccumulation in plant cells has been shown to be directly dependent on the system reduction potential (Haverkamp and Marshall, 2009). Thus, there are both positive and negative effects of nanomaterials on seed germination and root growth (Krishnaraj et al., 2012). Besides nanomaterials, the impact of AgNO₃ (precursor of AgNPs) is also very little explored in various plants such as *Cucumis sativus*

(Krishnaraj et al., 2012; Tripathi A. et al., 2017). In addition, there are only few studies that have reported the effect of these NPs on mustard plant (Pandey et al., 2014); however, they does not include their effect on cell death and cell morphology.

Therefore, the present study focuses on assessing biochemical changes caused by both silver nitrate (AgNO₃) and biosynthesized AgNPs at different concentrations on *Brassica* sp. by analyzing seedlings growth, accumulation, oxidative stress, DNA degradation and cell viability and activities of ascorbate peroxidase and catalase.

MATERIALS AND METHODS

Synthesis of Silver Nanoparticles Using Green Synthesis Approach

Aloe vera leaves that seem to be healthy were taken from Roxburgh Botanical Garden, University of Allahabad, India. Sterilized plastic bags were taken and leaves were stored in it. Twenty gram leaves were taken, subjected to washing with double distilled water, dried and chopped into fine pieces to prepare leaf extract. A glass beaker was used to boil the finely cut leaves in 100 ml de-ionized water for 20 min. The content obtained after boiling was subjected to filtration by using Whatman No.1 filter paper. To optimize the synthesis of AgNPs, its precursor (AgNO₃) at different concentrations (40, 60, 80, 100, and 120 mM) was freshly prepared in de-ionized water. To 90 ml of AgNO₃ solution, slow addition of 10 ml of leaf extract was performed with continuous stirring for 20 min followed by incubation for 24 h at room temperature (RT) with sample analysis at every 4 h. Initially, the colorless solution was turned into pale yellow color and after 24 h, the color was altered from pale yellow to reddish brown indicating formation of AgNPs. The solution was then subjected to centrifugation at 12,000 rpm for 15 min, and dispersed in de-ionized water to remove any other biological material with constant stirring at 50–60°C followed by three washings. Spectrum scan was taken on UV-Visible Spectrophotometer from 360 to 700 nm. After optimization, NPs so obtained were dried to get a fine powder. Estimation of Ag was performed as per method described in Tripathi D.K. et al. (2017).

Characterization of Silver Nanoparticles

Ultraviolet-visible spectrophotometer (Eppendorf BioSpectro-meter) was used to characterize AgNPs. Scanning was done in the range of 300–700 nm for the samples by taking de-ionized water as a blank. The shape and size of AgNPs were further determined by transmission electron microscopy (TEM).

Plant Material and Culture Conditions

Mustard (*Brassica* sp.) seeds were procured from the certified supplier of the Allahabad district. Surface sterilization for uniform-sized seeds was performed by using 10% (v/v) sodium hypochlorite solution for 10 min. Seeds were then washed with double-distilled water and soaked for 12 h to break dormancy. On next day, seeds were wrapped in cotton cloth to allow germination. Healthy looking uniform-sized seeds were kept in

Petri plates (150 mm, Riviera™) lined with Whatman No. 1 having half strength Hoagland solution followed by subsequent germination at $25 \pm 2°C$ for 4 days in the dark. After germination, five uniform length seedlings were placed in 40 ml half strength Hoagland solution per pot and subjected to the AgNP and $AgNO_3$ treatments at concentrations of 1 and 3 mM along with control seedlings.

Estimation of Growth

To estimate growth, 10 seedlings from control and treated samples were selected randomly and then weighed for fresh mass estimation. Length of root and shoot was determined by using centimeter scale. Fresh mass of root and shoot of treated and untreated *Brassica* seedlings was measured as described by Dwivedi et al. (2015).

Oxidative Stress

Histochemical staining of superoxide radical $(O_2^{\bullet -})$ was performed according to the method of Kariola et al. (2006). Root sections were kept in petri dishes followed by addition of 6 mM nitro blue tetrazolium (NBT) solution along with 50 mM sodium phosphate (pH: 7.5) and 10 mM sodium azide in the dark for 12 h. Further, root samples were soaked in lactate-glycerol-ethanol (1:1:4 v/v) to stop the $O_2^{\bullet -}$ reaction, followed by subsequent boiling in water for 5 min. Fifty percent ethanol was used to preserve the cleared roots which was further utilized for photography. The protocol of Thordal-Christensen et al. (1997) was utilized to histochemically stain the roots for the presence of hydrogen peroxide (H_2O_2) using 3, 30-diaminobenzidine (DAB) staining. In brief, roots were washed with double distilled water then kept in 1% DAB (pH 3.8; Sigma, United States) for 8 h at 25°C in the light. After staining, samples were subjected to washing and immediately followed by submerging and boiling in 95% ethanol two times for 10 min. Slides were prepared and photographed.

Biochemical Analyses

Catalase (CAT: EC 1.11.3.6) activity was determined in terms of reduction in optical density due to dissociation of H_2O_2 molecules which was recorded at 240 nm having an extinction coefficient of $39.4 \text{ mM}^{-1} \text{ cm}^{-1}$ (Aebi, 1984). A unit of enzyme activity was considered in terms of 1 nmol H_2O_2 dissociated min^{-1}. Ascorbate peroxidase (APX: EC 1.11.1.11) activity was measured as per the protocol of Nakano and Asada (1981). A unit of enzyme activity was considered in terms of 1 nmol ascorbate oxidized min^{-1}. Protein in each sample was determined as per method of Bradford (1976).

Measurement of Total Chlorophyll and Chlorophyll *a* Fluorescence

To determine total chlorophyll, 20 mg fresh leaves were taken from control and treated seedlings. Leaves were then subjected to crushing in 80% acetone followed by pigment extraction and centrifugation. Absorbance of the extract was taken at 663 and 646 nm by using Eppendorf UV-VIS Spectrophotometer. Total chlorophyll was measured according to the method of Lichtenthaler (1987). For measurement of photosynthetic performance of seedlings, chlorophyll *a* fluorescence was observed in the dark-adapted (30 min) state of leaves with the help of hand-held leaf fluorometer (Fluor Pen FP 100, Photon System Instrument, Czech Republic). The fluorescence parameters including maximum photochemical efficiency of PS II (F_v/F_m), photochemical quenching (qP) and non-photochemical quenching (NPQ) were then measured.

Plant DNA Extraction and Agarose Gel Electrophoresis

0.5 g plant tissue was taken on clean glass slide and chopped into a paste using clean single edge razor blade. The paste of plant tissue was used to extract DNA by the protocol suggested by Keb-Llanes et al. (2002). The pellet obtained after isolation procedure was suspended in 50 μL Tris-EDTA (TE) buffer. 1% (w/v) agarose gel was made in 1X Tris-acetate-EDTA (TAE) buffer followed by addition of 3 μL ethidium bromide. 5 μL of extracted DNA from each sample along with 1 μL DNA loading dye was loaded into the wells. Gel was run for 30 min at 85 V and visualized for degradation of DNA in the GelDoc (UVITEC Cambridge).

Measurement of Cell Viability by Fluorescence Spectroscopy and Flow Cytometry

Plant samples were prepared by using the method of Pozarowski and Darzynkiewicz (2004) after slight modifications. Cells (10^6) were suspended in 0.5 mL phosphate buffer saline (PBS) and fixed in 70% ethanol followed by incubation at 4°C for 30 min. Further, cells were washed in 1X PBS and centrifuged at 1200 rpm for 15 min. Supernatant was removed carefully and the pellet having cells was resuspended in 200 μL (Pozarowski and Darzynkiewicz, 2004).

Plant extract prepared above was subjected to staining with propidium iodide (PI which stain nucleic acids), and analyzed through fluorescence microscopy and flow cytometry (FCM). For fluorescence microscopy, addition of 2 μl PI (50 μg/ml) to 200 μl of plant extract was carried out and subjected to incubation for 20 min in the dark to allow cells to retain the stain. After incubation, cells were taken out from the dark for analysis by fluorescence microscopy. The fluorescence data indicates the basic outline with respect to cells' relative number and morphology that is based on cells' auto fluorescence and/or labeling with fluorescent dyes, further helping in characterization of plant cells, resolving them from electronic noise and debris, as well as indicating the cell viability and vitality.

Flow cytometry study was performed using a BD Accuri™ C6 Flow Cytometer having a red laser of 14.7 mW output with 640 nm excitation wavelength and a blue laser of 20 mW output having 488 nm excitation wavelength. For flow cytometry, the cell suspension was stained with 10 μL PI followed by 20 min incubation in dark followed by its analysis in FCM. FCM instrument is based on the detection of forward scattering (FSC) and side scattering (SSC) of light by four different types of fluorescence detectors with optical filters. In the present work, red color fluorescence detector (FL2, 585 nm; PE/PI) was utilized.

Statistical Analysis

All values are means of three independent experiments. The significance of data was confirmed by applying one-way analysis of variance (ANOVA). Comparison with the control and treatment's means was carried out by applying Duncan's Multiple Range Test (DMRT) at $p < 0.05$ significance level.

RESULTS

Synthesis of Nanoparticles by Green Synthesis Process

AgNPs were synthesized in the reaction mixture containing 120 mM AgNO$_3$ and a visible change in color from colorless to brown was observed within 24 h. Supplementary Figures S1a,b shows scanning spectra of concentration and time optimization and Supplementary Figures S2a,b shows difference in color of AgNO$_3$ and AgNPs after synthesis.

Characterization of Synthesized Nanoparticles

The absorbance maxima of formed AgNPs were noticed at 425 nm. The peak was observed in the range of 420–500 nm by UV-Visible spectrophotometer (**Figure 1a**). The TEM image of AgNPs synthesized by using plant extract confirmed the reduction of silver by *Aloe vera* and showed an average size of 47 nm for AgNPs (**Figure 1b**).

Effect on Growth Parameters

Root and shoot fresh mass and length were used to evaluate the impact of AgNO$_3$ and AgNPs on growth of *Brassica* seedlings. The results of root and shoot length clearly revealed that AgNO$_3$ treatments (1 and 3 mM) significantly reduced the length with increasing concentration whereas AgNPs showed slight decrease when both were compared with the control. AgNO$_3$ at 1 mM concentration showed 10.8 and 18.7% reduction in shoot and root length while AgNP at same concentration observed to cause reduction by only 5.3 and 10.9% for shoot and root, respectively (**Figure 2**). However, significant effect was recorded at higher concentration, i.e., 3 mM AgNO$_3$ with 24.8 and

28.1% reduction in shoot and root length as compared to its counterpart AgNPs that showed reduction only by 15.4 and 18.7% for shoot and root, respectively and thus, lesser than AgNO$_3$.

The fresh mass of shoot and root of seedlings treated with AgNO$_3$ and AgNPs was lower than seedlings grown under normal conditions (control) (**Figure 2**). 1 mM AgNO$_3$ caused reduction in shoot and root fresh mass by 11.4 and 28.7% respectively while same concentration of AgNPs led to a reduction by only 8.7 and 13% in shoot and root, respectively. The reduction percentage increased with increasing concentration as there was 18 and 24.2% reduction in shoot and root with 3 mM AgNPs, while it was 25.6 and 31.5%, respectively with 3 mM AgNO$_3$. The observed results indicate that AgNO$_3$ caused more harmful impact on growth characteristics of plant than the biosynthesized AgNPs.

Effect on Photosynthetic Pigments and Photosynthetic Performance

AgNO$_3$ and AgNP treatments (1 and 3 mM) decreased total chlorophyll and carotenoids concentration while the impact of AgNPs at respective concentrations was lesser as compared to the control (**Figures 3A,B**). However, it was clear that percentage reduction for carotenoids was lesser than chlorophylls.

A vision about the health of photosynthetic systems in the leaves is provided by chlorophyll *a* fluorescence. The results showed that reduction in F_v/F_m was significant when plants were treated with AgNO$_3$ as compared to the control (**Figure 4A**). The effect of AgNO$_3$ (1 and 3 mM) was greater than the effect of AgNPs in *Brassica* seedlings. Photochemical (qP) and non-photochemical quenching (NPQ) were also tested under stressed conditions (**Figures 4B,C**). Under given treatments, decrease in qP with increasing values of NPQ were observed in comparison to the control (**Figure 4C**). However, there were no significant changes in NPQ among AgNPs and AgNO$_3$ treatments.

Protein Content

Total protein content in *Brassica* treated with AgNO$_3$ and AgNPs was found to be lesser than protein level in control *Brassica* seedlings (**Figure 3C**). 3 mM of AgNO$_3$ showed greater reduction (16%) with respect to AgNPs at same concentration (7%), when

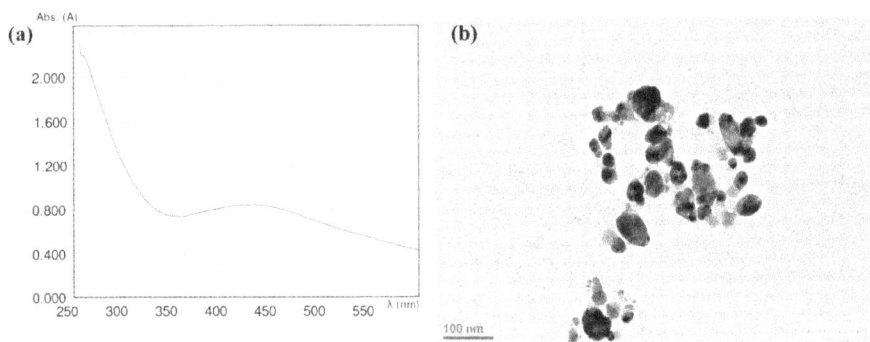

FIGURE 1 | (a) Absorption spectrum and **(b)** TEM image of AgNPs formed in the reaction media.

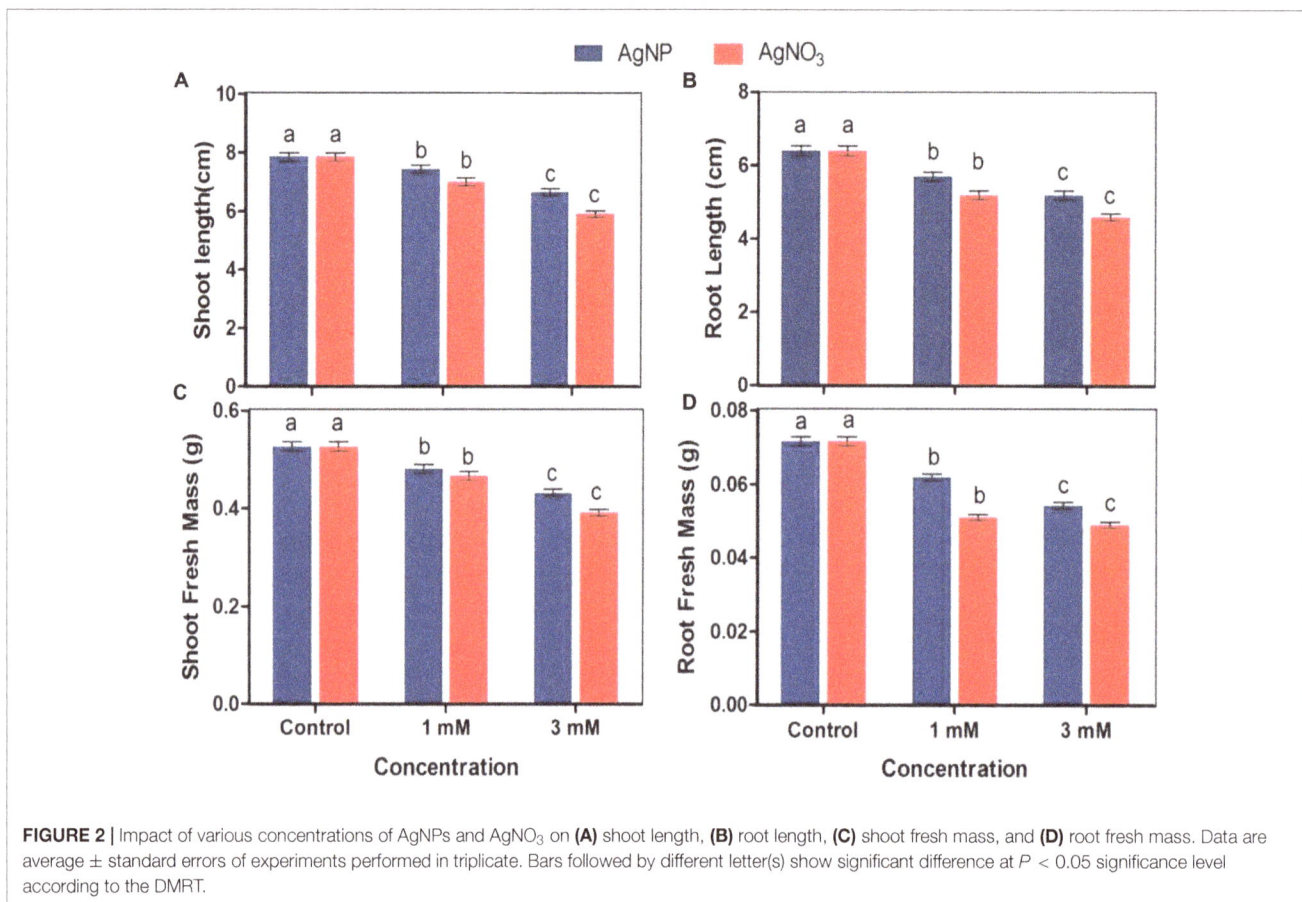

FIGURE 2 | Impact of various concentrations of AgNPs and AgNO$_3$ on **(A)** shoot length, **(B)** root length, **(C)** shoot fresh mass, and **(D)** root fresh mass. Data are average ± standard errors of experiments performed in triplicate. Bars followed by different letter(s) show significant difference at $P < 0.05$ significance level according to the DMRT.

both were compared with the control. In addition, at 1 mM concentration of AgNPs and AgNO$_3$, reduction in total protein content was observed (i.e., 3 and 6%, respectively). The data clearly indicate that damaging effect of AgNO$_3$ was more than AgNPs.

Enzyme Activities

Ascorbate peroxidase (APX) enzyme is involved in dissociation of H$_2$O$_2$ into water and oxygen by utilizing ascorbate as a particular electron donor, while catalase (CAT) also dissociates H$_2$O$_2$ into water and oxygen without any external use of reductant(s). The activities of APX and CAT enzymes were significantly inhibited when *Brassica* seedlings were treated with AgNO$_3$ while inhibition in their activities in the presence of AgNP was found to be lesser than AgNO$_3$ as compared to the control (**Figure 5**).

Accumulation

Brassica seedlings exposed to AgNO$_3$ and AgNPs showed their higher accumulation in roots than shoots. However, it was noticed that AgNO$_3$ at 1 and 3 mM accumulated more in both roots and shoots than AgNPs (**Figure 6**).

Oxidative Damage

The results showed that AgNO$_3$ enhanced accumulation of superoxide radical and hydrogen peroxide in *Brassica* more than

AgNPs as indicated by histochemical staining of roots by NBT (for superoxide radical) and DAB (for H$_2$O$_2$) dye reduction tests. In addition, the appearance of root hairs were reduced under AgNO$_3$ treated seedlings (**Figure 7**).

Degradation of DNA

DNA of the *Brassica* seedlings treated with AgNPs and AgNO$_3$ was isolated by CTAB method followed by its suspension in TE buffer and stored at −40°C till further use. Thereafter, the extracted and stored DNA of each sample was run in agarose gel and it was found that a degradation of DNA took place in samples treated with AgNP and AgNO$_3$ in comparison to the control (**Figure 8**). However, more degradation was observed at 3 mM AgNO$_3$ as well as 3 mM AgNPs concentration when compared to control where no degradation was observed.

Cell Viability Assessment with Fluorescence Microscopy

Brassica seedlings treated with different concentrations of AgNPs and AgNO$_3$ were taken and plant extract was prepared. Qualitative assessment of cell viability was done using fluorescence microscopy after staining the cells in the extract with PI. PI is considered as a marker of the cell death because of its exclusion by cell membrane of live cells. Hence, the fluorescence conferred by the dye generally indicates the

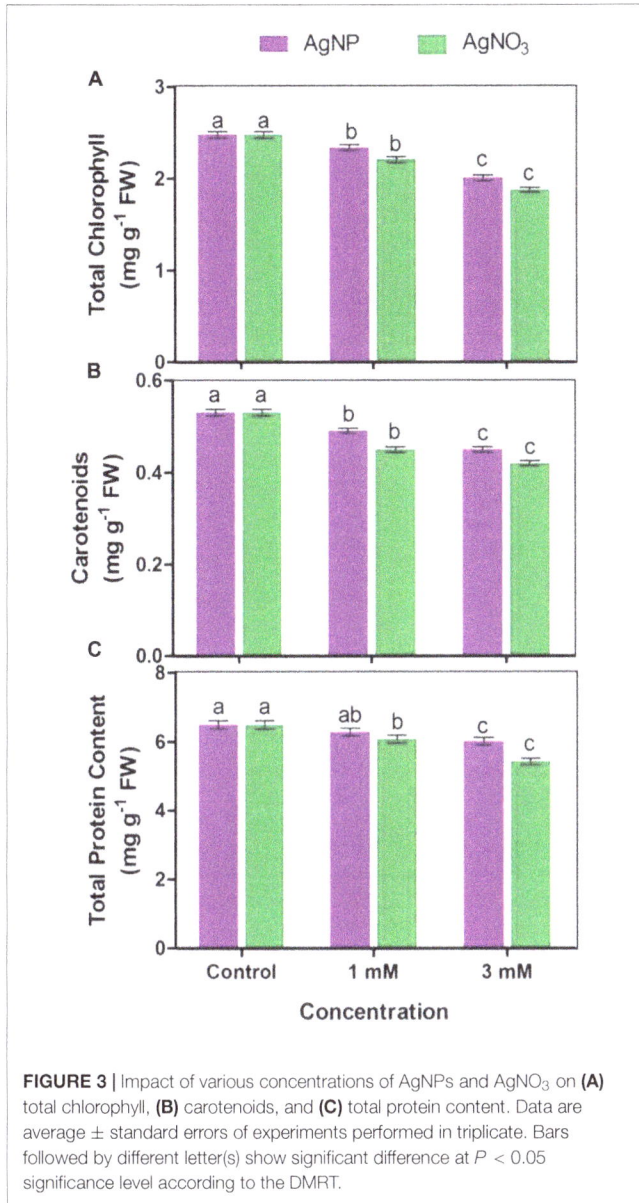

FIGURE 3 | Impact of various concentrations of AgNPs and $AgNO_3$ on (A) total chlorophyll, (B) carotenoids, and (C) total protein content. Data are average ± standard errors of experiments performed in triplicate. Bars followed by different letter(s) show significant difference at $P < 0.05$ significance level according to the DMRT.

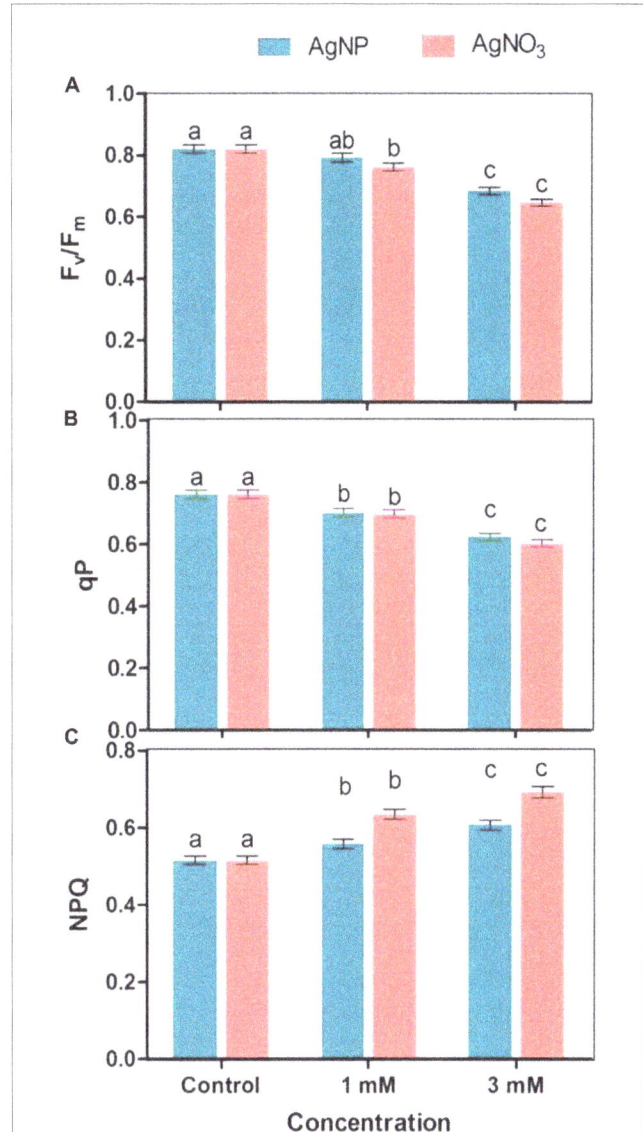

FIGURE 4 | Effect of different concentrations of AgNPs and $AgNO_3$ on photosynthetic parameters such as (A) F_v/F_m, (B) qP: photochemical quenching, and (C) NPQ, non-photochemical quenching. Data are average ± standard errors of experiments performed in triplicate. Bars followed by different letter(s) show significant difference at $P < 0.05$ significance level according to the DMRT.

presence of cells in which cell membrane is damaged or ruptured. After staining, dead cells take up the dye and appear as a red spots upon observation by fluorescence microscopy (**Figures 9b–f**). It is evident from the results that cells under $AgNO_3$ treatment showed more cell death (more stained cells) in comparison to the AgNPs. Number of cell death was higher in treated seedlings than the control (**Figures 9b–f**).

Cell Cycle Analysis by Flow Cytometry

In this study, extract of treated seedlings was stained with PI and observed by flow cytometry. Cells gave high PI fluorescence intensity that indicates the percent cell death in the presence of $AgNO_3$ and AgNPs (**Figure 9a**). The flow cytometric bar graph showed dominant G_0/G_1 profile. It was evident from the results that cell cycle dynamics

of various populations was influenced by treatments given. The plants exposed to 3 mM $AgNO_3$ and AgNP showed a decline in subpopulation of cells in G_0/G_1, S, G_2 phases (**Figure 9a**).

DISCUSSION

The present study was undertaken to investigate the differential impact of $AgNO_3$ and AgNPs on mustard (*Brassica* sp.) seedlings. $AgNO_3$ and AgNPs exposure significantly reduced growth

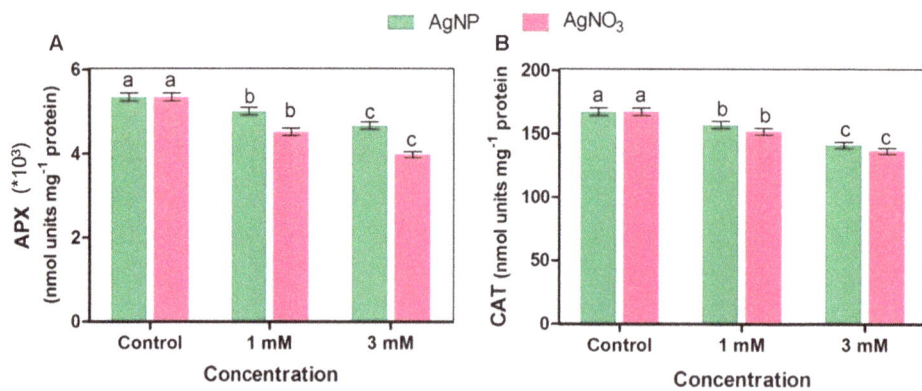

FIGURE 5 | Effect of different concentrations of AgNPs and AgNO$_3$ on enzymes **(A)** APX: ascorbate peroxidase, and **(B)** CAT, catalase. Data are average ± standard errors of experiments performed in triplicate. Bars followed by different letter(s) show significant difference at $P < 0.05$ significance level according to the DMRT.

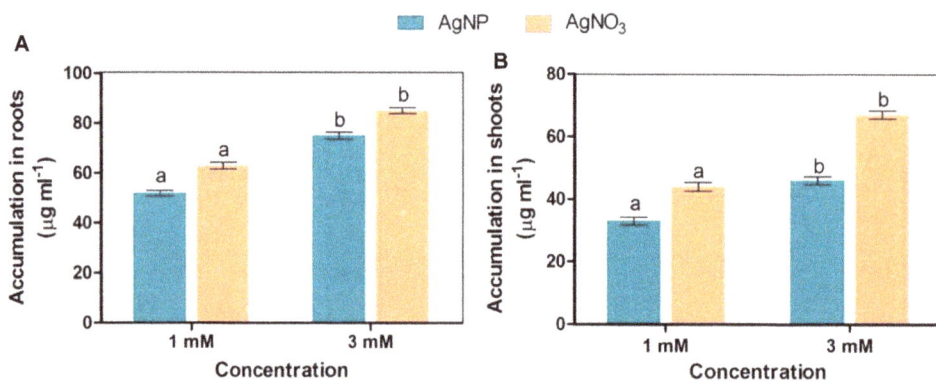

FIGURE 6 | Accumulation status of AgNPs and AgNO$_3$ in **(A)** roots and **(B)** shoots. Data are average ± standard errors of experiments performed in triplicate. Bars followed by different letter(s) show significant difference at $P < 0.05$ significance level according to the DMRT.

parameters in mustard seedlings that might be directly linked with reduction in photosynthetic performance.

More reduction in fresh mass and length of root and shoot, photosynthetic parameters, and protein content was observed under AgNO$_3$ treatment than that under AgNPs when impacts of both were compared with the control. Similarly, in a study by Krishnaraj et al. (2012) showed that plants when given AgNPs exhibited no severe toxic impacts on morphology studied by the scanning electron microscopy (SEM). However, growth retarded at high concentrations of AgNO$_3$ treatment. It was previously stated that reduction in growth characteristics under AgNO$_3$ is due to increased uptake of silver by plants (Harris and Bali, 2008). The above observations suggest that Ag in both forms, i.e., bulk silver and nano silver might show interaction with proteins and components of the plant system. There are only few reports that showed interaction of NPs with plant systems, but observations are mixed in nature (positive or negative). In a study, the impact of AgNPs on growth parameters of three plant species, i.e., watermelon, corn, and zucchini were examined and it was noticed that their germination rates enhanced in response to AgNPs (Almutairi and Alharbi, 2015).

Although, AgNPs imparted toxicity on corn root elongation, however, growth of watermelon and zucchini seedlings was positively influenced by AgNPs (Almutairi and Alharbi, 2015). Additionally, in the present study, some evident signs such as alterations in shoots and root growth were also observed (**Figure 2**). It might be because of uptake of AgNPs through seeds and changes in membrane and other cell structures, as well as defensive mechanisms (Blaser et al., 2008), and modification in cell division and/or cell elongation (Singh et al., 2014). The result that AgNO$_3$ caused more deleterious effects on growth of seedling than AgNPs was in agreement with the work carried out by Sarabi et al. (2015). Interestingly, in this research, there was no definite effect of high concentration of AgNP on mustard seeds. Also, the observed effects of AgNO$_3$ and AgNPs on root and shoot length were in accordance with the study by Pandey et al. (2014).

In our study, a minor decline in length of shoot and root of seedlings under AgNPs treatment was observed (**Figures 2A,B**). However, few other studies have noticed that association of Ag$^+$ ions with roots distorted its epidermis structure and changed anatomical properties of sunflower plant (Krizkova

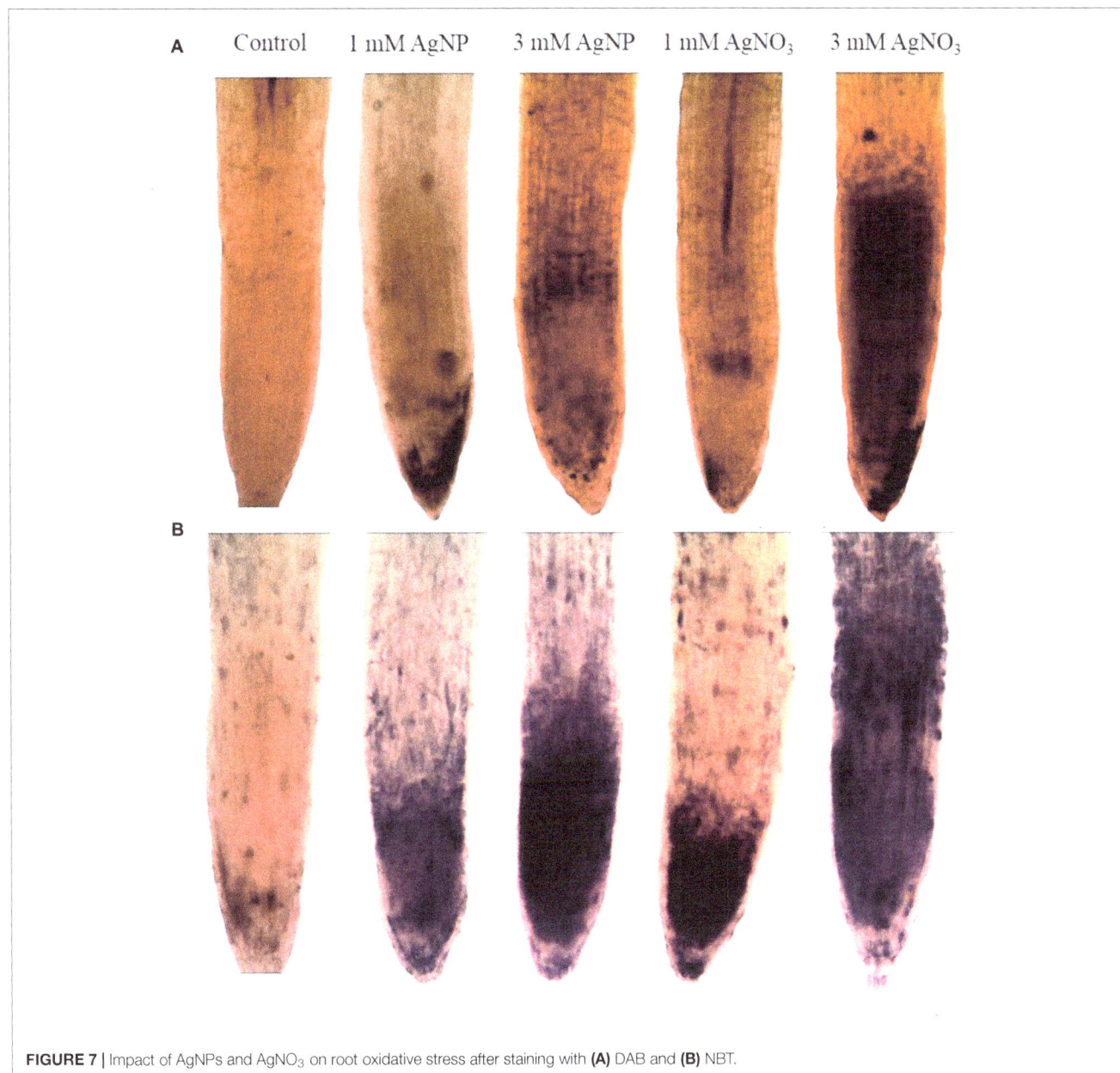

FIGURE 7 | Impact of AgNPs and AgNO$_3$ on root oxidative stress after staining with **(A)** DAB and **(B)** NBT.

et al., 2008). Furthermore, the same plant was observed, and arrested growth of various parts of plant along with damaged root hairs were noticed (Krizkova et al., 2008), while treatment with alumina (Yang and Watts, 2005) and copper (Lee et al., 2008) NPs have retarded the growth of seedlings. Interestingly, not much harmful impact like root hair distortion, and changes in root morphology were observed in the roots of AgNP treated plants, while plants that were given AgNO$_3$ showed more accumulation in roots, which retarded growth of plant (**Figures 2, 6**). This observation is in accordance with earlier studies which mentioned that maximum silver present in the plant remains associated within the roots and the translocation factor ([Ag]$_{shoot}$/[Ag]$_{root}$) is very low (Yin et al., 2012; Pandey et al., 2014).

Photosynthetic efficiency can be detected by measuring some parameters of chlorophyll a fluorescence in $vivo$ under normal and stressed conditions (Pandey et al., 2014). The results confirmed that under AgNO$_3$ exposure, F$_v$/F$_m$ (a reliable marker of photosynthetic efficiency) and qP were markedly declined (**Figures 4A,B**), which could be correlated with reduction in total chlorophyll (**Figure 3A**). However, these parameters when tested under AgNPs exposure showed lesser reduction when compared to AgNO$_3$ treatment, but were significant when compared to the control (**Figure 4**). However, enhanced quantum efficiency and more chlorophyll content were measured in mustard leaves treated with AgNPs (Sharma et al., 2012). Earlier studies have shown that the activity of photosystem II (PS II) is influenced by various stresses that resulted in decreased F$_v$/F$_m$ and qP (Xing et al., 2010; Singh et al., 2014). Substantial decrease in F_v/F_m (**Figure 4A**) and qP (**Figure 4B**) gives the sign of modifications in structure and function of photosynthetic process which might be linked with decrease in biomass accumulation in mustard seedlings. Furthermore, Genty et al. (1990) observed that rise in the values of NPQ when subjected to stressed situations might be due to the down-regulation of photosystem II to evade additional reduction of Q_A so as to equalize the decreased need for electrons through NADPH. In this study, NPQ was enhanced significantly under $AgNO_3$ treatment while its enhancement was lesser under AgNPs treatment, suggesting that AgNPs allow the proper working of electron transport chain than $AgNO_3$ (**Figure 4C**).

The decrease in growth of mustard seedlings subjected to $AgNO_3$ could be correlated with generation of more ROS as measured by histochemical staining by NBT and DAB (**Figure 7**), which are responsible for causing lipid peroxidation and protein oxidation. AgNPs treatment caused little effect on oxidative stress as compared to $AgNO_3$. The reason of such remarkable reduction in growth after treatments might be attributed to ROS mediated damages to the biomolecules linked with the photosynthetic apparatus, which mainly involved the membranes of energy transduction system as reported in a recent study (Tripathi D.K. et al., 2017). Earlier, it was studied that stress factor accelerates the generation of ROS (H_2O_2) (Sharma et al., 2012; Jiang et al., 2017). For regulation of ROS levels inside the cell, plants have fine-tuned antioxidant defense system. In the present study, noteworthy ($P < 0.05$) decrease in APX (**Figure 5A**) and CAT (**Figure 5B**) activity was observed in $AgNO_3$ treated seedlings, as compared to control and AgNPs. Other NPs like CuNP and its bulk $CuCl_2$ have also been found to decrease CAT activity because of direct involvement of copper to interact with thiol moieties of protein resulting in altered tertiary structure of CAT and ultimate inhibition (Atli et al., 2006). Possibly a similar mechanism may operate for Ag resulting in increased production of ROS. In a study by Sharma et al. (2012), it was shown that AgNPs support the growth of *Brassica juncea* seedlings by controlling their antioxidant level. Similarly, Mehta et al. (2016) observed that AgNPs application did not affect growth parameters in wheat seedlings. Likewise, in the present study, though AgNP enhanced the production of ROS but the increase was not significant when compared to the control. It was in support of a recent study, which showed that such effect of AgNPs was related to the blocking of electrons transfer that induces oxidative stress under light (Zou et al., 2016). As observed here that mustard seedlings under AgNPs treatment showed decrease in CAT and APX activities as compared to the control but relatively lesser reduction was observed than $AgNO_3$ (**Figure 5**), this might indicate the interaction of AgNPs with proteins present in cytosol and lipid bilayer; thereby, altering the configuration and damaging the antioxidant defense systems (Saptarshi et al., 2013; Hayashi et al., 2013; McShan et al., 2014; Mendes et al., 2015).

From the results, it was apparent that AgNPs synthesized biologically, were polydispersive in nature, decreased ability of penetrating and bioaccumulation, and posed relatively lesser toxic impact than silver nitrate (**Figure 6**). In general, it is

mentioned that AgNPs uptake and translocation across the root cells depend upon type and concentration of ions and plant species (Lin and Xing, 2007; Albanese and Chan, 2011; Dos Santos et al., 2011). Since proteins adsorb on the NP surface almost instantly and then NP-protein complex enters the cell (Saptarshi et al., 2013), therefore it might be probable that due to the reduction in total protein content (**Figure 3C**), reduced accumulation of AgNPs was observed with its increasing concentration.

The observation on impact of $AgNO_3$ and AgNPs on total protein content is shown in **Figure 3C**. The high concentrations of $AgNO_3$ decreased level of total protein in mustard seedlings as compared to AgNPs and control. It is known from earlier investigations that NPs induce conformational changes in proteins when present in high concentrations. $AgNO_3$ treatment at high concentration showed 16% reduction in total protein content when compared to the control and same concentration of AgNPs (6% reduction) (**Figure 3C**). It might be due to the reason that AgNP and $AgNO_3$ cause discrete modifications in the proteome of root and shoot (Vannini et al., 2013) and interfere with the protein interactions, cell-signaling and DNA transcription (Saptarshi et al., 2013).

The study showed that there was a substantial reduction in number of live plant cells at high concentration of $AgNO_3$ when observed under fluorescence microscopy. The notable increase in fluorescence intensity of PI at high concentrations (**Figures 9b–f**) reflects the increase in cell membrane damage. Furthermore, under AgNPs and $AgNO_3$ treatments, difference in cell death was witnessed by changing their concentrations

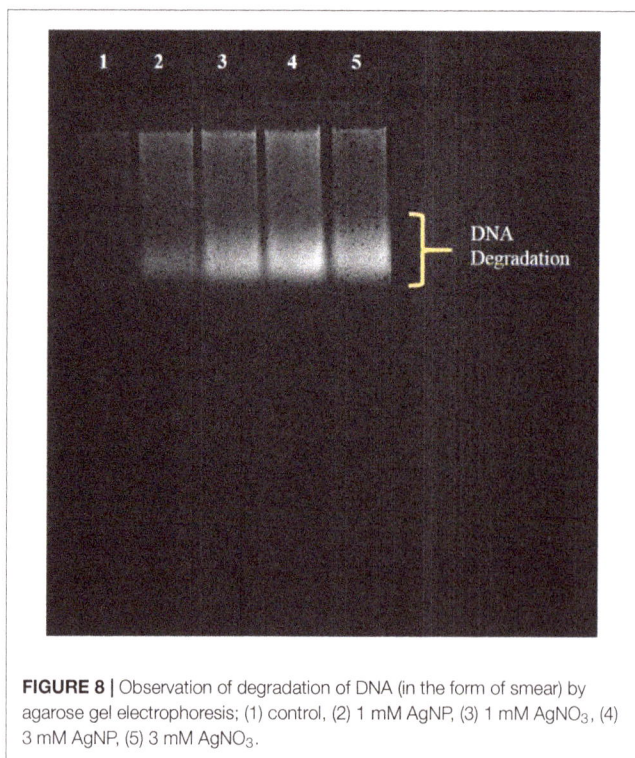

FIGURE 8 | Observation of degradation of DNA (in the form of smear) by agarose gel electrophoresis; (1) control, (2) 1 mM AgNP, (3) 1 mM $AgNO_3$, (4) 3 mM AgNP, (5) 3 mM $AgNO_3$.

using fluorescence microscopy. Maximum cell death in plant samples corresponds to the presence of greater number of cells that have acquired PI, which might be correlated with the decreased growth of mustard plants with AgNO$_3$ treatments as compared to the AgNPs and control. In addition, it was evident from **Figure 8** that degradation of DNA had occurred during treatments of AgNO$_3$ and AgNPs as a result of loss in membrane integrity due to which PI is able to intercalate with DNA of plant cells because of uptake and accumulation of silver inside the cells. Still, the results could not clarify the rise in PI fluorescence intensity at extreme stress conditions, whether it might be either due to the damage in membrane at localized

position or due to the complete loss of membrane integrity. Treatment of AgNO$_3$ and AgNPs on plant cells influenced the membrane potential (difference in internal and external electrical potential of a cell) of cytoplasm resulting in damaged membrane permeability. In control, very less intensity was observed, whereas a notable increase was observed at 3 mM AgNO$_3$ concentration. However, at 3 mM AgNPs treatment, the intensity was found to be less in comparison to the 3 mM AgNO$_3$ concentration (**Figures 9e,f**). Cell cycle dynamics studied by flow cytometry was also found to be affected under various treatments. Variations in the cell percentage occurred at different phases; decreasing the number of cells in S, G$_2$ phase

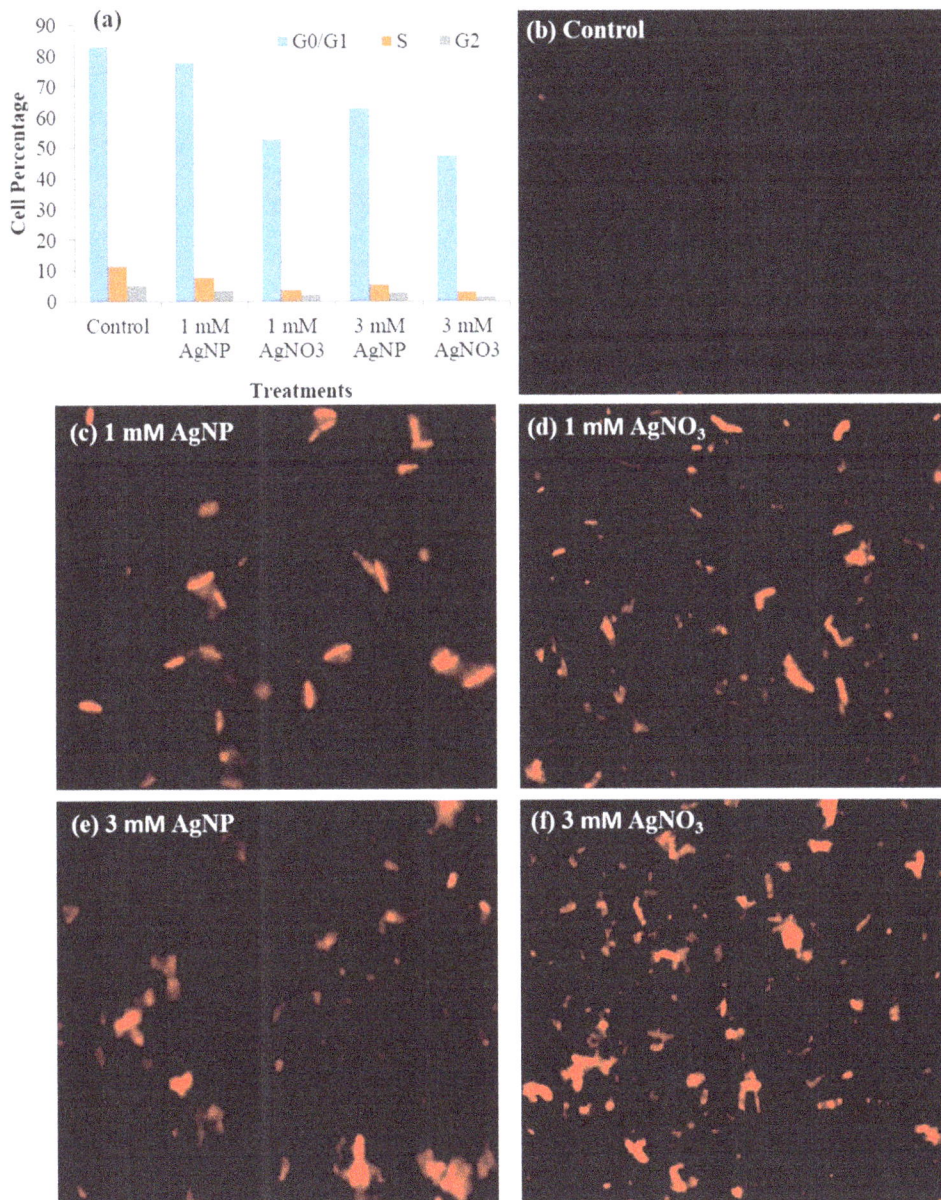

FIGURE 9 | (a) Cell cycle profile of plant analyzed by flow cytometry showing percent of cells in each phase, **(b–f)** Fluorescence microscopy images of plant cells treated with different concentrations of AgNP and AgNO$_3$, and stained with PI.

and G_0/G_1 phase indicating that synthesis of DNA was hampered during cell cycle under treatments. (**Figure 9a**). It was also supported by the result of gel electrophoresis that disruption of nucleic acids occurred in treatments; thereby resulting in degradation of DNA (**Figure 8**). Therefore, from the results, it is concluded that the toxicity of $AgNO_3$ was more prominent than AgNPs.

CONCLUSION

It is evident from the results that $AgNO_3$ posed more deleterious effects on growth of mustard seedling. Although, silver NPs did affect the growth at high concentrations, its impact is lower than that of $AgNO_3$. Hence, from this study, it can be concluded that the interacting ability of silver prompted the stressed condition for growth and metabolism of *Brassica* sp. Among the two treatment conditions, $AgNO_3$ treated plants had lower levels of total protein, carotenoids, CAT and APX activities while a somewhat higher level was observed in AgNPs treated seedlings. The cell viability assessment by fluorescence microscopy and cell cycle analysis by flow cytometry concluded that treatment with $AgNO_3$ posed harmful effects on structural and functional properties

of the cell. Therefore, in view of present findings, $AgNO_3$ interaction retards the growth of plants by imparting toxicity, whereas biologically synthesized AgNPs interaction on the growth and metabolism of mustard seedlings impose mild stress condition.

AUTHOR CONTRIBUTIONS

KV, DT, S, VS, SP, and NU designed the manuscript. KV, DT, S, NU, and JS performed the experiment and wrote the manuscript. SL, DT, DC, and SS critically evaluated the manuscript.

ACKNOWLEDGMENTS

Authors are thankful to Director MNNIT Allahabad for providing necessary facilities. The services and facilities provided by Centre for Medical Diagnostic and Research, MNNIT Allahabad, MHRD sponsored project "Design Innovation Centre" and Department of Botany, University of Allahabad are also acknowledged.

REFERENCES

Aebi, I. I. (1984). Catalase *in vitro*. *Methods Enzymol.* 105, 121–126. doi: 10.1016/S0076-6879(84)05016-3

Albanese, A., and Chan, W. C. W. (2011). Effect of gold nanoparticle aggregation on cell uptake and toxicity. *ACS Nano* 5, 5478–5489. doi: 10.1021/nn2007496

Almutairi, Z. M., and Alharbi, A. (2015). Effect of silver nanoparticles on seed germination of crop plants. *J. Adv. Agric.* 4, 283–288. doi: 10.1016/j.scitotenv.2013.02.059

Atli, G., Alptekin, Ö., Tükel, S., and Canli, M. (2006). Response of catalase activity to Ag^+, Cd^{2+}, Cr^{6+}, Cu^{2+} and Zn^{2+} in five tissues of freshwater fish *Oreochromis niloticus*. *Comp. Biochem. Phys. C. Toxicol Pharmacol.* 143, 218–224.

Austin, L. A., Mackey, M. A., Dreaden, E. C., and El-Sayed, M. A. (2014). The optical, photothermal, and facile surface chemical properties of gold and silver nanoparticles in biodiagnostics, therapy, and drug delivery. *Arch. Toxicol.* 88, 1391–1417. doi: 10.1007/s00204-014-1245-3

Blaser, S. A., Scheringer, M., Macleod, M., and Hungerbühler, K. (2008). Estimation of cumulative aquatic exposure and risk due to silver: contribution of nano-functionalized plastics and textiles. *Sci. Total. Environ.* 390, 396–409. doi: 10.1016/j.scitotenv.2007.10.010

Bradford, M. M. (1976). A rapid and sensitive method for the quantitation of microgram quantities of protein utilizing the principle of protein-dye binding. *Anal. Biochem.* 72, 248–254. doi: 10.1016/0003-2697(76)90527-3

Chaloupka, K., Malam, Y., and Seifalian, A. M. (2010). Nanosilver as a new generation of nanoproduct in biomedical applications. *Trends Biotechnol.* 28, 580–588. doi: 10.1016/j.tibtech.2010.07.006

Cheng, L., Shao, M., Zhang, M., and Ma, D. D. (2010). An ultrasensitive method to detect dopamine from single mouse brain cell: surface-enhanced Raman scattering on Ag nanoparticles from beta-silver vanadate and copper. *Sci. Adv. Mater.* 2, 386–389. doi: 10.1039/b802405g

Daniel, M. C., and Astruc, D. (2004). Gold nanoparticles: assembly, supramolecular chemistry, quantum-size-related properties, and applications toward biology,

catalysis, and nanotechnology. *Chem. Rev.* 104, 293–346. doi: 10.1021/cr030698+

Dos Santos, T., Varela, J., Lynch, I., Salvati, A., and Dawson, K. A. (2011). Quantitative assessment of the comparative nanoparticle-uptake efficiency of a range of cell lines. *Small* 7, 3341–3349. doi: 10.1002/smll.201101076

Dwivedi, R., Singh, V. P., Kumar, J., and Prasad, S. M. (2015). Differential physiological and biochemical responses of two *Vigna* species under enhanced UV-B radiation. *J. Radiat. Res. Appl. Sci.* 8, 173–181. doi: 10.1016/j.jrras.2014.12.002

Genty, B., Harbinson, J., Briantais, J. M., and Baker, N. R. (1990). The relationship between non-photochemical quenching of chlorophyll fluorescence and the rate of photosystem 2 photochemistry in leaves. *Photosynth. Res.* 25, 249–257. doi: 10.1007/BF00033166

Harris, A. T., and Bali, R. (2008). On the formation and extent of uptake of silver nanoparticles by live plants. *J. Nanopart. Res.* 10, 691–695. doi: 10.1007/s11051-007-9288-5

Haverkamp, R. G., and Marshall, A. T. (2009). The mechanism of metal nanoparticle formation in plants: limits on accumulation. *J. Nanopart. Res.* 11, 1453–1463. doi: 10.1007/s11051-008-9533-6

Hayashi, Y., Miclaus, T., Scavenius, C., Kwiatkowska, K., Sobota, A., Engelmann, P., et al. (2013). Species differences take shape at nanoparticles: protein corona made of the native repertoire assists cellular interaction. *Environ. Sci. Technol.* 47, 14367–14375. doi: 10.1021/es404132w

Husen, A., and Siddiqi, K. S. (2014). Phytosynthesis of nanoparticles: concept, controversy and application. *Nanoscale Res. Lett.* 9:229. doi: 10.1186/1556-276X-9-229

Jiang, H. S., Yin, L. Y., Ren, N. N., Zhao, S. T., Li, Z., Zhi, Y., et al. (2017). Silver nanoparticles induced reactive oxygen species via photosynthetic energy transport imbalance in an aquatic plant. *Nanotoxicology* 11, 157–167. doi: 10.1080/17435390.2017.1278802

Kaegi, R., Sinnet, B., Zuleeg, S., Hagendorfer, H., Mueller, E., Vonbank, R., et al. (2010). Release of silver nanoparticles from outdoor facades. *Environ. Pollut.* 158, 2900–2905. doi: 10.1016/j.envpol.2010.06.009

Kariola, T., Brader, G., Helenius, E., Li, J., Heino, P., and Palva, E. T. (2006). *Early responsive to dehydration 15*, a negative regulator of ABA-responses in *Arabidopsis*. *Plant Physiol.* 142, 1559–1573. doi: 10.1104/pp.106.086223

Keb-Llanes, M., González, G., Chi-Manzanero, B., and Infante, D. (2002). A rapid and simple method for small-scale DNA extraction in Agavaceae and other tropical plants. *Plant Mol. Biol. Rep.* 20, 299–300. doi: 10.1007/BF02782465

Krishnaraj, C., Jagan, E. G., Ramachandran, R., Abirami, S. M., Mohan, N., and Kalaichelvan, P. T. (2012). Effect of biologically synthesized silver nanoparticles on *Bacopa monnieri* (Linn.) Wettst. plant growth metabolism. *Process. Biochem.* 47, 651–658. doi: 10.1016/j.procbio.2012.01.006

Krizkova, S., Ryant, P., Krystofova, O., Adam, V., Galiova, M., and Beklova, M. (2008). Multi-instrumental analysis of tissues of sunflower plants treated with silver(I) ions—plants as bioindicators of environmental pollution. *Sensors* 8, 445–463. doi: 10.3390/s8010445

Lee, W. M., An, Y. J., Yoon, H., and Kweon, H. S. (2008). Toxicity and bioavailability of copper nanoparticles to the terrestrial plants mung bean (*Phaseolus radiatus*) and wheat (*Triticum aestivum*): plant agar test for water-insoluble nanoparticles. *Environ. Toxicol. Chem.* 27, 1915–1921. doi: 10.1897/07-481.1

Lichtenthaler, H. K. (1987). "Chlorophylls and carotenoids, the pigments of photosynthetic biomembranes," in *Methods in Enzymology*, eds R. Douce and L. Packer (New York, NY: Academic Press), 350–382.

Lin, D., and Xing, B. (2007). Phytotoxicity of nanoparticles: inhibition of seed germination and root growth. *Environ. Pollut.* 150, 243–250. doi: 10.1016/j.envpol.2007.01.016

Majdalawieh, A., Kanan, M. C., El-Kadri, O., and Kanan, S. M. (2014). Recent advances in gold and silver nanoparticles: synthesis and applications. *J. Nanosci. Nanotechnol.* 14, 4757–4780. doi: 10.1166/jnn.2014.9526

McShan, D., Ray, P. C., and Yu, H. (2014). Molecular toxicity mechanism of nanosilver. *J. Food Drug Anal.* 22, 116–127. doi: 10.1016/j.jfda.2014.01.010

Mehta, C. M., Srivastava, R., Arora, S., and Sharma, A. K. (2016). Impact assessment of silver nanoparticles on plant growth and soil bacterial diversity. *3 Biotech* 6, 254. doi: 10.1007/s13205-016-0567-7

Mendes, L. A., Maria, V. L., Scott-Fordsmand, J. J., and Amorim, M. J. (2015). Ag nanoparticles (Ag NM300K) in the terrestrial environment: effects at population and cellular level in *Folsomia candida* (Collembola). *Int. J. Environ. Res. Public Health* 12, 12530–12542. doi: 10.3390/ijerph121012530

Mueller, N. C., and Nowack, B. (2008). Exposure modeling of engineered nanoparticles in the environment. *Environ. Sci. Technol.* 42, 4447–4453. doi: 10.1021/es7029637

Nakano, Y., and Asada, K. (1981). Hydrogen peroxide is scavenged by Ascorbate specific peroxidase in spinach chloroplasts. *Plant Cell Physiol.* 22, 867–880.

Nigam, N., Kumar, S., Dutta, P. K., and Ghose, T. (2009). Preparation and characterization of chitosan based silver nanocomposites as bio-optical material. *Asian Chitin J.* 5, 97–100.

Nowack, B. (2010). Nanosilver revisited downstream. *Science* 330, 1054–1055. doi: 10.1126/science.1198074

Pandey, C., Khan, E., Mishra, A., Sardar, M., and Gupta, M. (2014). Silver nanoparticles and its effect on seed germination and physiology in *Brassica juncea* L. (Indian mustard) plant. *Adv. Sci. Lett.* 20, 1673–1676. doi: 10.1166/asl.2014.5518

Pozarowski, P., and Darzynkiewicz, Z. (2004). Analysis of cell cycle by flow cytometry. *Methods Mol. Biol.* 281, 301–311. doi: 10.1385/1-59259-811-0:301

Saptarshi, S. R., Duschl, A., and Lopata, A. L. (2013). Interaction of nanoparticles with proteins: relation to bio-reactivity of the nanoparticle. *J. Nanobiotechnol.* 11:26. doi: 10.1186/1477-3155-11-26

Sarabi, M., Afshar, A. S., and Mahmoodzadeh, H. (2015). Physiological analysis of silver nanoparticles and AgNO3 effect to *Brassica napus* L. *J. Chem. Health Risks* 5, 285–294.

Sharma, P., Bhatt, D., Zaidi, M. G. H., Saradhi, P. P., Khanna, P. K., and Arora, S. (2012). Silver nanoparticle-mediated enhancement in growth and antioxidant status of *Brassica juncea*. *Appl. Biochem. Biotechnol.* 167, 2225–2233. doi: 10.1007/s12010-012-9759-8

Singh, V. P., Kumar, J., Singh, S., and Prasad, S. M. (2014). Dimethoate modifies enhanced UV-B effects on growth, photosynthesis and oxidative stress in mung bean (*Vigna radiata* L.) seedlings: implication of salicylic acid. *Pestic. Biochem. Physiol.* 116, 13–23. doi: 10.1016/j.pestbp.2014.09.007

Siripattanakul-Ratpukdi, S., and Fürhacker, M. (2014). Review: issues of silver nanoparticles in engineered environmental treatment systems. *Water Air Soil Pollut.* 225, 1–18. doi: 10.1007/s11270-014-1939-4

Thordal-Christensen, H., Zhang, Z., Wei, Y., and Collinge, D. B. (1997). Subcellular localization of H_2O_2 in plants. H_2O_2 accumulation in papillae and hypersensitive response during the barley-powdery mildew interaction. *Plant J.* 11, 1187–1194. doi: 10.1046/j.1365-313X.1997.11061187.x

Tripathi, A., Liu, S., Singh, P. K., Kumar, N., Pandey, A. C., Tripathi, D. K., et al. (2017). Differential phytotoxic responses of silver nitrate (AgNO3) and silver nanoparticle (AgNps) in *Cucumis sativus* L. *Plant Gene.* (in press). doi: 10.1016/j.plgene.2017.07.005

Tripathi, D. K., Singh, S., Singh, V. P., Prasad, S. M., Dubey, N. K., and Chauhan, D. K. (2017). Silicon nanoparticles more effectively alleviated UV-B stress than silicon in wheat (*Triticum aestivum*) seedlings. *Plant Physiol. Biochem.* 110, 70–81. doi: 10.1016/j.plaphy.2016.06.026

Tripathi, D. K., Singh, V. P., Kumar, D., and Chauhan, D. K. (2012). Rice seedlings under cadmium stress: effect of silicon on growth, cadmium uptake, oxidative stress, antioxidant capacity and root and leaf structures. *Chem. Ecol.* 28, 281–291. doi: 10.1080/02757540.2011.644789

Vannini, C., Domingo, G., Onelli, E., Prinsi, B., Marsoni, M., Espen, L., et al. (2013). Morphological and proteomic responses of *Eruca sativa* exposed to silver nanoparticles or silver nitrate. *PLoS ONE* 8:68752. doi: 10.1371/journal.pone.0068752

Xing, W., Li, D., and Liu, G. (2010). Antioxidative responses of *Elodea nuttallii* (Planch.) H. St. John to short-term iron exposure. *Plant Physiol. Biochem.* 48, 873–878. doi: 10.1016/j.plaphy.2010.08.006

Yang, L., and Watts, D. J. (2005). Particle surface characteristics may play an important role in phytotoxicity of alumina nanoparticles. *Toxicol. Lett.* 158, 122–132. doi: 10.1016/j.toxlet.2005.03.003

Yin, L., Benjamin, P. C., Bonnie, M. M., Justin, P. W., and Emily, S. B. (2012). Effects of silver nanoparticle exposure on germination and early growth of eleven wetland plants. *PLoS ONE* 7:47674. doi: 10.1371/journal.pone.0047674

Yin, L., Cheng, Y., Espinasse, B., Colman, B. P., Auffan, M., Wiesner, M., et al. (2011). More than the ions: the effects of silver nanoparticles on *Lolium multiflorum*. *Environ. Sci. Technol.* 45, 2360–2367. doi: 10.1021/es103995x

Zou, X., Li, P., Huang, Q., and Zhang, H. (2016). The different response mechanisms of *Wolffia globosa*: light-induced silver nanoparticle toxicity. *Aquat. Toxicol.* 176, 97–105. doi: 10.1016/j.aquatox.2016.04.019

Conflict of Interest Statement: The authors declare that the research was conducted in the absence of any commercial or financial relationships that could be construed as a potential conflict of interest.

12

Uptake, Accumulation and Toxicity of Silver Nanoparticle in Autotrophic Plants, and Heterotrophic Microbes: A Concentric Review

Durgesh K. Tripathi[1,2]*, Ashutosh Tripathi[3], Shweta[3], Swati Singh[3], Yashwant Singh[3], Kanchan Vishwakarma[4], Gaurav Yadav[2,4], Shivesh Sharma[2,4], Vivek K. Singh[5,6], Rohit K. Mishra[2], R. G. Upadhyay[7], Nawal K. Dubey[1], Yonghoon Lee[8] and Devendra K. Chauhan[3]*

[1] Centre of Advanced Study in Botany, Banaras Hindu University, Varanasi, India, [2] Center for Medical Diagnostic and Research, Motilal Nehru National Institute of Technology Allahabad, Allahabad, India, [3] D. D. Pant Interdisciplinary Research Laboratory, Department of Botany, University of Allahabad, Allahabad, India, [4] Department of Biotechnology, Motilal Nehru National Institute of Technology Allahabad, Allahabad, India, [5] Department of Physics, Shri Mata Vaishno Devi University, Katra, India, [6] Lawrence Berkeley National Laboratory, Berkeley, CA, USA, [7] Veer Chand Singh Garhwali Uttarakhand University of Horticulture and Forestry, Tehri Garhwal, India, [8] Department of Chemistry, Mokpo National University, Mokpo, South Korea

Edited by:
Jayanta Kumar Patra,
Dongguk University, South Korea

Reviewed by:
Sanjay Kumar Bharti,
University of Virginia Hospital, USA
Santosh Kumar,
University of Coimbra, Portugal
Ishan Barman,
Johns Hopkins University, USA

*Correspondence:
Durgesh K. Tripathi
dktripathiau@gmail.com
Devendra K. Chauhan
dkchauhanau@yahoo.com

Nanotechnology is a cutting-edge field of science with the potential to revolutionize today's technological advances including industrial applications. It is being utilized for the welfare of mankind; but at the same time, the unprecedented use and uncontrolled release of nanomaterials into the environment poses enormous threat to living organisms. Silver nanoparticles (AgNPs) are used in several industries and its continuous release may hamper many physiological and biochemical processes in the living organisms including autotrophs and heterotrophs. The present review gives a concentric know-how of the effects of AgNPs on the lower and higher autotrophic plants as well as on heterotrophic microbes so as to have better understanding of the differences in effects among these two groups. It also focuses on the mechanism of uptake, translocation, accumulation in the plants and microbes, and resulting toxicity as well as tolerance mechanisms by which these microorganisms are able to survive and reduce the effects of AgNPs. This review differentiates the impact of silver nanoparticles at various levels between autotrophs and heterotrophs and signifies the prevailing tolerance mechanisms. With this background, a comprehensive idea can be made with respect to the influence of AgNPs on lower and higher autotrophic plants together with heterotrophic microbes and new insights can be generated for the researchers to understand the toxicity and tolerance mechanisms of AgNPs in plants and microbes.

Keywords: nanotoxicology, silver, uptake, autotrophic plants, heterotrophic microbes

INTRODUCTION

The ever-increasing indiscriminate anthropogenic activities worldwide together with the technological advances have led to the creation of huge waste material contaminating our biosphere and causing many ecological risks. Due to this, environmental stability is gradually diminishing thereby resulting in the damage to ecosystem facilities. In addition, the uncontrolled rise in human population will continue to intensify the ecosystem degradation in the near future (Lee, 2011). Due to the imbalanced population growth and simultaneous increase in ecological risk, the problems of food security and proliferation of pathogenic organisms may increase. Many scientists and pharmaceutical industries are working to develop antibacterial agents that can confer resistance against the attack of pathogens (Rai et al., 2009; Ahmed et al., 2013). In order to provide food for the increasing population, scientists are exploring new ways to increase the yield of the crops with the help of biotechnological techniques (Moose and Mumm, 2008). Presently, nanotechnology has proved to be an important tool in many industrial and agricultural applications such as raising productivity of many crops. The agricultural productivity can be increased by using nano-fertilizers or nanoparticles (NPs) in order to reduce the toxic effects of many metal pollutants (Anjum et al., 2013; Tripathi et al., 2015, 2016, 2017a). The naturally occurring NPs have always existed in the environment without any undesired properties (Murr et al., 2004; Handy et al., 2008; Macken et al., 2012). There are various modes of synthesis of Nps which include physical, chemical and biological methods. These smallest objects are referred to as the engineered NPs and may be counted as a whole unit in terms of its physiochemical or microscopic properties with a reduction of any one dimension (Donaldson and Poland, 2013). Such particles exhibit different behavior from their larger counterparts when reduced to nanoscale (Choi et al., 2008; Khanna, 2016). The production of engineered NPs will likely to increase from 2000 tons in 2004 to over 58,000 tons annually between 2011 and 2020 (Khanna, 2016). There are different varieties of nanoparticles and among them, silver nanoparticles (AgNPs) are fetching more attention because of their application or requirement in daily life (Chen and Schluesener, 2008; Aziz et al., 2015, 2016) as well as their toxic behavior (Tripathi et al., 2017b). In order to search for better solutions to the problems related to food security and occurrence of diseases, nanosilver is gaining priority as one of the leading solutions with more stability and surface area as compared to other nano-solutions (Donaldson and Poland, 2013; Khanna, 2016). Apart from this, AgNPs have wide range of applications in solar energy (Clavero, 2014), Raman scattering (Samal et al., 2013; Zheng et al., 2015), and antimicrobial applications (Rai et al., 2009). The effective antimicrobial properties and low toxicity of AgNPs toward mammalian cells have made them to be easily utilized in many consumer-based products. The silver nanoparticles also finds its use in biocidal coatings, shampoo, soap and toothpaste (Rai et al., 2009).

Owing to the increasing commercial production of NPs and their unregulated release into aquatic as well as terrestrial systems via number of pathways, there is a growing concern over their impending environmental effects (Choi et al., 2008; Mirzajani et al., 2013; Shweta et al., 2016; Singh et al., 2016). In a study by Nowack et al. (2011), it was observed that the potential concentration of AgNPs have increased in surface water up to 0.1 mg L^{-1} and in sludge up to 2.9 mg kg^{-1}. Despite its beneficial applications, numerous harmful effects of AgNPs have also been reported in plants and animals (Navarro et al., 2008; Tripathi et al., 2017a,b). The effluents having AgNPs are found to contaminate water bodies, soil and atmosphere (Benn and Westerhoff, 2008; Farkas et al., 2011; Nair and Chung, 2014). Cytotoxicity, genotoxicity and ecotoxicity of coated AgNPs have also been reported (Lee et al., 2007; Lima et al., 2012). It poses undesirable effects on plants such as inhibition of seed germination and growth (Yin et al., 2012; Dimkpa et al., 2013; Nair and Chung, 2014). From soil and water, they may penetrate into food crops (Mazumdar and Ahmed, 2011; Nair and Chung, 2014) and enter into heterotrophs or consumers by means of food chain. Studies have revealed that AgNPs show toxic behavior against mitochondria and generate reactive oxygen species (ROS) (Hsin et al., 2008; Kim et al., 2012). These ROS damage the cell membrane, disrupt ATP production pathway and DNA replication and alter gene expression (Moreno-Garrido et al., 2015). In algae and microbes also, it induces imbalanced generation of ROS and cause oxidative stress. There are various methods by which the affected plant or other organisms try to cope up with the problems induced by the NPs. Number of defense strategies are found in the organisms through which they avoid or lessen the possible impact of AgNPs. These defense mechanisms are important to understand as it may provide an exact understanding toward the amelioration of the problems arising due to the nanoparticle pollution and its impact on environment. However, the effect as well as the tolerance may vary across the organisms. The autotrophs show different response as compared to heterotrophs against NPs; thereby making it essential to understand such differences and related survival mechanisms. Hence, the present review details about the impact and tolerance of widely used nanomaterial, i.e., AgNP on both autotrophs and heterotrophs. It will lead to the enhancement of the knowledge in this regard and provide a differential approach towards the issue.

Sources of Silver Nanoparticle in the Environment

Engineered NPs may be found in the form of metals, other dust or various compounds where they are used (**Figure 1**). Synthesis of the NPs in laboratory or industry is one of the important sources of its release in the environment (Bhaduri et al., 2013). Physical and chemical methods of NP synthesis are not eco-friendly and may contaminate the surrounding environment (Bhaduri et al., 2013; Kuppusamy et al., 2015) whereas biological synthesis of NPs is rather eco-friendly (Bhaduri et al., 2013). By using strong reducing and stabilizing agents, the chemical methods have an undesirable effect on biotic components (Bhaduri et al., 2013; Kuppusamy et al., 2015). However, the NPs synthesized from plant extract do not include any reductants or stabilizing agents (Carlson et al., 2008; Kuppusamy et al., 2015). An outline of the various point and non-point sources of AgNPs has been given in **Figure 1**.

The sources of metallic AgNPs are not new. AgNPs could have been naturally occurred via natural reduction process from Ag^+ ions or produced anthropogenetically and then released into the environment (Nowack et al., 2012; Samal et al., 2013). Colloidal AgNPs had been produced and used as biocidal material in USA in 1954 (Nowack et al., 2011). The formation of AgNPs can be facilitated by photochemical reduction of

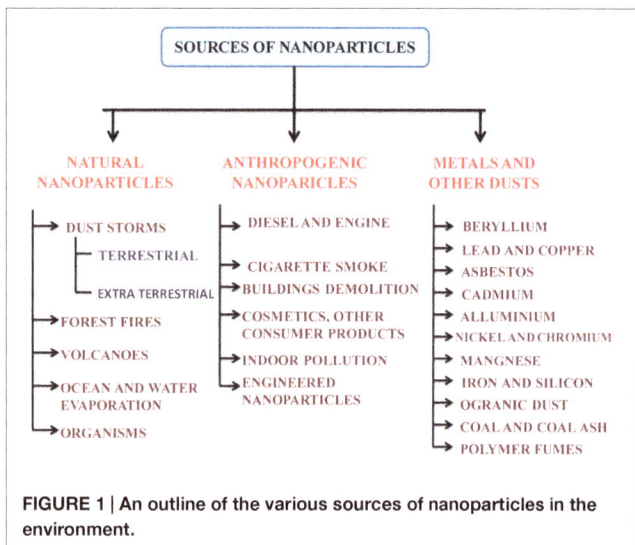

FIGURE 1 | An outline of the various sources of nanoparticles in the environment.

Ag^+ ions by dissolved organic matter in natural water under sunlight within several hours (Buzea et al., 2007; Samal et al., 2013). AgNPs may also be generated from silver objects through oxidative dissolution and subsequent reduction (Samal et al., 2013; Khanna, 2016). During washing, recycling, disposal and other manufacturing processes, they enter the surrounding environment (Nowack et al., 2011; Khanna, 2016) (**Figure 1**). The nanosilver species such as Ag^0, AgCl, and Ag_2S are frequently observed in various environmental compartments (Buzea et al., 2007; Wang et al., 2015; Khanna, 2016). There are various sources of AgNPs in the environment which could be point and non-point sources (Anjum et al., 2013) (**Figure 1**). AgNPs generated from anthropogenic activities are of greater concern as they are most widely incorporated in multidisciplinary applications. AgNPs are released in soil either from point sources that are suspended in surrounding environment after the application of NPs or organic matters in the forms of nano-fertilizers, sludge recycling, etc. in agricultural fields or from non-point sources such as products that contain AgNPs in themselves and directly contaminate the system (Mueller and Nowack, 2008; Benn et al., 2010).

Though AgNPs are found naturally, there should be no doubt that anthropogenic activities play a major role in pollution of silver nanoparticles in the environment. The widespread industrial uses of AgNPs have raised the chances of contamination. They are used in electronic devices, incorporated into textiles, dressing and medical devices, or directly added into disinfectants from where they could be directly released into the environment (Buzea et al., 2007; Khaydarov et al., 2009; Khanna, 2016). AgNPs may also appear from inappropriate disposal of biosolids or wastes, spills and other organic fertilizers or pesticides (Calder et al., 2012; Anjum et al., 2013). Despite these facts, the properties of AgNPs still enable them to be used in more than 250 consumer products in the world (Rai et al., 2009; Anjum et al., 2013). It is estimated that around 500 tons per annum AgNP is being produced (Mueller and Nowack, 2008), and is rapidly growing every year (Boxall et al., 2007). The generation of AgNPs

in United States has been reported to be up to 2,500 tons per year from which approximately 150 ton is released in sewage sludge and 80 ton in surface waters (Khaydarov et al., 2009; El-Temsah and Joner, 2012).

Chemistry of Silver Nanoparticles

In the periodic table, Silver (Ag) is an element of group 11 and period 5 with atomic number 47 and standard atomic weight 107.862. It has high electrical and thermal conductivity and also the reflectivity. This is considered as one of the main property of any metal. Silver belongs to the 'd' block in periodic table and its electronic configuration is $[Kr]4d^{10}5s^1$. It occurs in solid form with 2162°C and 961.78°C boiling and melting points, respectively. The density of silver is approximately 10.49 g/cm^3, its oxidation state is +1 and atomic radius about 145 pm.

The range of AgNPs lies between 1 and 100 nm (Graf et al., 2003) that contains around 20–15,000 silver atoms (Anjum et al., 2013). However, the bulk material of silver may be silver oxide NP and characterized by high ratio of their surface area to bulk silver atom. Beside this, AgNP has distinct chemical and physical properties such as catalytic activity and non-linear optical characteristics. They are also found in different shapes and sizes such as spherical, octagonal or in the shape of sheets (Graf et al., 2003), rod shaped, cylindrical shaped, wire like, plate like, and belt like etc. (Pal et al., 2007; Jana et al., 2012; Kim et al., 2012; Anjum et al., 2013). Furthermore, they have various dimensions such as, zero dimensions, one dimension (1D), two dimension (2D), and three dimension (3D) and accordingly they may be laments, surface films, strands and particles, respectively (Tiwari et al., 2012). AgNP can be characterized by different spectrophotometric and electroscopic techniques such as SEM (Scanning Electron Microscopy), TEM (Transmission Electron Microscopy), XRD (X-ray Diffractometer), and UV-VIS spectrophotometer.

However, Mura et al. (2015) reported that AgNP can become more hazardous when oxidized in water because they make bonds with anions and hence transform into the characteristics of heavy metals. This conversion of AgNP to a complex of anion or heavy metal causes toxic effect on various living organisms (Chen and Schluesener, 2008; Wijnhoven et al., 2009; Fabrega et al., 2011; Anjum et al., 2013). Another distinct trait of AgNPs is large surface area-to-volume ratio, on the basis of which they act as antibacterial agent on both types of bacteria, i.e., Gram-positive and Gram-negative (Kim et al., 2007; Marambio-Jones and Hoek, 2010; Anjum et al., 2013).

Applications of Silver Nanoparticles

Silver nanoparticles are intensively used in our daily life. AgNPs along with various other engineered NPs have wider application in many commercial and industrial sectors. It has also been used in the field of bioremediation and biomedicine because of its characteristic physiochemical properties (Chaloupka et al., 2010; Wong and Liu, 2010). Notably, they are used in antibiotics such as nanogels and nanolotions (Ma et al., 2010; Piccinno et al., 2012). These AgNPs are largely used in bedding, washers, toothpaste, waste water treatment, shampoo and fabrics, food packaging materials, food storage containers, water purifiers, odor-resistant socks and undergarments, room sprays, laundry detergents, etc. (Wijnhoven et al., 2009). Among other domestic uses, they are highly utilized for cleaning the bacteria from vacuum cleaner, refrigerators and ACs, laboratory coats, plastics, paints, textiles and other medical related applications such as in bandages, surgical gowns, wound dressings, female-hygiene products, bone cements and implantable devices etc. (Boxall et al., 2007; Kim et al., 2007; Klaine et al., 2008; Ma et al., 2010).

Due to some unique properties, AgNPs are used in sensing and imaging applications, including the detection of DNA (de la Escosura-Muñiz and Merkoçi, 2014), selective colorimetric sensing of cysteine, sensing purine nucleoside phosphorylase activity and selective colorimetric sensing of mercury(II) as well (Silver, 2003; Sapsford et al., 2013). Due to its antimicrobial activity, it inhibits the growth of both Gram-positive and Gram-negative bacteria and also its antibacterial activity is important for different drug-resistant pathogens (Samberg et al., 2011). Nanosilver is also used as an efficient fungicide against several ordinary fungal strains, such as *Aspergillus fumigatus, Mucor, Saccharomyces cerevisiae, and Candida tropicalis* (Velmurugan et al., 2009; Kim et al., 2012). AgNP also has antiviral properties which can be used against the HIV, hepatitis B and Herpes simplex virus (Galdiero et al., 2011). These are also used in many diagnostic and theranostic applications, such as in making nano-probes (Zheng et al., 2015; Li et al., 2016). However, we must understand that why silver nano differs from other nanomaterial in these applications. For example, gold NPs (AuNPs) are also widely used in medical science owing to their flow in endocytosis; they are diffused through lipid bilayer of the cell membrane and are mostly used in cancer treatments (Siddhanta et al., 2015; Alaqad and Saleh, 2016). Due to large surface-to-volume ratio, AuNPs functionalized with target specific biomolecules can efficiently destroy cancer cells or bacteria (Wang et al., 2010). AgNPs are commonly used due to their electrical conductivity, wide antimicrobial activity against various microorganisms and localized surface plasmon resonance effect (Raghavendra et al., 2014).

Autotrophic Plants and Heterotrophic Microbes and Their Importance

Autotrophs are organisms that produce organic compounds (carbohydrates, fats, and proteins) from simple substances present in the surrounding by using energy from sunlight via photosynthesis. They are plants on land or algae in water. Autotrophs can reduce CO_2 to make organic compounds and use water as the reducing agent, but some of them can also use other hydrogen compounds such as hydrogen sulfide for this purpose. However, the heterotrophs are organisms which are dependent on the autotrophs and cannot make their own food by fixing carbon rather they use organic carbon for their growth (Crane and Grover, 2010). The reduced carbon compounds in autotrophs provide the energy in food consumed by heterotrophs. All animals, fungi, most of the bacteria and protists are heterotrophs. Both kind of organisms have their own importance in an ecosystem in maintaining the food chain in which producers generate energy which is consumed by the consumers to degrade the organic compounds into simpler

form to be free in the environment to complete biogeochemical cycles. Any change in the physiology and biochemistry of these organisms can, thus, disrupt the ecological balance in many ways (Crane and Grover, 2010). Hence, it is important to understand the impacts pose by any such chemical pollutant which is new to the environment and for which more elaborative studies are needed to regulate their release to the environments. The nanosilver is widely used nowadays and regularly released therefore; its uptake, accumulation, and toxicity must be known with respect to autotrophic and heterotrophic organisms in order to better understand the impact of nano-pollution and to search future ways to combat the problems.

INTERACTION OF SILVER NANOPARTICLE WITH AUTOTROPHS

Interaction with Algae

Algae are considered as polyphyletic eukaryotic autotrophs which include many unicellular as well as multicellular forms and most of them are aquatic in nature and instead of lacking different tissues and cells like xylem and phloem, they make their own food. As most of the AgNP traces are released into the water after being used and are also employed for waste water treatment, it affects aquatic organisms in which algae are prime (Boxall et al., 2007). The toxicity of AgNPs toward algae can be estimated by means of many laboratory-based experiments and these studies demonstrate that AgNP is toxic to algae at different concentrations (Marshall et al., 2005; He et al., 2012; Moreno-Garrido et al., 2015). Due to different and variable growing conditions of these organisms, the amount of experiments and data on toxicity on various algal species are still sporadic. Since algal communities are important not only for the aquatic photosynthesis and food resource (Marshall et al., 2005) but also for industrial applications (Moreno-Garrido et al., 2015), therefore understanding the toxicity of nanosilver on this vital organism becomes necessary.

Uptake, Translocation, and Accumulation

The uptake, translocation, and accumulation of the AgNPs in the cells depend on the cellular structure, its permeability, size of the particles and other cell properties (Carlson et al., 2008; Li et al., 2015). The cell wall in algae is an important point for any type of reciprocal action with AgNPs as it acts as an obstruction or blocking point of the inflowing AgNPs from surrounding environment. The algal cell wall mainly comprise of carbohydrates, proteins, and cellulose (glycoproteins and polysaccharides) which organize a stiff elusive network (Navarro et al., 2008). Due to this, algal cell wall works as a semi permeable sieve and screens out larger NPs by allowing the transition of the smaller particles (Navarro et al., 2008). The smaller size and larger surface area of the AgNPs enable them to transit through the pores of cell wall and eventually reach to the plasma membrane (Samberg et al., 2011). Cellular reproduction may alter the permeability of cell wall and recently fabricated pores may become permeable for silver nanoparticles to a greater extent (Ovečka et al., 2005; Navarro et al., 2008).

It has been reported that due to the influence of AgNPs on algal cell, newly formed pores are larger than the prior ones and this may led to instigate the increase in uptake of the nanosilver in the cells of algae (Navarro et al., 2008). The sizes of pores in cell wall through which a single NP can be passed ranges from 5 to 20 nm. However, the interaction with NPs creates new and large-size pores in the cell wall and hence increases the internalization efficiency of cell (Carlson et al., 2008). After this transition through cell wall, AgNPs converge with plasma membrane. The possible mode of entry by lipid bilayer membrane has been discussed by some researchers (Navarro et al., 2008; Leonardo et al., 2015). AgNPs can encompass in cavities like the structure fabricated by plasma membrane and then can be imbibed into the cell through endocytic processes (Ovečka et al., 2005; Siddhanta et al., 2015). Apart from these, the ion channels or transport carrier proteins could also be used by AgNP as a mode of entry into the cell membrane (Mueller and Nowack, 2008). After entering into the cell, these NPs get attached with the various cell organelles (e.g., endoplasmic reticulum, Golgi bodies and endo-lysosomal system) and it shows some significant symptoms such as swelling of the endoplasmic reticulum and vacuolar changes (Miao et al., 2010). Navarro et al. (2008) reported that algal cell wall contains some barriers to create hindrance as well as some primary sites for interaction with NPs. Moreover, their bimolecular system contains many functional groups such as hydroxyl, carboxylate, imidazole, sulfhydryl, phosphate, and amine which are associated with many active sites of the AgNP interaction (Cao and Liu, 2010). After reaching to the specific cell organelle, they start disturbing the metabolic processes by enhancing the production of ROS and affect the biochemical processes in the cell (Miao et al., 2010).

Toxicity

Silver nanoparticles induces physical and chemical substructure alterations by means of its toxicity in the algal cells (**Table 1**). AgNP shows toxic effects as it releases silver ions and poses adverse effects on algal community at varied concentrations. The structural and functional properties of the algal cell could be affected by severe alterations induced by these NPs. The toxicity is induced through decrease in chlorophyll content, viable cell counts, increased ROS generation and lipids peroxidation (MDA) (Marshall et al., 2005; Miao et al., 2010; Dewez and Oukarroum, 2012; He et al., 2012; Li et al., 2015). It was noticed that AgNPs in association with light alter the oxygen evolution complex, inhibit the electron transport activity as well as induce some structural deterioration (He et al., 2012; Oukarroum et al., 2012; Leonardo et al., 2015). There are reports showing increased toxicity of AgNPs as compared to metallic silver ions which means silver ions are more toxic if present in the form of NPs in environments (Roh et al., 2009; Fabrega et al., 2010). The negative impact of AgNPs are also seen on the algal reproduction as well as on the subsequent stimulated imposition of oxidative stress (Roh et al., 2009; Fabrega et al., 2010; Ribeiro et al., 2014).

Various properties of the released Ag^+ ion (such as preparation, stability, aggregation, and speciation) have differing impacts on algae (**Table 1**). Burchardt et al. (2012) have also demonstrated this difference in Thalassiosira pseudonana and

cyanobacterium Synechococcus. Various algal species have been tested for the toxicity of AgNPs at various concentrations (**Table 1**) such as, Dunaliella tertiolecta and Chlorella vulgaris (Oukarroum et al., 2012), T. pseudonana and cyanophyte Synechococcus (Burchardt et al., 2012) and Euglena gracilis (Li et al., 2015). While Ribeiro et al. (2014) compared the effects and found that AgNPs were more toxic than silver nitrate. AgNPs have been reported to enhance the biotic generation of superoxide in Chattonella marina (He et al., 2012). The AgNPs also affect photosystem II (PSII) photochemistry, alternation of the oxygen evolution complex, inhibition of electron transport activity, and structural deterioration of PSII reaction of the green algal species (Navarro et al., 2015; Huang et al., 2016). AgNP acts as a catalyst for redox reactions when they get in touch with organic molecules and they also affect photosynthetic and respiratory processes (Navarro et al., 2008) which is an outcome of the impacts on photo-induced electron transfer capacity by AgNP (Navarro et al., 2008).

In various algal species, the toxicity mechanisms for AgNPs depend on various processes occurring in the cell such as adhesion to membranes and altering their permeability or ion transport properties, disturbing cellular phosphate management, and inhibition of DNA synthesis and DNA damage by breaking the H-bonding; crumpling proton pump; ROS generation; denaturation of ribosome; and inactivation of proteins and enzymes by bonding on active sites (Moreno-Garrido et al., 2015; Kwok et al., 2016; Taylor et al., 2016). Smaller AgNPs (<80 nm) are shown to be able to enter into bacterial cells (Klaine et al., 2008), but there are contradictions on the entry of bigger NPs into the cells of different algae (He et al., 2012). Several reports have indicated the "Trojan horse" effect of AgNPs in which NPs start releasing ionic Ag^+ after its entry and damage the cellular structure (Huang et al., 2016). However, due to the ability of some microalgae to produce internal NPs from dissolved metals (Moreno-Garrido et al., 2015), the intracellular NPs observed in certain studies should be carefully investigated.

Tolerance Mechanism

Algal cells have specific mechanisms to cope up and reduce the toxic effects produced by AgNPs. Algal cell egests certain compounds to tolerate the toxic effects of AgNPs. The discharge of metal chelates from root system may either repress the availability of toxic metal ions excreted through AgNPs or increase its intake of metals (Dong et al., 2007; Navarro et al., 2008). These excreted compounds may regulate the dissolution rate of metals released from AgNPs. Certain compounds released from algal cells may also increase the AgNP flocculation and repress its bioavailability (Soldo et al., 2005). Many exopolymeric substances are released upon the introduction of AgNPs into the cell and this lead to their detoxification mechanisms (Miao et al., 2009).

Although AgNPs affect the algal population, these algal populations can also affect the potential toxicity and release of Ag from AgNPs by producing extracellular dissolved organic carbon compounds (DOCs) in order to inactivate AgNPs toxicity (Taylor et al., 2016; Xu et al., 2016). Hence, this is certain that feedback response by the algae against the presence of NPs seems to occur

TABLE 1 | Inimical effects of silver nanoparticles (AgNPs) on different algal varieties.

Algae	Size of AgNPs	Concentrations	Effect of NPs	Reference
Chlamydomonas reinhardtii	10 nm	10, 50, 100, and 500 μM	Reduction in photosynthetic yield of algae	Navarro et al., 2008
Ceramium tenuicorne	<5, 5–10 nm	26.6 μg L^{-1}	AgNPs induce toxic effects in organism	Macken et al., 2012
Chlorella vulgaris, Dunaliella tertiolecta	50 nm	0–10 mg L^{-1}	Strong decrease in chlorophyll content as well as formation of ROS and lipid peroxidation takes place	Oukarroum et al., 2012
Pseudokirchneriella subcapitata	20–30 nm	LC_{50} 0.19 mg L^{-1}	Low toxicity of AgNPs observed than silver ions	Griffitt et al., 2009; Fabrega et al., 2010
Chlamydomonas reinhardtii	25 ± 13 nm	EC_{50} 1H: 3300 nM; EC_{50} 5h: 829 nM	Toxicity of silver ions observed released from AgNp accumulated in cell.	Navarro et al., 2008 Fabrega et al., 2010
Thalassiosira weissflogii	60–70 nm	0.02–0.0002 nM	Decreased production of chlorophyll and low photosynthesis rate. Reduced cell growth observed.	Fabrega et al., 2010
Chara vulgaris	10–15 nm	0.9 mM	Green colored thalli turned yellow due to progressive loss of chlorophyll	Das et al., 2012
Pithophora oedogonia	10–15 nm	1.5 mM	Fragmented and disintegrated chloroplasts; thin and ruptured cell wall; condensed and clumped chromosomes at metaphase stage	Das et al., 2012
Ochromonas danica	1–10 nm	More than 10 μM	Showed inhibiting effect even after supplementation of glutathione	Miao et al., 2010
Thalassiosira weissflogii	60–70 nm	0.02–0.0002 nM	Suppressed chlorophyll production, photosynthetic activity and hence growth of the cell	Miao et al., 2009
Pseudokirchneriella subcapitata	80 nm	Nominal EC_{50} - 5.25 ± 1.82	Growth inhibited in size dependent manner	Ivask et al., 2014
Chlorella sp.	<100 nm	10 ppm	Shown to cause adverse effect on chloroplasts and finally death of cells	Zaidi et al., 2014
Chlamydomonas acidophila	50 nm	1, 10, and 100 mg L^{-1}	Altered chlorophyllous contents, cellular and parameters like cellular viability, generation of intracellular ROS	Oukarroum et al., 2014
Chattonella marina	50 nm	10 μM	Generation of ROS	He et al., 2012

in the cells which can alter bioavailability and chemical behavior of the NPs (He et al., 2012). Therefore, it may be understood that algal species have various tolerance mechanisms for the initial impacts posed by the AgNPs while concentration and exposure duration are the significant factors determining the longevity of the effects and also their intensities on the algal species. However, it still seems a bit complex in the arena of research to understand comprehensive tolerance mechanisms in algal cells possibly due to greater diversity in them and complex ecological conditions in which they live which, further, have certain effects on the adaptation and tolerance mechanisms of the algal cells toward AgNPs.

Interaction with Plants

Plants as producers are the building blocks of the basic structure of any ecosystem. Plants uptake, translocate and accumulate AgNPs from their surrounding growing medium (Monica and Cremonini, 2009). When AgNP is released in the environment, they find their way into the plants through food chain and then impart toxicity to them. Various studies show marked positive and negative impacts of AgNPs on plants (Siddiqui et al., 2015; Tripathi et al., 2017b) which depend on various factors

regulating the uptake and accumulation in plants (Wang et al., 2015). Uptake of AgNPs depends upon the cellular permeability of the concerned plant and also on the different size and shape of AgNPs (Tripathi et al., 2017b). After entry into cells and sub cells, they create biological alterations and essential macrobiotic elements such as protein are affected (Griffitt et al., 2009; Pham et al., 2012). After their entry into the roots, AgNPs have been found to regulate the accumulation of protein, such as CDK-2 (cell division cycle kinase-2), 1,6-bisphosphate aldolase, protochlorophyllide oxidoreductase (Siddiqui et al., 2015). They also regulate the expression of some genes involved in cellular metabolism such as expression of IAA-8 (Indole acetic acid protein 8), RD22 (dehydration responsive), and NCED3 (9-cis-epoxycarotenoid dioxygenase) (Siddiqui et al., 2015). In Arabidopsis thaliana, inhibition of root elongation of seedling by activation of ACC (aminocyclopropane- 1-carboxylic acid) declines the expression of aminocyclopropane- 1-carboxylic acid synthase 7 and aminocyclopropane- 1-carboxylic acid oxidase 2, that ultimately inhibits biosynthesis of ethylene under the effect of AgNPs uptake in roots (Siddiqui et al., 2015). AgNPs also affect plants by producing ROS together with DNA destruction (Roh et al., 2009; Kim et al., 2013).

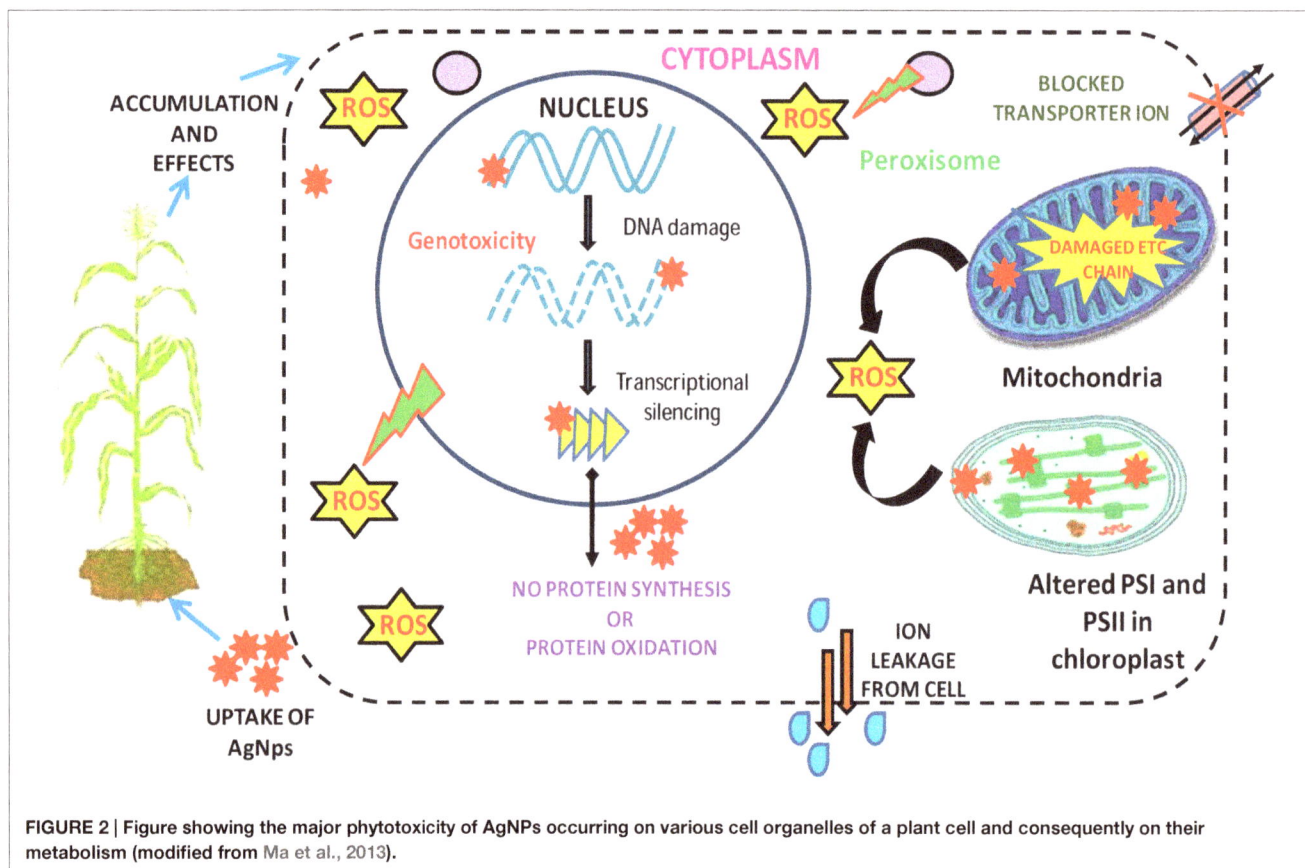

FIGURE 2 | Figure showing the major phytotoxicity of AgNPs occurring on various cell organelles of a plant cell and consequently on their metabolism (modified from Ma et al., 2013).

Uptake, Translocation, and Accumulation

Plant cell walls are mainly composed of cellulose which act as semi-permeable layer precisely permitting the entry of smaller particles and inhibiting the larger ones. The cell wall of the root cells is the main site through which AgNPs enter in plant cells (**Figure 2**). After entering into the plant, they penetrate the cell wall and plasma membranes of epidermal layer of roots, and then enter inside the vascular tissues (**Figure 2**). The AgNPs come in the plant together with the plant's uptake of water and other solutes. The cell wall consists of pores which are smaller than the NPs (Ma et al., 2010) and the cell wall serve as natural sieves (Navarro et al., 2008). The small sized AgNPs transit through the pores and enter into plasma membrane whereas large sized AgNPs are sieved out. They are further translocated to the stems and then to the leaves.

Sometimes, AgNPs influence the creation of new pores which permits the internalization of large AgNPs through cell wall (Navarro et al., 2008). Large leaf area and static plants enhance the accumulation of AgNPs from the ecosphere (Dietz and Herth, 2011). Translocation of AgNPs is aided by endocytosis (Ovečka et al., 2005; Fabrega et al., 2010) which include the creation of vesicle that enfold the material and finally transport AgNPs from plasma membrane to the cells. The AgNPs that eventually reach to the cell wall may also be translocated through plasmodesmata (Heinlein and Epel, 2004; Lucas and Lee, 2004; Ma et al., 2010).

Plant's acquirement of AgNPs usually occur via intercellular spaces and translocated within the cells of plant through the plasmodesmata process. After getting accumulated in the plant cells, AgNPs pose many gregarious impacts on plants including physiological, biochemical, and structural as well (Tripathi et al., 2017b).

Toxicity

Silver nanoparticle causes phytotoxicity in plants to a great extent which can be observed variably by analyzing different physical, physiological, biochemical, and structural traits (Tripathi et al., 2017b) (**Table 2**). They damage the cell membranes; interrupt ATP production as well as DNA replication (**Figure 2**). The enhanced production of ROS and subsequent generation of oxidative stress lead to various toxic impacts and may also affect the gene expressions and the demolition of DNA due to enhanced generation of ROS. Toxicity of AgNPs can be seen from seedling growth stage up to a full developed stage of the plants (Yin et al., 2012). It generally gives negative impact on the root growth of germinating seedlings and reduces the fresh biomass of the plant through reduction in root elongation and weight (Tripathi et al., 2017b). They also induce morphological modifications not only on the contact parts of the roots but also in the stem and leaves (Tripathi et al., 2017b). AgNPs modify the expression of several proteins of primary metabolism and cell defense system (Ma et al., 2010). AgNPs also affect the reproductive structure of the plant and

TABLE 2 | Impact of different concentration of AgNPs in plants.

Plants	Size	Concentration	Inimical effects	Reference
Cucurbita pepo	>100 nm	500 mg L^{-1}	Rate of transcription declined up to 66–84%. Biomass reduction was also reported	Musante and White, 2012
Triticum aestivum	10 nm	0–5 mg kg^{-1}	Reduction in root and shoot length occur in dose dependent manner	Dimkpa et al., 2013
Triticum aestivum	10 nm	0–5 mg kg^{-1}	Accumulation of oxidized GSSG in dose dependent manner	Dimkpa et al., 2013
Cucurbita pepo	NA	250 and 750 mg L^{-1}	49–91% decreased rate of transpiration and biomass as compared to silver compound	Hawthorne et al., 2012
Cucumis sativus; Lactuca sativa	2 nm	62, 100, and 116 mg L^{-1}	Negotiable toxicity	Barrena et al., 2009
Linum usitatissimum	20 nm	20, 40, 60, 80, and 100 mg L^{-1}	No effect seen on germination	El-Temsah and Joner, 2012
Lolium perenne	0.6–2 nm (Colloidal)	10 mg L^{-1}	20% reduction in germination percentage	El-Temsah and Joner, 2012
Lolium perenne	0.6–2 nm (Colloidal)	20 mg L^{-1}	50% reduction in germination percentage	El-Temsah and Joner, 2012
Lolium perenne; Linum usitatissimum	0.6–2 nm (Colloidal)	10 mg L^{-1}	Reduction in length of shoot	El-Temsah and Joner, 2012
Hordeum vulgare; Lolium perenne; Linum usitatissimum	0.6–2 nm (Colloidal)	20 mg L^{-1}	Reduction in length of shoot	El-Temsah and Joner, 2012
Hordeum vulgare	5 nm	10 mg L^{-1}	Reduced rate of germination	El-Temsah and Joner, 2012
Linum usitatissimum; Hordeum Vulgare	5 nm	10 mg L^{-1}	Reduction in length of shoot	El-Temsah and Joner, 2012
Hordeum vulgare	20 nm	10 mg L^{-1}	Reduction in rate of germination and shoot length	El-Temsah and Joner, 2012
Hordeum vulgare; Lolium perenne	20 nm	20 mg L^{-1}	Declined shoot length	El-Temsah and Joner, 2012
Cucurbita pepo	100 nm	100, 500, and 1,000 mg L^{-1}	41–79% of reduction in rate of transpiration	Stampoulis et al., 2009
Lolium multiflorum	6 nm (Gum arabic-coated)	1–40 mg L^{-1}	Dose dependent toxicity Undeveloped root hairs Crumpled cortical cells Ruptured epidermis Undeveloped root cap Declined biomass Decreased root length	Yin et al., 2011
Populus deltoides nigra	25 nm	100 mg L^{-1}	87% declined evapotranspiration that result in decreased fresh biomass of leaves, stem, and roots.	Wang et al., 2013
Arabidopsis thaliana	5 and 10 nm	1 mg L^{-1}	Growth of root completely inhibited	Wang et al., 2013
Oryza sativa	NA	1,000 mg L^{-1}	Vacuolar damage in root cells Cell wall breakage	Mazumdar and Ahmed, 2011
Allium cepa	70 nm	0–80 mg L^{-1}	Cytotoxicity seen at LC$_{50}$, i.e., up to 10 mg L^{-1} concentration DNA damage at 10 mg L^{-1} concentration	Panda et al., 2011
Allium cepa	24–55 nm	0–80 mg L^{-1}	Generation ROS that causes damage in structure of DNA and ultimately death of the cell	Panda et al., 2011
Allium cepa	<100 nm	100 mg L^{-1}	Sticky chromosomes led to chromosome breakage and disturbance in metaphase, that result in disruption of cell wall	Kumari et al., 2009
Vicia faba	60 nm	12.5, 25, 50, and 100 mg L^{-1}	Increased chromosomal aberrations	Patlolla et al., 2012

destruction of DNA involve the creation of chromatin bridges, stickiness, disarranged metaphase and multiple chromosomal breaks (Panda et al., 2011; Patlolla et al., 2012; Anjum et al., 2013).

Silver nanoparticles also affect the photosynthetic system of the plants (Tripathi et al., 2017b) through reducing total chlorophyll, affecting fluorescence parameters, and enhancing proline content (Monica and Cremonini, 2009). The main reason behind the dreadful toxicity of AgNPs in the plants is its impact on the biochemical properties of plants and inducing free radical generation resulting in induced oxidative stress in plant cells (Nair et al., 2010). The increased generation of the hydrogen peroxide (H_2O_2) in the plants cells is also an important toxic effect to be considered which affect the growth and development of the plants and kill the cells (Monica and Cremonini, 2009; Tripathi et al., 2017a,b). AgNPs may also affect the mitochondrial membrane potential (DWm) of roots with increasing concentrations (Hsin et al., 2008). The toxicity of AgNPs is more noticeable in roots as compared to shoots because roots are the main site of interaction while plant's self-defense mechanism involve translocation of the AgNPs from roots to shoots and thus restrict its accumulation in above ground parts completely or partially (Yin et al., 2012; Vannini et al., 2014). The research is needed to understand the effects of NPs on cellular level and how to reduce NPs' inherent toxicity by modifying some cellular processes. One way could be the modification in the osmolyte concentration in the environment for which researches should concern for plasmonic NP–cell interaction and internalization dealing with the NP surface composition and aggregation behavior in the cellular environment (Siddhanta et al., 2016). Some researchers have shown the osmolyte-based approach to reduce the toxicity of NPs by surface aggregation on the plasma membrane of the cells without changing the specific surface functionalization. The toxicity may also be reduced by inhibiting protein aggregation through lysozyme–AgNP interaction (Siddhanta et al., 2015).

Tolerance Mechanism

The toxicity of AgNPs leads to the cellular damage as well as affects metabolic activities which lead to the phytotoxicity in plants. Thus, activation of tolerance mechanism is very important so that the plant cells should be protected from stress conditions. The different stresses of plant cells require varied tolerance mechanisms to eliminate their toxic effects. The enhanced concentrations of cellular metabolite proline as well as oxidative stress controlling genes indicates the readiness of plant's antioxidative defense mechanism for the termination of oxidative stress factors (Apel and Hirt, 2004; Nair et al., 2010). According to Hsin et al. (2008), the cells should be given pretreatment of cyanide which suppresses the mitochondrial electron transferring process of cytochrome C oxidase that intercepts the generation of ROS through AgNPs. For the protection of cells against induced generation of ROS, plants involve many processes such as regulation of genes in which oxidative stress responses lead to the production of antioxidant enzymes (Apel and Hirt, 2004). There are various types of enzymatic scavengers present in cells of plants such as SOD, CAT, and APX which are ready to protect the cells from stress conditions (Nair and Chung, 2014). These toxic effects are dependent on various factors of plants, i.e., species, seeds, seedlings, and cell suspensions; and AgNPs, i.e., its concentration, size, aggregation, and functionalization. Also, the surrounding factors like temperature, time, and method of exposure can inhibit the AgNP phytotoxicity (Navarro et al., 2008; Siddhanta et al., 2016).

Interaction with Microbes

Microbes include bacteria, molds, yeasts, and viruses that are present in the environment and may induce several diseases. All having a very simple morphological structure perform different types of metabolic functions. For studies regarding AgNPs and their interaction with microbes, bacteria are among the most important organisms due to their small size and simple cell structure (Pal et al., 2007; Samberg et al., 2011; Prasad et al., 2016). As they are pathogenic in nature and result in serious infections for all life forms, a new antimicrobial agent is required to suppress the formation of pathogens. Silver compounds have been used as an inorganic antimicrobial agent to combat contagion of different pathogens since ancient days (Shrivastava et al., 2007; Lee et al., 2007). AgNPs act as an antibacterial agent toward bacterial stresses and eliminate its atrocious effects (Lee et al., 2007). Studies have also been conducted on the interaction of AgNPs with fungi and viruses and they have also been found to be affected by AgNPs at various concentrations (Velmurugan et al., 2009; Galdiero et al., 2011).

Uptake, Translocation, and Accumulation

Beside the simple shape or structure of bacteria, they possess a well developed structure of cell that performs many biological functions. Intracellular distribution of any solute or AgNP depends on their surface area to volume ratio (Pal et al., 2007). Studies have demonstrated that some small granular (electron dense) structures either accumulate in the cells or adhere near the cell wall (Feng et al., 2000). Furthermore, Feng et al. (2000) also demonstrated that accumulation of the sulfur, silver ions and dense electron granules occurs in the cytoplasm. This process disrupts the bacterial membrane and makes the entrance of AgNPs in the cell easy. Moreover, it also leads to the alteration in integrity of cell by continuous leakage of intracellular potassium from the cell (Navarro et al., 2008). The probable mechanism for the target and interaction of silver species might also be the thiol groups found in proteins (Lok et al., 2006). Similarly, another site for interaction of silver species to the bacterial cell could be phospholipid membrane (Lok et al., 2006). In the same way, fungal cells comprise of cell wall which inhibit the transition from AgNP in cells. Fungi cell wall consists of some significant constituents like carbohydrates which form a stiff and elusive structure (Navarro et al., 2008). The main component of fungal cell is their chitinous cell wall that is semipermeable in nature and acts as a sieve to allow the transition of small particles while inhibiting the larger ones. However, sometimes the pore size increases during reproduction period due to the effect of AgNPs and recently formed pores permit the translocation of the larger AgNPs (Ovećka et al., 2005; Navarro et al., 2008). Due to substantial alteration during exposure of AgNPs, "pits" are formed on the cell wall surface leading into the creation of pores and result in the destruction of cell metabolism (Navarro et al., 2008). The membrane barriers may collapse due to AgNPs by outflow of

TABLE 3 | Effect of different concentrations of AgNPs on microbes.

Microbes	Size	Concentration	Effect	Reference
Escherichia coli	12 nm	50–60 μg cm^{-3}	Inhibition of bacterial growth Increased permeability due to formation of "pits"	Sondi and Salopek-Sondi, 2004
Aspergillus sp.	30–45 nm	50 μg mL^{-1}	AgNPs shows antifungal activity Suppress the growth of fungal cells	Kuppusamy et al., 2015
Pneumocystis sp.	30–45 nm	50 μg mL^{-1}		
Yeast	13.5 nm	13.2 nM	Generation of free radicals Loss in permeability of membrane	Kim et al., 2007
Escherichia coli	13.5 nm	3.3 and 6.6 nM		
Staphylococcus aureus	13.5 nm	>33 nM		
Escherichia Coli	3 nm	40–140 μg mL^{-1}	Inhibitory effect	Ruparelia et al., 2008
Bacillus subtilis	3 nm	40 μg mL^{-1}		
Staphylococcus aureus	3 nm	120 μg mL^{-1}		
Escherichia coli	40 nm	40 μg mL^{-1}	On interaction of bacterial cell with AgNPs causes Proton Motive Force dissipation and finally death of the cell	Yoon et al., 2007
Bacillus subtilis	40 nm	20 μg/mL		
Escherichia coli	10 nm	0.1–1 mg L^{-1}	Damage occur in protein and membranes	Hwang et al., 2008
Escherichia coli	From 39 to 41 nm	0.1–10 μg mL^{-1}	Truncated AgNPs possess biocidal effect	Pal et al., 2007
Escherichia coli	9.3 ± 2.8 nm	0.4–0.8 nM	Unstable outer membrane Disintegrated plasma membrane	Lok et al., 2006
Escherichia coli	9.3 ± 2.8 nm	0–100 μg mL^{-1}	Small sized AgNPs showed more detrimental effect than larger ones	Lok et al., 2006
Nitrifying bacteria	21 nm	0.05–1 mg L^{-1}	Generation of Reactive Oxygen Species	Choi et al., 2008
Autotrophic nitrifying bacteria	14 ± 6 nm	1 mg L^{-1}	Respiration declined by 87% (calculated)	Choi and Hu, 2008
Pseudomonas fluorescens	65 ± 30 nm	0–2000 ppb	Toxicity of AgNPs varies according to the pH	Fabrega et al., 2009
Pseudomonas putida biofilm	65 ± 30 nm	0–2000 ppb	Toxicity of silver nanoparticles enhanced in combination of organic matter	Fabrega et al., 2010
Escherichia coli, Staphylococcus aureus	26 nm	MIC range of 1.69–6.75 μg mL^{-1}	Enhanced antibacterial activity	Kvitek et al., 2008
Escherichia coli, Salmonella typhi, Pseudomonas aeruginosa and *Vibrio cholerae*	16 ± 8 nm	0–100 μg mL^{-1}	AgNPs of less than 10 nm attached with membrane and cause toxicity	Morones et al., 2005
Escherichia coli, Ampicillin-resistant *Escherichia coli*, Multi-drug resistant *Salmonella typhi, Staphylococcus aureus*.	10–15 nm	5–35 μg mL^{-1}	More detrimental for Gram-negative bacteria as compared to Gram-positive.	Shrivastava et al., 2007

ions and other materials which alters the electrical potential of the membrane.

Toxicity

Silver ions and the related compounds show high toxicity to the microorganisms (**Table 3**). Choi et al. (2008) described the inhibitory mechanism of silver or the toxicity created by silver in microbes and in the context of the toxicity of AgNPs, free silver ions were found to be more toxic than silver nitrate. Moreover, when silver ions react with SH functional group of proteins, they cause inactivation in bacterial cell (Morones et al., 2005). Also, concentration of silver ions in micromole level has been found to inhibit the process of microbial respiration by uncoupling the electrons involved in phosphorylation and thus disrupt the permeability of membrane (Feng et al., 2000). Both Gram-negative (*Escherichia coli*) and Gram-positive (*Staphylococcus aureus*) types of the bacteria are found to be affected by the silver

ions (Feng et al., 2000; Jung et al., 2008). AgNPs may interact with nucleic acid and lead to the destruction in DNA replication in bacteria (Feng et al., 2000) (**Figure 2**). From many studies and researches, it is proved that AgNPs or silver ions may be toxic to some species of bacteria like *E. coli* (Gogoi et al., 2006; Kim et al., 2007; Mohan et al., 2007) and yeast (Kim et al., 2007). It was also found that after interaction of AgNPs with *E. coli*, the membrane integrity completely disrupts due to high surface area to volume ratio of AgNPs facilitating more interaction with *E. coli* (Raffin et al., 2008). However, Morones et al. (2005) demonstrated that penetration of AgNPs in Gram-negative bacteria (*Vibrio cholera, Salmonella enterica typhi, E. coli,* and *Pseudomonas aeruginosa*) depend on their size. The most preferable size they observed was in between 1 and 10 nm. The proteomic analysis in *E. coli* revealed the change in expression of HSPs (heat shock proteins) due to the impact of AgNPs (Lok et al., 2006) which disrupts the bacterial membrane and entrance of the smaller particles

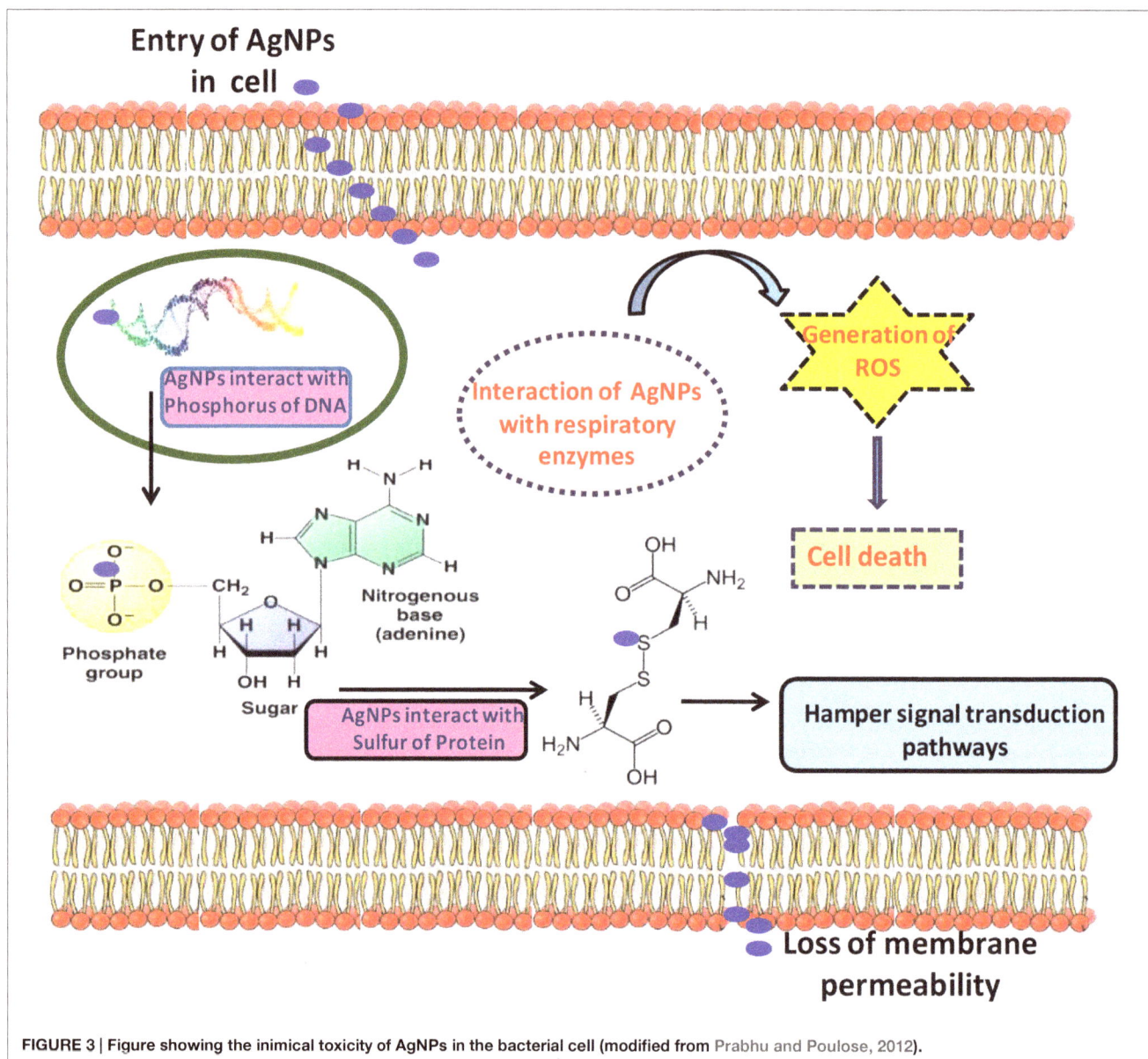

FIGURE 3 | Figure showing the inimical toxicity of AgNPs in the bacterial cell (modified from Prabhu and Poulose, 2012).

in the cell membrane becomes easy. This process also leads to the alteration in cellular integrity by continuous leakage of intracellular potassium from the cell which reduces the ATP and damages the cell viability. A hypothetical toxicity mechanism has been given in **Figure 3**.

While, in fungal cell wall, formation of new larger pores takes place due to effect of AgNPs and thus the transition of large AgNPs becomes easier. The perturbation of membrane by AgNPs leads to the generation of glucose and trehalose which indicate that they are the intracellular components of the membrane (Siddhanta et al., 2016). Reidy et al. (2013) have reported the mechanisms involved in antimicrobial actions of AgNPs which start with the adhesion of AgNPs on the surface of bacterial cell and changes the properties of membrane. AgNPs (smaller than 5 nm) have also been reported to suppress the growth of nitrifying bacteria (Choi and Hu, 2008; Reidy et al., 2013). After

the disintegration of AgNPs, the released Ag ions interact with bacterial cell wall which chiefly consists of sulfur protein resulting in compromised functionality (Levy and Marshall, 2004; Reidy et al., 2013). Silver ions also interact with the cytoplasmic proteins of bacterial cell wall (Cao and Liu, 2010; Reidy et al., 2013) and also affect the thiol group leading into the improper functioning or inhibition of bacterial cell. Uptake and accumulation of silver ions in bacterial cell disrupt hereditary biomolecules such as DNA and may lead to many unwanted changes in the genetic makeup of the bacteria (Feng et al., 2000).

Tolerance Mechanism

Bacterial cells also adopt some defense mechanisms to save themselves from the harmful effects of AgNPs. It has been reported that the peptidoglycan membrane thickness and their component (phospholipids) participate in the defense mechanism against AgNPs as their first line of defense (Sedlak et al., 2012). In order to protect bacterial cells from the toxic effects of AgNPs, many proteins such as heat shock proteins also get activated (Sedlak et al., 2012). Generally, a bacterium shows tolerance mechanism against the high concentrations of AgNPs, and preferably use efflux pumps resistant toward silver ions in natural environment (Jung et al., 2008). The encoding of this efflux pump is carried out by the plasmid-borne cassettes and it can also transmit to other strains of bacteria. Beside this, for the production of periplasmic silver ion-binding protein along with pumps responsible for efflux of ions, i.e., P-type ATPase and chemiosmotic silver ion/H$^+$ exchange protein, a sensor or transcriptional regulatory system play a key role behind it (Kvitek et al., 2008). According to Khan et al. (2011), *Bacillus pumilus* is tolerant toward AgNPs' antibacterial activity at high concentrations. The growth of bacteria stays the same while the diminution of extracellular polymeric substances (EPS) has been recorded (Taylor et al., 2016). According to Feng et al. (2000) and Jung et al. (2008), both types of Gram-negative (*E. coli*) and Gram-positive (*S. aureus*) bacteria accumulate dense electron light particles or granules as a defense strategy in the center of the cell. This region actually consists of thicker DNA (deoxyribonucleic acid) molecules and this thickness provides them security against the attack of silver ions.

Jung et al. (2008) also reported that in the presence of silver ions, some morphological changes also occur in the bacterial cells and these bacterial cells attain a non-culturable position and at last, lead to death. The peptidoglycan layer in Gram-positive bacteria is also very important in providing protection against AgNPs due to their thickness and shows a high degree of inhibitory effect against the adverse effects of silver ions, especially in *E. coli* (Jung et al., 2008). Bacterial defense mechanism also works even at molecular level (Silver, 2003). After the exposure of bacteria against the silver salts or silver ions, genes of plasmid and chromosomes have shown high degree of resistance mechanism (Silver, 2003). However, studies have been able to solve some of the questions relating to the defense mechanisms against AgNPs in the bacterial cells, such explorations regarding other microbial forms are still sporadic and a great deal of work related to genetic transformations and molecular markers in these microbes is yet to be done.

CONCLUSION AND FUTURE RESEARCH APPROACHES

It is quite evident that the technological interventions related to the nanotechnology have immense use and importance in modern times; however, somehow they are leading to the destruction or imbalance in ecosystem with their unregulated release posing toxic impacts on plants, algae and micro-organisms. Although researches are being carried on the beneficial as well as harmful impacts of AgNPs, there is a need to work in order to understand the toxicity of AgNPs at the cellular level of these organisms and their further impacts. Studies have yet not been able to fix any conclusive results on the effective/lethal/sub-lethal/optimum concentrations of NPs/AgNPs as a whole or/and the organism wise on which some regulatory framework can be made. The data regarding this are inadequate and researches must be carried on looking these things into consideration. The toxicity of AgNPs is translocated from plants to other communities through food chain and leads to the disruption of balanced ecosystem. However, the food chain analysis and health effects on trophic structure on this regard is sporadic and must be considered in the studies. The differences existing among experimental results of toxicity are thus creating problems in interpretation of the toxicological data. The studies on the toxicity and tolerance in plants, algae and microbes on molecular level are yielding some good results though, studies regarding fungi, yeasts, and viruses are very few on these aspects.

There must be some microcosm studies involving ecosystem based studies on this regard. The molecular marker approach, the evaluation of the tolerance mechanism and their use to develop artificial tolerance in the organisms may pave the significant pathways in this research field not only for developing new nanomaterials of use but also for formulating some regulatory concentration in various components of the environment. There is a strong need for the appropriate association between analytical techniques and toxicological studies through which more understanding towards this subject may be developed for the future research projects. Studies are generating good amount of data to be interpreted. A common research platform is needed to agglomerate all the data and to reach out to a logical conclusion to safeguard the ecosystem functioning and humankind as well.

AUTHOR CONTRIBUTIONS

DKT, AT, S, ShS, KV, GY, RM, and YS designed the manuscript. DKT, AT, S, KV, and SwS wrote the manuscript. DKT, DC, ShS, VKS, YL, RU, and ND critically evaluated the manuscript.

ACKNOWLEDGMENTS

Authors are grateful to the University Grants Commission, New Delhi for financial assistance. The support provided by MHRD sponsored project DIC (Design and Innovation Centre) is also acknowledged. DKT is also grateful to the UGC for providing the D. S. Kothari Fellowship. AT is thankful to NASI, Allahabad for providing Ganga Research fellowship. VKS is also thankful to Shri Mata Vaishno Devi (SMVD) University, Jammu and Kashmir, India for study leave to pursue Post Doctoral Research in Lawrence Berkeley National Laboratory (LBNL), Berkeley, California, USA during "Raman Fellowship (2015–2016)" awarded by University Grants Commission (UGC), Govt. of India.

REFERENCES

Ahmed, F., Arshi, N., Kumar, S., Gill, S. S., Gill, R., Tuteja, N., et al. (2013). "Nanobiotechnology: scope and potential for crop improvement," in *Crop Improvement Under Adverse Conditions*, eds N. Tuteja and S. S. Gill (New York: Springer), 245–269. doi: 10.1007/978-1-4614-4699-0-11

Alaqad, K., and Saleh, T. A. (2016). Gold and silver nanoparticles: synthesis methods, characterization routes and applications towards drugs. *J. Environ. Anal. Toxicol.* 6:384. doi: 10.4172/2161-0525.1000384

Anjum, N. A., Gill, S. S., Duarte, A. C., Pereira, E., and Ahmad, I. (2013). Silver nanoparticles in soil–plant systems. *J. Nano Res.* 15, 1–26. doi: 10.1007/s11051-013-1896-7

Apel, K., and Hirt, H. (2004). Reactive oxygen species: metabolism, oxidative stress, and signal transduction. *Annu. Rev. Plant Biol.* 55, 373–399. doi: 10.1146/annurev.arplant.55.031903.141701

Aziz, N., Faraz, M., Pandey, R., Shakir, M., Fatma, T., Varma, A., et al. (2015). Facile algae-derived route to biogenic silver nanoparticles: synthesis, antibacterial, and photocatalytic properties. *Langmuir* 16, 11605–11612. doi: 10.1021/acs.langmuir.5b03081

Aziz, N., Pandey, R., Barman, I., and Prasad, R. (2016). Leveraging the attributes of *Mucor hiemalis*-derived silver nanoparticles for a synergistic broad-spectrum antimicrobial platform. *Front. Microbiol.* 7:1984. doi: 10.3389/fmicb.2016.01984

Barrena, R., Casals, E., Colón, J., Font, X., Sánchez, A., and Puntes, V. (2009). Evaluation of the ecotoxicity of model nanoparticles. *Chemosphere* 75, 850–857. doi: 10.1016/j.chemosphere.2009.01.078

Benn, T., Cavanagh, B., Hristovski, K., Posner, J. D., and Westerhoff, P. (2010). The release of nanosilver from consumer products used in the home. *J. Environ. Qual.* 39, 1875–1882. doi: 10.2134/jeq2009.0363

Benn, T. M., and Westerhoff, P. (2008). Nanoparticle silver released into water from commercially available sock fabrics. *Environ. Sci. Technol.* 42, 4133–4139. doi: 10.1021/es7032718

Bhaduri, G. A., Little, R., Khomane, R. B., Lokhande, S. U., Kulkarni, B. D., Mendis, B. G., et al. (2013). Green synthesis of silver nanoparticles using sunlight. *J. Photochem. Photobiol. A Chem.* 258, 1–9. doi: 10.1016/j.jphotochem.2013.02.015

Boxall, A. B., Tiede, K., and Chaudhry, Q. (2007). Engineered nanomaterials in soils and water: How do they behave and could they pose a risk to human health? *Nanomedicine (Lond.)* 2, 919–927. doi: 10.2217/17435889.2.6.919

Burchardt, A. D., Carvalho, R. N., Valente, A., Nativo, P., Gilliland, D., and García, C. P. (2012). Effects of silver nanoparticles in diatom *Thalassiosira pseudonana* and Cyanobacterium *Synechococcus* sp. *Environ. Sci. Technol.* 46, 11336–11344. doi: 10.1021/es300989e

Buzea, C., Pacheco, I. I., and Robbie, K. (2007). Nanomaterials and nanoparticles: sources and toxicity. *Biointerphases* 2, MR17–MR71. doi: 10.1116/1.2815690

Calder, A. J., Dimkpa, C. O., McLean, J. E., Britt, D. W., Johnson, W., and Anderson, A. J. (2012). Soil components mitigate the antimicrobial effects of silver nanoparticles towards a beneficial soil bacterium, *Pseudomonas chlororaphis* O6. *Sci. Total Environ.* 429, 215–222. doi: 10.1016/j.scitotenv.2012.04.049

Cao, H., and Liu, X. (2010). Silver nanoparticles-modified films versus biomedical device-associated infections. *Wiley Interdiscip. Rev. Nanomed. Nanobiotechnol.* 2, 670–684. doi: 10.1002/wnan.113

Carlson, C., Hussain, S. M., Schrand, A. M. K., Braydich-Stolle, L., Hess, K. L., Jones, R. L., et al. (2008). Unique cellular interaction of silver nanoparticles: size-dependent generation of reactive oxygen species. *J. Phys. Chem. B* 112, 13608–13619. doi: 10.1021/jp712087m

Chaloupka, K., Malam, Y., and Seifalian, A. M. (2010). Nanosilver as a new generation of nanoproduct in biomedical applications. *Trends Biotechnol.* 28, 580–588. doi: 10.1016/j.tibtech.2010.07.006

Chen, X., and Schluesener, H. J. (2008). Nanosilver: a nanoproduct in medical application. *Toxicol. Lett.* 176, 1–12. doi: 10.1016/j.toxlet.2007.10.004

Choi, O., Deng, K. K., Kim, N. J., Ross, L., Surampalli, R. Y., and Hu, Z. (2008). The inhibitory effects of silver nanoparticles, silver ions, and silver chloride colloids on microbial growth. *Water Res.* 42, 3066–3074. doi: 10.1016/j.watres.2008.02.021

Choi, O., and Hu, Z. (2008). Size dependent and reactive oxygen species related nanosilver toxicity to nitrifying bacteria. *Environ. Sci. Technol.* 42, 4583–4588. doi: 10.1021/es703238h

Clavero, C. (2014). Plasmon-induced hot-electron generation at nanoparticle/metal-oxide interfaces for photovoltaic and photocatalytic devices. *Nat. Photonics* 8, 95–103. doi: 10.1038/nphoton.2013.238

Crane, K. W., and Grover, J. P. (2010). Coexistence of mixotrophs, autotrophs, and heterotrophs in planktonic microbial communities. *J. Theor. Biol.* 262, 517–527. doi: 10.1016/j.jtbi.2009.10.027

Das, A., Singh, A. P., Chaudhary, B. R., Singh, S. K., and Dash, D. (2012). Effect of silver nanoparticles on growth of eukaryotic green algae. *Nano Micro Lett.* 4, 158–165. doi: 10.1007/BF03353707

de la Escosura-Muñiz, A., and Merkoçi, A. (2014). "Application of nanomaterials for DNA sensing," in *Nucleic Acid Nanotechnology*, eds J. Kjems, E. Ferapontova, and K. V. Gothelf (Berlin: Springer), 305–332.

Dewez, D., and Oukarroum, A. (2012). Silver nanoparticles toxicity effect on photosystem II photochemistry of the green alga *Chlamydomonas reinhardtii* treated in light and dark conditions. *Toxicol. Environ. Chem.* 94, 1536–1546. doi: 10.1080/02772248.2012.712124

Dietz, K. J., and Herth, S. (2011). Plant nanotoxicology. *Trends Plant Sci.* 16, 582–589. doi: 10.1016/j.tplants.2011.08.003

Dimkpa, C. O., McLean, J. E., Martineau, N., Britt, D. W., Haverkamp, R., and Anderson, A. J. (2013). Silver nanoparticles disrupt wheat (*Triticum aestivum* L.) growth in a sand matrix. *Environ. Sci. Technol.* 47, 1082–1090. doi: 10.1021/es302973y

Donaldson, K., and Poland, C. A. (2013). Nanotoxicity: challenging the myth of nano-specific toxicity. *Curr. Opin. Biotechnol.* 24, 724–734. doi: 10.1016/j.copbio.2013.05.003

Dong, J., Mao, W. H., Zhang, G. P., Wu, F. B., and Cai, Y. (2007). Root excretion and plant tolerance to cadmium toxicity-a review. *Plant Soil Environ.* 53, 193–200.

El-Temsah, Y. S., and Joner, E. J. (2012). Impact of Fe and Ag nanoparticles on seed germination and differences in bioavailability during exposure in aqueous suspension and soil. *Environ. Toxicol.* 27, 42–49. doi: 10.1002/tox.20610

Fabrega, J., Fawcett, S. R., Renshaw, J. C., and Lead, J. R. (2009). Silver nanoparticle impact on bacterial growth: effect of pH, concentration, and organic matter. *Environ. Sci. Technol.* 43, 7285–7290. doi: 10.1021/es803259g

Fabrega, J., Fawcett, S. R., Renshaw, J. C., and Lead, J. R. (2010). Silver nanoparticle impact on bacterial growth: effect of pH, concentration, and organic matter. *Environ. Sci. Technol.* 43, 7285–7290. doi: 10.1021/es803259g

Fabrega, J., Luoma, S. N., Tyler, C. R., Galloway, T. S., and Lead, J. R. (2011). Silver nanoparticles: behaviour and effects in the aquatic environment. *Environ. Int.* 37, 517–531. doi: 10.1016/j.envint.2010.10.012

Farkas, J., Peter, H., Christian, P., Urrea, J. A. G., Hassellöv, M., Tuoriniemi, J., et al. (2011). Characterization of the effluent from a nanosilver producing washing machine. *Environ. Int.* 37, 1057–1062. doi: 10.1016/j.envint.2011.03.006

Feng, Q. L., Wu, J., Chen, G. Q., Cui, F. Z., Kim, T. N., and Kim, J. O. (2000). A mechanistic study of the antibacterial effect of silver ions on *Escherichia coli* and *Staphylococcus aureus*. *J. Biomed. Mater. Res.* 52, 662–668. doi: 10.1002/1097-4636(20001215)52:4<662::AID-JBM10>3.0.CO;2-3

Galdiero, S., Falanga, A., Vitiello, M., Cantisani, M., Marra, V., and Galdiero, M. (2011). Silver nanoparticles as potential antiviral agents. *Molecules* 16, 8894–8918. doi: 10.3390/molecules16108894

Gogoi, S. K., Gopinath, P., Paul, A., Ramesh, A., Ghosh, S. S., and Chattopadhyay, A. (2006). Green fluorescent protein-expressing *Escherichia coli* as a model system for investigating the antimicrobial activities of silver nanoparticles. *Langmuir* 22, 9322–9328. doi: 10.1021/la060661v

Graf, C., Vossen, D. L., Imhof, A., and van Blaaderen, A. (2003). A general method to coat colloidal particles with silica. *Langmuir* 19, 6693–6700. doi: 10.1021/la0347859

Griffitt, R. J., Hyndman, K., Denslow, N. D., and Barber, D. S. (2009). Comparison of molecular and histological changes in zebrafish gills exposed to metallic nanoparticles. *Toxicol. Sci.* 107, 404–415. doi: 10.1093/toxsci/kfn256

Handy, R. D., Owen, R., and Valsami-Jones, E. (2008). The ecotoxicology of nanoparticles and nanomaterials: current status, knowledge gaps, challenges, and future needs. *Ecotoxicology* 17, 315–325. doi: 10.1007/s10646-008-0206-0

Hawthorne, J., Musante, C., Sinha, S. K., and White, J. C. (2012). Accumulation and phytotoxicity of engineered nanoparticles to *Cucurbita pepo*. *Int. J. Phytoremed.* 14, 429–442. doi: 10.1080/15226514.2011.620903

He, D., Dorantes-Aranda, J. J., and Waite, T. D. (2012). Silver nanoparticle-algae interactions: oxidative dissolution, reactive oxygen species generation and synergistic toxic effects. *Environ. Sci. Technol.* 46, 8731–8738. doi: 10.1021/es300588a

Heinlein, M., and Epel, B. L. (2004). Macromolecular transport and signaling through plasmodesmata. *Int. Rev. Cytol.* 235, 93–164. doi: 10.1016/S0074-7696(04)35003-5

Hsin, Y. H., Chen, C. F., Huang, S., Shih, T. S., Lai, P. S., and Chueh, P. J. (2008). The apoptotic effect of nanosilver is mediated by a ROS-and JNK-dependent mechanism involving the mitochondrial pathway in NIH3T3 cells. *Toxicol. Lett.* 179, 130–139. doi: 10.1016/j.toxlet.2008.04.015

Huang, J., Cheng, J., and Yi, J. (2016). Impact of silver nanoparticles on marine diatom *Skeletonema costatum*. *J. Appl. Toxicol.* 36, 1343–1354. doi: 10.1002/jat.3325

Hwang, E. T., Lee, J. H., Chae, Y. J., Kim, Y. S., Kim, B. C., Sang, B. I., et al. (2008). Analysis of the toxic mode of action of silver nanoparticles using stress-specific bioluminescent bacteria. *Small* 4, 746–750. doi: 10.2134/jeq2009.0479

Ivask, A., Kurvet, I., Kasemets, K., Blinova, I., Aruoja, V., Suppi, S., et al. (2014). Size-dependent toxicity of silver nanoparticles to bacteria, yeast, algae, crustaceans and mammalian cells in vitro. *PLoS ONE* 9:e102108. doi: 10.1371/journal.pone.0102108

Jana, D., Mandal, A., and De, G. (2012). High Raman enhancing shape-tunable Ag nanoplates in alumina: a reliable and efficient SERS technique. *ACS Appl. Mater. Interfaces* 4, 3330–3334. doi: 10.1021/am300781h

Jung, W. K., Koo, H. C., Kim, K. W., Shin, S., Kim, S. H., and Park, Y. H. (2008). Antibacterial activity and mechanism of action of the silver ion in *Staphylococcus aureus* and *Escherichia coli*. *Appl. Environ. Microbiol.* 74, 2171–2178. doi: 10.1128/AEM.02001-07

Khan, S., Mukherjee, A., and Chandrasekaran, N. (2011). Silver nanoparticles tolerant bacteria from sewage environment. *J. Environ. Sci.* 23, 346–352. doi: 10.1016/S1001-0742(10)60412-3

Khanna, V. K. (2016). "Nanomaterials and their Properties," in *Integrated Nanoelectronics*, (New Delhi: Springer), 25–41.

Khaydarov, R. R., Khaydarov, R. A., Estrin, Y., Evgrafova, S., Scheper, T., Endres, C., et al. (2009). "Silver nanoparticles," in *Nanomaterials: Risks and Benefits*, eds I. Linkov and J. Steevens (Dordrecht: Springer), 287–297.

Kim, J. S., Kuk, E., Yu, K. N., Kim, J. H., Park, S. J., Lee, H. J., et al. (2007). Antimicrobial effects of silver nanoparticles. *Nanomedicine* 3, 95–101. doi: 10.1016/j.nano.2006.12.001

Kim, J. Y., Kim, K. T., Lee, B. G., Lim, B. J., and Kim, S. D. (2013). Developmental toxicity of Japanese medaka embryos by silver nanoparticles and released ions in the presence of humic acid. *Ecotoxicol. Environ. Saf.* 92, 57–63. doi: 10.1016/j.ecoenv.2013.02.004

Kim, S. W., Jung, J. H., Lamsal, K., Kim, Y. S., Min, J. S., and Lee, Y. S. (2012). Antifungal effects of silver nanoparticles (AgNPs) against various plant pathogenic fungi. *Mycobiology* 40, 53–58. doi: 10.5941/MYCO.2012.40.1.053

Klaine, S. J., Alvarez, P. J., Batley, G. E., Fernandes, T. F., Handy, R. D., Lyon, D. Y., et al. (2008). Nanomaterials in the environment: behavior, fate, bioavailability, and effects. *Environ. Toxicol. Chem.* 27, 1825–1851. doi: 10.1897/08-090.1

Kumari, M., Mukherjee, A., and Chandrasekaran, N. (2009). Genotoxicity of silver nanoparticles in *Allium cepa*. *Sci. Total Environ.* 407, 5243–5246. doi: 10.1016/j.scitotenv.2009.06.024

Kuppusamy, P., Ichwan, S. J., Parine, N. R., Yusoff, M. M., Maniam, G. P., and Govindan, N. (2015). Intracellular biosynthesis of Au and Ag nanoparticles using ethanolic extract of *Brassica oleracea* L. and studies on their physicochemical and biological properties. *J. Environ. Sci.* 29, 151–157. doi: 10.1016/j.jes.2014.06.050

Kvitek, L., Panáček, A., Soukupova, J., Kolar, M., Vecerova, R., Prucek, R., et al. (2008). Effect of surfactants and polymers on stability and antibacterial activity of silver nanoparticles (NPs). *J. Phys. Chem. C* 112, 5825–5834. doi: 10.1021/jp711616v

Kwok, K. W., Dong, W., Marinakos, S. M., Liu, J., Chilkoti, A., Wiesner, M. R., et al. (2016). Silver nanoparticle toxicity is related to coating materials and disruption of sodium concentration regulation. *Nanotoxicology* 10, 1–46. doi: 10.1080/17435390.2016.1206150

Lee, H. Y., Park, H. K., Lee, Y. M., Kim, K., and Park, S. B. (2007). A practical procedure for producing silver nanocoated fabric and its antibacterial evaluation for biomedical applications. *Chem. Commun.* 2959–2961. doi: 10.1039/B703034G

Lee, R. (2011). The outlook for population growth. *Science* 333, 569–573. doi: 10.1126/science.1208859

Leonardo, T., Farhi, E., Pouget, S., Motellier, S., Boisson, A. M., Banerjee, D., et al. (2015). Silver accumulation in the green microalga *Coccomyxa actinabiotis*: toxicity, in situ speciation, and localization investigated using synchrotron XAS, XRD, and TEM. *Environ. Sci. Technol.* 50, 359–367. doi: 10.1021/acs.est.5b03306

Levy, S. B., and Marshall, B. (2004). Antibacterial resistance worldwide: causes, challenges and responses. *Nat. Med.* 10, S122–S129. doi: 10.1038/nm1145

Li, M., Banerjee, S. R., Zheng, C., Pomper, M. G., and Barman, I. (2016). Ultrahigh affinity Raman probe for targeted live cell imaging of prostate cancer. *Chem. Sci.* 7, 6779–6785. doi: 10.1039/C6SC01739H

Li, X., Schirmer, K., Bernard, L., Sigg, L., Pillai, S., and Behra, R. (2015). Silver nanoparticle toxicity and association with the alga *Euglena gracilis*. *Environ. Sci. Nano* 2, 594–602.

Lima, R., Seabra, A. B., and Durán, N. (2012). Silver nanoparticles: a brief review of cytotoxicity and genotoxicity of chemically and biogenically synthesized nanoparticles. *J. Appl. Toxicol.* 32, 867–879. doi: 10.1002/jat.2780

Lok, C. N., Ho, C. M., Chen, R., He, Q. Y., Yu, W. Y., Sun, H., et al. (2006). Proteomic analysis of the mode of antibacterial action of silver nanoparticles. *J. Proteome Res.* 5, 916–924. doi: 10.1021/pr0504079

Lucas, W. J., and Lee, J. Y. (2004). Plasmodesmata as a supracellular control network in plants. *Nat. Rev. Mol. Cell Biol.* 5, 712–726. doi: 10.1038/nrm1470

Ma, N. L., Rahmat, Z., and Lam, S. S. (2013). A review of the "omics" approach to biomarkers of oxidative stress in *oryza sativa*. *Int. J. Mol. Sci.* 14, 7515–7541. doi: 10.3390/ijms14047515

Ma, X., Geiser-Lee, J., Deng, Y., and Kolmakov, A. (2010). Interactions between engineered nanoparticles (ENPs) and plants: phytotoxicity, uptake and accumulation. *Sci. Total Environ.* 408, 3053–3061. doi: 10.1016/j.scitotenv.2010.03.031

Macken, A., Byrne, H. J., and Thomas, K. V. (2012). Effects of salinity on the toxicity of ionic silver and Ag-PVP nanoparticles to Tisbe battagliai and *Ceramium tenuicorne*. *Ecotoxicol. Environ. Saf.* 86, 101–110. doi: 10.1016/j.ecoenv.2012.08.025

Marambio-Jones, C., and Hoek, E. M. (2010). A review of the antibacterial effects of silver nanomaterials and potential implications for human health and the environment. *J. Nanopart. Res.* 12, 1531–1551. doi: 10.1007/s11051-010-9900-y

Marshall, J., De Salas, M., Oda, T., and Hallegraeff, G. (2005). Superoxide production in marine microalgae: I. Survey of 37 species from 6 classes. *Mar. Biol.* 47, 533–540.

Mazumdar, H., and Ahmed, G. U. (2011). Phytotoxicity effect of silver nanoparticles on Oryza sativa. *Int. J. Chem. Tech Res.* 3, 1494–1500.

Miao, A. J., Luo, Z., Chen, C. S., Chin, W. C., Santschi, P. H., and Quigg, A. (2010). Intracellular uptake: a possible mechanism for silver engineered nanoparticle toxicity to a freshwater alga *Ochromonas danica*. *PLoS ONE* 5:e15196. doi: 10.1371/journal.pone.0015196

Miao, A. J., Schwehr, K. A., Xu, C., Zhang, S. J., Luo, Z., Quigg, A., et al. (2009). The algal toxicity of silver engineered nanoparticles and detoxification by exopolymeric substances. *Environ. Pollut.* 157, 3034–3041. doi: 10.1016/j.envpol.2009.05.047

Mirzajani, F., Askari, H., Hamzelou, S., Farzaneh, M., and Ghassempour, A. (2013). Effect of silver nanoparticles on Oryza sativa L. and its rhizosphere bacteria. *Ecotoxicol. Environ. Saf.* 88, 48–54. doi: 10.1016/j.ecoenv.2012.10.018

Mohan, Y. M., Lee, K., Premkumar, T., and Geckeler, K. E. (2007). Hydrogel networks as nanoreactors: a novel approach to silver nanoparticles for antibacterial applications. *Polymer* 48, 158–164. doi: 10.1016/j.polymer.2006.10.045

Monica, R. C., and Cremonini, R. (2009). Nanoparticles and higher plants. *Caryologia* 62, 161–165. doi: 10.1080/00087114.2004.10589681

Moose, S. P., and Mumm, R. H. (2008). Molecular plant breeding as the foundation for 21st century crop improvement. *Plant Physiol.* 147, 969–977. doi: 10.1104/pp.108.118232

Moreno-Garrido, I., Pérez, S., and Blasco, J. (2015). Toxicity of silver and gold nanoparticles on marine microalgae. *Mar. Environ. Res.* 111, 60–73. doi: 10.1016/j.marenvres.2015.05.008

Morones, J. R., Elechiguerra, J. L., Camacho, A., Holt, K., Kouri, J. B., Ramírez, J. T., et al. (2005). The bactericidal effect of silver nanoparticles. *Nanotechnology* 16, 2346–2353. doi: 10.1088/0957-4484/16/10/059

Mueller, N. C., and Nowack, B. (2008). Exposure modeling of engineered nanoparticles in the environment. *Environ. Sci. Technol.* 42, 4447–4453. doi: 10.1021/es7029637

Mura, S., Greppi, G., and Irudayaraj, J. (2015). "Latest developments of nanotoxicology in plants," in *Nanotechnology and Plant Sciences*, eds M. H. Siddiqui, M. H. Al-Whaibi, and F. Mohammad (Cham: Springer International Publishing), 125–151.

Murr, L. E., Esquivel, E. V., Bang, J. J., De La Rosa, G., and Gardea-Torresdey, J. L. (2004). Chemistry and nanoparticulate compositions of a 10,000 year-old ice core melt water. *Water Res.* 38, 4282–4296. doi: 10.1016/j.watres.2004.08.010

Musante, C., and White, J. C. (2012). Toxicity of silver and copper to *Cucurbita pepo*: differential effects of nano and bulk-size particles. *Environ. Toxicol.* 27, 510–517. doi: 10.1002/tox.20667

Nair, P. M. G., and Chung, I. M. (2014). Assessment of silver nanoparticle-induced physiological and molecular changes in *Arabidopsis thaliana*. *Environ. Sci. Pollut. Res.* 21, 8858–8869. doi: 10.1007/s11356-014-2822-y

Nair, R., Varghese, S. H., Nair, B. G., Maekawa, T., Yoshida, Y., and Kumar, D. S. (2010). Nanoparticulate material delivery to plants. *Plant Sci.* 179, 154–163. doi: 10.1016/j.plantsci.2010.04.012

Navarro, E., Baun, A., Behra, R., Hartmann, N. B., Filser, J., Miao, A. J., et al. (2008). Environmental behavior and ecotoxicity of engineered nanoparticles to algae, plants, and fungi. *Ecotoxicology* 17, 372–386. doi: 10.1007/s10646-008-0214-0

Navarro, E., Wagner, B., Odzak, N., Sigg, L., and Behra, R. (2015). Effects of differently coated silver nanoparticles on the photosynthesis of *Chlamydomonas reinhardtii*. *Environ. Sci. Technol.* 49, 8041–8047. doi: 10.1021/acs.est.5b01089

Nowack, B., Krug, H. F., and Height, M. (2011). 120 years of nanosilver history: implications for policy makers. *Environ. Sci. Technol.* 45, 1177–1183. doi: 10.1021/es103316q

Nowack, B., Ranville, J. F., Diamond, S., Gallego-Urrea, J. A., Metcalfe, C., Rose, J., et al. (2012). Potential scenarios for nanomaterial release and subsequent alteration in the environment. *Environ. Toxicol. Chem.* 31, 50–59. doi: 10.1002/etc.726

Oukarroum, A., Bras, S., Perreault, F., and Popovic, R. (2012). Inhibitory effects of silver nanoparticles in two green algae, Chlorella vulgaris and *Dunaliella tertiolecta*. *Ecotoxicol. Environ. Saf.* 78, 80–85. doi: 10.1016/j.ecoenv.2011.11.012

Oukarroum, A., Samadani, M., and Dewez, D. (2014). Influence of pH on the toxicity of silver nanoparticles in the green alga *Chlamydomonas acidophila*. *Water Air Soil Pollut.* 225, 1–8. doi: 10.1007/s11270-014-2038-2

Ovečka, M., Lang, I., Baluška, F., Ismail, A., Illeš, P., and Lichtscheidl, I. K. (2005). Endocytosis and vesicle trafficking during tip growth of root hairs. *Protoplasma* 226, 39–54. doi: 10.1007/s00709-005-0103-9

Pal, S., Tak, Y. K., and Song, J. M. (2007). Does the antibacterial activity of silver nanoparticles depend on the shape of the nanoparticle? A study of the gram-negative bacterium *Escherichia coli*. *Appl. Environ. Microbiol.* 73, 1712–1720. doi: 10.1128/AEM.02218-06

Panda, K. K., Achary, V. M. M., Krishnaveni, R., Padhi, B. K., Sarangi, S. N., Sahu, S. N., et al. (2011). In vitro biosynthesis and genotoxicity bioassay of silver nanoparticles using plants. *Toxicol. In Vitro* 25, 1097–1105. doi: 10.1016/j.tiv.2011.03.008

Patlolla, A. K., Berry, A., May, L., and Tchounwou, P. B. (2012). Genotoxicity of silver nanoparticles in *Vicia faba*: a pilot study on the environmental monitoring of nanoparticles. *Int. J. Environ. Res. Public Health* 9, 1649–1662. doi: 10.3390/ijerph9051649

Pham, C. H., Yi, J., and Gu, M. B. (2012). Biomarker gene response in male Medaka (*Oryzias latipes*) chronically exposed to silver nanoparticle. *Ecotoxicol. Environ. Saf.* 78, 239–245. doi: 10.1016/j.ecoenv.2011.11.034

Piccinno, F., Gottschalk, F., Seeger, S., and Nowack, B. (2012). Industrial production quantities and uses of ten engineered nanomaterials in Europe and the world. *J. Nanopart. Res.* 14, 1–11. doi: 10.1007/s11051-012-1109-9

Prabhu, S., and Poulose, E. K. (2012). Silver nanoparticles: mechanism of antimicrobial action, synthesis, medical applications, and toxicity effects. *Int. Nano Lett.* 2:32. doi: 10.1186/2228-5326-2-32

Prasad, R., Pandey, R., and Barman, I. (2016). Engineering tailored nanoparticles with microbes: Quo vadis? *Wiley Interdiscip. Rev. Nanomed. Nanobiotechnol.* 8, 316–330. doi: 10.1002/wnan.1363

Raffin, M., Hussain, F., Bhatti, T. M., Akhter, J. I., Hameed, A., and Hasan, M. M. (2008). Antibacterial characterization of silver nanoparticles against *E. coli* ATCC-15224. *J. Mater. Sci. Technol.* 24, 192–196.

Raghavendra, R., Arunachalam, K., Annamalai, S. K., and Arunachalam, A. M. (2014). Diagnostics and therapeutic application of gold nanoparticles. *Int. J. Pharm. Pharm. Sci.* 6(Suppl. 2), 74–87. doi: 10.1016/bs.irn.2016.06.007

Rai, M., Yadav, A., and Gade, A. (2009). Silver nanoparticles as a new generation of antimicrobials. *Biotechnol. Adv.* 27, 76–83. doi: 10.1016/j.biotechadv.2008.09.002

Reidy, B., Haase, A., Luch, A., Dawson, K. A., and Lynch, I. (2013). Mechanisms of silver nanoparticle release, transformation and toxicity: a critical review of current knowledge and recommendations for future studies and applications. *Materials* 6, 2295–2350. doi: 10.3390/ma6062295

Ribeiro, F., Gallego-Urrea, J. A., Jurkschat, K., Crossley, A., Hassellöv, M., Taylor, C., et al. (2014). Silver nanoparticles and silver nitrate induce high toxicity to *Pseudokirchneriella subcapitata, Daphnia magna* and *Danio rerio*. *Sci. Total Environ.* 466, 232–241. doi: 10.1016/j.scitotenv.2013.06.101

Roh, J. Y., Sim, S. J., Yi, J., Park, K., Chung, K. H., Ryu, D. Y., et al. (2009). Ecotoxicity of silver nanoparticles on the soil nematode *Caenorhabditis elegans* using functional ecotoxicogenomics. *Environ. Sci. Technol.* 43, 3933–3940. doi: 10.1021/es803477u

Ruparelia, J. P., Chatterjee, A. K., Duttagupta, S. P., and Mukherji, S. (2008). Strain specificity in antimicrobial activity of silver and copper nanoparticles. *Acta Biomater.* 4, 707–716. doi: 10.1016/j.actbio.2007.11.006

Samal, A. K., Polavarapu, L., Rodal-Cedeira, S., Liz-Marzaìn, L. M., Peìrez-Juste, J., and Pastoriza-Santos, I. (2013). Size Tunable Au@ Ag core–shell nanoparticles: synthesis and surface-enhanced raman scattering properties. *Langmuir* 29, 15076–15082. doi: 10.1021/la403707j

Samberg, M. E., Orndorff, P. E., and Monteiro-Riviere, N. A. (2011). Antibacterial efficacy of silver nanoparticles of different sizes, surface conditions and synthesis methods. *Nanotoxicology* 5, 244–253. doi: 10.3109/17435390.2010.525669

Sapsford, K. E., Algar, W. R., Berti, L., Gemmill, K. B., Casey, B. J., Oh, E., et al. (2013). Functionalizing nanoparticles with biological molecules: developing chemistries that facilitate nanotechnology. *Chem. Rev.* 113, 1904–2074. doi: 10.1021/cr300143v

Sedlak, R. H., Hnilova, M., Grosh, C., Fong, H., Baneyx, F., Schwartz, D., et al. (2012). Engineered *Escherichia coli* silver-binding periplasmic protein that promotes silver tolerance. *Appl. Environ. Microbiol.* 78, 2289–2296. doi: 10.1128/AEM.06823-11

Shrivastava, S., Bera, T., Roy, A., Singh, G., Ramachandrarao, P., and Dash, D. (2007). Characterization of enhanced antibacterial effects of novel silver nanoparticles. *Nanotechnology* 18:225103. doi: 10.1088/0957-4484/18/22/225103

Shweta, Tripathi, D. K., Singh, S., Singh, S., Dubey, N. K., and Chauhan, D. K. (2016). Impact of nanoparticles on photosynthesis: challenges and opportunities. *Mater. Focus* 5, 405–411. doi: 10.1166/mat.2016.1327

Siddhanta, S., Barman, I., and Narayana, C. (2015). Revealing the trehalose mediated inhibition of protein aggregation through lysozyme–silver nanoparticle interaction. *Soft Matter* 11, 7241–7249. doi: 10.1039/c5sm01896j

Siddhanta, S., Zheng, C., Narayana, C., and Barman, I. (2016). An impediment to random walk: trehalose microenvironment drives preferential endocytic uptake of plasmonic nanoparticles. *Chem. Sci.* 7, 3730–3736. doi: 10.1039/C6SC00510A

Siddiqui, M. H., Al-Whaibi, M. H., and Mohammad, F. (eds) (2015). *Nanotechnology and Plant Sciences: Nanoparticles and Their Impact on Plants.* Cham: Springer. doi: 10.1007/978-3-319-14502-0

Silver, S. (2003). Bacterial silver resistance: molecular biology and uses and misuses of silver compounds. *FEMS Microbiol. Rev.* 27, 341–353. doi: 10.1016/S0168-6445(03)00047-0

Singh, S., Tripathi, D. K., Dubey, N. K., and Chauhan, D. K. (2016). Effects of nano-materials on seed germination and seedling growth: striking the slight balance between the concepts and controversies. *Mater. Focus* 5, 195–201. doi: 10.1166/mat.2016.1329

Soldo, D., Hari, R., Sigg, L., and Behra, R. (2005). Tolerance of Oocystis nephrocytioides to copper: intracellular distribution and extracellular complexation of copper. *Aquat. Toxicol.* 71, 307–317. doi: 10.1016/j.aquatox.2004.11.011

Sondi, I., and Salopek-Sondi, B. (2004). Silver nanoparticles as antimicrobial agent: a case study on *E. coli* as a model for Gram-negative bacteria. *J. Colloid Interface Sci.* 275, 177–182. doi: 10.1016/j.jcis.2004.02.012

Stampoulis, D., Sinha, S. K., and White, J. C. (2009). Assay-dependent phytotoxicity of nanoparticles to plants. *Environ. Sci. Technol.* 43, 9473–9479. doi: 10.1021/es901695c

Taylor, C., Matzke, M., Kroll, A., Read, D. S., Svendsen, C., and Crossley, A. (2016). Toxic interactions of different silver forms with freshwater green algae and cyanobacteria and their effects on mechanistic endpoints and the production of extracellular polymeric substances. *Environ. Sci. Nano* 3, 396–408. doi: 10.1039/C5EN00183H

Tiwari, J. N., Tiwari, R. N., and Kim, K. S. (2012). Zero-dimensional, one-dimensional, two-dimensional and three-dimensional nanostructured materials for advanced electrochemical energy devices. *Prog. Mater. Sci.* 57, 724–803. doi: 10.1016/j.pmatsci.2011.08.003

Tripathi, D. K., Singh, S., Singh, S., Pandey, R., Singh, V. P., Sharma, N. C., et al. (2017a). An overview on manufactured nanoparticles in plants: uptake, translocation, accumulation and phytotoxicity. *Plant Physiol. Biochem.* 110, 2–12. doi: 10.1016/j.plaphy.2016.07.030

Tripathi, D. K., Singh, S., Singh, S., Srivastava, P. K., Singh, V. P., Singh, S., et al. (2017b). Nitric oxide alleviates silver nanoparticles (AgNps)-induced phytotoxicity in *Pisum sativum* seedlings. *Plant Physiol. Biochem.* 110, 167–177. doi: 10.1016/j.plaphy.2016.06.015

Tripathi, D. K., Singh, S., Singh, V. P., Prasad, S. M., Chauhan, D. K., and Dubey, N. K. (2016). Silicon nanoparticles more efficiently alleviate arsenate toxicity than silicon in maize cultiver and hybrid differing in arsenate tolerance. *Front. Environ. Sci.* 4:46. doi: 10.3389/fenvs.2016.00046

Tripathi, D. K., Singh, V. P., Prasad, S. M., Chauhan, D. K., and Dubey, N. K. (2015). Silicon nanoparticles (SiNp) alleviate chromium (VI) phytotoxicity in *Pisum sativum* (L.) seedlings. *Plant Physiol. Biochem.* 96, 189–198. doi: 10.1016/j.plaphy.2015.07.026

Vannini, C., Domingo, G., Onelli, E., De Mattia, F., Bruni, I., Marsoni, M., et al. (2014). Phytotoxic and genotoxic effects of silver nanoparticles exposure on germinating wheat seedlings. *J. Plant Physiol.* 171, 1142–1148. doi: 10.1016/j.jplph.2014.05.002

Velmurugan, N., Kumar, G. G., Han, S. S., Nahm, K. S., and Lee, Y. S. (2009). Synthesis and characterization of potential fungicidal silver nano-sized particles and chitosan membrane containing silver particles. *Iran. Polymer J.* 18, 383–392.

Wang, G., Stender, A. S., Sun, W., and Fang, N. (2010). Optical imaging of non-fluorescent nanoparticle probes in live cells. *Analyst* 135, 215–221. doi: 10.1039/b916395f

Wang, J., Koo, Y., Alexander, A., Yang, Y., Westerhof, S., Zhang, Q., et al. (2013). Phytostimulation of poplars and *Arabidopsis* exposed to silver nanoparticles and Ag+ at sublethal concentrations. *Environ. Sci. Technol.* 47, 5442–5449. doi: 10.1021/es4004334

Wang, P., Menzies, N. W., Lombi, E., Sekine, R., Blamey, F. P. C., Hernandez-Soriano, M. C., et al. (2015). Silver sulfide nanoparticles (Ag2S-NPs) are taken up by plants and are phytotoxic. *Nanotoxicology* 9, 1041–1049. doi: 10.3109/17435390.2014.999139

Wijnhoven, S. W., Peijnenburg, W. J., Herberts, C. A., Hagens, W. I., Oomen, A. G., Heugens, E. H., et al. (2009). Nano-silver–a review of available data and knowledge gaps in human and environmental risk assessment. *Nanotoxicology* 3, 109–138. doi: 10.1080/17435390902725914

Wong, K. K., and Liu, X. (2010). Silver nanoparticles—the real "silver bullet" in clinical medicine? *Med. Chem. Commun.* 1, 125–131. doi: 10.1039/C0MD00069H

Xu, H., Pan, J., Zhang, H., and Yang, L. (2016). Interactions of metal oxide nanoparticles with extracellular polymeric substances (EPS) of algal aggregates in an eutrophic ecosystem. *Ecol. Eng.* 94, 464–470. doi: 10.1016/j.ecoleng.2016.06.019

Yin, L., Cheng, Y., Espinasse, B., Colman, B. P., Auffan, M., Wiesner, M., et al. (2011). More than the ions: the effects of silver nanoparticles on *Lolium multiflorum*. *Environ. Sci. Technol.* 45, 2360–2367. doi: 10.1021/es103995x

Yin, L., Colman, B. P., McGill, B. M., Wright, J. P., and Bernhardt, E. S. (2012). Effects of silver nanoparticle exposure on germination and early growth of eleven wetland plants. *PLoS ONE* 7:e47674. doi: 10.1371/journal.pone.0047674

Yoon, K. Y., Byeon, J. H., Park, J. H., and Hwang, J. (2007). Susceptibility constants of *Escherichia coli* and *Bacillus subtilis* to silver and copper nanoparticles. *Sci. Total Environ.* 373, 572–575. doi: 10.1016/j.actbio.2007.11.006

Zaidi, S., Maurya, C., Shankhadarwar, S., and Pius, J. (2014). "Silver nanoparticles-chlorella interaction: effect on metabolites," in *Proceeding of the National Conference on Conservation of Natural Resources & Biodiversity for Sustainable Development*, (Matunga: Ramnarain Ruia College).

Zheng, C., Shao, W., Paidi, S. K., Han, B., Fu, T., Wu, D., et al. (2015). Pursuing shell-isolated nanoparticle-enhanced Raman spectroscopy (SHINERS) for concomitant detection of breast lesions and microcalcifications. *Nanoscale* 7, 16960–16968. doi: 10.1039/C5NR05319F

Conflict of Interest Statement: The authors declare that the research was conducted in the absence of any commercial or financial relationships that could be construed as a potential conflict of interest.

13

Cytotoxicity, Intestinal Transport, and Bioavailability of Dispersible Iron and Zinc Supplements

Hyeon-Jin Kim[1†], Song-Hwa Bae[1†], Hyoung-Jun Kim[2†], Kyoung-Min Kim[2], Jae Ho Song[2], Mi-Ran Go[1], Jin Yu[1], Jae-Min Oh[2*] and Soo-Jin Choi[1*]

[1] Division of Applied Food System, Major of Food Science and Technology, Seoul Women's University, Seoul, South Korea, [2] Department of Chemistry and Medical Chemistry, College of Science and Technology, Yonsei University, Wonju, South Korea

Edited by:
Jayanta Kumar Patra,
Dongguk University, South Korea

Reviewed by:
Yong-Ro Kim,
Seoul National University,
South Korea
Dae-Hwan Park,
University of South Australia, Australia

*Correspondence:
Jae-Min Oh
jaemin.oh@yonsei.ac.kr
Soo-Jin Choi
sjchoi@swu.ac.kr

[†] These authors have contributed
equally to this work.

Iron or zinc deficiency is one of the most important nutritional disorders which causes health problem. However, food fortification with minerals often induces unacceptable organoleptic changes during preparation process and storage, has low bioavailability and solubility, and is expensive. Nanotechnology surface modification to obtain novel characteristics can be a useful tool to overcome these problems. In this study, the efficacy and potential toxicity of dispersible Fe or Zn supplement coated in dextrin and glycerides (SunActive Fe™ and SunActive Zn™) were evaluated in terms of cytotoxicity, intestinal transport, and bioavailability, as compared with each counterpart without coating, ferric pyrophosphate (FePP) and zinc oxide (ZnO) nanoparticles (NPs), respectively. The results demonstrate that the cytotoxicity of FePP was not significantly affected by surface modification (SunActive Fe™), while SunActive Zn™ was more cytotoxic than ZnO-NPs. Cellular uptake and intestinal transport efficiency of SunActive Fe™ were significantly higher than those of its counterpart material, which was in good agreement with enhanced oral absorption efficacy after a single-dose oral administration to rats. These results seem to be related to dissolution, particle dispersibility, and coating stability of materials depending on suspending media. Both SunActive™ products and their counterpart materials were determined to be primarily transported by microfold (M) cells through the intestinal epithelium. It was, therefore, concluded that surface modification of food fortification will be a useful strategy to enhance oral absorption efficiency at safe levels.

Keywords: surface modification, SunActive™, toxicity, uptake, oral absorption, dispersibility

INTRODUCTION

Iron deficiency is the most common nutritional disorder often occurs in infants, children, and premenopausal women, causing iron deficiency anemia. Approximately, 20–50% of the world population are reported to be affected by iron deficiency (Beard and Stoltzfus, 2001; Hider and Kong, 2013). Zinc deficiency is another important nutrition-related health problem, which affects about 25% of population around the world (Prasad, 2012; Raya et al., 2016). Since a small portion of iron or zinc in the diet can be absorbed, the deficiencies are mainly associated with low bioavailability of these elements.

Food fortification refers to the addition of essential trace elements and vitamins to food, so as to improve the nutritional quality of the food and to provide a public health benefit with minimal risk to health [World Health Organization (WHO) and Food and Agriculture Organization (FAO)]. This is a relatively simple and efficient way to prevent mineral deficiency. However, mineral fortification in food has challenges because it often causes unacceptable organoleptic changes during preparation process and storage, has a low bioavailability, and is expensive (Hurrell, 2002). For example, water-soluble ferrous sulfate has a good bioavailability, but induces unacceptable color or flavor changes in food (Hurrell, 1997). In contrast, ferric pyrophosphate (FePP) does not induce serious organoleptic change, although it has low bioavailability coming from poor water solubility (Kloots et al., 2004). In case of Zn fortification, zinc sulfate, zinc oxide (ZnO), zinc acetate, and zinc gluconate have been frequently used (Brown and Wuehler, 2000; Brown et al., 2004). Recently, zinc nanoparticle (NP) formulation, like ZnO-NP, has also attracted much attention to enhance intestinal absorption and bioavailability and to reduce undesirable side effects, taking advantage of high surface area to volume ratio and great reactivity as nano-sized materials (Srinivas et al., 2010; Torabi et al., 2013; Zhao et al., 2014; Faiz et al., 2015; Raya et al., 2016). However, ZnO-NP itself tends to be easily precipitated or aggregated, thereby possessing low stability under physiological condition. Moreover, most forms for Zn fortification have unpleasant zinc flavor when applied in the diet.

A modification of surface characteristics can be a fascinating strategy to overcome such disadvantages in food fortification, for example, by using colloidal techniques, emulsification, surface coating, and etc. (Rossi et al., 2014). In this regard, innovative SunActive^{TM} compounds (Taiyo Kagaku, Yokkaichi, Japan), in which FePP or ZnO-NPs are emulsified by dextrin, surfactants, and lecithin (Nanbu et al., 2000; Honda et al., 2010), are currently available on the market. These techniques provide high dispersion stability in aqueous media, and thus, can be also used for application in liquid foods or drinks without compromising taste. It was reported that SunActive Fe^{TM}, an emulsified FePP with dextrin and glycerides, has a similar bioavailability to ferrous sulfate when added to a wheat-milk infant cereal and a yogurt drink (Fidler et al., 2004). Similarly, SunActive Zn^{TM}, an emulsion of ZnO-NPs with maltodextrin and glycerides, was reported to enhance Zn bioavailability in zinc-deficient rats (Ishihara et al., 2008).

Nanotechnology application to food sectors covers techniques or tools used during cultivation, production, processing, or packaging of the food, as defined by the Nanoforum (Joseph and Morrison, 2006). Surface modification of Fe or Zn moiety by SunActive^{TM} coating, which induces novel characteristics, can be an important nanotechnology strategy, although an average particle size of SunActive^{TM} products have been reported to be relatively large (~300 nm) (Fidler et al., 2004; Sakaguchi et al., 2004; Ishihara et al., 2008; Roe et al., 2009). Nevertheless, SunActive^{TM} coating can be classified as a nano-coating, considering general definition of nanomaterials as "particles with dimensions less than one micrometer (i.e., <1000 nm) that exhibit unique properties not recognized in micron or larger sized particles," provided by US Food and Drug Administration (Food and Drug Administration [FDA], 2014).

Relatively little information is currently available on physicochemical characteristics and *in vitro* and *in vivo* efficacy as well as toxicity of commercially available, nanotechnology applied mineral fortification products. The aim of the present study was, therefore, to answer the question as to whether they have enhanced efficacy, without affecting potential toxicity. We evaluated the cytotoxicity of SunActive Fe^{TM} and SunActive Zn^{TM} in human intestinal cells. Furthermore, the efficacy and bioavailability of SunActive^{TM} products were assessed in terms of cellular uptake, intestinal transport mechanism, and *in vivo* biokinetics. The physicochemical properties of SunActive^{TM} products, such as solubility, dispersibility, and coating stability, were also analyzed to explain their biological behaviors. In all experiments, FePP and ZnO-NPs with similar primary particle size compared to SunActive^{TM} products were used as counterpart materials for comparative study.

MATERIALS AND METHODS

Materials

SunActive Fe^{TM} (FePP 26.6%, dextrin 54.7%, glycerides 5.3%, lecithin 1.3%, and etc.) and SunActive Zn^{TM} (ZnO 30.0%, maltodextrin 62.1%, glyceride 6.0%, lecithin 1.9%, and etc.) were purchased from Taiyo Kagaku (Yokkaichi, Japan). According to the certificate of analysis provided by the manufacturer, both Sunactive Fe^{TM} and Sunactive Zn^{TM} were prepared by materials mixing, pasteurization at 80°C, homogenization, filtration, and spray drying. FePP and ZnO-NPs possessing similar primary particle sizes to corresponding SunActive^{TM} product were purchased from Dr. Paul Lohmann GmbH KG (Emmerthal, Germany) and American Elements, Co. Ltd (Los Angeles, CA, USA), respectively.

Characterization

Crystalline phase of each sample was analyzed by powder X-ray diffractometry (XRD, Bruker AXS D2 Phaser, Billerica, MA, USA) with Cu Kα radiation ($\lambda = 1.5418$ Å). The powder sample mounted on poly (methyl methacrylate) holder was scanned from 5 to 80° (2θ) with 0.1 mm equatorial slit and 1.0 mm air scattering slit. For measurement of hydrodynamic radius of materials, aqueous suspension of each sample (1 mg/ml) was subjected to dynamic light scattering (DLS) instrument (ELSZ-1000, Otsuka, Japan). The particle size and morphology of powder samples were examined with field emission-scanning electron microscope (SEM; Quanta 250 FEG, FEI Company, Hillsboro, OR, USA). Each powder was loaded on carbon tape and coated with Pt/Pd for 60 s before SEM measurement.

Solubility and Hydrodynamic Radii in Cell Culture Medium and Simulated Body Fluids

Dissolution of metal species from each sample was evaluated in cell culture minimum essential medium (MEM; Welgene,

Gyeongsangbuk-do, South Korea) and simulated body fluids, such as gastric and intestinal solutions. For preparation of simulated gastric solution, 4 g NaCl (Daejung Chemicals & Metals, Gyeonggi-do, South Korea) and 6.4 g pepsin (Sigma–Aldrich, St. Louis, MO, USA) were dissolved in 1,900 ml of distilled and deionized water (DDW), and then the solution was titrated with 1 M HCl until pH 1.5. In order to prepare simulated intestinal solution, pH of simulated gastric solution was adjusted to 6.8 with 0.95 M NaHCO$_3$ (Sigma–Aldrich) solution, and then, 175 mg of bile salt (Sigma–Aldrich) and 50 mg pancreatin (Sigma–Aldrich) were added.

Each sample with designated mass was dispersed in MEM containing 10% fetal bovine serum (FBS), and simulated gastric or intestinal solution. The final concentrations of suspension were 3.72, 1, 3.24, and 1 mg/ml for SunActive Fe^{TM}, FePP, SunActive Zn^{TM}, and ZnO-NPs, respectively, to have an equivalent Zn or Fe amount. Each SunActive^{TM} suspension was prepared to contain an equivalent Fe or Zn amount compared to its counterpart, FePP or ZnO-NPs. After incubation for 24 h, the supernatants were collected by centrifugation (9,240 × g), pre-treated with ultrapure nitric acid, filtered through syringe filter (0.45 μm, Advantec MFS, Inc., Dublin, CA, USA), and then subjected to inductively coupled plasma-atomic emission spectroscopy (JY2000 Ultrace, HORIBA Jobin Yvon, Longjumeau, France).

In order to investigate dispersibility of samples under physiological conditions, hydrodynamic radii in MEM and simulated body fluids were measured. Each suspension was prepared to have 1 mg/ml concentration, and then subjected to DLS measurement (ELSZ-1000).

Cell Culture

Human intestinal epithelial INT-407 cells were provided by Dr. Tae-Sung Kim at Korea University (Seoul, South Korea) and cultured in MEM (Welgene). Caco-2 cells were purchased from the Korean Cell Line Bank (Seoul, South Korea) and cultured in Dulbecco's Modified Eagle Medium (DMEM; Welgene), under a humidified atmosphere (5% CO$_2$/95% air) at 37°C. The media were supplemented with 10% heat inactivated FBS, 100 units/ml of penicillin, and 100 μg/ml of streptomycin (Welgene).

Cell Proliferation

The effect of materials on cell proliferation was measured with the water soluble tetrazolium salt-1 (WST-1; Roche, Molecular Biochemicals, Mannheim, Germany). Cells (5 × 10^3 cells/100 μl) were incubated with SunActive^{TM} products or equivalent amounts of FePP or ZnO-NPs based on Fe or Zn content for 24 h. Then, 10 μl of WST-1 solution (Roche) was added to each well, and cells were further incubated for 4 h. Absorbance was measured using a plate reader at 440 nm (SpectraMax® M3, Molecular Devices, Sunnyvale, CA, USA). Cells incubated in the absence of materials were used as controls.

Clonogenic Assay

Cells (5 × 10^2 cells/2 ml) were seeded in 6-well plates and incubated overnight at 37°C under a 5% CO$_2$ atmosphere. The medium in the plates was then replaced with fresh medium containing various concentrations of SunActive^{TM} materials or equivalent amounts of FePP or ZnO-NPs based on Fe or Zn content, and incubation was continued for 7 days. For colony counting, cells were washed with phosphate buffered saline (PBS) and fixed with 90% crystal violet solution (Sigma–Aldrich) for 1 h. After cell washing with DDW and air-drying, colonies consisted of more than 50 cells were counted. Each experiment was done in triplicate and colony numbers in the absence of test materials were used as controls.

LDH Leakage Assay

The release of lactate dehydrogenase (LDH) was monitored with the CytoTox 96 Non-Radioactive Cytotoxicity assay (Promega, Madison, WI, USA). Cells (2 × 10^4 cells/1 ml) were exposed to SunActive^{TM} products or equivalent amounts of FePP or ZnO-NPs based on Fe or Zn content for 24 h. The cells were detached with trypsin- ethylenediaminetetraacetic acid (EDTA) treatment, centrifuged, and aliquots (50 μl) of cell culture medium were collected from each well and placed in new microplates. Then, 50 μl of substrate solution was added to each well and the plates were further incubated for 30 min at room temperature. Finally, after adding 50 μl of stop solution, the absorbance at 490 nm was measured with a microplate reader (SpectraMax® M3, Molecular Devices). Cytotoxicity is expressed relative to the basal LDH release from untreated control cells.

Intracellular ROS Production

Intracellular reactive oxygen species (ROS) levels were monitored using a peroxide-sensitive fluorescent probe, 2′,7′-dichlorofluorescein diacetate (H$_2$DCFDA) (Molecular Probes, Inc., Eugene, OR, USA), according to the manufacturer's guidelines. Briefly, cells (1 × 10^4 cells/100 μl) were incubated with SunActive^{TM} products or equivalent amounts of FePP or ZnO-NPs based on Fe or Zn content for 24 h under standard condition as described above, washed with PBS, and incubated with 5 μM H$_2$DCFDA for 60 min at 37°C. After washing with PBS, DCF fluorescence was immediately measured using a fluorescence microplate reader (SpectraMax® M3, Molecular Devices). Excitation and emission wavelengths were 495 and 518 nm, respectively. Cells not treated with test materials were used as controls.

Cellular Uptake

Cells (1 × 10^6 cells/2 ml) were incubated overnight at 37°C, and then, the medium was replaced with fresh medium containing SunActive^{TM} products or an equivalent amount of FePP or ZnO-NPs (50 μg/ml Fe or Zn content). After 6 h of incubation, the cells were washed three times with PBS, treated with 5 mM EDTA in PBS for 40 s to detach adsorbed particles on the cell membrane, and further washed with PBS. The cells were harvested by scraping and re-suspended in DDW. After centrifugation, the cell pellets were digested and Fe or Zn concentrations were determined by ICP-AES (JY2000 Ultrace, HORIBA Jobin Yvon), as described in "ICP-AES analysis" below.

Intestinal Transport Mechanism

3D Cell Culture for Intestinal Epithelial Monolayers

The monoculture system of Caco-2 cells, representing the intestinal epithelium monolayers of tight junctions, was prepared as follow; after coating Transwell® inserts (SPL Life Science, Gyeonggi-do, South Korea) with Matrigel™ matrix (Becton Dickinson, Bedford, MA, USA) for 1 h, supernatants were removed, and then inserts were washed with DMEM. Caco-2 cells (4.5×10^5 cells/well) were grown on upper insert sides, cultured for 21 days, and treated with SunActive™ products or an equivalent amount of FePP or ZnO-NPs (50 μg/ml Fe or Zn content) for 6 h. The concentrations of transported Fe or Zn in basolateral solutions were determined by ICP-AES (JY2000 Ultrace, HORIBA Jobin Yvon).

3D Cell Culture for FAE Model

Non-adherent human Burkitt's lymphoma Raji B cells were purchased from the Korean Cell Line Bank and grown in RPMI 1640 medium, supplemented with 10% FBS, 1% non-essential amino acids, 1% L-glutamine, 100 units/ml of penicillin, and 100 μg/ml of streptomycin at 37°C in 5% CO_2 atmosphere. An *in vitro* model of human intestinal follicle-associated epithelium (FAE) was prepared according to the method described by des Rieux et al. (2007) to study material transport by microfold (M) cells; Caco-2 cells (1×10^6 cells/well) were grown on upper insert sides in the same manner as described in Caco-2 monoculture system and incubated for 14 days. Raji B cells (1×10^6 cells/well) in DMEM were then added to basolateral insert compartments, and these co-cultures were maintained for 5 days. Apical medium of cell monolayers was then replaced with a particle suspension containing SunActive™ products or an equivalent amount of FePP or ZnO-NPs (50 μg/ml Fe or Zn content), and treated for 6 h. The concentrations of transported Fe or Zn in basolateral solutions were determined by ICP-AES (JY2000 Ultrace, HORIBA Jobin Yvon).

Animals

Five-week-old female Sprague Dawley (SD) rats weighing 130–150 g were purchased from Nara Biotech, Co., Ltd (Seoul, South Korea). Animals were housed in plastic laboratory animal cages in a ventilated room maintained at $20 \pm 2°C$ and $60 \pm 10\%$ relative humidity under a 12 h light/dark cycle. Water and commercial laboratory complete food were provided *ad libitum*. Animals were allowed to acclimate to the environment for 7 days before treatment. All animal experiments were performed in compliance with the guideline issued by the Animal and Ethics Review Committee of Seoul Women's University and Ethics Review Committee of Seoul Women's University approved the procedures used in this study (IACUC-2014A-3).

Biokinetics and Tissue Distribution

Six female rats per group were administered a single dose of SunActive Fe™ (1000 mg/kg) or SunActive Zn™ (500 mg/kg) and an equivalent Fe or Zn amount of FePP (320 mg/kg) or ZnO-NPs (156.25 mg/kg) by oral gavage; controls ($n = 6$) received an equivalent volume of distilled water (DW). To determine plasma Fe or Zn concentrations, blood samples were collected *via* tail vein at 0, 0.25, 0.5, 1, 2, 4, 6, 10, and 24 h. Blood samples were centrifuged at $3,000 \times g$ for 15 min at 4°C to obtain plasma. Plasma Fe or Zn concentration-time profiles were determined by ICP-AES (JY2000 Ultrace, HORIBA Jobin Yvon). The following biokinetic parameters were calculated using Kinetica version 4.4 (Thermo Fisher Scientific, Waltham, MA, USA): maximum concentration (C_{max}), time to maximum concentration (T_{max}), area under the plasma concentration-time curve (AUC), half-life ($T_{1/2}$), and mean residence time (MRT).

For the tissue distribution study, the same doses were orally administered. Tissue samples of kidneys, liver, lungs, and spleen were collected after CO_2 euthanasia at 24 h post-administration. Fe or Zn biodistribution was determined by ICP-AES (JY2000 Ultrace, HORIBA Jobin Yvon).

ICP-AES Analysis

Biological samples were pre-digested with 10 ml of ultrapure nitric acid (HNO_3) overnight, heated to ~160°C, and 1 ml of hydrogen peroxide (H_2O_2) solution was added. The mixtures were heated until the samples were colorless and clear. After dilution with 3 ml of DDW, total Fe or Zn concentrations were determined by ICP-AES (JY2000 Ultrace, HORIBA Jobin Yvon).

Statistical Analysis

Results were expressed as means ± standard deviations. Experimental values were compared with corresponding untreated control values. One-way analysis of variance (ANOVA) with Tukey's test in SAS Version 9.4 (SAS Institute, Inc., Cary, NC, USA) was used to determine the significances of intergroup differences. Statistical significance was accepted for p values of less than 0.05.

RESULTS

Physicochemical Properties

Physicochemical properties of SunActive Fe™ and SunActive Zn™ are displayed in **Figures 1**, **2**, respectively. XRD results show that both SunActive Fe™ and FePP had amorphous phase (**Figure 1A**). Hydrodynamic radii, representing the secondary particle sizes in DDW suspension, were 240 and 410 nm for SunActive Fe™ and FePP (**Figure 1B**), respectively. The SEM images in low magnification exhibited sphere-like morphology for SunActive Fe™, which seems to be attributed to the organic moiety, such as dextrin and glycerides, while FePP showed irregular aggregation (**Figure 1C**). Magnified images taken at the surface of materials showed comparable primary particle size of ~50 nm for both SunActive Fe™ and FePP (insets in **Figure 1C**). On the other hand, both SunActive Zn™ and ZnO-NPs had the typical Wurtzite crystal phase [**Figure 2A**, Joint Committee on Powder Diffraction Standards (JCPDS) No. 36-1451], indicating the existence of ZnO-NPs in SunActive Zn™. Hydrodynamic radii of SunActive Zn™ and ZnO-NPs were 350 and 750 nm, respectively, indicating higher dispersion degree of SunActive Zn™ (**Figure 2B**). Similar to SunActive Fe™, SunActive Zn™ showed spherical morphology in low magnification, while ZnO-NPs had irregular and agglomerated

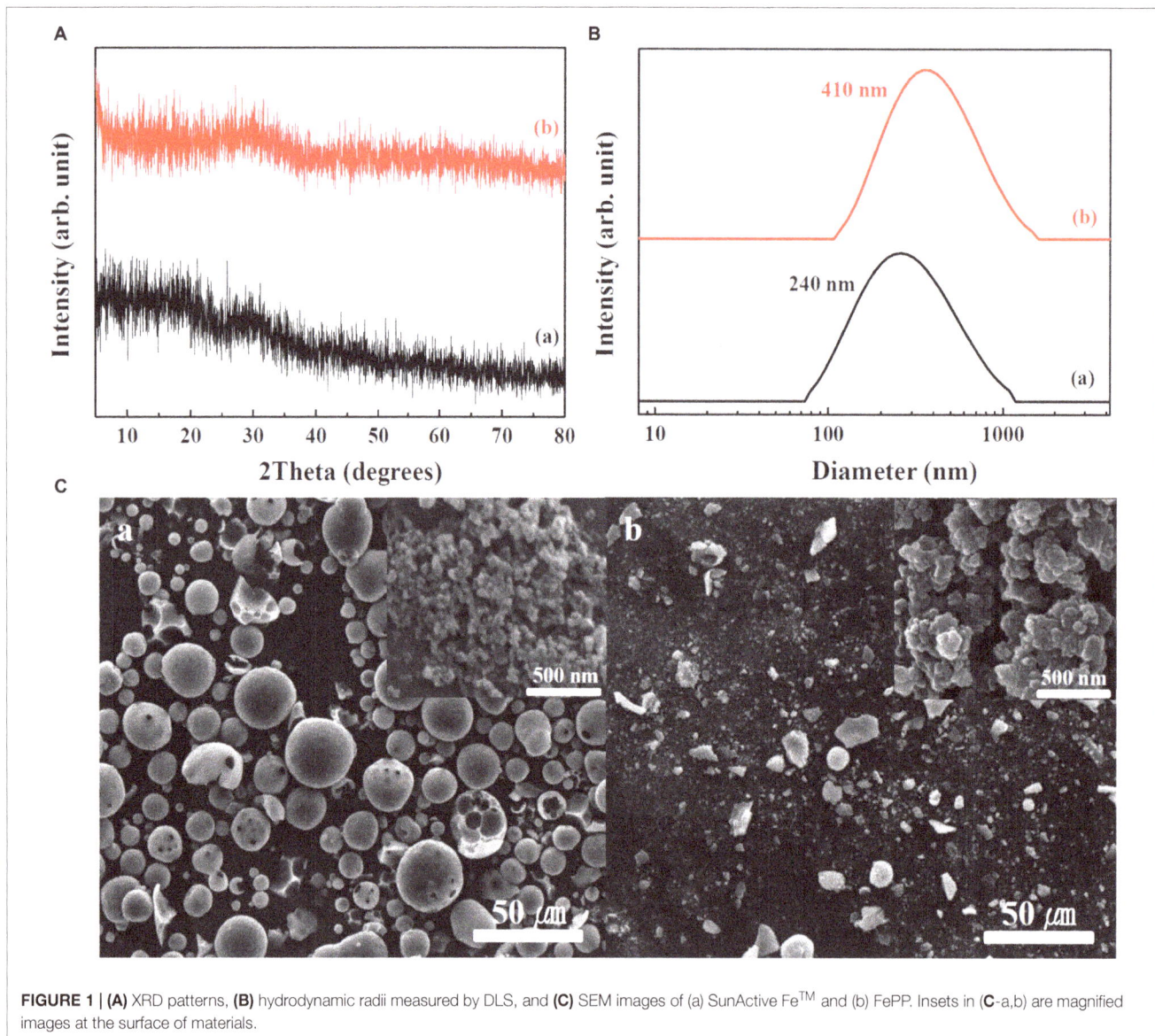

FIGURE 1 | (A) XRD patterns, **(B)** hydrodynamic radii measured by DLS, and **(C)** SEM images of (a) SunActive Fe^{TM} and (b) FePP. Insets in (**C**-a,b) are magnified images at the surface of materials.

shape (**Figure 2C**). Magnified SEM images at the surface of materials exhibited fairly similar primary particle size of 112 ± 42 and 103 ± 20 nm for SunActive Zn^{TM} and ZnO-NPs, respectively (insets in **Figure 2C**).

Solubility and Hydrodynamic Radii in Cell Culture Medium and Simulated Body Fluids

Dissolution properties of Fe or Zn from SunActive Fe^{TM} or SunActive Zn^{TM} in MEM and simulated gastric or intestinal solutions were compared to those of their counterpart materials (FePP or ZnO-NPs) and summarized in **Table 1**. There were no significant differences in Fe dissolution between SunActive Fe^{TM} and FePP in MEM, although slightly high solubility was found for FePP in simulated gastric fluid. On the other hand, Fe solubility from SunActive Fe^{TM} significantly and remarkably increased

compared to FePP under intestinal condition. Similar to Fe samples, Zn dissolution from SunActive Zn^{TM} was similar to that from ZnO-NPs in both MEM and gastric solution. Whereas, Zn solubility was slightly higher for ZnO-NPs than SunActive Zn^{TM} in intestinal solution, and both Zn samples showed the highest solubility in gastric solution. In other words, solubilities between SunActive^{TM} products and their counterparts were similar in MEM, while statistically different solubility was found in simulated intestinal solution.

Hydrodynamic radii of SunActive^{TM} products and their counterpart materials in MEM and simulated gastric and intestinal solutions were measured by DLS and displayed in **Figure 3**. The hydrodynamic radii were found to be highly dependent on the type of medium. The hydrodynamic radii of SunActive Fe^{TM}, FePP, and SunActive Zn^{TM} in MEM were 340, 460, and 300 nm, respectively (**Figure 3A**), which were comparable to the values obtained in DDW (**Figures 1B, 2B**). The

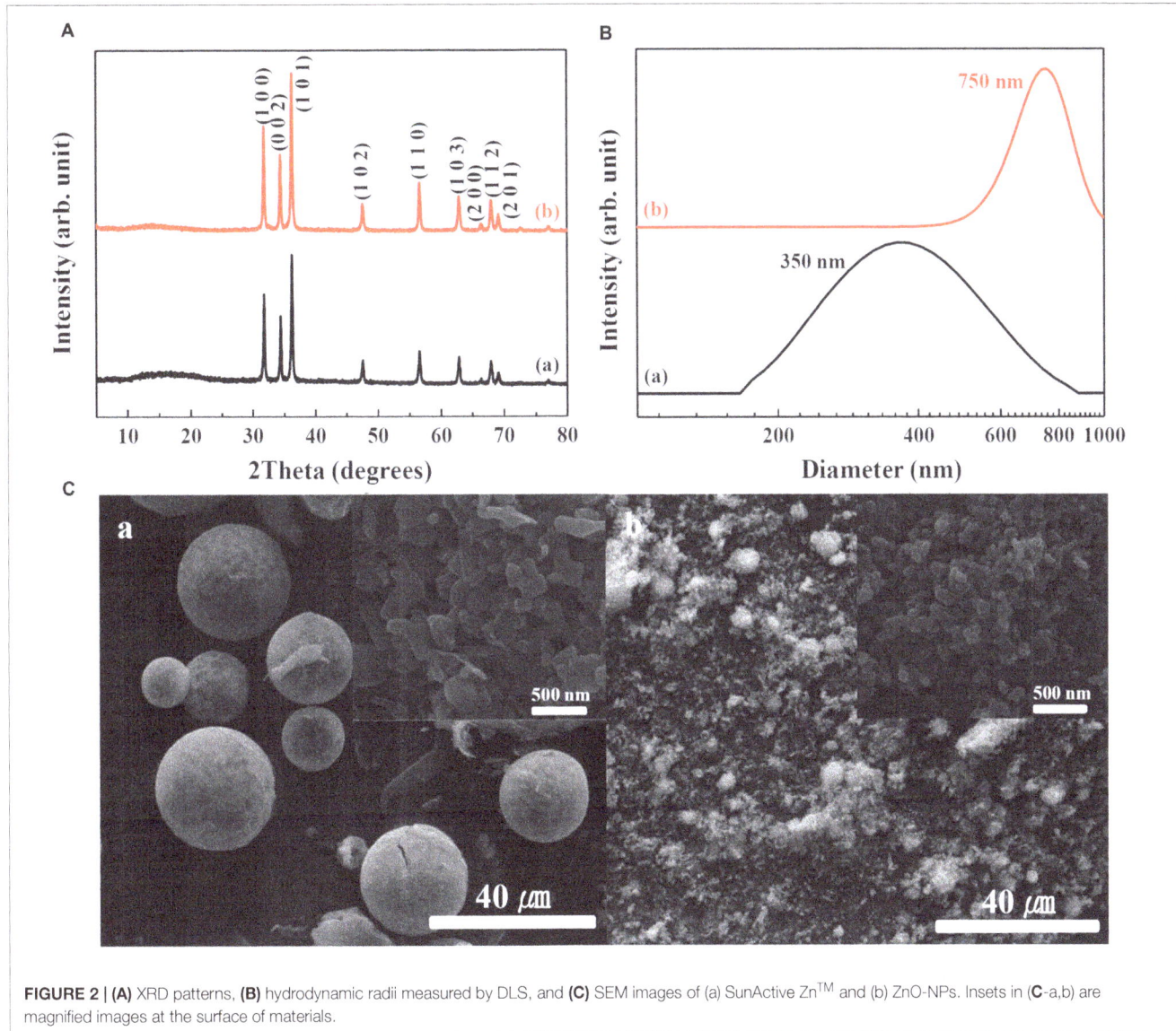

FIGURE 2 | (A) XRD patterns, (B) hydrodynamic radii measured by DLS, and (C) SEM images of (a) SunActive Zn™ and (b) ZnO-NPs. Insets in (C-a,b) are magnified images at the surface of materials.

hydrodynamic radius of ZnO-NPs in MEM considerably reduced compared to that in DDW, showing 380 nm. Under simulated gastric condition, the hydrodynamic radii of all materials significantly increased and more prominent increase in size was

found for Fe samples than Zn samples (**Figure 3B**). In simulated intestinal fluid (**Figure 3C**), both SunActive Fe™ and FePP had increased hydrodynamic radii and reduced homogeneity compared to those in MEM (**Figure 3A**). Furthermore, SunActive

TABLE 1 | Dissolution properties of Fe or Zn from SunActive™ products and their counterpart materials in cell culture medium (MEM) and simulated body fluids.

Materials	Unit	MEM		Simulated gastric solution		Simulated intestinal solution	
		Fe	Zn	Fe	Zn	Fe	Zn
SunActive Fe™	% (w/w)	0.067 ± 0.75		2.47 ± 0.20^a		25.6 ± 2.68^a	
FePP	% (w/w)	N.D.		4.02 ± 0.64^b		0.27 ± 0.14^b	
SunActive Zn™	% (w/w)		2.24 ± 0.42		86.0 ± 0.58		1.48 ± 0.11^a
ZnO-NP	% (w/w)		1.87 ± 0.77		84.1 ± 1.19		1.90 ± 0.03^b

Mean values with different superscripts (a,b) in the same column indicate significant difference between SunActive™ product and its counterpart ($p < 0.05$).
N.D., not determined; MEM, minimum essential medium; FePP, ferric pyrophosphate.

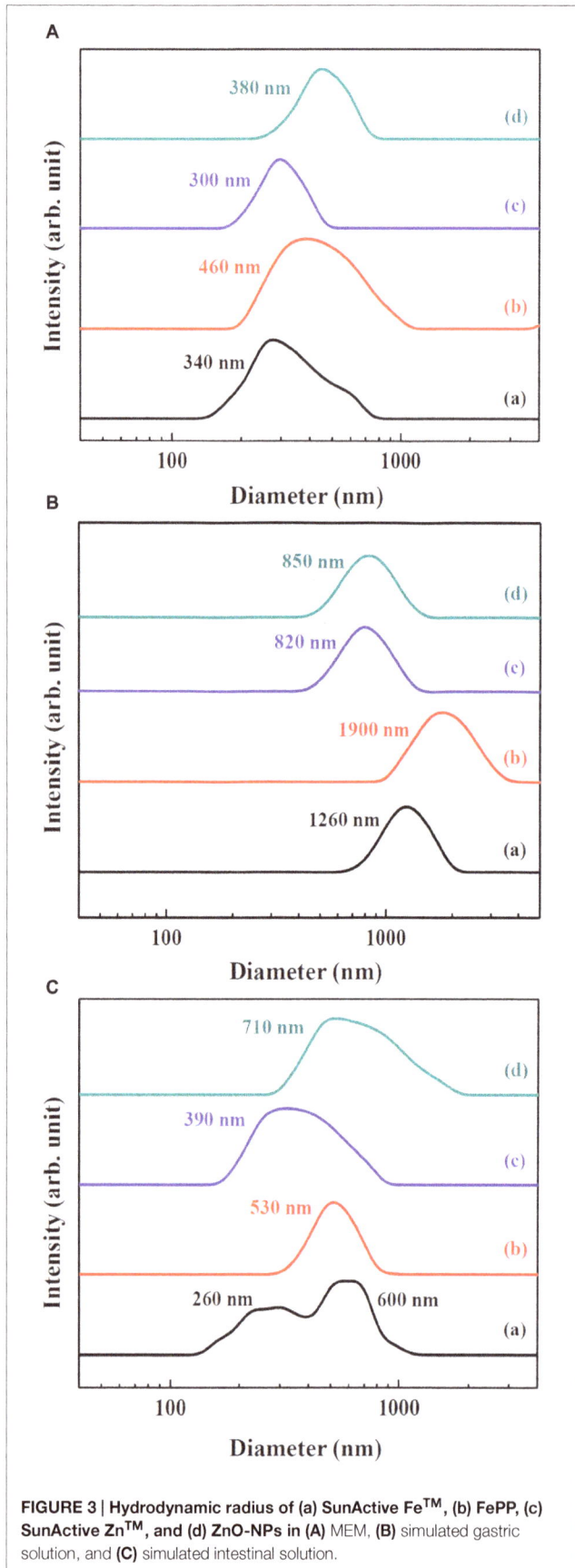

FIGURE 3 | Hydrodynamic radius of (a) SunActive Fe[TM], (b) FePP, (c) SunActive Zn[TM], and (d) ZnO-NPs in (A) MEM, (B) simulated gastric solution, and (C) simulated intestinal solution.

Fe[TM] showed split distribution in the range from 150 to 1000 nm. In case of SunActive Zn[TM] and ZnO-NPs, the single distribution patterns and similar hydrodynamic radii were found in intestinal fluid, compared to those in DDW.

Short- and Long-Term Cytotoxicity

Cytotoxicity of SunActive Fe[TM] or SunActive Zn[TM] was evaluated in human intestinal INT-407 cells in terms of inhibition of cell proliferation and colony-forming ability, and compared to that of each counterpart material. **Figure 4A** demonstrates that both SunActive Fe[TM] and FePP did not affect cell proliferation and viability up to the highest concentration tested after exposure for 24 h. On the other hand, SunActive Zn[TM] more remarkably inhibited cell proliferation than ZnO-NPs (**Figure 4B**). It is worth noting that all cytotoxicity tests were performed based on equivalent Fe or Zn amounts, and the counterpart materials had similar primary particle size to corresponding SunActive[TM] products. It was determined that Fe and Zn contents in SunActive Fe[TM] and SunActive Zn[TM] are 8 and 25%, respectively, and therefore, the highest concentrations as total compounds for the former and the latter were 1000 µg/ml (equivalent to 80 µg/ml Fe content) and 800 µg/ml (equivalent to 200 µg/ml Zn content), respectively.

Long-term cytotoxicity was also evaluated by clonogenic assay after treatment for 7 days, which investigates the ability of colony formation of cells. Interestingly, SunActive Fe[TM] and FePP did not affect colony-forming ability at all (**Figure 4C**). Whereas, SunActive Zn[TM] remarkably inhibited colony formation at the highest concentration tested (25 µg/ml), as compared to ZnO-NPs (**Figure 4D**). The same tendency was observed in human intestinal Caco-2 cells exposed to SunActive[TM] products (data not shown).

Membrane Damage and ROS Generation

Membrane damage caused by SunActive[TM] products was evaluated using LDH leakage assay, which measures released amount of cytosolic LDH into extracellular medium. The same tendency obtained from WST-1 assay was found; neither SunActive Fe[TM] nor FePP did induce membrane damage (**Figure 5A**), while significantly high LDH leakage from SunActive Zn[TM]-exposed cells compared to ZnO-NP exposure was found (**Figure 5B**).

When intracellular ROS generation was measured with H_2DCFDA, slightly but significantly increased ROS levels were detected by SunActive Fe[TM] treatment, whereas, no elevated ROS levels were found in FePP-exposed cells (**Figure 5C**). Meanwhile, both SunActive Zn[TM] and ZnO-NPs significantly induced ROS at Zn concentrations of more than 50 µg/ml, but much high ROS levels were induced by SunActive Zn[TM] at concentration range from 12.5 to 50 µg/ml (**Figure 5D**). The same tendency was found in Caco-2 cells (data not shown).

Cellular Uptake

Cellular uptake efficacy of SunActive[TM] products was evaluated in both INT-407 cells and Caco-2 cells by measuring intracellular Fe or Zn amount by ICP-AES, and compared with that of each

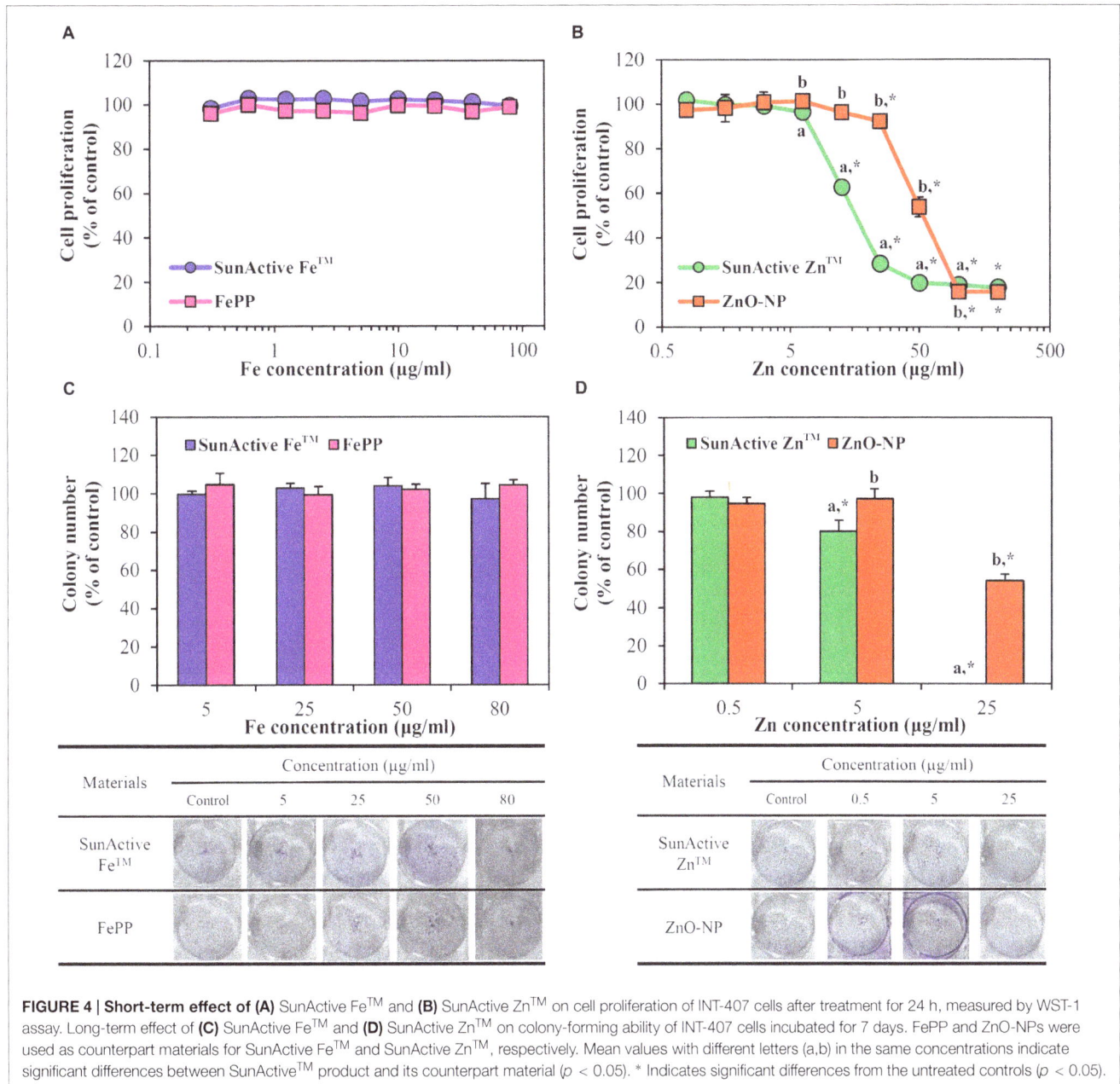

FIGURE 4 | Short-term effect of (A) SunActive Fe™ and (B) SunActive Zn™ on cell proliferation of INT-407 cells after treatment for 24 h, measured by WST-1 assay. Long-term effect of (C) SunActive Fe™ and (D) SunActive Zn™ on colony-forming ability of INT-407 cells incubated for 7 days. FePP and ZnO-NPs were used as counterpart materials for SunActive Fe™ and SunActive Zn™, respectively. Mean values with different letters (a,b) in the same concentrations indicate significant differences between SunActive™ product and its counterpart material ($p < 0.05$). * Indicates significant differences from the untreated controls ($p < 0.05$).

counterpart material. **Figure 6** shows that both SunActive Fe™ and SunActive Zn™ were more efficiently taken up by cells than FePP and ZnO-NPs, respectively. Significantly, high uptake of both materials was found in INT-407 cells (**Figures 6C,D**) than in Caco-2 cells (**Figures 6A,B**).

Intestinal Transport Mechanism

Intracellular transport pathway was assessed using *in vitro* models of human FAE and Caco-2 monolayers, which represent M cells in Peyer's Patches and intestinal tight junctions, respectively. The results demonstrate that both SunActive™ materials were primarily transported by M cells, but their significantly elevated transports through Caco-2 monolayers

were also found (**Figures 7A,B**). FePP and ZnO-NPs were also found to be translocated across both M cells and Caco-2 tight junctions. Total intestinal transport amounts of SunActive™ materials through both M cells and Caco-2 monolayers were pooled and presented in **Figures 7C,D**. Significantly higher translocation of SunActive Fe™ than FePP was found, contrary to SunActive Zn™ showing similar total transport amount compared to ZnO-NPs.

Biokinetics and Tissue Distribution

Plasma concentration versus time curves after a single-dose oral administration to rats demonstrate that SunActive Fe™ entered the bloodstream more rapidly and greatly than FePP, showing

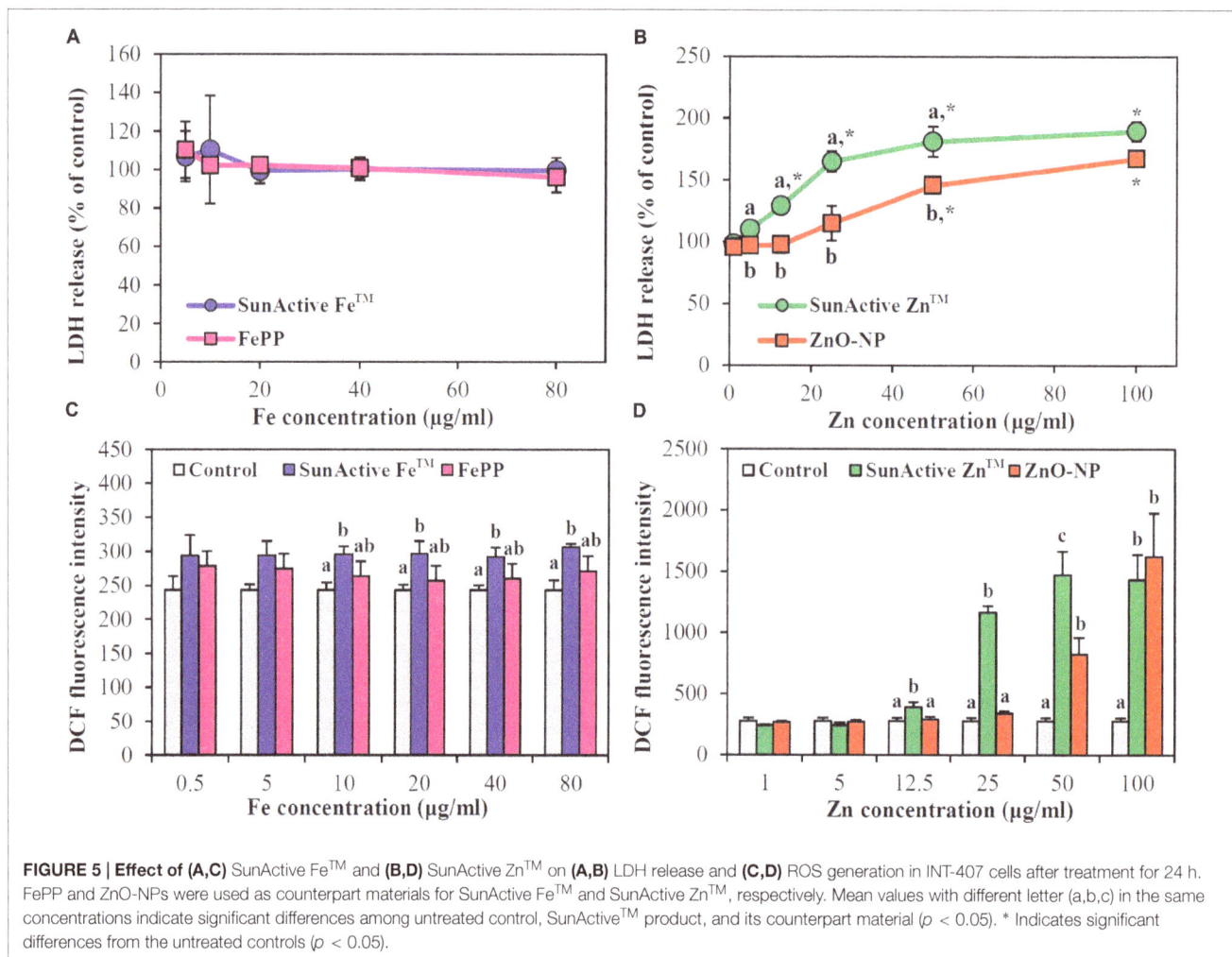

FIGURE 5 | Effect of (A,C) SunActive Fe^{TM} and **(B,D)** SunActive Zn^{TM} on **(A,B)** LDH release and **(C,D)** ROS generation in INT-407 cells after treatment for 24 h. FePP and ZnO-NPs were used as counterpart materials for SunActive Fe^{TM} and SunActive Zn^{TM}, respectively. Mean values with different letter (a,b,c) in the same concentrations indicate significant differences among untreated control, SunActive^{TM} product, and its counterpart material ($p < 0.05$). * Indicates significant differences from the untreated controls ($p < 0.05$).

peak concentration at 0.5 h for the former versus 2 h for the latter (**Figure 8A**). When biokinetic parameters were calculated (**Table 2**), all kinetic parameters between SunActive Fe^{TM} and FePP were significantly different, showing high C$_{max}$, AUC, T$_{1/2}$, and MRT values, but low T$_{max}$ values for SunActive Fe^{TM}. On the other hand, almost similar plasma concentration-time profiles between SunActive Zn^{TM} and ZnO-NPs were found (**Figure 8B**). Biokinetic parameters reveal only slightly high T$_{1/2}$ values for SunActive Zn^{TM} compared to ZnO-NPs, but there were no statistical significances in C$_{max}$, T$_{max}$, AUC, and MRT values between two materials ($p > 0.05$). When oral absorption efficiency was calculated based on AUC values (**Table 2**), absorption of SunActive Fe^{TM} (21.1 $\pm$ 4.6%) significantly increased, in comparison with that of FePP (9.7 $\pm$ 2.4%), while SunActive Zn^{TM} and ZnO-NPs were determined to have 9.1 $\pm$ 0.4 and 8.9 $\pm$ 1.0% of oral absorptions, respectively, without statistical difference ($p > 0.05$).

Tissue distributions of SunActive^{TM} products and their counterpart materials were also investigated after a single-dose oral administration to rats. **Figures 8C,D** show no significantly increased Fe or Zn levels in all organs analyzed compared to untreated control groups ($p > 0.05$).

DISCUSSION

The aim of this study was to investigate the efficacy and potential toxicity of commercially available, surface modified Fe or Zn fortification, coated with low-molecular-carbohydrates (dextrin for SunActive Fe^{TM} and maltodextrin for SunActive Zn^{TM}) and glycerides, which enables to have stable and high dispersibility in water, thereby expecting high bioavailability. Hydrodynamic radii obtained from DLS measurement in various media (DDW, MEM, and simulated gastric and intestinal solutions) clearly showed that SunActive^{TM} coating enhanced dipersibility of pristine materials, FePP and ZnO-NPs (**Figures 1B, 2B, 3A–C**). However, SunActive^{TM} products measured from SEM showed micron sizes (**Figures 1C, 2C**), which seems to be attributed to the agglomeration among coating moieties, such as dextrin (for SunActive Fe^{TM}) and maltodextrin (SunActive Zn^{TM}), during drying process for SEM measurement. The sizes of SunActive Fe^{TM} and SunActive Zn^{TM} can be, therefore, considered to be 240 and 350 nm, respectively, measured by DLS, rather than particle sizes from SEM images. Furthermore, SunActive^{TM} coating ensured long-term dispersion stability, while remarkable sedimentation of non-coated FePP and

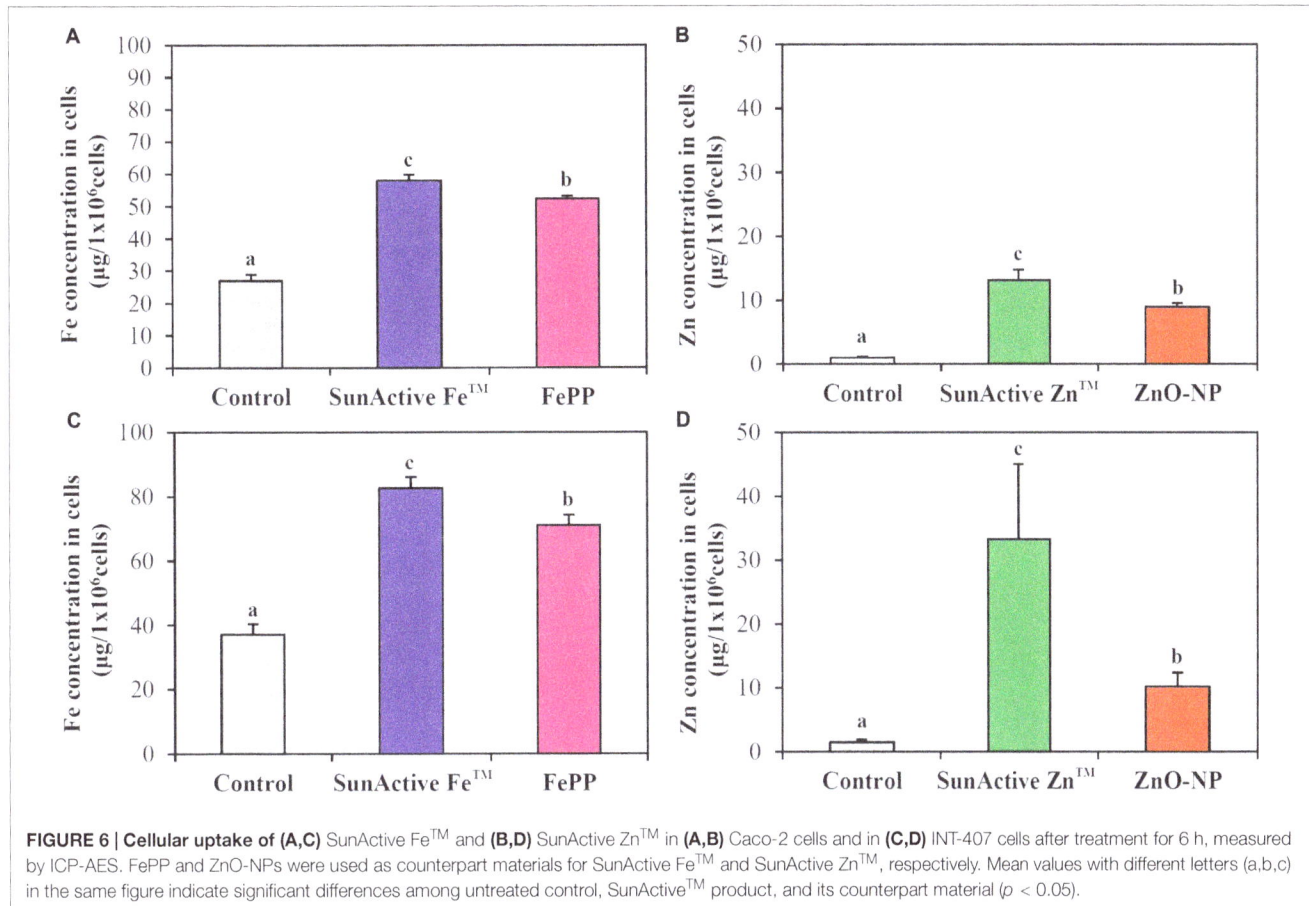

FIGURE 6 | Cellular uptake of (A,C) SunActive Fe[TM] and **(B,D)** SunActive Zn[TM] in **(A,B)** Caco-2 cells and in **(C,D)** INT-407 cells after treatment for 6 h, measured by ICP-AES. FePP and ZnO-NPs were used as counterpart materials for SunActive Fe[TM] and SunActive Zn[TM], respectively. Mean values with different letters (a,b,c) in the same figure indicate significant differences among untreated control, SunActive[TM] product, and its counterpart material ($p < 0.05$).

ZnO-NPs was observed (Supplementary Figure 1). It should be noted that the counterparts, FePP and ZnO-NPs were chosen to have the same primary particle size compared to SunActive[TM] products. Thus, the reduced hydrodynamic radii of SunActive[TM] products compared to counterpart materials are attributed to the organic coating, which reduces particle aggregation. It was also found that the hydrodynamic radius and dispersibility were highly dependent on dispersing medium. For instance, SunActive Fe[TM] and FePP showed significant agglomeration in simulated gastric (**Figure 3B**) and intestinal fluids (**Figure 3C**) compared to DDW (**Figure 1A**) and MEM (**Figure 3A**). The degree of agglomeration was relatively higher for FePP than for SunActive Fe[TM]. Agglomeration was also strongly related to dispersion stability; FePP showed clear sedimentation in DDW and simulated intestinal solution after 48 h of dispersion, while SunActive Fe[TM] showed retarded sedimentation under all conditions (Supplementary Figure 1). On the other hand, DLS results revealed that SunActive Zn[TM] showed less agglomeration than ZnO-NPs in MEM (**Figure 3A**) and simulated gastric and intestinal fluids (**Figures 3B,C**). This was in good agreement with sedimentation result (Supplementary Figure 1), which was attributed to the agglomeration of particles in each medium.

The cytotoxicity of SunActive Fe[TM] was determined to be not affected by surface coating in terms of short- and long-term inhibition of cell proliferation (**Figures 4A,C**) as well as

membrane damage (**Figure 5A**). Interestingly, SunActive Fe[TM] did not inhibit colony-forming ability even after exposure for 7 days, implying its low cytotoxicity. It is worth noting that all biological experiments were performed based on the same Fe or Zn content, in order to compare *in vivo* oral Fe or Zn absorption efficiency of SunActive[TM] products with that of each counterpart material. Thus, the highest concentrations tested as total compounds for cytotoxicity experiments were 1000 and 320 μg/ml for SunActive Fe[TM] (8% Fe) and FePP (25% Fe), respectively. Slightly increased intracellular ROS generation was detected in cells exposed to SunActive Fe[TM] compared to FePP (**Figure 5C**). The slight difference in ROS generation may be related to different particle size in aqueous suspension. Indeed, hydrodynamic radius of SunActive Fe[TM] was smaller than FePP in MEM (**Figure 3A-a,b**). It is still controversial, but generally accepted that smaller sized NPs induce higher ROS generation (Manke et al., 2013). Nevertheless, overall toxicity results on SunActive Fe[TM] suggest its low cytotoxicity potential, indicating that low cytotoxic nature of FePP was not influenced by surface coating. On the other hand, SunActive Zn[TM] exhibited higher inhibition of cell proliferation and colony formation (**Figures 4B,D**), induction of membrane damage (**Figure 5B**), and ROS generation (**Figure 5D**) than ZnO-NPs, when cytotoxicity was expressed based on Zn contents. The IC$_{50}$ values obtained by WST-1 assay were 18.11 and 61.99 μg/ml for

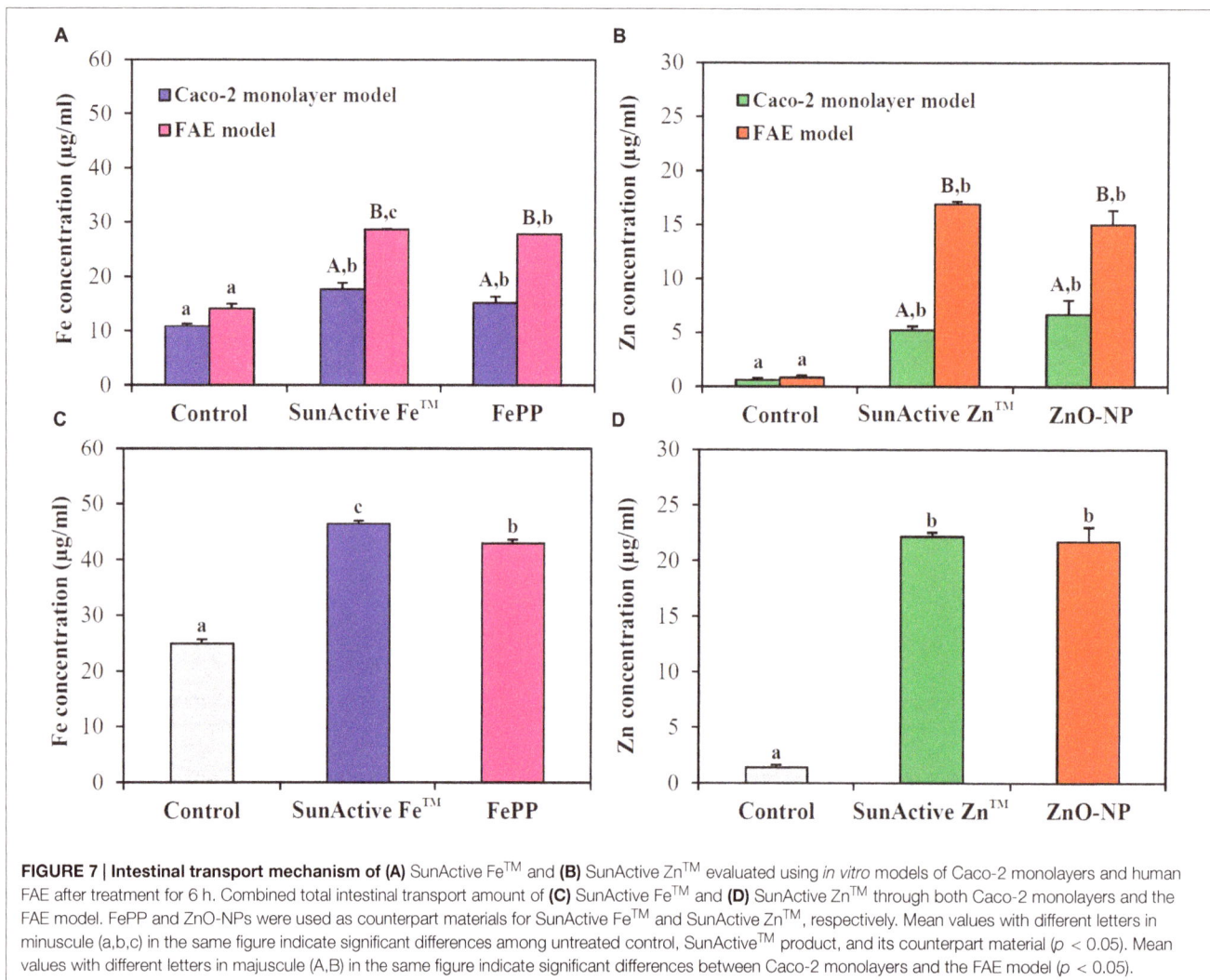

FIGURE 7 | Intestinal transport mechanism of (A) SunActive Fe[TM] and **(B)** SunActive Zn[TM] evaluated using *in vitro* models of Caco-2 monolayers and human FAE after treatment for 6 h. Combined total intestinal transport amount of **(C)** SunActive Fe[TM] and **(D)** SunActive Zn[TM] through both Caco-2 monolayers and the FAE model. FePP and ZnO-NPs were used as counterpart materials for SunActive Fe[TM] and SunActive Zn[TM], respectively. Mean values with different letters in minuscule (a,b,c) in the same figure indicate significant differences among untreated control, SunActive[TM] product, and its counterpart material ($p < 0.05$). Mean values with different letters in majuscule (A,B) in the same figure indicate significant differences between Caco-2 monolayers and the FAE model ($p < 0.05$).

SunActive Zn[TM] and ZnO-NPs as Zn contents, corresponding to 72.45 (25% Zn) and 77.48 μg/ml (80% Zn) as total compounds for the former and the latter, respectively. Hence, the cytotoxicity of SunActive Zn[TM] is much higher than ZnO-NPs based on Zn content, but similar toxicity between two materials was found when the toxicity was expressed as total compounds. Meanwhile, another critical factor affecting the cytotoxicity is the fate of materials under cell culture condition. It has been well-reported that Zn^{2+} ions cause high toxicity (Song et al., 2010; Yin et al., 2012; Wang et al., 2014; Jo et al., 2016). Furthermore, Zn^{2+} ion release from ZnO-NPs was reported to facilitate ROS generation and LDH release (Song et al., 2010). Indeed, the IC$_{50}$ values for ZnCl$_2$ as Zn^{2+} ions were 9.54 μg/ml based on Zn content and 19.89 μg/ml as a total compound (data not shown), which are different from those for SunActive Zn[TM] or ZnO-NPs. In the present study, SunActive Zn[TM] was more toxic than ZnO-NPs based on Zn content, in spite of similar Zn dissolution property between two materials in MEM (**Table 1**). Moreover, the solubility of both materials was $\sim$2% in MEM, indicating their primarily particulate fate. Hence, it is strongly likely that

high toxicity of SunActive Zn[TM] can be related to its small hydrodynamic radius and high dispersibility (**Figure 3A**), but not to their solubility or fate.

When intracellular uptake efficiency was evaluated by measuring total Fe or Zn concentration using ICP-AES, uptake efficiency in both INT-407 and Caco-2 cells was significantly enhanced by SunActive[TM] coatings, in comparison with their counterpart materials (**Figure 6**). Enhanced cellular uptake of SunActive[TM] products might be attributed to their reduced hydrodynamic radii compared to counterpart materials (**Figure 3A**), which probably causes high toxicity as well (**Figures 4, 5**). However, this type of experiment reflects only the amount taken up by cells, which can be different from absorption efficacy through the intestinal epithelium after oral administration. Thus, intestinal transport efficacy and mechanism were also assessed using *in vitro* 3D culture systems, human FAE model and Caco-2 monolayers. The FAE model represents M cells found in the Peyer's patches, which involve in the intestinal transport of a variety of materials and immune system (Kernéis et al., 1997; Neutra et al., 2001; Jang et al., 2004).

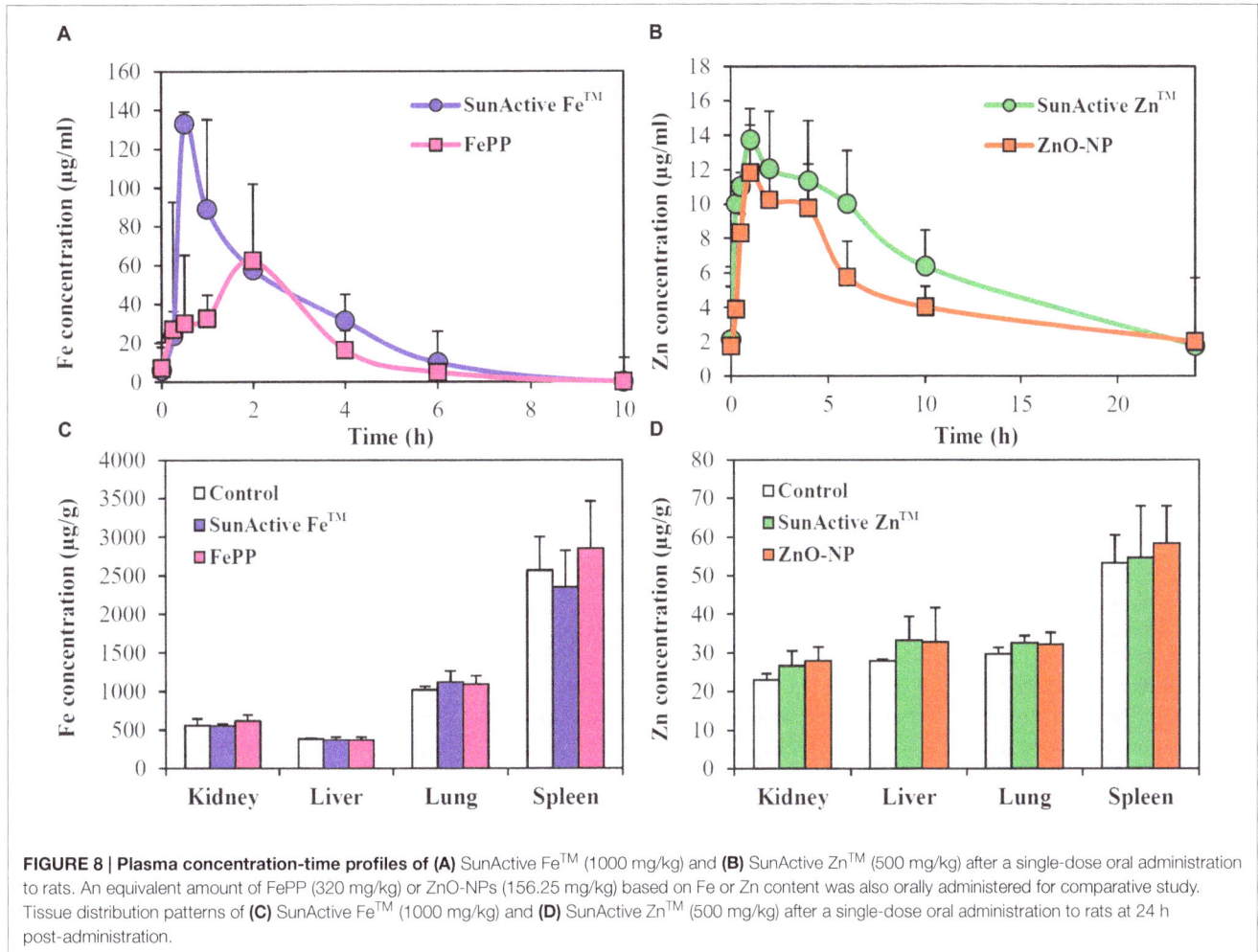

FIGURE 8 | Plasma concentration-time profiles of (A) SunActive Fe™ (1000 mg/kg) and **(B)** SunActive Zn™ (500 mg/kg) after a single-dose oral administration to rats. An equivalent amount of FePP (320 mg/kg) or ZnO-NPs (156.25 mg/kg) based on Fe or Zn content was also orally administered for comparative study. Tissue distribution patterns of **(C)** SunActive Fe™ (1000 mg/kg) and **(D)** SunActive Zn™ (500 mg/kg) after a single-dose oral administration to rats at 24 h post-administration.

Caco-2 monolayers were often used to represent a dense network of intestinal tight junctions (Ma et al., 2005; Hubatsch et al., 2007). The results show that both SunActive Fe™ and SunActive Zn™ were primarily transported by M cells, but slight transports by Caco-2 monolayers were also found (**Figure 7**). FePP and ZnO-NPs were also found to be translocated in the same manner. In other words, intestinal transport mechanism was not affected by surface coating. Since *in vivo* intestinal transport occurs simultaneously through M cells and tight junctions, it is more reasonable that combined translocation amounts obtained by two different models are compared. As shown in **Figures 7C,D**, significantly high intestinal transport of SunActive Fe™ compared to FePP suggests that surface coating can contribute to enhance intestinal transport of mineral fortification.

TABLE 2 | Biokinetic parameters and oral absorption of SunActive™ products and their counterpart materials after oral administration of a single-dose to rats.

Biokinetic parameters	SunActive Fe™	FePP	SunActive Zn™	ZnO-NP
C_{max} (mg/l)	133.07 ± 6.79[a]	63.53 ± 21.03[b]	13.81 ± 1.67[a]	12.82 ± 1.41[a]
T_{max} (h)	0.5[a]	2[b]	1[a]	1[a]
AUC (h × mg/l)	286.60 ± 62.32[a]	156.97 ± 38.38[b]	153.79 ± 7.32[a]	150.69 ± 16.63[a]
$T_{1/2}$	2.69 ± 1.34[a]	1.01 ± 0.31[b]	7.24 ± 0.14[a]	5.57 ± 0.12[b]
MRT (h)	4.31 ± 1.34[a]	2.49 ± 0.26[b]	11.74 ± 0.36[a]	11.78 ± 0.26[a]
Absorption (%)	21.07 ± 4.58[a]	9.69 ± 2.37[b]	9.05 ± 0.43[a]	8.86 ± 0.98[a]

Mean values with different superscripts (a,b) in the same row are significantly different between SunActive™ product and its counterpart material (p < 0.05). Absorption (%) was calculated based on AUC values.
FePP, ferric pyrophosphate; C_{max}, maximum concentration; T_{max}, time to maximum concentration; AUC, area under the plasma concentration-time curve; $T_{1/2}$, half-life; MRT, mean residence time.

However, increased translocation of SunActive Zn^{TM} compared to ZnO-NPs was not found. These results indicate that surface modification does not always result in transport efficiency, and that the intestinal transport of materials is governed by various factors, such as particle size, dispersibility, and coating stability.

Plasma concentration-time profiles after a single-dose oral administration to rats were highly consistent with the intestinal transport results. SunActive Fe^{TM} remarkably increased oral absorption rate and absorption efficiency, as shown in rapid T_{max} and high AUC values (**Figure 8A** and **Table 2**). Oral absorption efficacy of SunActive Fe^{TM} significantly increased by about 2.2 fold compared to its counterpart material, FePP. This result indicates that surface coating by dextrin and glycerides can enhance oral Fe bioavailability. This seems to be strongly associated with hydrodynamic radius, dispersibility, and solubility. SunActive Fe^{TM} had smaller hydrodynamic radius than FePP in all media tested (**Figure 3C-a,b**), showing better dispersibility than FePP (Supplementary Figure 1), which is likely to facilitate oral absorption of Fe moiety. Furthermore, intestinal solubility of SunActive Fe^{TM} dramatically enhanced compared to FePP, and thus, total gastrointestinal solubility of the former significantly increased in comparison with the latter, which surely affects Fe bioavailability (Zariwala et al., 2013). Meanwhile, almost similar biokinetic behaviors between SunActive Zn^{TM} and ZnO-NPs were found, as observed by similar plasma concentration versus time curves and biokinetic parameters (**Figure 8B** and **Table 2**). Moreover, oral absorption of SunActive Zn^{TM} did not significantly increase (**Table 2**), although reduced hydrodynamic radius and enhanced dispersibility compared to ZnO-NPs were observed (**Figure 3** and Supplementary Figure 1). It should be noted that major surface coating materials for SunActive Fe^{TM} and SunActive Zn^{TM} are dextrin (54.7%) and maltodextrin (62.1%), respectively, and maltodextrin is more easily digestible than dextrin and rapidly absorbed as glucose (Chronakis, 1998). It is, therefore, probable that SunActive Zn^{TM} is easily degradable after oral administration, resulting in its similar oral absorption compared to ZnO-NPs. Contrary to SunActive Zn^{TM}, SunActive Fe^{TM} with dextrin coating tends to be less degraded under gastrointestinal condition, which can contribute to enhance Fe bioavailability. It is interesting to note that SunActive Zn^{TM} had similar hydrodynamic radius compared to ZnO-NPs in simulated gastric fluid, while significantly reduced hydrodynamic radius was observed for SunActive Fe^{TM} compared to FePP (**Figure 3B**). Moreover, no significant difference in total gastrointestinal solubility between SunActive Zn^{TM} and ZnO-NPs was found (**Table 1**), which may explain similar oral Zn absorption between two materials. On the other hand, much higher total gastrointestinal solubility of SunActive Zn^{TM} than SunActive Fe^{TM} (**Table 1**) might also support coating stability of the latter. It is worth noting that biokinetic study was performed to have an equivalent amount of Fe or Zn, based on Fe or Zn contents in all tested materials. Taken together, it is likely that oral absorption of Fe or Zn fortification can be enhanced by surface coating to some extent, but highly dependent on the nature of fortified minerals and surface coating materials.

High oral absorption efficacy of SunActive Fe^{TM} is likely to be attributed to its increased intestinal transport through M cells and Caco-2 monolayers, which might be related to the size, dispersibility, and coating stability. Interestingly, we found that surface coating did not affect tissue distribution patterns, showing no accumulation in all organs analyzed after a single-dose oral administration of both SunActive^{TM} products. It can be strongly suggested that enhanced oral Fe absorption by surface coating did not influence on tissue accumulation, and surface modified Fe or Zn fortification has low toxicity potential.

CONCLUSION

Surface modified Fe or Zn fortification with low-molecular-carbohydrates and glycerides had stable and high dispersibility as well as small hydrodynamic radius in various aqueous media. However, cytotoxicity, uptake behaviors, and oral absorption were highly dependent on the nature of fortified minerals and surface coating materials; SunActive Fe^{TM} with dextrin coating enhanced cellular uptake, intestinal transport efficacy, and bioavailability, but did not cause cytotoxicity, which was not found for SunActive Zn^{TM} with maltodextrin coating. The discrepancy can be attributed to the stability of coating materials under physiological condition. On the other hand, surface modification of Fe or Zn fortification did not affect intestinal transport mechanism and tissue distribution pattern. Taken together, bioavailability of mineral fortification can be enhanced by surface modification to some extent, without affecting potential toxicity. These findings will provide a useful strategy to enhance oral absorption efficacy of food fortification at safe levels.

AUTHOR CONTRIBUTIONS

Conceived and designed the experiments and wrote the manuscript: S-JC and J-MO, performed biological experiments: H-JiK, S-HB, performed physicochemical experiments: H-JuK, K-MK, JHS, performed statistical and data analysis: M-RG and JY.

ACKNOWLEDGMENT

This research was supported by a grant (14162MFDS135) from Ministry of Food and Drug Safety in 2014 and by a research grant from Seoul Women's University (2017).

REFERENCES

Beard, J., and Stoltzfus, R. J. (2001). Iron deficiency anemia: reexamining the nature and magnitude of the public health problem. *J. Nutr.* 131, 563S–703S.

Brown, K. H., Rivera, J. A., Bhutta, Z., Gibson, R. S., King, J. C., Lönnerdal, B., et al. (2004). International Zinc Nutrition Consultative Group (IZiNCG) technical document #1. Assessment of the risk of zinc deficiency in populations and options for its control. *Food Nutr. Bull.* 25, S91–S204.

Brown, K. H., and Wuehler, S. E. (2000). *Zinc and Human Health: Results of Recent Trials and Implications for Program Interventions and Research.* Ottawa, ON: International Development Research Centre.

Chronakis, I. S. (1998). On the molecular characteristics, compositional properties, and structural-functional mechanisms of maltodextrins: a review. *Crit. Rev. Food Sci.* 38, 599–637. doi: 10.1080/10408699891274327

des Rieux, A., Fievez, V., Théate, I., Mast, J., Préat, V., and Schneider, Y. J. (2007). An improved *in vitro* model of human intestinal follicle-associated epithelium to study nanoparticle transport by M cells. *Eur. J. Pharm. Sci.* 30, 380–391. doi: 10.1016/j.ejps.2006.12.006

Faiz, H., Zuberi, A., Nazir, S., Rauf, M., and Younus, N. (2015). Zinc oxide, zinc sulfate and zinc oxide nanoparticles as source of dietary zinc: comparative effects on growth and hematological indices of juvenile grass carp (*Ctenopharyngodon idella*). *Int. J. Agric. Biol.* 17, 568–574. doi: 10.17957/IJAB/17.3.14.446

Fidler, M. C., Walczyk, T., Davidsson, L., Zeder, C., Sakaguchi, N., Juneja, L. R., et al. (2004). A micronised, dispersible ferric pyrophosphate with high relative bioavailability in man. *Br. J. Nutr.* 91, 107–112. doi: 10.1079/BJN20041018

Food and Drug Administration [FDA] (2014). *Guidance for Industry Assessing the Effects of Significant Manufacturing Process Changes, Including Emerging Technologies, on the Safety and Regulatory Status of Food Ingredients and Food Contact Substances, Including Food Ingredients that are Color Additives.* Silver Spring, MD: Food and Drug Administration.

Hider, R. C., and Kong, X. (2013). "Effect of overload and deficiency," in *Interrelations between Essential Metal Ions and Human Diseases*, ed. A. Sigel (Dordrecht: Springer), 229–294.

Honda, N., Sakaguchi, K., Nakata, K., and Nanbu, H. (2010). *Mineral Composition E.P. Patent No 1,649,762.* Munich: European Patent Office.

Hubatsch, I., Ragnarsson, E. G. E., and Artursson, P. (2007). Determination of drug permeability and prediction of drug absorption in Caco-2 monolayers. *Nat. Protoc.* 2, 2111–2119. doi: 10.1038/nprot.2007.303

Hurrell, R. F. (1997). Preventing iron deficiency through food fortification. *Nutr. Rev.* 55, 210–222. doi: 10.1111/j.1753-4887.1997.tb01608.x

Hurrell, R. F. (2002). Fortification: overcoming technical and practical barriers. *J. Nutr.* 132, 806S–812S.

Ishihara, K., Yamanami, K., Takano, M., Suzumura, A., Mita, Y., Oka, T., et al. (2008). Zinc bioavailability is improved by the micronised dispersion of zinc oxide with the addition of L-histidine in zinc-deficient rats. *J. Nutr. Sci. Vitaminol.* 54, 54–60.

Jang, M. H., Kweon, M. N., Iwatania, K., Yamamoto, M., Terahara, K., Sasakawa, C., et al. (2004). Intestinal villous M cells: an antigen entry site in the mucosal epithelium. *Proc. Natl. Acad. Sci. U.S.A.* 101, 6110–6115. doi: 10.1073/pnas.0400969101

Jo, M. R., Chung, H. E., Kim, H. J., Bae, S. H., Go, M. R., Yu, J., et al. (2016). Effects of zinc oxide nanoparticle dispersants on cytotoxicity and cellular uptake. *Mol. Cell. Toxicol.* 12, 281–288. doi: 10.1007/s13273-016-0033-y

Joseph, T., and Morrison, M. (2006). *Nanoforum Report: Nanotechnology in Agriculture and Food.* Wolverhampton: Institute of Nanotechnology.

Kernéis, S., Bogdanova, A., Kraehenbuhl, J. P., and Pringault, E. (1997). Conversion by Peyer's patch lymphocytes of human enterocytes into M cells that transport bacteria. *Science* 277, 949–952. doi: 10.1126/science.277.5328.949

Kloots, W., den Kamp, D. O., and Abrahamse, L. (2004). *In vitro* iron availability from iron-fortified whole-grain wheat flour. *J. Agric. Food Chem.* 52, 8132–8136. doi: 10.1021/jf040010+

Ma, T. Y., Boivin, M. A., Ye, D., Pedram, A., and Said, H. M. (2005). Mechanism of TNF-{alpha} modulation of Caco-2 intestinal epithelial tight junction barrier: role of myosin light-chain kinase protein expression. *Am. J. Physiol. Gastrointest. Liver Physiol.* 288, G422–G430. doi: 10.1152/ajpgi.00412.2004

Manke, A., Wang, L., and Rojanasakul, Y. (2013). Mechanisms of nanoparticle-induced oxidative stress and toxicity. *Biomed. Res. Int.* 2013:942916. doi: 10.1155/2013/942916

Nanbu, H., Nakata, K., Sakaguchi, N., and Yamazaki, Y. (2000). Mineral composition. US 6,074,675.

Neutra, M. R., Mantis, N. J., and Kraehenbuhl, J. P. (2001). Collaboration of epithelial cells with organized mucosal lymphoid tissues. *Nat. Immunol.* 2, 1004–1009. doi: 10.1038/ni1101-1004

Prasad, A. S. (2012). Discovery of human zinc deficiency: 50 years later. *J. Trace Elem. Med. Biol.* 26, 66–69. doi: 10.1016/j.jtemb.2012.04.004

Raya, S. D. H. A., Hassan, M. I., Farroh, K. Y., Hashim, S. A., and Salaheldin, T. A. (2016). Zinc oxide nanoparticles fortified biscuits as a nutritional supplement for zinc deficient rats. *J. Nanomed. Res.* 4:00081. doi: 10.15406/jnmr.2016.04.00081

Roe, M. A., Collings, R., Hoogewerff, J., and Fairweather-Tait, S. J. (2009). Relative bioavailability of micronized, dispersible ferric pyrophosphate added to an apple juice drink. *Eur. J. Nutr.* 48, 115–119. doi: 10.1007/s00394-008-0770-3

Rossi, L., Velikov, K. P., and Philipse, A. P. (2014). Colloidal iron(III) pyrophosphate particles. *Food Chem.* 151, 243–247. doi: 10.1016/j.foodchem.2013.11.050

Sakaguchi, N., Rao, T. P., Nakata, K., Nanbu, H., and Juneja, L. R. (2004). Iron absorption and bioavailability in rats of micronized dispersible ferric pyrophosphate. *Int. J. Vitam. Nutr. Res.* 74, 3–9. doi: 10.1024/03009831.74.1.3

Song, W., Zhang, J., Guo, J., Zhang, J., Ding, F., Li, L., et al. (2010). Role of the dissolved zinc ion and reactive oxygen species in cytotoxicity of ZnO nanoparticles. *Toxicol. Lett.* 199, 389–397. doi: 10.1016/j.toxlet.2010.10.003

Srinivas, P. R., Philbert, M., Vu, T. Q., Huang, Q., Kokini, J. L., Saltos, E., et al. (2010). Nanotechnology research: applications in nutritional sciences. *J. Nutr.* 140, 119–124. doi: 10.3945/jn.109.115048

Torabi, M., Kesmati, M., Harooni, H. E., and Varzi, H. N. (2013). Effects of nano and conventional zinc oxide on anxiety-like behavior in male rats. *Indian J. Pharmacol.* 45, 508–512. doi: 10.4103/0253-7613.117784

Wang, B., Zhang, Y., Mao, Z., Yu, D., and Gao, C. (2014). Toxicity of ZnO nanoparticles to macrophages due to cell uptake and intracellular release of zinc ions. *J. Nanosci. Nanotechnol.* 14, 5688–5696. doi: 10.1166/jnn.2014.8876

Yin, Y., Lin, Q., Sun, H., Chen, D., Wu, Q., Chen, X., et al. (2012). Cytotoxic effects of ZnO hierarchical architectures on RSC96 Schwann cells. *Nanoscale Res. Lett.* 7:439. doi: 10.1186/1556-276X-7-439

Zariwala, M. G., Somavarapu, S., Farnaud, S., and Renshaw, D. (2013). Comparison study of oral iron preparations using a human intestinal model. *Sci. Pharm.* 81, 1123–1139. doi: 10.3797/scipharm.1304-03

Zhao, C. Y., Tan, S. X., Xiao, X. Y., Qiu, X. S., Pan, J. Q., and Tang, Z. X. (2014). Effects of dietary zinc oxide nanoparticles on growth performance and antioxidative status in broilers. *Biol. Trace Elem. Res.* 160, 361–367. doi: 10.1007/s12011-014-0052-2

Conflict of Interest Statement: The authors declare that the research was conducted in the absence of any commercial or financial relationships that could be construed as a potential conflict of interest.

14

Biocombinatorial Synthesis of Novel Lipopeptides by COM Domain-Mediated Reprogramming of the Plipastatin NRPS Complex

Hongxia Liu[1], Ling Gao[1], Jinzhi Han[1], Zhi Ma[1], Zhaoxin Lu[1], Chen Dai[2], Chong Zhang[1] and Xiaomei Bie[1*]

[1] College of Food Science and Technology, Nanjing Agricultural University, Nanjing, China, [2] College of Life Sciences, Nanjing Agricultural University, Nanjing, China

Edited by:
Jayanta Kumar Patra,
Dongguk University, South Korea

Reviewed by:
Sean Doyle,
Maynooth University, Ireland
Vani Mishra,
Nanotechnology Application Centre,
India

*Correspondence:
Xiaomei Bie
bxm43@njau.edu.cn

Both donors and acceptors of communication-mediating (COM) domains are essential for coordinating intermolecular communication within nonribosomal peptides synthetases (NRPSs) complexes. Different sets of COM domains provide selectivity, allowing NRPSs to utilize different natural biosynthetic templates. In this study, novel lipopeptides were synthesized by reprogramming the plipastatin biosynthetic machinery. A Thr-to-Asp point mutation was sufficient to shift the selectivity of the donor COM domain of ppsB toward that of ppsD. Deletion and/or interchangeability established donor and acceptor function. Variations in acceptor COM domain did not result in novel product formation in the presence of its partner donor, whereas plipastatin formation was completely abrogated by altering donor modules. Five novel lipopeptides (cyclic pentapeptide, linear hexapeptide, nonapeptide, heptapeptide, and cyclic octapeptide) were identified and verified by high-resolution LC-ESI-MS/MS. In addition, we demonstrated the potential to generate novel strains with the antimicrobial activity by selecting compatible COM domains, and the novel lipopeptides exhibited antimicrobial activity against five of the fungal species at a contention of 31.25–125 μg/ml.

Keywords: plipastatin, biosynthetic complex, COM domain, lipopeptide, NRPS

INTRODUCTION

Bacillus subtilis is a Gram-positive bacterium that produces broad spectrum amphiphilic lipopeptides with excellent biosurfactant properties and antifungal, antibacterial and antiviral activities (Marahiel et al., 1997; Schwarzer et al., 2003; Batool et al., 2011). Iturins (Hiradate et al., 2002; Yu et al., 2002; Jin et al., 2014), fengycins (Vanittanakom et al., 1986; Guo et al., 2014) and surfactins (Bonmatin et al., 2003; Liu et al., 2015) are the three most well-known families of lipopeptides, which produced by *Bacilli*. And all of them are synthesized by nonribosomal peptide synthetases via a thioesterase chain release mechanism (Tosato et al., 1997; Finking and Marahiel, 2004; Calcott and Ackerley, 2014). These lipopeptides contain a variable cyclic amino acid portion attached to a variable β-amino or β-hydroxy fatty acid (de Faria et al., 2011). The fengycin is actually identical compounds to plipastatin that display slightly structural variations at different salty conditions (Honma et al., 2012). Therefore, the term plipastatin is used throughout this study.

Plipastatin is composed of 10 α-amino acids linked to one unique β-hydroxy fatty acid chain, and two variants (plipastatin A and B) with Val or Ala at position 6 have been reported (Vater et al., 2002; Sun et al., 2006). Plipastatin synthetase contains five distinct NRPSs subunits which assemble to form a co-linear chain in the order of ppsA-ppsB-ppsC-ppsD-ppsE (**Figure 1A**) that incorporates two (ppsA, B and C), three (ppsD), and one (ppsE) amino acid residues (**Figure 1B**) into the growing peptide, respectively (Marahiel et al., 1997).

Based on the molecular mechanisms employed by NRPSs, NRPS assembly lines have an enormous potential for biocombinatorial synthesis. Biosynthesis of a defined, full-length product relies on the selectivity of individual modules and their coordinated interplay with donor and acceptor communication-mediating (COM) domains (Chiocchini et al., 2006). In most bacterial NRPS systems, a donor COM domain (COM_X^D) situated at the C terminus of an aminoacyl- or peptidyl-donating NRPS (X) and an acceptor COM domain (COM^A) located at the N terminus of the accepting partner enzyme (Y) form a matching (compatible) set that forms productive interactions between adjacent modules (i.e., ppsA ppsB1 and ppsB2 ppsC1) (Hahn and Stachelhaus, 2006). Biochemical investigations on the selectivity of NRPSs have helped define sets of compatible inter-module linkers. In principle and in practice, enzymes of a NRPS complex can form other biosynthetic templates, making possible the synthesis of a vast array of novel peptide products via a process that is tantamount to combinatorial synthesis (Sieber and Marahiel, 2005; Fischbach and Walsh, 2006). The COM domain swapping experiments verified the decisive role of COM domains for the control of protein-protein interactions between surfactin NRPS (Chiocchini et al., 2006). The research had demonstrated that point mutation of one of these key residues within the COM domain of TycC1 was sufficient to shift its selectivity from the cognate donor COM of TycB3 toward the noncognate donor COM domain of TycB1 in tyrocidine NRPSs (Hahn and Stachelhaus, 2006). The fragments of COM domains were sequenced in previous study (Hahn and Stachelhaus, 2006), but the interaction between enzymes of a nonribosomal peptide has hardly been studied, which relies on the interplay of COM domains. Therefore, we attempted to

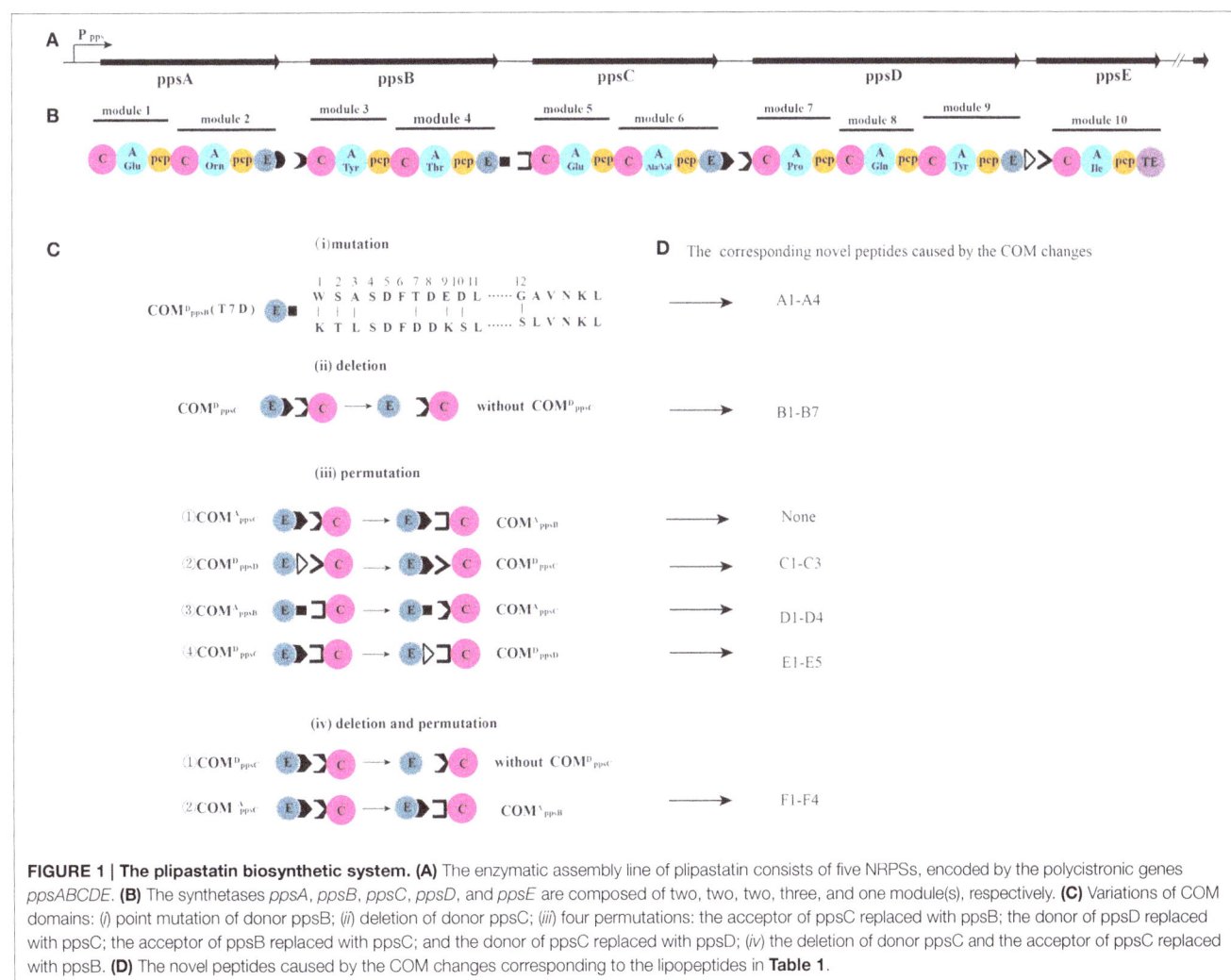

FIGURE 1 | The plipastatin biosynthetic system. (A) The enzymatic assembly line of plipastatin consists of five NRPSs, encoded by the polycistronic genes *ppsABCDE*. **(B)** The synthetases *ppsA*, *ppsB*, *ppsC*, *ppsD*, and *ppsE* are composed of two, two, two, three, and one module(s), respectively. **(C)** Variations of COM domains: (*i*) point mutation of donor ppsB; (*ii*) deletion of donor ppsC; (*iii*) four permutations: the acceptor of ppsC replaced with ppsB; the donor of ppsD replaced with ppsC; the acceptor of ppsB replaced with ppsC; and the donor of ppsC replaced with ppsD; (*iv*) the deletion of donor ppsC and the acceptor of ppsC replaced with ppsB. **(D)** The novel peptides caused by the COM changes corresponding to the lipopeptides in **Table 1**.

investigate the biocombinatorial synthesis of novel lipopeptides by COM domain-mediated reprogramming of the plipastatin NRPS complex.

The goal of this study was to investigate the influence on protein–protein communication based on the converting donor and acceptor of COM domains sequences, which maintain or prevent the selective interaction between partner NRPSs, furthermore the influence on the novel lipopeptides formation. In this work we began with the *B. subtilis* PB2-L strain from our previous work, which produces plipastatin (Figure S5) following integration of a gene expression frame composed of a constitutive promoter P43, functional gene *sfp*, and regulatory gene *degQ* into the chromosomal *amyE* locus of PB2. We studied the importance of COM domains in mediating the specific channeling of reaction intermediates between partner enzymes. First, the effect of site directed mutagenesis of module ppsB on protein-protein communication with the partner elongation module ppsC was explored. Subsequently, the importance and generality of COM domains was further substantiated by COM domain deletion and permutation experiments. This research may provide the theoretical basement for the biocombinatorial synthesis of NRPS complex and the exploitation of novel lipopeptides.

MATERIALS AND METHODS
Strains, Media, and General Methods

Bacterial strains and plasmids used in this work are listed in Table S1. *Escherichia coli* and *Bacillus subtilis* were grown in LB medium supplemented with 100 mg/ml ampicillin, 20 mg/ml kanamycin, or 5 mg/ml erythromycin (final concentrations) where applicable. All enzymes were commercial preparations and were used as specified by the suppliers (Vazyme Biotech Co.,Ltd, Nanjing, China and Takara Shuzo Co., Ltd, Dalian, China). *E. coli* transformation was performed as described previously (Sambrook et al., 1989). The plipastatin biosynthetic complex was reprogrammed using a homologous recombination approach (Figure S1) with upstream (A1) and downstream (A2) primers to construct plasmid pks2A1A2. Similarly, upstream B1 (targeting the *tgf* gene fragments) and downstream B2 were used to generate plasmid pks2B1B2. The constructs was verified by PCR analysis and used for deletion and substitutions, respectively. After transformation of the plasmids into the host, plasmids were inserted into the chromosome by homologous recombination between the target gene and homologous sequences, or between the homologous sequences alone via the well-established two-step exchange method (Chiocchini et al., 2006). Integration into the chromosome by a single crossover event was selected during growth at the nonpermissive temperature (37°C) while maintaining selective pressure. Subsequent growth of the cointegrates at the permissive temperature (30°C) leads to a second recombination event, resulting in their resolution (Arnaud et al., 2004). The novel *B.subtilis* strains were verified by PCR analysis and the sequences were sequenced by Genscript biotechnology co., LTD (Nanjing, China) (data not shown).

Cloning

Bacillus subtilis DNA was isolated from enrichment cultures using a bacterial genomic DNA extraction kit (Omega, USA) according to the manufacturer's instructions. Each 50 μl reaction contained 35 μl of ddH$_2$O, 5 μl of 2 × PCR buffer, 4 μl of dNTP, 2 μl of F-primer, 2 μl of R-primer, 1 μl of Phanta Super-Fidelity DNA Polymerase and 1 μl of template DNA. The PCR program consisted of an initial denaturation at 94°C for 3 min, followed by 30 amplification cycles of 94°C for 30 s, 60°C for 30 s, and 72°C for 30 s, then 72°C for 10 min. Primers (Supplemental Data) were purchased from Sangon Biotech (Shanghai, China). Standard procedures were applied for all DNA manipulations (Sambrook et al., 1989).

A 1382 bp product containing the COM$_{ppsB}$ fragment was amplified using oligonucleotides ppsB-F and ppsB-R. The resulting product was digested with *Sph*I and *Sal*I and cloned into the *E. coli* expression vector PMD-19 that was digested with the same enzymes to give pCC42. After digestion with *Sph*I and *Sal*I, the fragment was ligated into pKS2 to give pKSA1. Based on the 21 bp overlapping region in M-F and M-R, pKSA1 was used as template for the subsequent PCR with oligonucleotides M-F and M-R to generate point mutations.

Two 600 bp ppsC fragments were amplified using oligonucleotides K-up-F and K-up-R, and K-down-F and K-down-R, respectively. After cloning into the *E. coli* expression vector PMD-19 and digestion with *Cla*I, *Sal*I, *Sal*I, and *Kpn*I, the fragments were ligated into pKS2 to give pKSK1, which was used as template for PCR with oligonucleotides P1-F and P1-R to gain the pKSP1 for use in the initial substitution step. Subsequently, 650 bp upstream and downstream gene fragments were amplified by PCR and the target COM$_{ppsC}^D$ (the donor COM of ppsC) fragment was combined with the downstream fragment by fusion PCR using the oligonucleotides P2-down1-F and P2-down2-F. The resultant product was digested with *Sal*I and *Kpn*I and ligated into pKS2 that was also digested with these enzymes, and oligonucleotides P2-up-F and P2-down-R were used to generate pKSP2. Next, PCR was performed using oligonucleotides P3-F and P3-R and pKSK1 as template to generate pKSP3 for permutation in the third step of the procedure. To this end, 650 bp upstream and downstream fragments were amplified using oligonucleotides P4-up-F, P4-up-R, P4-down3-F, and P4-down-R, while a 72 bp fragment of COM$_{ppsD}^D$ was amplified using oligonucleotides P4-down1-F and P4-down2-F. After purification, DNA fragments were combined and used as template for PCR amplification with oligonucleotides P4-up-F and P4-down-R, and subsequent cloning into pKS2 yielded the final disruption vector pKSP4 (Table S1). The maps of plasmids used to construct the deletion mutant, point mutation and substitution were shown in Figures S3, S4, respectively.

B. subtilis Strain Construction

Transformations were carried out as described previously (Anagnostopoulos and Spizizen, 1961), and genotypes were verified by PCR. The pKSA1, pKSK1, and pKSP1 plasmids were transformed into *B. subtilis* PB2-L, *B. subtilis* PB2-LP1 was transformed with pKSP2, and the resulting

TABLE 1 | Lipopeptides, retention times, m/z values of protonated forms and peptide sequences.

Approach of COM changes	Homologues of peptides	RT (min)	Molecular formula	Mass (m/z)	Peptide sequence
Point mutation of donor ppsB (T7D)	A1	13.94	$n = 12$; $C_{45}H_{74}N_6O_{11}$	875.5498	β-OHFA-ABE
	A2	14.15	$n = 13$; $C_{46}H_{76}N_6O_{11}$	889.5688	Cyclo pentapeptide
	A3	14.15	$n = 14$; $C_{47}H_{78}N_6O_{11}$	903.5841	CH3(CH2)n1CHOHCH2CO-Glu-Orn-
	A4	14.15	$n = 15$; $C_{48}H_{80}N_6O_{11}$	917.6087	Tyr-Thr –Ile; n1 = 12-15
Deletion of donor ppsC	B1	13.07	$n = 10$; $C_{47}H_{77}N_7O_{15}$	980.5580	
	B2	13.15	$n = 11$; $C_{48}H_{79}N_7O_{15}$	994.5712	
	B3	13.20	$n = 12$; $C_{49}H_{81}N_7O_{15}$	1008.5873	β-OHFA-ABC
	B4	13.68	$n = 13$; $C_{50}H_{83}N_7O_{15}$	1022.6026	Linear hexapeptide
	B5	14.15	$n = 14$; $C_{51}H_{85}N_7O_{15}$	1036.6179	CH3(CH2)n2CHOHCH2CO-Glu-Orn-Tyr-Thr -Glu-Val; n2 = 10–16
	B6	14.15	$n = 15$; $C_{52}H_{87}N_7O_{15}$	1050.6338	
	B7	14.15	$n = 16$; $C_{53}H_{89}N_7O_{15}$	1064.6493	
the donor of ppsD replaced with ppsC	C1	15.63	$n = 11$; $C_{67}H_{103}N_{11}O_{20}$	1382.7556	β-OHFA-ABCD
	C2	15.63	$n = 12$; $C_{68}H_{105}N_{11}O_{20}$	1396.7715	Linear nonapeptide
	C3	15.98	$n = 13$; $C_{69}H_{107}N_{11}O_{20}$	1410.7876	CH3(CH2)n3CHOHCH2CO-Glu-Orn-Tyr-Thr-Glu-Val-Pro-Gln-Tyr; n3 = 11–13
the acceptor of ppsB replaced with ppsC	D1	12.15	$n = 10$; $C_{66}H_{101}N_{11}O_{20}$	1368.7387	β-OHFA-ABCD
	D2	12.33	$n = 11$; $C_{67}H_{103}N_{11}O_{20}$	1382.7517	Linear nonapeptide
	D3	13.15	$n = 12$; $C_{68}H_{105}N_{11}O_{20}$	1396.7667	CH3(CH2)n4CHOHCH2CO-Glu-Orn-Tyr-Thr -Glu-Val -Pro-Gln-Tyr;n4 = 11–14
	D4	13.68	$n = 13$; $C_{69}H_{107}N_{11}O_{20}$	1410.7817	
the donor of ppsC replaced with ppsD	E1	13.52	$n = 12$; $C_{53}H_{88}N_8O_{16}$	1093.6393	
	E2	13.68	$n = 13$; $C_{54}H_{90}N_8O_{16}$	1107.6598	β-OHFA-ABCE
	E3	14.15	$n = 14$; $C_{55}H_{92}N_8O_{16}$	1121.6753	Linear heptapeptide
	E4	14.69	$n = 15$; $C_{56}H_{94}N_8O_{16}$	1135.6871	CH3(CH2)n5CHOHCH2CO-Glu-Orn-
	E5	14.69	$n = 16$; $C_{57}H_{96}N_8O_{16}$	1149.7043	Tyr-Thr-Glu-Ala –Ile; n5 = 12–16
Deletion of donor ppsC and the acceptor of ppsC replaced with ppsB	F1	17.05	$n = 12$; $C_{64}H_{98}N_{10}O_{16}$	1263.7235	β-OHFA-ABDE
	F2	17.12	$n = 13$; $C_{65}H_{100}N_{10}O_{16}$	1277.7412	Cyclo octapeptide
	F3	17.68	$n = 14$; $C_{66}H_{102}N_{10}O_{16}$	1291.7504	CH3(CH2)n6CHOHCH2CO-Glu-Orn
	F4	18.06	$n = 15$; $C_{67}H_{104}N_{10}O_{16}$	1305.7682	-Tyr-Thr-Pro-Gln-Tyr -Ile; n6 = 12-15

β-OHFA-ABE, The lipopeptides produced by NRPS complex assembly line ppsA, ppsB, and ppsE; β-OHFA-ABC, The lipopeptides produced by NRPS complex assembly line ppsA, ppsB, and ppsC; β-OHFA-ABCD, The lipopeptides produced by NRPS complex assembly line ppsA, ppsB, ppsC, and ppsD; β-OHFA-ABCE, The lipopeptides produced by NRPS complex assembly line ppsA, ppsB, ppsC, and ppsE; β-OHFA-ABDE, The lipopeptides produced by NRPS complex assembly line ppsA, ppsB, ppsD, and ppsE.

strain was transformed with pKSP3 and pKSP4, generating *B. subtilis* mutants PB2-LA1 (COM^D_{ppsB}T::D), PB2-LK1 (ΔCOM^D_{ppsC}), PB2-LP1 (COM^A_{ppsC}:: COM^A_{ppsB}), PB2-LP2 (COM^A_{ppsC}:: COM^A_{ppsB} and COM^D_{ppsD}:: COM^D_{ppsC}), PB2-LP3 (COM^A_{ppsC}:: COM^A_{ppsB}, COM^D_{ppsD}:: COM^D_{ppsC} and COM^A_{ppsB}:: COM^A_{ppsC}), and PB2-LP4 (COM^A_{ppsC}:: COM^A_{ppsB}, COM^D_{ppsD}:: COM^D_{ppsC}, COM^A_{ppsB}:: COM^A_{ppsC} and COM^D_{ppsC}:: COM^D_{ppsD}). The pKSK1 and pKSP1 plasmids were transformed into *B. subtilis* PB2-L in two steps to give *B. subtilis* mutants PB2-LK1 (ΔCOM^D_{ppsC}) and PB2-LKP1 (ΔCOM^D_{ppsC}, COM^A_{ppsC}:: COM^A_{ppsB}) (Table S1).

Production of Antimicrobial Extracts

Modified strains were cultivated in Landy medium (20 g/L glucose, 1 g/L yeast extract, 5 g/L L-glutamic acid, 1 g/L KH_2PO_4, 0.16 mg/L $CuSO_4$, 0.5 g/L $MgSO_4 \cdot 7H_2O$, 0.15 mg/L $FeSO_4$, 0.5 g/L KCl, 5 mg/L $MnSO_4 \cdot H_2O$, pH 7.0) at 33°C in a rotary shaker at 180 rpm for 3 days to produce antimicrobial substances (Landy et al., 1947; Deng et al., 2011b). Antimicrobial extracts were obtained using an organic solvent (methanol) extraction method.

Product Formation Assays

The crude methanol extract was fractioned by the high resolution LC-ESI-MS and LC-ESI-MS/MS analysis, which was performed with a Thermo Finnigan Surveyor-LCQ DECA XP Plus (Thermo Electron Corporation, San Jose, CA, USA). The flow rate was maintained at 0.2 mL/min with a gradient of 22 min (0–95 %, vol/vol acetonitrile for 18 min and 95–5%, vol/vol acetonitrile for 4 min). The electrospray needle was operated at a spray voltage of 5 kV, a capillary voltage of 32 V and a capillary temperature of 300°C. For the HCD experiment, helium was used as the collision gas and the collision energy was set at 35%. Elution was monitored by UV detection at 214 nm (Mootz et al., 2000; Deng et al., 2011b).

Determination of Antimicrobial Activity

The linear products were synthetized by KareBay BioChem, Inc. Antimicrobial activity was evaluated by determining MIC values using the standard broth microdilution method (Nedorostova et al., 2009) with some modifications. Tested microorganisms were incubated at 37 or 30°C to approximately 10^6–10^7 CFU/ml in lysogeny broth (LB) medium. Serial dilutions of antimicrobial compounds were prepared to obtain final concentrations of 1000, 500, 250, 125, 62.5, 31.25, 15.63, and 7.81 µg/ml in LB or PDA medium. The ability of cyclic products strains to inhibit the growth of various indicator organisms was determined qualitatively by agar-well diffusion method (Deng et al., 2011a). Plates were incubated for up to 24 (bacteria) and 48 h (fungi) before the antimicrobial activities were determined (Oh et al., 2011). Samples incubated without antimicrobials were used as controls.

RESULTS

Influence of a ppsB Donor Point Mutation on Plipastatin Synthesis

To demonstrate the biocombinatorial potential of COM domains, we investigated crosstalk between ppsB and ppsC modules through point mutations. Seven positions in COM^D_{ppsB} (Figure 1C) were mutated as follows: W1K, S2T, A3L, T7D, E9K, D10S, and G12S. We evaluated the impact of COM^D_{ppsB} point mutants on lipopeptide biosynthesis.

Based on the high-resolution LC-ESI–MS spectra, all the point mutants did not produce novel lipopeptides (data not shown) except for the T7D mutation. In high-resolution ESI–MS spectra, a series of charge 1 ($z = 1$) molecular ions with m/z values at 875.5488, 889.5644, 903.5801, and 917.5963 (Figure 2A) led us to predict the formation of a β-OH fatty acid (β-OHFA)-ABE product. And the predicted formula was $C_{46}H_{76}N_6O_{11}$ (ppm = 4.831) at m/z 889.5644. Using the HCD–MS/MS spectrum of the precursor ion [M + H]$^+$ at m/z of 889.5644 (Figure 2B), -b and -y productions (Pathak et al., 2013) were assigned (Figure 2C), which enabled us to infer the sequence cyclo-[β-OHFA-Glu-Orn- Tyr-Thr -Ile]. The formula ($C_{46}H_{76}N_6O_{11}$) was consistent with the results of mass spectrometry (Table 1). The molecular mass ions at m/z 875.5488, 903.5801, and 917.5963 were found to be derived from a β-OHFA chain variant at m/z 889.5644 sharing the same peptide sequence (Figure 1D), based on the fragment ion profile obtained by high-resolution HCD–MS/MS. These lipopeptide precursor ions with a 14 Da (Yang et al., 2015) mass difference from

m/z 889.5644 ion were assigned as new pentapeptide variants differing only in β-OHFA chain length.

The T7D mutation resulted in a loss in the ability of the COM^D domain to form a productive complex with its native partner COM^A_{ppsB} (the acceptor COM of ppsB). However, at the same time, ppsB point mutant gained the ability to interact with the ppsE. Meanwhile, a novel NRPS complex assembly line ppsA, ppsB and ppsE was formed. This indicated that the mutation of threonine in the donor COM domain of ppsB to aspartic acid was sufficient to transfer selectivity from the cognate acceptor COM toward the noncognate accepter COM of ppsD, and highlights the importance of the threonine residue at this position in protein-protein communication.

Influence of ppsC Donor Deletion Mutant on Plipastatin Synthesis

The deletion of ppsC donor on the ability to interact with the partner elongation module ppsD was investigated in vivo (Figure 1C). As shown in Table 1, the high-resolution ESI–MS revealed a series of charge 1 ($z = 1$) ions with m/z at 980.5560, 994.5712, 1008.5873, 1022.6026, 1036.6179, 1050.6338, and 1064.6493 (Figure 3A) that eluted with a retention time between 13.07 and 14.15 min. And the predicted formula was $C_{49}H_{81}N_7O_{15}$ (ppm = 0.673) at m/z 1008.5873. And the high-resolution HCD-MS/MS spectrum of the precursor ion (Figure 3B) was analyzed, and -b and -y ions were consistent with a linear [β-OHFA-Glu-Orn- Tyr-Thr-Glu-Val] peptide (Figure 3C), which lacked four amino acid residues (Pro, Gln, Tyr, and Ile) that are present in plipastatin. The formula of linear peptide was in accord with the MS spectra analysis. Furthermore, other molecular ions were clearly derived from this peptide moiety, and the 14 Da differences again indicated varying chain length of the β-OHFA (Figure 1D). The COM^D_{ppsC} deletion largely resulted in a loss in the ability of the COM^D domain to form a productive complex with its native partner COM^A_{ppsC}. The deletion mutant was completely inactive in a product formation assay with ppsD, resulting in formation of the novel NRPS complex assembly line ppsA, ppsB, and ppsC.

Influence of Interchangeabilities in COM Domain on Plipastatin Synthesis

Indeed, deletion of the donor module ppsC prevents plipastatin biosynthesis. However, the inability of mutant systems to synthesize plipastatin and the formation of novel products is due to either the loss of protein–protein interaction between donor and acceptor NRPS partners, or inactivity of donor modules causing by the deletion of the most C-terminal amino acid residues.

To challenge these potential explanations, crosstalk experiments were performed, which confirmed that members of the compatible pairs between ppsB and ppsC, ppsC and ppsD, as well as ppsD and ppsE were interchangeable in vivo. First, acceptor ppsC was initially swapped with ppsB, and donor ppsD was subsequently substituted with donor ppsC. Next, acceptor ppsB was mutated to acceptor ppsC, and donor ppsC was replaced by donor ppsD. Plipastatin was observed in the

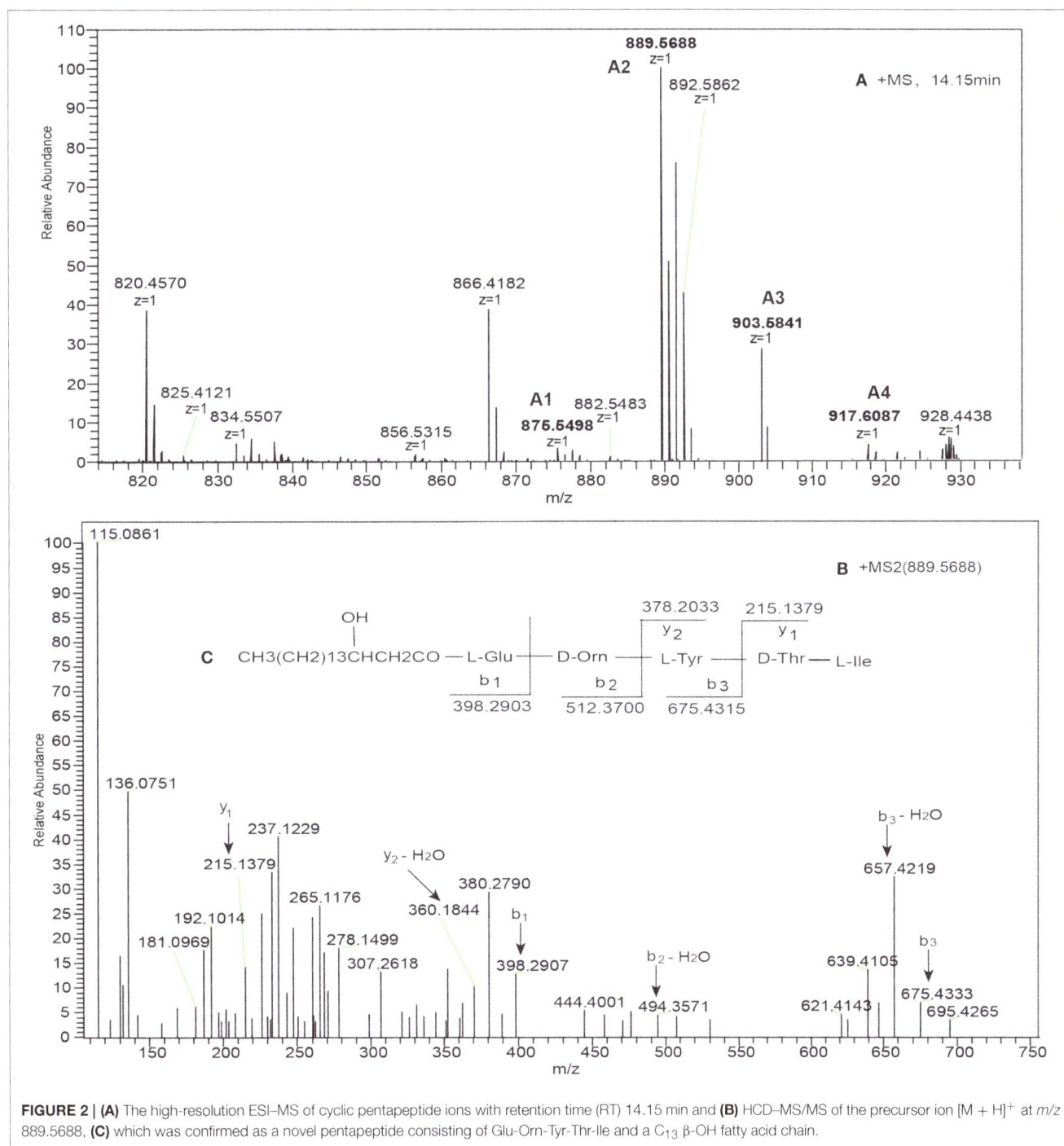

FIGURE 2 | (A) The high-resolution ESI–MS of cyclic pentapeptide ions with retention time (RT) 14.15 min and **(B)** HCD–MS/MS of the precursor ion [M + H]⁺ at m/z 889.5688, **(C)** which was confirmed as a novel pentapeptide consisting of Glu-Orn-Tyr-Thr-Ile and a C_{13} β-OH fatty acid chain.

methanolic extract when acceptor ppsC was replaced by acceptor ppsB (data not shown), and the anticipated novel product was not formed. However, when COM^D_{ppsD} was replaced by COM^D_{ppsC}, the high-resolution ESI–MS gave the charge 1 molecular mass ions of m/z 1382.7556, 1396.7715, and 1410.7876 (**Figure 4A**), which were assigned as hydrogen adducts of corresponding protonated novel nonapeptide ions. The charge 2 (z = 2) of the mass ions was 691.8796, 698.8879, and 705.8953, which verified the predicted formula was $C_{68}H_{105}N_{11}O_{20}$ (ppm = 4.333).

In order to confirm the assignments, the m/z 1396.7715 ion was subjected to the high-resolution HCD-MS/MS (**Figure 4B**), and the -y ion series at m/z 736.3506 (−H₂O, 718.3523)→508.2406→407.1911→ 310.1407→ 181.0986 was consistent with Glu-Val-Pro-Gln-Tyr from the N-terminus, while the -b ion series 762.4486→661.4158→ 498.3586→ 384.2732 suggested that the precursor ion possessed a Glu-Orn-Tyr-Thr at the C-terminus and a C12 β-OH fatty acid that lacked the Ile residue present in plipastatin, and the fragmentation

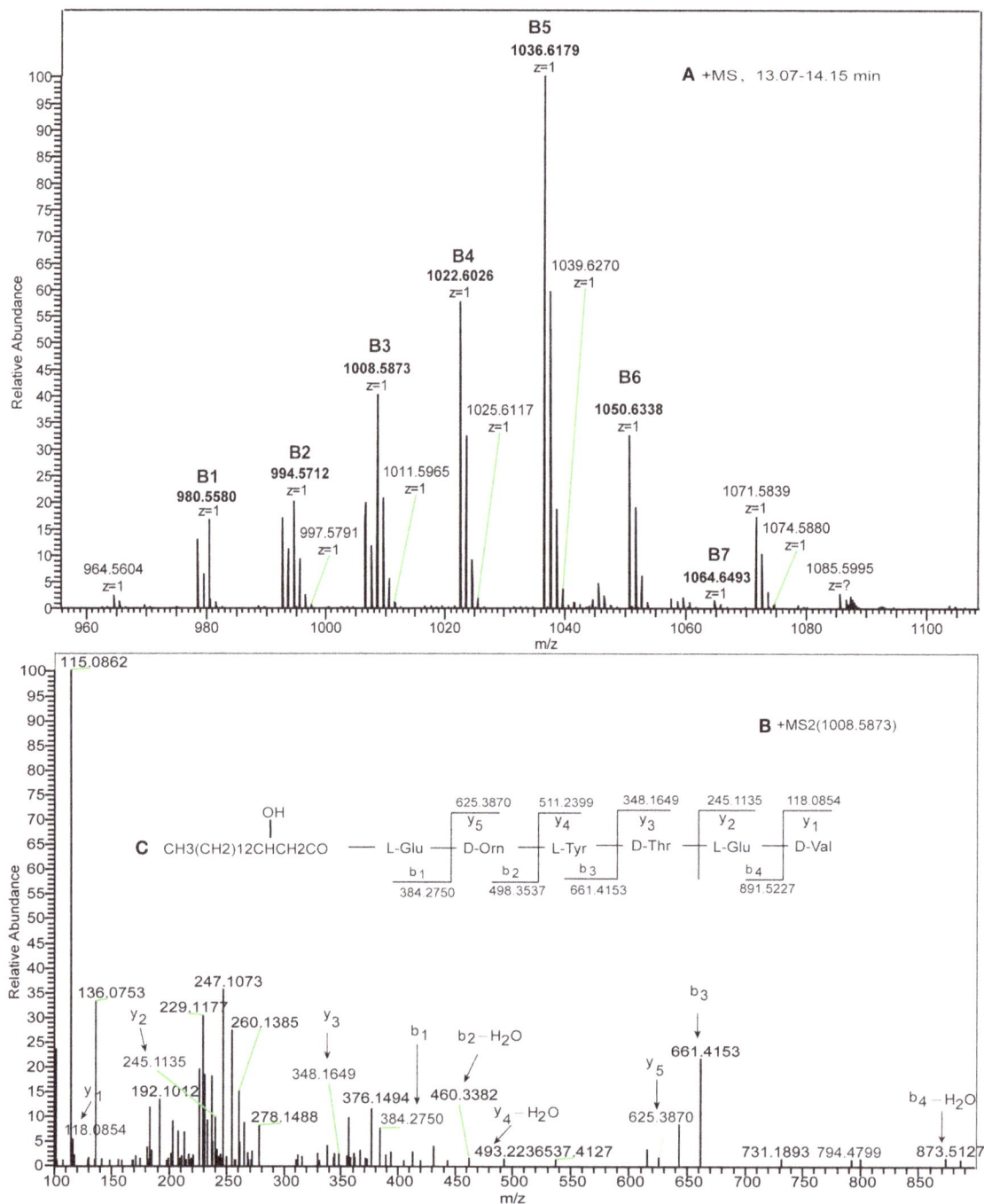

FIGURE 3 | (A) The high-resolution ESI–MS of linear hexapeptide ions with RT in the range 13.07–14.15 min and **(B)** HCD–MS/MS of the precursor ion [M + H]⁺ at m/z 1008.5873, **(C)** which was confirmed as a novel hexapeptide consisting of Glu-Orn-Tyr-Thr-Glu-Val and a C_{12} β-OH fatty acid chain.

pattern for the entire m/z 1396.7715 is consistent with MS/MS analysis (**Figure 4C**). Again, other molecular mass ions differing by 14 Da ($-CH_2-$) were clearly derived from the same peptide sequence but with C11 and C13 β-OH fatty acids, respectively (**Table 1, Figure 1D**). This indicated that the donor of ppsD loss the ability of forming a productive complex with its native partner, and resulting the formation of the NRPS complex assembly line ppsA, ppsB, ppsC and ppsD.

Subsequent mutation of acceptor ppsB to acceptor ppsC was predicted to result in a product with one further modification. The charge 1 ($z = 1$) molecular mass ions at m/z 1368.7387, 1382.7517, 1396.7667, and 1410.7817 and the charge 2 ($z = 2$) of the mass ions at m/z 684.8743, 691.8776, 698.8860, and 705.8932 were observed in the high-resolution ESI–MS spectrum (Figure S6A), which verified the predicted formula at m/z 1396.7667 was $C_{68}H_{105}N_{11}O_{20}$(ppm = 4.181).

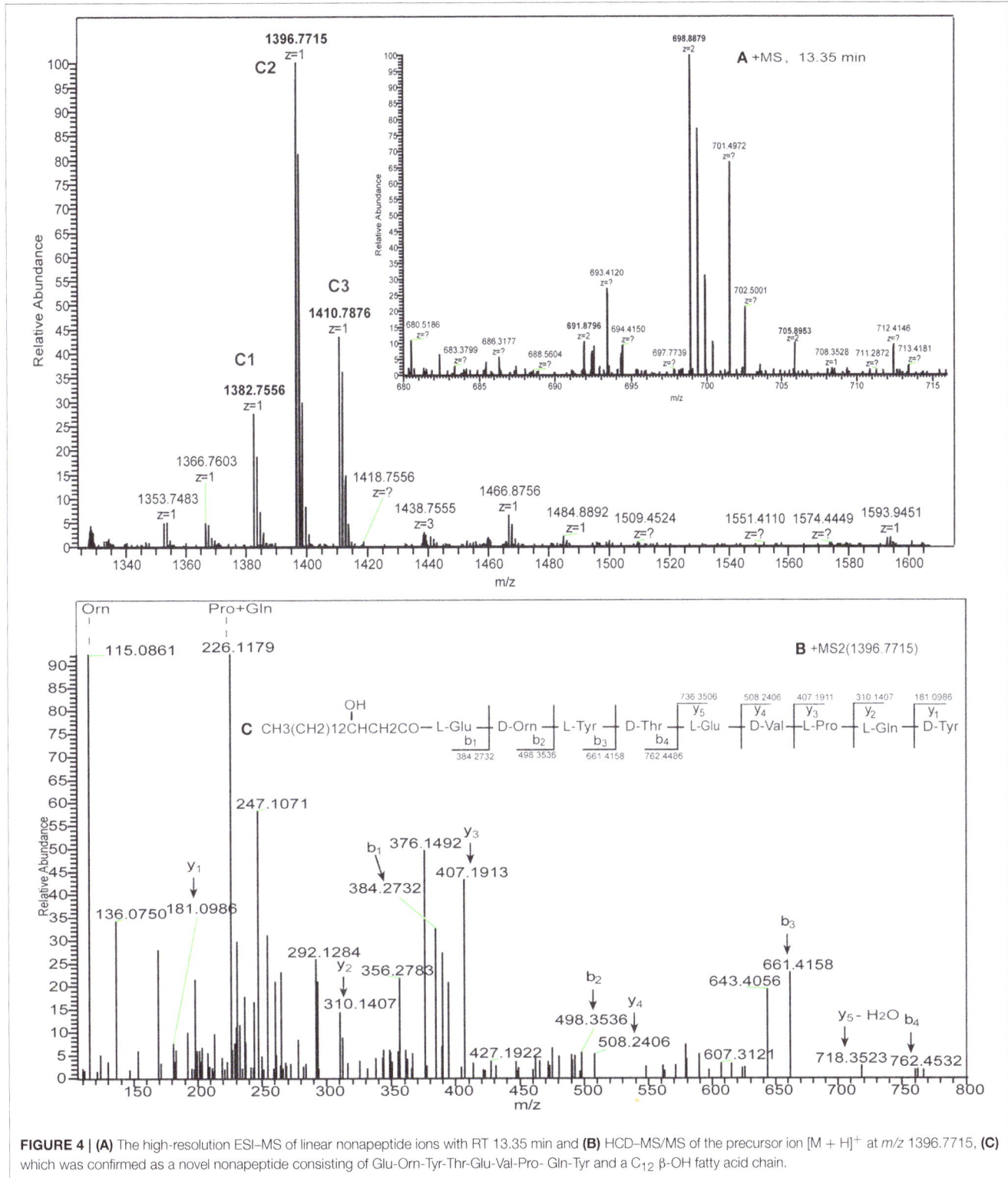

FIGURE 4 | (A) The high-resolution ESI–MS of linear nonapeptide ions with RT 13.35 min and **(B)** HCD–MS/MS of the precursor ion [M + H]$^+$ at m/z 1396.7715, **(C)** which was confirmed as a novel nonapeptide consisting of Glu-Orn-Tyr-Thr-Glu-Val-Pro- Gln-Tyr and a C$_{12}$ β-OH fatty acid chain.

The high-resolution HCD-MS/MS fragmentation of the m/z 1396.7667 ion (Figure S6B) yielded a series of -b fragment ions at m/z 891.5045 762.4468→ 661.4145→498.3531→384.2725 and -y fragment ions at m/z 736.3506 (−H$_2$O, 718.3517) →508.2404→ 407.1911 →310.1393→ 181.0977 that were consistent with

hydrogen adducts of a C12 β-OH FA and an nonapeptide with the sequence Glu-Orn-Tyr-Thr-Glu-Val- Pro-Gln- Tyr that is the same as that obtained with the modification described above, and is also lacking the Ile residue comparing the original product plipastatin. The result showed that the NRPS complex generated the assembly line ppsA, ppsB, ppsC, and ppsD. Finally, COM_{ppsB}^{D} was replaced with COM_{ppsD}^{D}, this caused the disability of COM_{ppsB}^{D} operon and generation of COM_{ppsB}^{D} mutated to COM_{ppsD}^{D} as expected. The high-resolution ESI–MS spectrum again included hydrogen adduct charge 1 ($z = 1$) mass ions differing by multiples of 14 Da that eluted between 13.52 and 14.69 min (**Figure 5A**). The formula of the $[M + H]^{+}$

ion at m/z 1107.6598 was $C_{54}H_{90}N_{8}O_{16}$(ppm = 3.553) with the mass spectrum analysis. The high-resolution HCD-MS/MS fragmentation of m/z 1107.6598 ion eluting at 13.68 min resulted in -b and -y ion assignments (**Figure 5C**) consistent with a linear heptapeptide with a linear C13 β-OH FA and the sequence Glu-Orn-Tyr-Thr-Glu-Ala-Ile (**Figure 5B**). This sequence was produced by the NRPS complex assembly line ppsA, ppsB, ppsC,

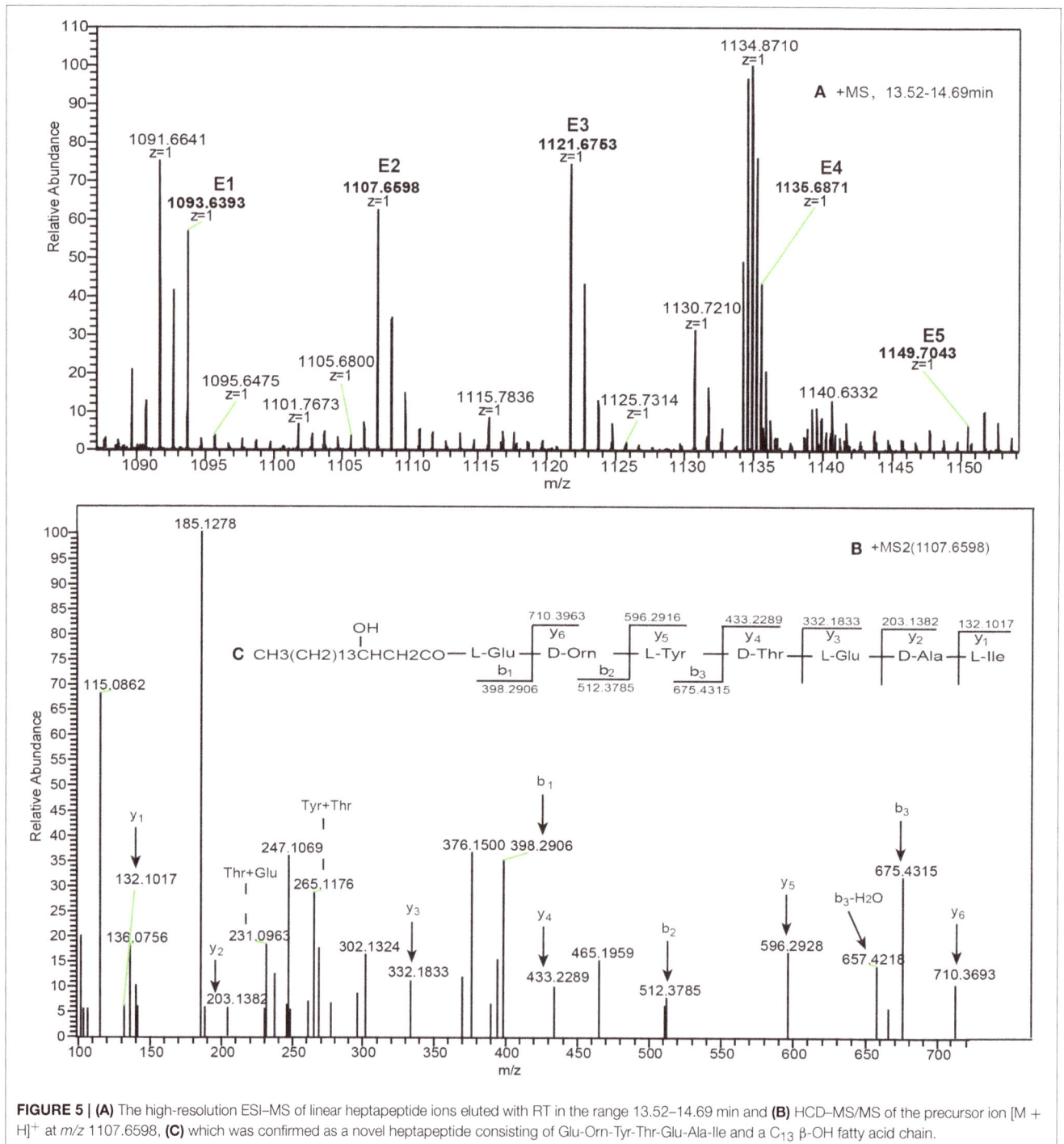

FIGURE 5 | **(A)** The high-resolution ESI–MS of linear heptapeptide ions eluted with RT in the range 13.52–14.69 min and **(B)** HCD–MS/MS of the precursor ion [M + H]⁺ at m/z 1107.6598, **(C)** which was confirmed as a novel heptapeptide consisting of Glu-Orn-Tyr-Thr-Glu-Ala-Ile and a C_{13} β-OH fatty acid chain.

and ppsE, and lacks three amino acid residues (Pro, Gln and Tyr) that are present in plipastatin. Other linear heptapeptides identified in this study are summarized in **Table 1**, **Figure 1D**. Experiments unequivocally confirmed that the acceptor modules did not contribute to the formation of novel products, however, all tested donor modules were equally able to change the native system synthesizing plipastatin.

Influence of Combined Deletion and Substitution of COM Domains on Plipastatin Synthesis

The second experiment demonstrated that the ppsC module could not couple with the next ppsD module following deletion of the ppsC donor. However, when the partner donor was present, altering the acceptor module had no effect on the formation of novel lipopeptides in the permutation experiment. We therefore decided to investigate whether the acceptor module retains the ability to interact when its partner donor module is deleted. The donor of ppsC was first deleted, and the acceptor of ppsC was then substituted with the acceptor of ppsB (**Figure 1C**). The high-resolution ESI–MS analysis of the resulting product gave a charge 1 ($z = 1$) molecular mass ion series of m/z 1263.7235, 1277.7412, 1291.7504, and 1305.7682 (**Figure 6A**) that corresponded with hydrogen adducts of ions derived from a novel cyclic octapeptide. The predicted formula was $C_{65}H_{100}N_{10}O_{16}$ (ppm = 1.132) at m/z 1277.7412. The high-resolution HCD-MS/MS fragmentation of the m/z 1277.7412 ion (**Figure 6B**) yielded a -y ion series at m/z 102.0553→766.3765→881.4631 from the N-terminus, and a -b ion series at m/z 226.1178→389.4315→502.2661 →398.2906→512.3785 from the C-terminus, consistent with a cyclic Glu-Orn-Tyr- Thr-Pro-Gln-Tyr-Ile peptide and a C13 β-OH fatty acid lacking Glu and Ala/Val amino acid residues that are present in plipastatin and the fragmentation pattern is illustrated in **Figure 6C**. Again, other molecular mass ions (**Table 1**, **Figure 1D**) differing by 14 Da ($-CH_2-$) were derived from the same peptide but with C11, C12, and C14 β-OH fatty acids. This indicated that a novel NRPS complex assembly line ppsA, ppsB, ppsD and ppsE directed the production of cyclic octapeptide.

Deletion of donor of ppsC largely abrogated the ability of the COM^D domain to form a productive complex with its native partner COM$^A_{ppsC}$, but formation of plipastatin was not affected when the acceptor of ppsC was substituted with the acceptor of ppsB. However, the including of both modifications mutant resulted in the ability of ppsB to interact with the ppsC acceptor (COM$^A_{ppsC}$). Therefore, the acceptor module retained selectivity when its partner donor module was deleted.

Antimicrobial Activity

The antimicrobial activity of the linear lipopeptide products was tested against one Gram-positive bacteria, one Gram-negative bacteria and five fungi using a standard dilution approach. The lipopeptides exhibited antimicrobial activity against five of the fungal species at a contention of 31.25–125 µg/ml (**Table 2**). Interestingly, the novel lipopeptides inhibited both tested bacteria. *Penicillium notatum* exhibited the least resistance

against linear heptapeptide and hexapeptide compared with all fungi, showing a MIC of 31.25 µg/ml, which was the value of the linear heptapeptide against *Aspergillus ochraceus*. The MIC values of all linear lipopeptides against *Escherichia coli* and *Staphylococcus aureus* were much higher than against the fungi, and the MIC values were, respectively, 250 and 500 µg/ml. The cyclic lipopeptides exhibited the inhibitory activity as shown in Figure S7.

DISCUSSION

Based on the characterization of NRPS systems, each of the large modular enzymes is deemed responsible for the incorporation of a monomeric precursor into a specific amino acid (Walsh, 2002; Schwarzer et al., 2003). In multienzymatic NRPS complexes, synthesis by most known NRP assembly lines requires appropriate communication between partner enzymes including A-PCP minimal modules (Schneider et al., 1998), C-A-PCP elongation modules (Mootz et al., 2002), A domains (Eppelmann et al., 2002), and translocation of the terminal, product-releasing Te domains (de Ferra et al., 1997). A recent study has examined the interaction between NRPS modules by introducing the photocrosslinking unnatural amino acidBpF (Chin et al., 2002) into aminimal, dimodular NRPS system. Further, the crosslinks were Photocrosslinked and mapped of by MS (Dehling et al., 2016). In addition, protein-protein communication is controlled by the interplay between linker (COM) domains that comprise only 20–30 or 15–25 amino acid residues, and that are located at the C and N termini between NRPS enzyme modular domains (Hahn and Stachelhaus, 2004). The selectivity provided by different sets of compatible COM^D and COM^A domains results in NRPS complexes that utilize defined biosynthetic templates and synthesize specific peptide products (Hahn and Stachelhaus, 2004, 2006).

In the present study, we investigated the selectivity of COM domains by direct reprogramming of the plipastatin biosynthetic complex, and achieved a biocombinatorial synthesis of new lipopeptides possessing significant antimicrobial activity. Novel lipopeptides sequences were identified by the structure of plipastatin (Deleu et al., 2008) coupled with HCD-MS/MS fragmentation (Pathak et al., 2013). In the first part of the study, the amino acid substitutions can be classified into four groups: (i) conservative exchanges between hydrophobic amino acids (A3L) and polar uncharged residues (S2T and G12S), (ii) substitution of a hydrophobic with a polar residue (W1K), (iii) exchange of a polar uncharged residue for a negatively charged residue (T7D), and (iv) substitution of a polar negatively charged residue with a polar positively charged (E9K) or uncharged (D10S) amino acid. The key residues that mediate the interaction between donor ppsB and acceptor ppsC COM domains were mutated, and the resultant NRPS system was able to form a novel assembly line (ppsA/ppsB/ppsE) that could synthesize lipopentapeptides. Indeed, a Thr-to-Asp mutation resulted in a series of homologous shortened pentapeptide products (**Table 1**). The previous research demonstrated that the selective

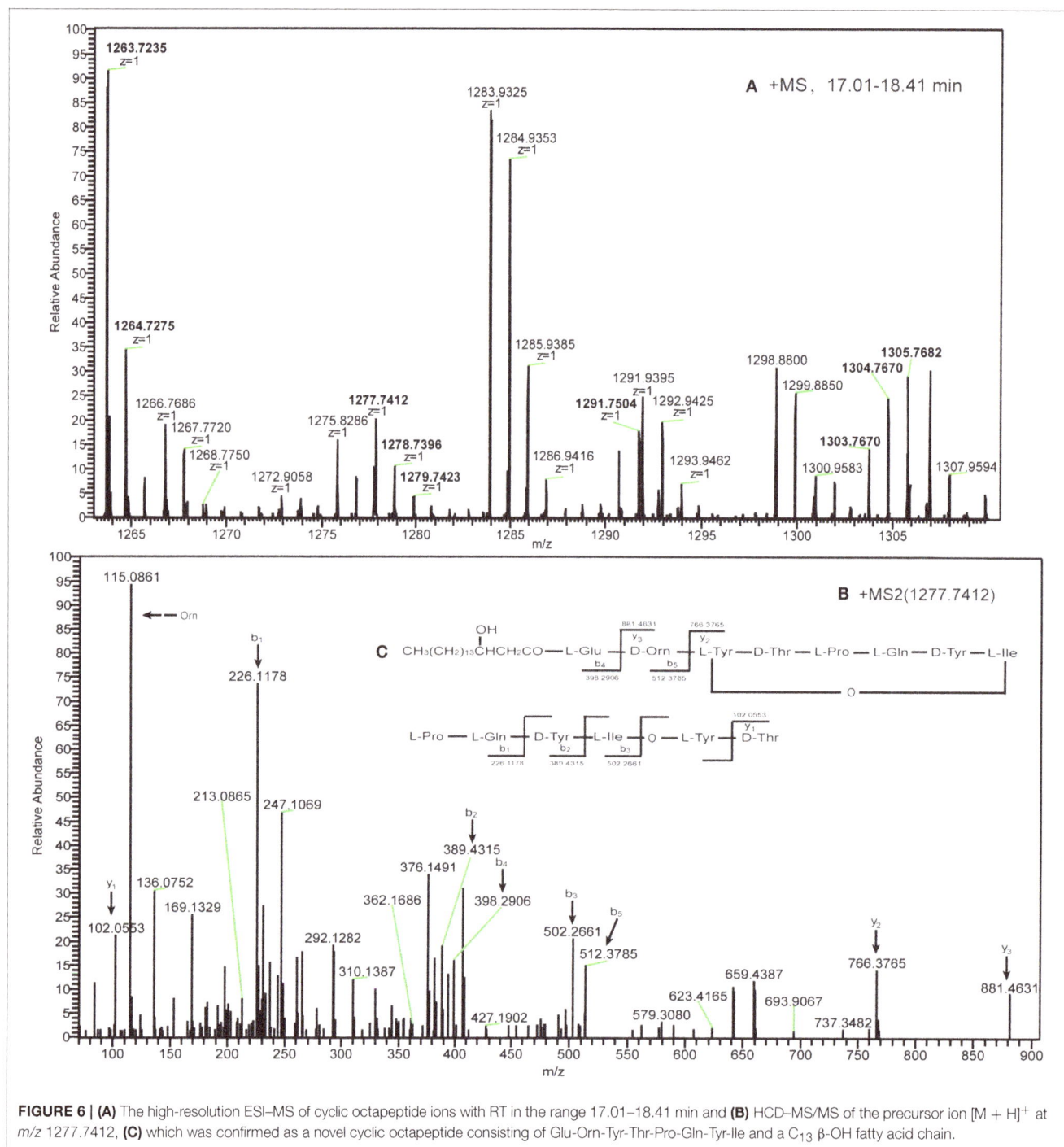

FIGURE 6 | (A) The high-resolution ESI–MS of cyclic octapeptide ions with RT in the range 17.01–18.41 min and (B) HCD–MS/MS of the precursor ion [M + H]$^+$ at m/z 1277.7412, (C) which was confirmed as a novel cyclic octapeptide consisting of Glu-Orn-Tyr-Thr-Pro-Gln-Tyr-Ile and a C$_{13}$ β-OH fatty acid chain.

communication is predominantly established by polar and/or electrostatic interactions in tyrocidine NRPS complex (Hahn and Stachelhaus, 2006). However, the substitution of a hydrophobic with a polar residue (W1K) did not change the selectivity. Therefore, the electrostatic interactions might play a key role in the selective communication. This approach can therefore be used to identify key residues that are important for maintaining (or preventing) the interaction between COM^D and COM^A in NRPSs. Deletion of the ppsC donor prevented selectivity

of the acceptor of ppsC, and resulted in a novel assembly line (ppsA/ppsB/ppsC) that synthesized a lipohexapeptide. It was also demonstrated that the COM^A domain alone is insufficient for interacting with its upstream partner enzyme in the enzyme complex with specificity (Cheng et al., 2016). We next questioned whether both the donor and acceptor COM domains influence the selectivity by switching the acceptors of ppsC and ppsB. The differences between respective COM domain pairs were minimal (Figure S2) (Linne et al., 2003), and the mutation of acceptor

TABLE 2 | Minimum inhibitory concentration (MIC) of lipopeptides against fungal species (μg/mL).

Fungal species	Linear hexapeptide	Linear nonapeptide	Linear heptapeptide
Rhizopus stolonifer	62.5	62.5	62.5
Fusarium oxysporum	62.5	62.5	62.5
Aspergillus ochraceus	62.5	62.5	31.25
Penicillium notatum	31.25	62.5	31.25
Aspergillus flavus	62.5	125	125
Escherichia coli	250	500	250
Staphylococcus aureus	250	500	500

COM domains impairs rather than halts product formation. In contrast, formation of plipastatin was not only impaired but completely abrogated when donor modules of COM domains were changed, and resulted in formation of new assembly lines (ppsA/ppsB/ppsC/ppsD and ppsA/ppsB/ppsC/ppsE) that synthesized novel lipononapeptide and lipoheptapeptide products. These results further demonstrated that donor COM domains influence interactions in the biosynthetic assembly line. Deletion of the ppsC donor ceased the interaction with its partner acceptor module, but when the partner acceptor module was mutated to the acceptor of ppsB, which facilitated the synthesis of a novel assembly line (ppsA/ppsB/ppsD/ppsE). Therefore, the acceptor COM domains retain selectivity when their partner donor is deleted. It was previously shown that when all donor and acceptor were replaced with the same (compatible) COM^D and COM^A domains in the surfactin biosynthetic complex, different assembly lines (SrfA-A/SrfA-B/SrfA-C and SrfA-A/SrfA-C) were formed that led to the synthesis of lipoheptapeptide and the lipotetrapeptide products (Chiocchini et al., 2006). A universal communication system for the plipastatin biosynthetic complex is therefore likely to result in additional novel lipopeptides.

Nonribosomal peptides synthetases (NRPSs) synthesis structurally diverse peptide-based natural products (Marahiel and Essen, 2009). Manipulation of their modular organization provides enormous potential for generating novel bioactive compounds (Miao et al., 2006; Butz et al., 2008; Doekel

et al., 2008). Matching pairs of COM domains form protein–protein interactions (Linne et al., 2003; Hahn and Stachelhaus, 2006), and the present study demonstrated the decisive role of COM domain pairs in NRP biosynthetic complexes. The need for new antibiotics to fight the emergence of resistant pathogenic microorganisms is growing (Marr et al., 2006), and the novel lipopeptides proved to have good antimicrobial activity.

CONCLUSIONS

Our results and experimental approaches pave the way for (1) identifying key residues in COM domains (2) characterizing the interactions between donor and acceptor COM domains, and (3) the preparation of novel lipopeptides possessing antimicrobial activity. Donor or acceptor modules could therefore influence the selectivity of NRPS systems. We demonstrated that (1) a single point mutation of a COM domain can alter the selectivity of the biosynthetic assembly line, (2) donor COM domains influence interactions in the biosynthetic assembly line, (3) mutation of acceptor COM domains impairs rather than halts product formation, (4) acceptor COM domains retain selectivity when their partner donor is deleted, and (5) the lipopeptides possess strong antifungal and some antibacterial activity. Our experimental approaches provide the theoretical basis for the production of the novel lipopeptides, which has potential for use in food preservation.

AUTHOR CONTRIBUTIONS

XB conceived the project and revising it critically for important intellectual content, HL and LG designed the experiments, HL analyzed the data and wrote the manuscript. CD, JH, and ZM assisted in data analysis, XB, ZL, and CZ supervised the study.

ACKNOWLEDGMENTS

This work was supported by grants from the National Natural Science Foundation of China (grant no. 31271828).

REFERENCES

Anagnostopoulos, C., and Spizizen, J. (1961). Requirements for transformation in Bacillus subtilis. J. Bacteriol. 81, 741.

Arnaud, M., Chastanet, A., and Débarbouillé, M. (2004). New vector for efficient allelic replacement in naturally nontransformable, low-GC-content, gram-positive bacteria. Appl. Environ. Microbiol. 70, 6887–6891. doi: 10.1128/AEM.70.11.6887-6891.2004

Batool, M., Khalid, M. H., Hassan, M. N., and Hafeez, F. Y. (2011). Homology modeling of an antifungal metabolite plipastatin synthase from the bacillus subtilis 168. Bioinformation 7, 384–387. doi: 10.6026/97320630007384

Bonmatin, J.-M., Laprévote, O., and Peypoux, F. (2003). Diversity among microbial cyclic lipopeptides: iturins and surfactins. Activity-structure relationships to

design new bioactive agents. Comb. Chem. High Throughput Screen. 6, 541–556. doi: 10.2174/138620703106298716

Butz, D., Schmiederer, T., Hadatsch, B., Wohlleben, W., Weber, T., and Süssmuth, R. D. (2008). Module extension of a non-ribosomal peptide synthetase of the glycopeptide antibiotic balhimycin produced by amycolatopsis balhimycina. Chembiochem 9, 1195–1200. doi: 10.1002/cbic.2008 00068

Calcott, M. J., and Ackerley, D. F. (2014). Genetic manipulation of non-ribosomal peptide synthetases to generate novel bioactive peptide products. Biotechnol. Lett. 36, 2407–2416. doi: 10.1007/s10529-014-1642-y

Cheng, Y. C., Ke, W. J., and Liu, S. T. (2016). Regions involved in fengycin synthetases enzyme complex formation. J. Microbiol. Immunol. Infect. doi: 10.1016/j.jmii.2015.12.001. [Epub ahead of print].

Chin, J. W., Martin, A. B., King, D. S., Wang, L., and Schultz, P. G. (2002). Addition of a photocrosslinking amino acid to the genetic code of *Escherichia coli*. *Proc. Natl. Acad. Sci. U.S.A.* 99, 11020–11024. doi: 10.1073/pnas.172226299

Chiocchini, C., Linne, U., and Stachelhaus, T. (2006). *In vivo* biocombinatorial synthesis of lipopeptides by COM domain-mediated reprogramming of the surfactin biosynthetic complex. *Chem. Biol.* 13, 899–908. doi: 10.1016/j.chembiol.2006.06.015

de Faria, A. F., Stéfani, D., Vaz, B. G., Silva, Í. S., Garcia, J. S., Eberlin, M. N., et al. (2011). Purification and structural characterization of fengycin homologues produced by *Bacillus subtilis* LSFM-05 grown on raw glycerol. *J. Ind. Microbiol. Biotechnol.* 38, 863–871. doi: 10.1007/s10295-011-0980-1

de Ferra, F., Rodriguez, F., Tortora, O., Tosi, C., and Grandi, G. (1997). Engineering of peptide synthetases key role of the thioesterase-like domain for efficient production of recombinant peptides. *J. Biol. Chem.* 272, 25304–25309. doi: 10.1074/jbc.272.40.25304

Dehling, E., Volkmann, G., Matern, J. C., Dörner, W., Alfermann, J., Diecker, J., et al. (2016). Mapping of the communication-mediating interface in nonribosomal peptide synthetases using a genetically encoded photocrosslinker supports an upside-down helix-hand motif. *J. Mol. Biol.* 428, 4345–4360. doi: 10.1016/j.jmb.2016.09.007

Deleu, M., Paquot, M., and Nylander, T. (2008). Effect of fengycin, a lipopeptide produced by Bacillus subtilis, on model biomembranes. *Biophys. J.* 94, 2667–2679. doi: 10.1529/biophysj.107.114090

Deng, Y., Lu, Z., Lu, F., Wang, Y., and Bie, X. (2011a). Study on an antimicrobial protein produced by *Paenibacillus polymyxa* JSa-9 isolated from soil. *World J. Microbiol. Biotechnol.* 27, 1803–1807. doi: 10.1007/s11274-010-0638-6

Deng, Y., Lu, Z., Lu, F., Zhang, C., Wang, Y., Zhao, H., et al. (2011b). Identification of LI-F type antibiotics and di-n-butyl phthalate produced by *Paenibacillus polymyxa*. *J. Microbiol. Methods* 85, 175–182. doi: 10.1016/j.mimet.2011.02.013

Doekel, S., Coëffet-Le Gal, M. F., Gu, J. Q., Chu, M., Baltz, R. H., and Brian, P. (2008). Non-ribosomal peptide synthetase module fusions to produce derivatives of daptomycin in *Streptomyces roseosporus*. *Microbiology* 154, 2872–2880. doi: 10.1099/mic.0.2008/020685-0

Eppelmann, K., Stachelhaus, T., and Marahiel, M. A. (2002). Exploitation of the selectivity-conferring code of nonribosomal peptide synthetases for the rational design of novel peptide antibiotics. *Biochemistry* 41, 9718–9726. doi: 10.1021/bi0259406

Finking, R., and Marahiel, M. A. (2004). Biosynthesis of nonribosomal peptides 1. *Annu. Rev. Microbiol.* 58, 453–488. doi: 10.1146/annurev.micro.58.030603.123615

Fischbach, M. A., and Walsh, C. T. (2006). Assembly-line enzymology for polyketide and nonribosomal peptide antibiotics: logic, machinery, and mechanisms. *Chem. Rev.* 106, 3468–3496. doi: 10.1021/cr0503097

Guo, Q., Dong, W., Li, S., Lu, X., Wang, P., Zhang, X., et al. (2014). Fengycin produced by Bacillus subtilis NCD-2 plays a major role in biocontrol of cotton seedling damping-off disease. *Microbiol. Res.* 169, 533–540. doi: 10.1016/j.micres.2013.12.001

Hahn, M., and Stachelhaus, T. (2004). Selective interaction between nonribosomal peptide synthetases is facilitated by short communication-mediating domains. *Proc. Natl. Acad. Sci. U.S.A.* 101, 15585–15590. doi: 10.1073/pnas.0404932101

Hahn, M., and Stachelhaus, T. (2006). Harnessing the potential of communication-mediating domains for the biocombinatorial synthesis of nonribosomal peptides. *Proc. Natl. Acad. Sci. U.S.A.* 103, 275–280. doi: 10.1073/pnas.0508409103

Hiradate, S., Yoshida, S., Sugie, H., Yada, H., and Fujii, Y. (2002). Mulberry anthracnose antagonists (iturins) produced by *Bacillus amyloliquefaciens* RC-2. *Phytochemistry* 61, 693–698. doi: 10.1016/S0031-9422(02)00365-5

Honma, M., Tanaka, K., Konno, K., Tsuge, K., Okuno, T., and Hashimoto, M. (2012). Termination of the structural confusion between plipastatin A1 and fengycin IX. *Bioorg. Med. Chem.* 20, 3793–3798. doi: 10.1016/j.bmc.2012.04.040

Jin, H., Zhang, X., Li, K., Niu, Y., Guo, M., Hu, C., et al. (2014). Direct bio-utilization of untreated rapeseed meal for effective iturin A production by *Bacillus subtilis* in submerged fermentation. *PLoS ONE* 9:e111171. doi: 10.1371/journal.pone.0111171

Landy, M., Rosenman, S. B, and Warren, G. H. (1947). An antibiotic from *Bacillus subtilis* active against pathogenic fungi. *J. Bacteriol.* 54, 24–24.

Linne, U., Stein, D. B., Mootz, H. D., and Marahiel, M. A. (2003). Systematic and quantitative analysis of protein-protein recognition between nonribosomal peptide synthetases investigated in the tyrocidine biosynthetic template. *Biochemistry* 42, 5114–5124. doi: 10.1021/bi034223o

Liu, Q., Lin, J., Wang, W., Huang, H., and Li, S. (2015). Production of surfactin isoforms by *Bacillus subtilis* BS-37 and its applicability to enhanced oil recovery under laboratory conditions. *Biochem. Eng. J.* 93, 31–37. doi: 10.1016/j.bej.2014.08.023

Marahiel, M. A., Stachelhaus, T., and Mootz, H. D. (1997). Modular peptide synthetases involved in nonribosomal peptide synthesis. *Chem. Rev.* 97, 2651–2674. doi: 10.1021/cr960029e

Marahiel, M. A., and Essen, L. O. (2009). Nonribosomal peptide synthetases: mechanistic and structural aspects of essential domains. *Meth. Enzymol.* 458, 337–351. doi: 10.1016/S0076-6879(09)04813-7

Marr, A. K., Gooderham, W. J., and Hancock, R. E. (2006). Antibacterial peptides for therapeutic use: obstacles and realistic outlook. *Curr. Opin. Pharmacol.* 6, 468–472. doi: 10.1016/j.coph.2006.04.006

Miao, V., Coëffet-Le Gal, M.-F., Nguyen, K., Brian, P., Penn, J., Whiting, A., et al. (2006). Genetic engineering in *Streptomyces roseosporus* to produce hybrid lipopeptide antibiotics. *Cell Chem. Biol.* 13, 269–276. doi: 10.1016/j.chembiol.2005.12.012

Mootz, H. D., Kessler, N., Linne, U., Eppelmann, K., Schwarzer, D., and Marahiel, M. A. (2002). Decreasing the ring size of a cyclic nonribosomal peptide antibiotic by in-frame module deletion in the biosynthetic genes. *J. Am. Chem. Soc.* 124, 10980–10981. doi: 10.1021/ja027276m

Mootz, H. D., Schwarzer, D., and Marahiel, M. A. (2000). Construction of hybrid peptide synthetases by module and domain fusions. *Proc. Natl. Acad. Sci. U.S.A.* 97, 5848–5853. doi: 10.1073/pnas.100075897

Nedorostova, L., Kloucek, P., Kokoska, L., Stolcova, M., and Pulkrabek, J. (2009). Antimicrobial properties of selected essential oils in vapour phase against foodborne bacteria. *Food Control.* 20, 157–160. doi: 10.1016/j.foodcont.2008.03.007

Oh, I., Yang, W-Y, Park, J., Lee, S., Mar, W., Oh, K-B, et al. (2011). *In vitro* Na+/K+-ATPase inhibitory activity and antimicrobial activity of sesquiterpenes isolated from Thujopsis dolabrata. *Arch. Pharm. Res.* 34, 2141–2147. doi: 10.1007/s12272-011-1218-5

Pathak, K. V., Bose, A., and Keharia, H. (2013). Characterization of novel lipopeptides produced by *Bacillus tequilensis* P15 Using Liquid Chromatography Coupled Electron Spray Ionization Tandem Mass Spectrometry (LC-ESI-MS/MS). *Int. J. Pept. Res. Ther.* 20, 133–143. doi: 10.1007/s10989-013-9375-7

Sambrook, J., Fritsch, E. F., and Maniatis, T. (1989). Molecular *Cloning*. New York, NY: Cold spring harbor laboratory press.

Schneider, A., Stachelhaus, T., and Marahiel, M. (1998). Targeted alteration of the substrate specificity of peptide synthetases by rational module swapping. *Mol. Gen. Genet.* 257, 308–318. doi: 10.1007/s004380050652

Schwarzer, D., Finking, R., and Marahiel, M. A. (2003). Nonribosomal peptides: from genes to products. *Nat. Prod. Rep.* 20, 275. doi: 10.1039/b111145k

Sieber, S. A., and Marahiel, M. A. (2005). Molecular mechanisms underlying nonribosomal peptide synthesis: approaches to new antibiotics. *Chem. Rev.* 105, 715–738. doi: 10.1021/cr0301191

Sun, L., Lu, Z., Bie, X., Lu, F., and Yang, S. (2006). Isolation and characterization of a co-producer of fengycins and surfactins, endophytic Bacillus amyloliquefaciens ES-2, from Scutellaria baicalensis Georgi. *World J. Microbiol. Biotechnol.* 22, 1259–1266. doi: 10.1007/s11274-006-9170-0

Tosato, V., Albertini, A. M., Zotti, M., Sonda, S., and Bruschi, C. V. (1997). Sequence completion, identification and definition of the fengycin operon in Bacillus subtilis 168. *Microbiology* 143, 3443–3450. doi: 10.1099/00221287-143-11-3443

Vanittanakom, N., Loeffler, W., Koch, U., and Jung, G. (1986). Fengycin-A novel antifungal lipopeptide antibiotic produced by *Bacillus subtilis* F-29-3. *J. Antibiot.* 39, 888–901. doi: 10.7164/antibiotics.39.888

Vater, J., Kablitz, B., Wilde, C., Franke, P., Mehta, N., and Cameotra, S. S. (2002). Matrix-assisted laser desorption ionization-time of flight mass spectrometry of lipopeptide biosurfactants in whole cells and culture filtrates of *Bacillus subtilis* C-1 isolated from petroleum sludge. *Appl. Environ. Microbiol.* 68, 6210–6219. doi: 10.1128/AEM.68.12.6210-6219.2002

Walsh, C. T. (2002). Combinatorial biosynthesis of antibiotics: challenges and opportunities. *Chembiochem* 3, 124–134. doi: 10.1002/1439-7633 (20020301)3:2/3<124::AID-CBIC124>3.0.CO;2-J

Yang, H., Li, X., Li, X., Yu, H., and Shen, Z. (2015). Identification of lipopeptide isoforms by MALDI-TOF-MS/MS based on the simultaneous purification of iturin, fengycin, and surfactin by RP-HPLC. *Anal. Bioanal. Chem.* 407, 2529–2542. doi: 10.1007/s00216-015-8486-8

Yu, G., Sinclair, J., Hartman, G., and Bertagnolli, B. (2002). Production of iturin A by *Bacillus amyloliquefaciens* suppressing *Rhizoctonia solani*. *Soil Biol. Biochem.* 34, 955–963. doi: 10.1016/S0038-0717(02)00027-5

Conflict of Interest Statement: The authors declare that the research was conducted in the absence of any commercial or financial relationships that could be construed as a potential conflict of interest.

15

Development of Multiple Cross Displacement Amplification Label-Based Gold Nanoparticles Lateral Flow Biosensor for Detection of *Shigella* spp.

Yi Wang, Yan Wang, Jianguo Xu and Changyun Ye*

State Key Laboratory of Infectious Disease Prevention and Control, National Institute for Communicable Disease Control and Prevention, Collaborative Innovation Center for Diagnosis and Treatment of Infectious Diseases, Chinese Center for Disease Control and Prevention, Beijing, China

Edited by:
*Spiros Paramithiotis,
Agricultural University of Athens,
Greece*

Reviewed by:
*Yean Yean Chan,
University of Science, Malaysia,
Malaysia
Boris B. Dzantiev,
A.N.Bakh Institute of Biochemistry
of Russian Academy of Sciences,
Russia*

*Correspondence:
*Changyun Ye
yechangyun@icdc.cn*

Shigella spp., the etiological agent of shigellosis or "bacillary dysentery," are responsible for considerable morbidity and mortality in excess of a million deaths globally per year. Although PCR-based techniques (such as PCR-based dipstick biosensors) have been used for the molecular diagnosis of infectious disease, these assays were restricted due to the need for a sophisticated thermal cycling apparatus to denature target templates. To facilitate simple and rapid detection of target pathogens, we successfully devised an inexpensive, reliable and nearly instrument-free molecular technique, which incorporates multiple cross displacement amplification (MCDA) combined with a newly designed lateral flow biosensor (LFB) for visual, sensitive and specific detection of *Shigella*. The MCDA-LFB assay was conducted at 65°C for only 20 min during the amplification stage, and then products were directly analyzed on the biosensor, alleviating the use of special reagents, electrophoresis equipment and amplicon detection instruments. The entire process, including specimen processing (35 min), amplification (20) and detection (2–5 min), can be finished within 1 h. The MCDA-LFB assay demonstrated high specificity for *Shigella* detection. The analytical sensitivity of the assay was 10 fg of genomic templates per reaction in pure culture and 5.86 CFU per tube in human fecal samples, which was consistent with MCDA by colorimetric indicator, gel electrophoresis, real time turbidity and fluorescence detection. Hence, the simplicity, rapidity and nearly instrument-free platform of the MCDA-LFB assay make it practical for 'on-site' diagnosis, point-of-care testing and more. Moreover, the proof-of-concept approach can be reconfigured to detect a wide variety of target sequences by re-designing the specific MCDA primers.

Keywords: *Shigella* spp., multiple cross displacement amplification, lateral flow biosensor, MCDA-LFB, limit of detection

INTRODUCTION

Shigella spp. are exquisitely fastidious gram-negative pathogens that are responsible for as many as 167 million cases of shigellosis worldwide, resulting in a million deaths annually (Schroeder and Hilbi, 2008). Four *Shigella* species, including *S. sonnei*, *S. boydii*, *S. flexneri*, and *S. dysenteriae*, are considered as pathogenic to humans, particularly in young children (Koh et al., 2012). The typical symptoms of *Shigella* infection include dysentery and/or diarrhea with frequent mucoid boldly stools, fever, abdominal pain, tenesmus and malaise (Khan et al., 2013). The individuals, including young children, older adults and immune-compromised populations, may be at more risk for *Shigella* infection (Njuguna et al., 2013). The low infective dose (10 cells) of *Shigella* permits the disease to be effectively spread by contaminated food or water, and also by person-to-person contact, thus the foodborne or waterborne outbreaks of *Shigella* are common (Haley et al., 2010; Nygren et al., 2013; Baker et al., 2015). Herein, a reliable detection tool is needed to offer accurate diagnosis of *Shigella* to achieve infection control, clinical care and epidemiologic investigations.

The traditional detection of *Shigella* relies on culture-based methods, while only a small fraction of the actual shigellosis cases can be identified (Echeverria et al., 1991). Moreover, the growth, and thus the identification of these pathogens is frequently further impaired by ongoing antimicrobial therapy prior to specimen collection. The molecular detection techniques, such as PCR-based protocols, which overcome some of disadvantages posed by culture methods, are employed for the diagnostic of *Shigella* spp. (McKillip and Drake, 2004; Warren et al., 2006; Mandal et al., 2011; Villalobo and Torres, 1998). These methodologies require a sophisticated thermal cycling apparatus to denature target templates and still analysis of the amplified products with either agarose gel electrophoresis or probe hybridization techniques, which significantly hampered its application in the laboratories with limited resources settings (Wang et al., 2015a,b, 2016a). Although, newer approaches, including chemical and biological sensors, have been reported to be very rapid, sensitive and specific for detecting PCR amplicons of different target, thermal cycling of PCR-based methods during the amplification stage imposed instrumental constraints, limiting these assays to a low-resource setting (Chua et al., 2011; Liao et al., 2016). As such, the suitable detection assays using a simple, rapid, sensitive and specific technique are continuously required for the effective control and prevention of *Shigella*.

The growing use of molecular diagnostic methods has emphasized speed, simplicity and inexpensiveness as key criteria for adoption in 'on-site' analysis, field diagnosis and point-of-care testing and more, and the isothermal amplification technologies were well-suited for these application. Among dozens of isothermal nucleic acid amplification technologies, a few of these techniques (e.g., RCA, rolling circle amplification; LAMP, loop-mediated isothermal amplification; CPA, cross priming amplification) can efficiently achieve amplification using only one enzyme (Zhao et al., 2015). However, RCA was limited to amplify the circular target DNA, and a ligation process before amplification was always conducted for the specific recognition of a sequence. Although LAMP and CPA assays displayed high amplification efficiency comparable to that of the PCR method, the marginal amounts of nucleic acid sequences were still difficultly to analyze in various samples (Wang et al., 2016b).

More recently, multiple cross displacement amplification (MCDA) (Chinese IP Office Patent Application CN201510280765.X) was successfully established to overcome the technical barriers posed by current isothermal amplification strategies, and the mechanism and rationale of MCDA technique have been described in details (Wang et al., 2015c). MCDA has exhibited unique advantages of simplicity, rapidity, sensitivity, specificity and repeatability, generating amplicons from as few as three bacterial cells. The gold nanoparticle-based immunochromatographic technique is another strategy that has been widely used for the detection of amplicons yielded by various nucleic acid amplification-based assays (Vikesland and Wigginton, 2010). Here, the amplicon detection using gold nanoparticle-based dipstick biosensor was employed to simplify and accelerate the process of interpreting MCDA approach results. In the current report, we devised a MCDA assay combined with lateral flow biosensor (MCDA-LFB) for simple, rapid, sensitive and accurate visual detection of target sequence. As a proof of concept, *Shigella* was detected by MCDA-LFB assay to demonstrate the capability of target analysis. The performance of the MCDA-LFB methodology in detecting *Shigella* from pure culture and practical sample was successfully evaluated.

MATERIALS AND METHODS

Reagents and Instruments

The sample pad, conjugate pad, nitrocellulose membrane (NC), absorbent pad and backing card were purchased from the Jie Yi Biotechnology Co., Ltd. (Shanghai, China). The streptavidin-immobilized gold nanoparticles (SA-G), rabbit anti-fluorescein antibody (anti-FITC) and biotinylated bovine serum albumin (biotin-BSA) were purchased from the Resenbio Co., Ltd. (XiAn, China). The QIAamp DNA Stool Mini Kit and QIAamp DNA Mini Kit (QIAamp DNA minikits; Qiagen, Hilden, Germany) were purchased from Qiagen (Beijing, China). Loopamp™ Fluorescent Detection Reagent (FD) and the Loopamp kits were purchased from Eiken Chemical (Beijing, China).

Preparation of Gold Nanoparticle-Based Dipstick Biosensor

The dry-reagent dipstick (5 mm × 70 mm), illustrated in **Figure 1**, consisted of an absorbent pad, a NC membrane, a conjugate pad and an immersion pad assembled on a plastic adhesive backing card. The capture reagents, including anti-FITC (0.15 mg/ml) and biotin-BSA (4 mg/ml) in 0.01 M phosphate-buffered saline (PBS, PH 7.4), were dispensed onto the reaction regions. On the NC membrane, there are two zones as the test zone (conjugated with anti-FITC) and control zone (conjugated with biotin-BSA), with each line separated by 5 mm. SA-G in 0.01M PBS (PH 7.4) was deposited on the conjugate pad of the biosensor. Then, the assembled cards were cut at 5 mm widths, and the biosensors were dryly stored at the room temperature until use.

Visual Detection of MCDA Products Using the Biosensor

A 0.5 μl aliquot of MCDA amplicons was deposited to the sample application area of the biosensor. Then, the strip was directly

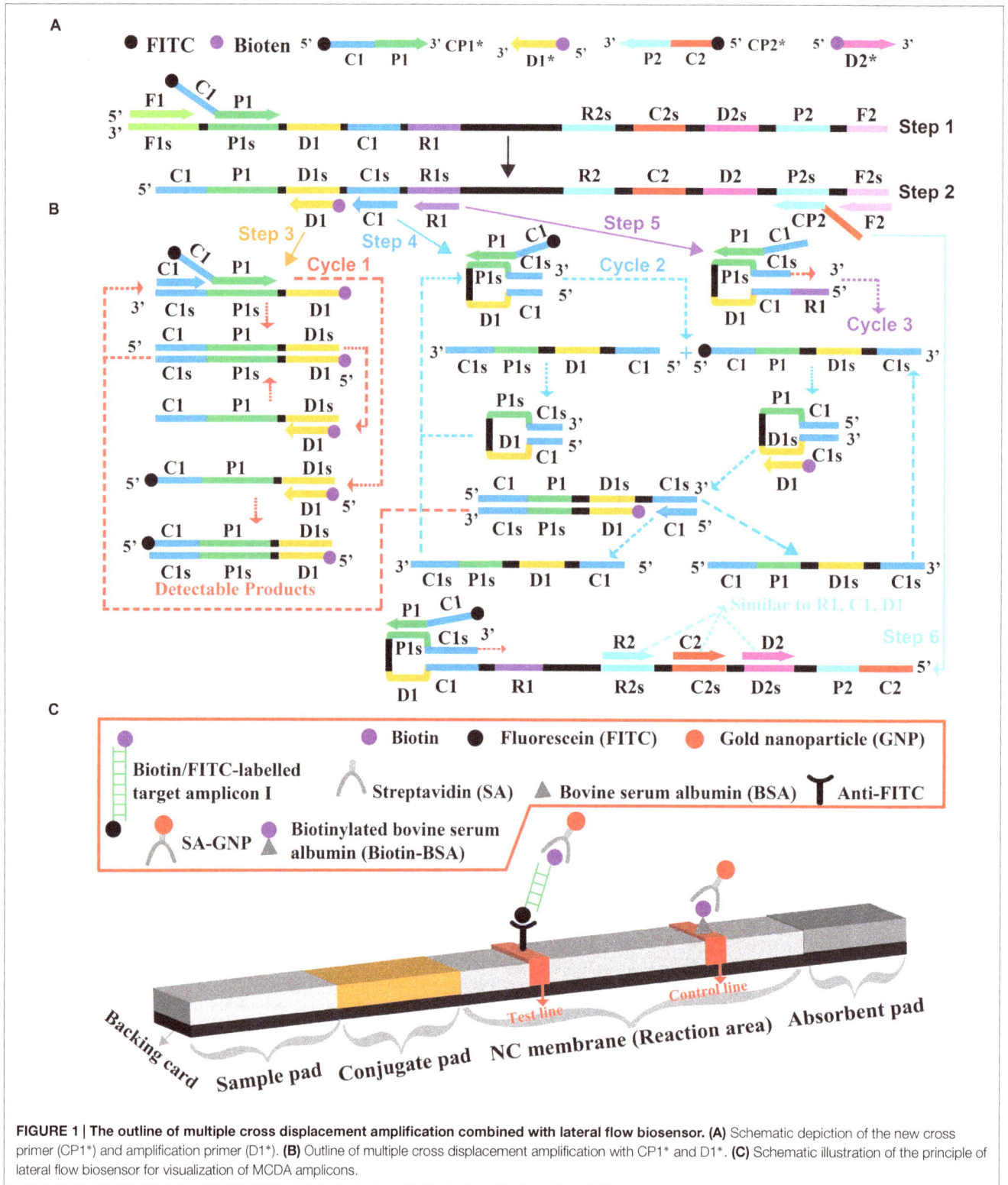

FIGURE 1 | The outline of multiple cross displacement amplification combined with lateral flow biosensor. (A) Schematic depiction of the new cross primer (CP1*) and amplification primer (D1*). (B) Outline of multiple cross displacement amplification with CP1* and D1*. (C) Schematic illustration of the principle of lateral flow biosensor for visualization of MCDA amplicons.

immersed into 120 µl of running buffer (10 mM PBS, PH 7.4 with 1% Tween 20) and the biosensor allowed absorbing the whole running buffer. After 2 min, the MCDA product detection was visualized in the form of red lines on the NC membrane.

Primer Design for MCDA Approach

In order to design *Shigella* spp. specific MCDA primers, the nucleotide sequence of the specific *ipaH* gene (GenBank accession no. M32063) was downloaded from the NCBI Genbank database, and a set of MCDA primers was designed by PrimerExplorer V4 (Eiken Chemical, Japan) and primer software PRIMER PREMIER 5.0 (Thiem et al., 2004). Blast analysis demonstrated that the MCDA primer set was specific for *Shigella* spp. strains. The details of primer design, primers sequences, locations and modifications of MCDA primers were displayed in **Figure 2** and **Table 1**. All of the oligomers were synthesized and purified by TsingKe Biological Technology (Beijing, China) at HPLC purification grade.

Bacterial Strains and Genomic Template Preparation

A total of 60 bacterial strains were used in this study (**Table 2**). Twenty-three *Shigella* strains and 37 non-*Shigella* strains were included to test the specificity of the MCDA-LFB assay. All bacterial strains were stored in 10% (w/v) glycerol broth at −70°C and then were refreshed three times on nutrient agar plate at 37°C. The genomic DNA templates were extracted from all culture strains using DNA extraction kits according to the manufacturer's instructions. The extracted templates were

examined with ultraviolet spectrophotometer (Nano drop ND-1000, Calibre, Beijing, China) at A260/280 and stored under at −20°C before the templates were used. The strains of *S. flexneri* serovar 1d (ICDC-NPS001) were applied for confirmation performance, optimal temperature, sensitivity analysis and practical application conducted in the report. Moreover, the genomic templates of *S. flexneri* serovar 1d (ICDC-NPS001) were serially diluted (10 ng, 10 pg, 10 fg, 1 fg, and 0.1 fg) for sensitivity evaluation of MCDA-LFB detection.

The Standard MCDA Assay

In order to assess the feasibility of *ipaH*-MCDA primers, the MCDA reaction was conducted as the standard MCDA condition, which has been reported in previous report (Wang et al., 2015c). In brief, the MCDA assay was carried out in 25-μl amplification mixtures containing the following components: 0.4 μM each of displacement primers F1 and F2, 0.8 μM each of amplification primers C1 and C2, 1.2 μM each of amplification primers R1, R2, D1* and D2, 1.2 μM each of cross primers CP1* and CP1, 2.4 μM cross primer CP2, 12.5 μl 2× reaction mix (Loopamp kits), 1.25 μl of *Bst* DNA polymerase (10 U)

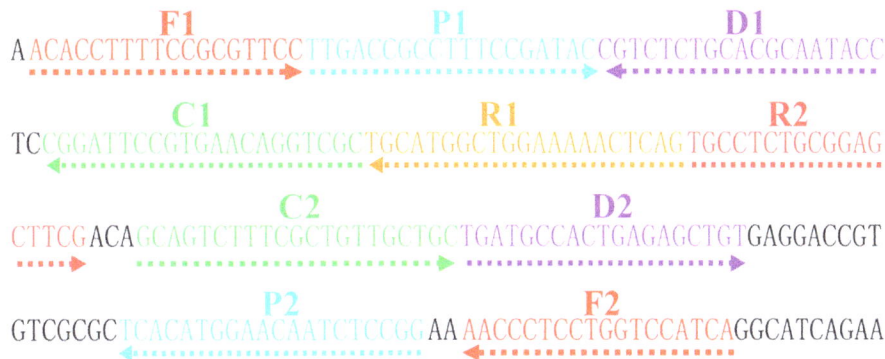

FIGURE 2 | Location and sequence of *ipaH* gene (*Shigella* app.-specific gene) used to design multiple cross displacement amplification primers. The nucleotide sequence of the sense strand of *ipaH* was exhibited. Right arrows and left arrows indicate sense and complementary sequences that were used.

TABLE 1 | The primers used in this study.

Primers name[a]	Sequences and modifications[b]	Length[c]	Gene
Shi-F1	5′-ACACCTTTTCCGCGTTCC-3′	18 nt	*ipaH*
Shi-F2	5′-TGATGGACCAGGAGGGTT-3′	18 nt	
Shi-CP1	5′-GCGACCTGTTCACGGAATCCG-TTGACCGCCTTTCCGATAC-3′	40 mer	
Shi-CP1*	5′-FITC-GCGACCTGTTCACGGAATCCG-TTGACCGCCTTTCCGATAC-3′	40 mer	
Shi-E-CP1	5′-Hex-TGCAATG-GCGACCT(BHQ1)GTTCACGGAATCCG-TTGACCGCCTTTCCGATAC-3′	47 mer	
Shi-CP2	5′-GCAGTCTTTCGCTGTTGCTGC-CCGGAGATTGTTCCATGTGA-3′	41 mer	
Shi-C1	5′-GCGACCTGTTCACGGAATCCG-3′	21 nt	
Shi-C2	5′-GCAGTCTTTCGCTGTTGCTGC-3′	21 nt	
Shi-D1	5′-GGTATTGCGTGCAGAGACG-3′	19 nt	
Shi-D1*	5′-Biotin-GGTATTGCGTGCAGAGACG-3	19 nt	
Shi-D2	5′-TGATGCCACTGAGAGCTGT-3′	19 nt	
Shi-R1	5′-CTGAGTTTTTCCAGCCATGCA-3′	21 nt	
Shi-R2	5′-TGCCTCTGCGGAGCTTCG-3′	18 nt	

[a]*Shi, shigella; Shi-CP1*, 5′-labeled with FITC when used in MCDA-LFB assay; Shi-D1*, 5′-labeled with biotin when used in MCDA-LFB assay; Shi-E-CP1, 5′-labeled with Hex when used in ET-MCDA assay;* [b]*Hex, hexachloro-fluorescein; FITC, fluorescein isothiocyanate.* [c]*mer, monomeric unit; nt, nucleitide.*

TABLE 2 | Bacterial strains used in this study.

Bacteria	Serovar/ Species	Strain no. (source of strain)[a]	No. of strains
Shigella flexneri	1d	ICDC-NPS001	1
	4a	ICDC-NPS002	1
	5a	ICDC-NPS003	1
	2b	ICDC-NPS004	1
	1b	ICDC-NPS005	1
	3a	ICDC-NPS006	1
	4av	ICDC-NPS007	1
	3b	ICDC-NPS008	1
	5b	ICDC-NPS009	1
	Y	ICDC-NPS010	1
	Yv	ICDC-NPS011	1
	1a	ICDC-NPS012	1
	X	ICDC-NPS013	1
	Xv	ICDC-NPS014	1
	F6	ICDC-NPS015	1
	7b	ICDC-NPS016	1
	$2a_1$	ICDC-NPS017	1
	4b	ICDC-NPS018	1
Shigella boydii	U	Isolated strains (ICDC)	1
Shigella dysenteriae	U	Isolated strains (ICDC)	2
Shigella sonneri	U	Isolated strains (ICDC)	2
Salmonella	Choleraesuis	ICDC-NPSa001	1
	U	Isolated strains (ICDC)	10
Listeria seeligeri	U	ATCC35967	1
Listeria grayii	U	Isolated strains (ICDC)	1
Listeria monocytogenes	4a	ATCC19114	1
Listeria welshimeri	U	ATCC35897	1
Listeria ivanovii	U	Isolated strains (ICDC)	1
Bacillus cereus	U	Isolated strains (ICDC)	1
Enteropathogenic Escherichia coli	U	Isolated strains (ICDC)	1
Enterotoxigenic Escherichia coli	U	Isolated strains (ICDC)	1
Enteroaggregative Escherichia coli	U	Isolated strains (ICDC)	1
Enteroinvasive Escherichia coli	U	Isolated strains (ICDC)	1
Enterohemorrhagic Escherichia coli	U	EDL933	1
Plesiomonas shigelloides	U	ATCC51903	1
Campylobacter jejuni	U	ATCC33291	1
Enterobacter cloacae	U	Isolated strains (ICDC)	1
Enterococcus faecalis	U	ATCC35667	1
Enterococcus faecium	U	Isolated strains (ICDC)	
Yersinia enterocolitica	U	ATCC23715	1
Streptococcus pneumoniae	U	Isolated strains (ICDC)	1
Aeromonas hydrophila	U	ATCC7966	1
Vibrio vulnificus	U	Isolated strains (ICDC)	1
Proteus vulgaris	U	Isolated strains (ICDC)	1
Vibrio fluvialis	U	Isolated strains (ICDC)	1
Streptococcus bovis	U	Isolated strains (ICDC)	1
Vibrio parahaemolyticus	U	ATCC17802	1
Klebsiella pneumoniae	U	ATCC700603	1
Bntorobater sakazakii	U	Isolated strains (ICDC)	1

[a]*U, unidentified serotype; ATCC, American Type Culture Collection; ICDC, National Institute for Communicable Disease Control and Prevention, Chinese Center for Disease Control and Prevention.*

and 1 μl DNA template. Four monitoring techniques, including colorimetric indicator (FD), gel electrophoresis, turbidimeters (LA-320C) and LFB detection, were employed to analyze the MCDA products. Furthermore, the endonuclease restriction-mediated real-time multiple cross displacement amplification (ET-MCDA), which was reported in a recent study, was employed to achieve real time fluorescence measurement of MCDA reaction (Wang et al., 2015c, 2016b).

Then, we tested the optimal reaction temperature of *ipaH*-MCDA primers. The MCDA reaction mixtures were performed at a constant temperature ranging from 60°C to 67°C for 1 h and then incubated at 85°C for 5 min to stop the amplification. Mixtures with 1 μl genomic template of *Listeria monocytogens* strain (*L. monocytogenes*, ATCC19114) and *Salmonella* strain (ICDC-NPsa001) were used as negative controls, and mixtures with 1 μl double distilled water (DW) were used as a blank control.

The Analytical Sensitivity of the *Shigella*-MCDA by Five Monitoring Techniques

The templates of *S. flexneri* serovar 1d (ICDC-NPS001) were serially diluted to confirm the limit of detection (LoD), which was defined by genomic DNA amount of the template. The analytical sensitivity of MCDA by colorimetric indicator (FD reagent), real time turbidity, 2% agarose gel electrophoresis, real time fluorescence and LFB detection was determined as described above. At least three replicates of each dilution were examined to test the analytical sensitivity.

The Analytical Specificity of the MCDA-LFB Approach

In order to assess the analytical specificity of MCDA-LFB methodology, the MCDA reactions were carried out under the conditions described above with purely genomic templates from 23 *Shigella* strains and 37 non-*Shigella* strains (**Table 2**). The MCDA products were tested using 2.5% agarose gel electrophoresis and LFB detection. Analysis of each sample was examined in at least two independent experiments.

Examination of MCDA-LFB Assay Using Simulated Human Fecal Specimens

Human fecal samples were acquired from a healthy donor with the written informed consent. Our study was reviewed and approved by the ethics committee of the National Institute for Communicable Disease Control and Prevention, China CDC, according to the medical research regulations of the Ministry of Health China (Approval No. ICDC2014003).

In order to evaluate the suitability of MCDA-LFB technique as a surveillance tool for *Shigella*, the MCDA-LFB assay was applied to rapidly diagnose the target pathogens in human fecal samples. Firstly, the human fecal samples were confirmed as being *Shigella*-negative by culture-based methods and PCR detection. Then, to test the minimal detectable colony forming units (CFUs), the cultures with *S. flexneri* strains were serially diluted (10^{-1} to 10^{-9}), and the aliquots of 100 μl appropriate dilution

(10^{-6}) was spread in triplicate onto brain heart infusion (BHI) agar. The CFUs were counted after 24 h at 37°C. Simultaneously, the aliquots of 100 μl appropriate dilution (10^{-3} to 10^{-8}) with *S. flexneri* strains were inoculated into the fecal samples (0.2 g), and the number of *Shigella* was adjusted to approximate 1.42×10^6, 1.42×10^5, 1.42×10^4, 1.42×10^3, 1.42×10^2 and 1.41×10^1 CFU/g. Then, the artificially contaminated stool samples were applied to extract the genomic DNA templates, and the supernatants (2 μl) were used for MCDA detections. Non-contaminated fecal sample was used as negative control and this analysis was independently conducted in triplicate. The MCDA products were also analyzed by colorimetric indicator (FD reagent), real time turbidity, 2% agarose gel electrophoresis, real time fluorescence and LFB detection as described above.

RESULTS

Development of the MCDA-LFB Assay

A schematic of MCDA-LFB technique was shown in **Figure 1**. In the MCDA-LFB system, the cross primer (CP1 or CP2) involved in MCDA reaction were labeled at the 5' end with FITC, and the amplification primers (D1 or D2) were modified at the 5' end with biotin (**Figure 1A**). The new CP1, CP2, D1, and D2 primers were named as CP1*, CP2*, D1*, and D2*, respectively. For clarity, the CP2* and D2* primers were not displayed in outline of MCDA reaction during the reaction stage (**Figure 1B**). The CP1* primer initiated MCDA reaction at the P1s site of the target sequence, and the newly synthesized strand was displaced by upstream synthesis from the primer F1 (Step 1). Five primers (D1*, C1, R1, CP2, and F2) annealed to the newly generated strand, and then the Bst polymerase extended in tandem producing four different products (Step 2). The D1* product was used as the template by C1 and CP1* primers, enter a cyclic process (Step 3, Cycle 1). In the cycle, a larger amounts of double-labled detectable amplicons, which contained biotin-labeled D1* primer and a FITC-labeled CP1* primer, were successfully yielded. The details of the reaction process for C1, R1, and CP2 products (Step 4, 5, 6) has been reported in previous study (Wang et al., 2015c). In addition, a double-labeled detectable product (CP2*/D2* product), which was similar to the detectable CP1*/D1* product, could be formed when the CP2 primer was modified with a FITC at the 5' end and D2 primer for biotin.

The principle of LFB for visualization of MCDA amplicons was exhibited in **Figure 1C**. The LFB detected MCDA amplicons through specific recognition of the FITC labels at the end of products, which were formed by using FITC labeled primers (CP1* primer). The other end, the amplicons labeled with biotin binded streptavidin-conjugated gold nanoparticles for visualization. The MCDA products were deposited onto on the sample application region of the biosensor, and then the biosensor was directly immersed in the running buffer. The running buffer moved along the biosensor by capillary action, which rehydrated the immobilized detector reagents (SA-G). The target amplicons was specifically captured by the immobilized anti-FITC at the first test zone and detector reagents rapidly accumulate in the reaction zone of the strip through biotion/streptavidin interaction, resulting in a visual red colored line on the test region. The proper function of the strip is demonstrated by the control line formation which contained biotinylated bovine serum albumin that captured excess detector reagent.

Confirmation and Detection *Shigella*-MCDA-LFB Products

In order to verify the feasibility of *Shigell*-MCDA primers, the MCDA reactions were carried out in the presence or absence of genomic DNA templates within 60 min at a constant temperature (65°C). Three monitoring techniques, including colorimetric indicator (FD reagent), gel electrophoresis analysis and LFB detection, were employed to confirm the *Shigella*-MCDA products. A color shift of positive amplification in *Shigella*-MCDA tubes was directly observed from light gray to green (**Figure 3A**). The positive MCDA products were seen many bands of different sizes in a typical ladder-like pattern on ethidium bromide-stained 2% agarose gel electrophoresis, but not in the negative and blank control (**Figure 3B**). It was also observed that two visible red bands (Test line, TL; Control line, CL) were seen in positive amplifications, and only the CL were seen in negative and blank controls (**Figure 3C**). Therefore, the MCDA primer set was a good candidate for establishment of the MCDA-LFB method for *Shigella* detection.

The Optimal Amplification Temperature of the MCDA-LFB Assay

In order to examine the optimal assay temperature during the amplification stage, the *Shigella*-MCDA reactions were conducted at eight distinct temperatures (60°C–67°C) with 1°C intervals. The strain *S. flexneri* serovar 1d (ICDC-NPS001) was employed as the positive control to evaluate the optimal amplification temperature at the level of 10 pg genomic templates per reaction. The reactions were analyzed by means of real time turbidity detection and the typical kinetics graphs corresponding to eight temperatures were obtained (**Figure 4**). Eight reaction temperatures provided a robust signal, with the faster amplifications generated from assay temperature of 63°C–67°C, which were recommended as the standard temperature for *Shigella*-MCDA-LFB assay during the amplification stage. The assay temperature of 65°C was used for the rest of MCDA-LFB tests conducted in this study.

Analytical Sensitivity of MCDA-LFB Technique in Pure Culture

The analytical sensitivity of MCDA-LFB technique on *Shigella* was determined by analyzing the products yielded from the serial dilutions (10 ng, 10 pg, 10 fg, 1 fg, and 0.1 fg per microliter) of *Shigella* genomic DNA in triplicate (**Figure 5**). The *Shigell*-MCDA reactions were monitored by real time measurement of turbidity and the LoD of MCDA-LFB assay for *Shigella* detection was 10 fg of genomic templates per reaction (**Figure 5A**). By the FD reagent, a color shift of positive amplification in *Shigella*-MCDA tubes was directly observed from light gray to green (**Figure 5B**). Then, the *Shigella*-MCDA products were analyzed

by 2% agarose gel electrophoresis and positive products were observed as the ladder-like patterns but not in negative reactions, negative control and blank control (**Figure 5C**). The biosensor was also subjected to detect the *Shigella*-MCDA products (**Figure 5E**). As expected, the biosensor exhibited clear visible red bands for both TL and CL when the products came from positive MCDA amplifications, and only the CL were generated from for negative MCDA amplifications, negative control and blank control. The LoD of MCDA-LFB assay for detecting *ipaH* gene was also 10 fg of genomic templates per reaction. Moreover, the LoD of ET-MCDA assay for *Shigella* detection was also 10 fg of genomic DNA in pure culture (**Figure 5D**). These results indicated that the analytical sensitivity by FD reagent, real time turbidity, real time fluorescence and agarose gel electrophoresis detection for *Shigella*-MCDA amplifications was conformity with biosensor analysis.

Then, we assessed the optimal duration of time require for the MCDA-LFB assay during the amplification stage, and four different reaction times (10, 15, 20, and 25 min) were compared at 65°C according to the standard MCDA conditions. The lowest genomic DNA level (10 fg of *Shigella* templates per tube) showed two red bands when the reaction only lasted for 20 min at 65°C (**Figure 6**). A reaction time of 20 min was used as the optimal time for the MCDA-LFB assay during the reaction stage. Hence, the whole procedure, including specimen (such as fecal sample) processing (35 min), isothermal reaction (20 min), and result reporting (5 min), could be completed within 60 min.

The Analytical Specificity of MCDA-LFB Assay

The analytical specificity of the MCDA-LFB technique was evaluated by MCDA-LFB amplification of genomic DNA extracted from 23 *Shigella* strains and 37 non-*Shigella* strains (roughly 10 ng of genomic templates for each pathogen). The detection was positive only for the four *Shigella* species, and was negative for non-*Shigella* species and blank control (**Figure 7**). As shown in **Figure 7**, two red bands, including TL and CL, appeared on the biosensor from the positive test, and only a red band at the control line appeared, indicating negative results for non-*Shigella* strains and blank control. The results demonstrated that the MCDA-LFB assay has a 100% analytical specificity for *Shigella* detection.

MCDA-LFB Assay for Artificially Contaminated Fecal Samples

In order to determine the suitability of the MCDA-LFB assay as a nucleic acid detection tool, the MCDA-LFB approach was examined by the artificially inoculating *Shigella* strains into human fecal samples. As shown in **Figure 8A**, the MCDA-LFB assay could generate positive results when the contaminated numbers of *Shigella* were more than 1.42×10^3 CFU/g (~5.68 CFU/reaction). The MCDA-LFB approach produced the negative results when the contaminated numbers of *Shigella* were lower than 1.42×10^2 CFU/g (~0.568 CFU/reaction). Only a red band at the control line appeared, indicating negative results for

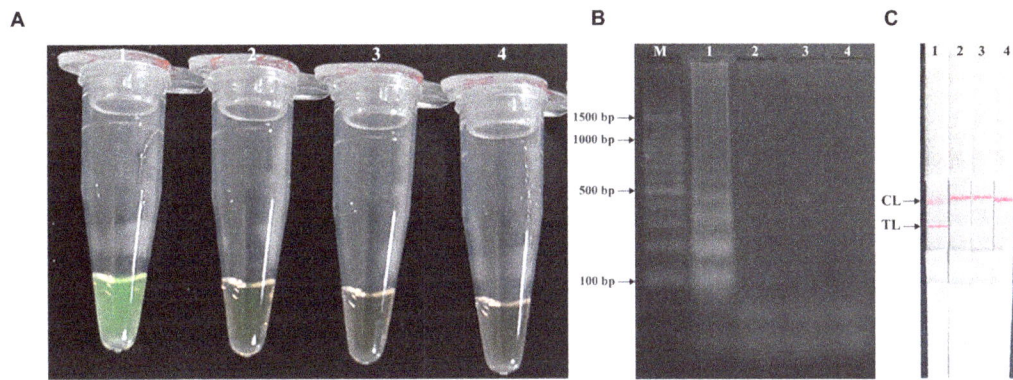

FIGURE 3 | Detection and confirmation of *Shigella*-MCDA products. (A) Amplification products of *Shigella*-MCDA assay were visually detected by observation of the color change: tube 1, positive amplification of *Shigella flexneri* strain (ICDC-NPS001); negative control of *Listeria monocytogenes* strain (ATCC19114); negative control of *Salmonella* strain (ICDC-NPSa001); blank control (DW). (B) Agarose gel electrophoresis of *Shigella*-MCDA products was shown: lane M, DNA marker DL100; lane 1, *Shigella*-MCDA products of *Shigella flexneris* (ICDC-NPS001); lane 2, negative control (*Listeria monocytogenes*, ATCC19114); negative control (*Salmonella*, ICDC-NPSa001); lane 4, blank control (DW); (C) Lateral flow biosensor applied for visual detection of *Shigella* MCDA products: strip 1, positive amplification of *Shigella flexneris* (ICDC-NPS001); strip 2, negative control (*Listeria monocytogenes*, ATCC19114); strip 3, negative control (*Salmonella*, ICDC-NPSa001); strip 4, blank control (DW).

negative control and blank control. Thus, the LoD of MCDA-LFB method was 5.68 CFU per tube, which was consistent with MCDA-FD, MCDA-turbidity and MCDA-gel electrophoresis assays (**Figures 8B,D,E**). In contrast, the analytical sensitivity of ET-MCDA assay for detection of *Shigella* in fecal samples was also 5.68 CFU per reaction, which was as sensitive as MCDA-LFB detection (**Figures 8A,C**). The results indicated that the analytical sensitivity of MCDA-LFB assay was in complete accordance with MCDA-FD, MCDA-turbidity, MCDA-gel electrophoresis and ET-MCDA assays.

DISCUSSION

Species of the genus *Shigella* were the causative agents of shigellosis or "bacillary dysentery," and responsible for 5–15% of all diarrheal episodes worldwide, disproportionately affecting children 5 years of age living in developing countries (Von Seidlein et al., 2006; Schroeder and Hilbi, 2008). Thus, a simple, rapid and accurate detection assay, which can be used in clinical laboratories, primary care facilities and resource-poor settings, is necessary. In this study, we successful developed a MCDA-LFB technique for simple, rapid, sensitive and specific detection of *Shigella* spp. as a valuable screening tool. Comparing with the currently existent PCR-based technologies, the MCDA-LFB assay during the reaction stage was preceded at a uniform temperature, alleviating the use of a sophisticated thermal cycling instrument, and only a water bath or heat block was need to conduct the reaction. Hence, the MCDA-LFB method developed here had the potential for point-of-care testing, field detection, 'on-site' diagnosis and more. Furthermore, only a reaction time of 20 min was required for the MCDA-LFB assay during the amplification stage. Consequently, the entire procedure, including specimen (such as stool sample) processing (35 min), isothermal reaction (20 min), and result reporting (5 min), could be completed within

60 min (**Figure 6**). The rapid detection of *Shigella* was valuable for determining the choice of treatment in clinical laboratories, especially in acute-care settings.

In the MCDA assay, CP1 and D1 primers, which involved in isothermal amplification, were labeled at the 5′ end with FITC and biotin, respectively (**Figure 1**). During the amplification stage, the double-labeled detectable amplicons were constructed, which were generated from FITC-labeled CP1 primers and biotin-labeled D1 primers. The end of the detectable products labeled with FITC could be captured by the anti-FITC body fixed on the first line of the biosensor, known as the test line. The other end of the amplicons labeled with biotin could bind streptavidin-conjugated gold nanoparticles for visualization. The excess streptavidin-conjugated color particles were captured by biotinylated bovine serum albumin located on the second line of strip, known as the control line, which validated the working condition of the biosensor. Importantly, the test results are displayed as colored bands visible by the naked eye about 2 min, thus the whole process of detection could be finished within 5 min.

In the MCDA-LFB system, the interpretation of test results is based on the appearance of red bands on the reaction pad. The presence of two red lines (TL and CL) on the biosensor indicated a positive result for *Shigella*, whereas only a red line appeared in the CL zone, indicating the negative result, negative control and blank control. Several other monitoring techniques, including colorimetric indicator (such as FD reagent), real time turbidity, gel electrophoresis and fluorescence detection, were employed to analyze the MCDA products. Firstly, the assessment of color shift with naked eye was potentially subjective, thus there was the possibility that a sample was somewhat ambiguous to the unaided eye when the concentration of target sequences was low. Secondly, due to use of ten primers, MCDA could produce a complex mixture of various amplicons, and thus these detection techniques (such as colorimetric indicator, real time

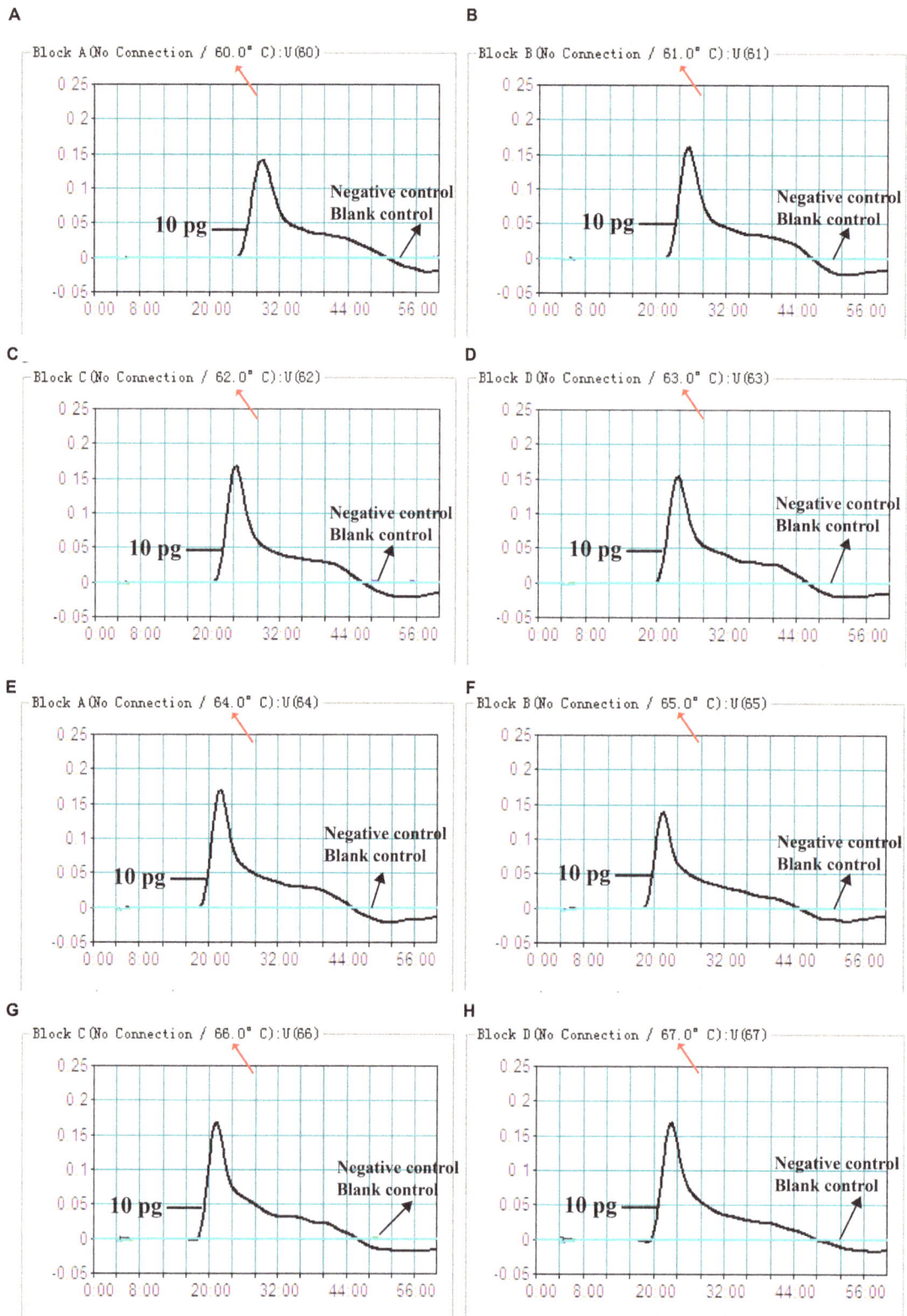

FIGURE 4 | The optimal amplification temperature for *Shigella*-MCDA primer sets. The standard MCDA reactions for detection of *Shigella* were monitored by real-time measurement of turbidity and the corresponding curves of concentrations of DNA were marked in the figures. The threshold value was 0.1 and the turbidity of >0.1 was considered to be positive. Eight kinetic graphs **(A–H)** were generated at various temperatures (60–67°C, 1°C intervals) with target pathogens DNA at the level of 10 pg per reaction. The graphs from **(D–H)** showed robust amplification.

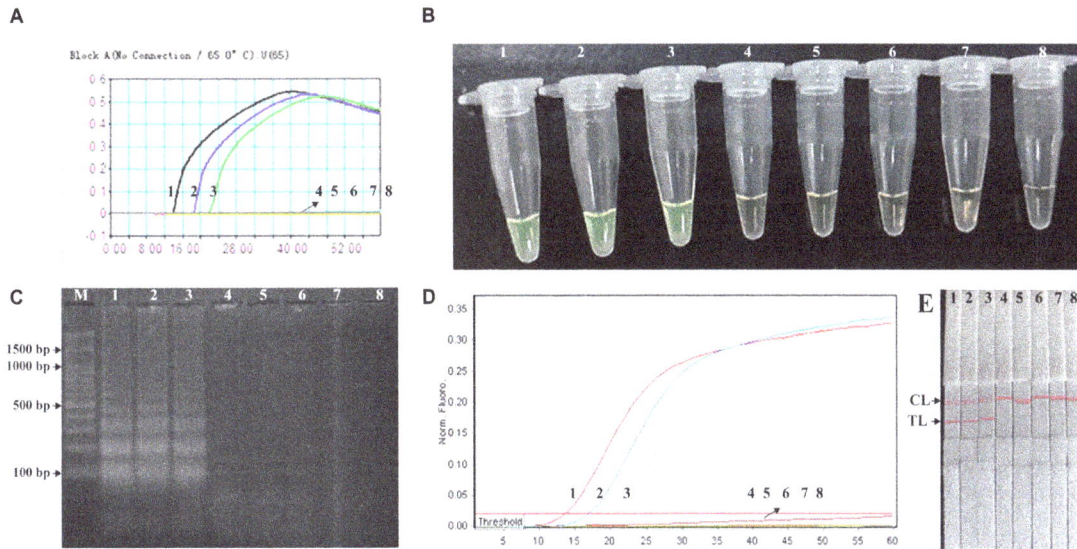

FIGURE 5 | Analytical sensitivity of *Shigella*-MCDA assay using serially diluted genomic DNA with *Shigella flexneris* strain ICDC-NPs001. Five monitoring techniques, including real time turbidity **(A)**, colorimetric indicator (FD) **(B)**, gel electrophoresis **(C)**, real time fluorescence **(D)** and lateral flow biosensor **(E)**, were applied for analyzing the amplification products. The serial dilutions (10 ng, 10 pg, 10 fg, 1 fg, and 0.1 fg) of target templates were subjected to standard MCDA or ET-MCDA reactions. Turbidity signals **(A)**/Tubes **(B)**/Lanes **(C)**, Fluorescence signals **(D)**/Strips **(E)** 1–8 represented the DNA levels of 10 ng, 10 pg, 10 fg, 1 fg, and 0.1 fg per reaction, negative control (10 pg of *Listeria monocytogenes* genomic DNA), negative control (10 pg of *Salmonella* genomic DNA) and blank control (DW). The genomic DNA levels of 10 ng, 10 pg, and 10 fg per reaction produced the positive reactions.

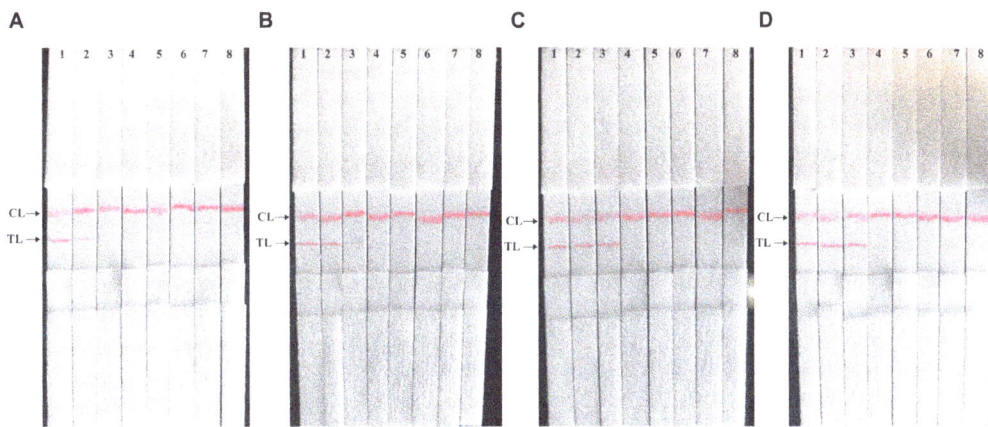

FIGURE 6 | The optimal duration of time required for MCDA-LFB assay. Four different reaction times (**A**, 10 min; **B**, 15 min; **C**, 20 min; **D**, 25 min) were examined and compared at 65°C. Strips 1, 2, 3, 4, 5, 6, 7, and 8 represent DNA levels of 10 ng of *Shigella* templates, 10 pg of *Shigella* templates, 10 fg of *Shigella* templates, 1 fg of *Shigella* templates, 0.1 fg *Shigella* templates per tube, negative control (*L. monocytogenes*, 10 pg per reaction), negative control (*Salmonella*, 10 pg per reaction) and blank control (DW). The best sensitivity was seen when the amplification lasted for 20 min (**C**).

turbidity and gel electrophoresis) could not distinguish the non-specific and specific products (Ge et al., 2013). Furthermore, these detection methods required a post detection procedure (gel electrophoresis), turbidimeter (real time turbidity detection), or a fluorescence instrument (real time fluorescence detection), and the resultant instrumental restraint could hamper the uptake of MCDA analysis in point-of use and field settings. In our report, the MCDA technique coupled a lateral flow strip offered a simple, rapid, cost-effective and nearly instrument-free platform for molecular testing with easily interpretable results. Moreover,

the proof-of-concept method may be reconfigured to detect a wide variety of nucleic acid sequences by re-designing the specific MCDA primers.

The newly developed MCDA-LFB approach could detect as little as 10 fg of *Shigella* DNA per reaction in pure culture and 5.86 CFU per tube in human fecal samples, and the results were further confirmed by FD, real time turbidity, gel electrophoresis and real time fluorescence detection (**Figure 5** and **8**). The results showed the LFB technique was as sensitive as FD, real time turbidity, gel electrophoresis and real time florescence detection.

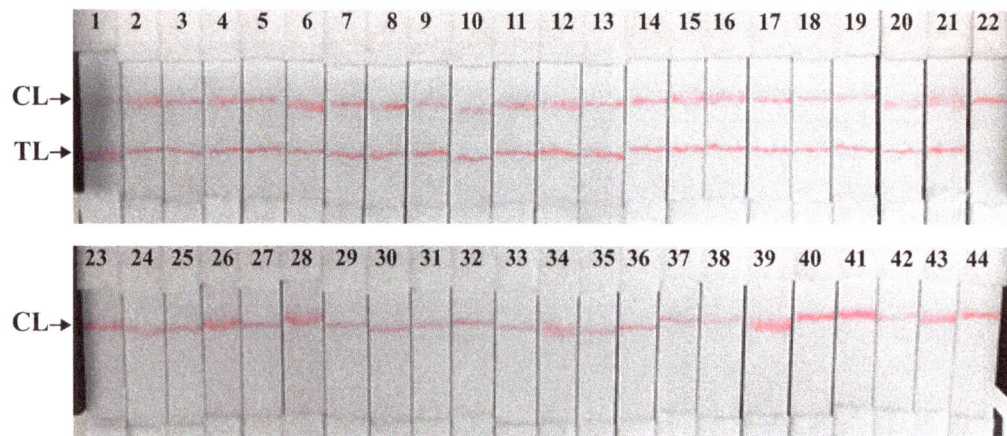

FIGURE 7 | The specificity of MCDA-LFB assay for different strains. The MCDA reactions were conducted using different genomic DNA templates and were monitored by means of visual format. Biosensor 1–18, *Shigella flexneri* strains of serovar 1d (ICDC-NPS001), 4a (ICDC-NPS002), 5a (ICDC-NPS003), 2b (ICDC-NPS004), 1b (ICDC-NPS005), 3a (ICDC-NPS006), 4av (ICDC-NPS007), 3b (ICDC-NPS008), 5b (ICDC-NPS007), Y (ICDC-NPS010), Yv (ICDC-NPS011), 1a (ICDC-NPS012), X (ICDC-NPS013), Xv (ICDC-NPS014), F6 (ICDC-NPS015), 7b (ICDC-NPS016), 2a$_1$ (ICDC-NPS017), 4b (ICDC-NPS018); biosensor 19–21, *Shigella boydii*, *Shigella sonneri* and *Shigella dysenteriae*; biosensor 22–43, Enteropathogenic *E. coli*, Enterotoxigenic *E. coli*, Enteroaggregative *E. coli*, Enteroinvasive *E. coli*, Enterohemorrhagic *E. coli*, *Plesiomonas shigelloides*, *Campylobacter jejuni*, *Enterobacter cloacae*, *Enterococcus faecalis*, *Enterococcus faecium*, *Yersinia enterocolitica*, *Streptococcus pneumonia*, *Aeromonas hydrophil*, *Vibrio vulnificus*, *Vibrio fluvialis*, *Vibrio parahaemolyticus*, *Klebsiella pneumonia*, *Bntorobater sakazakii*, *Bacillus cereus*, *Listeria grayii*, *Listeria welshimeri*, and *Listeria ivanovii*; biosensor 44, blank control (DW).

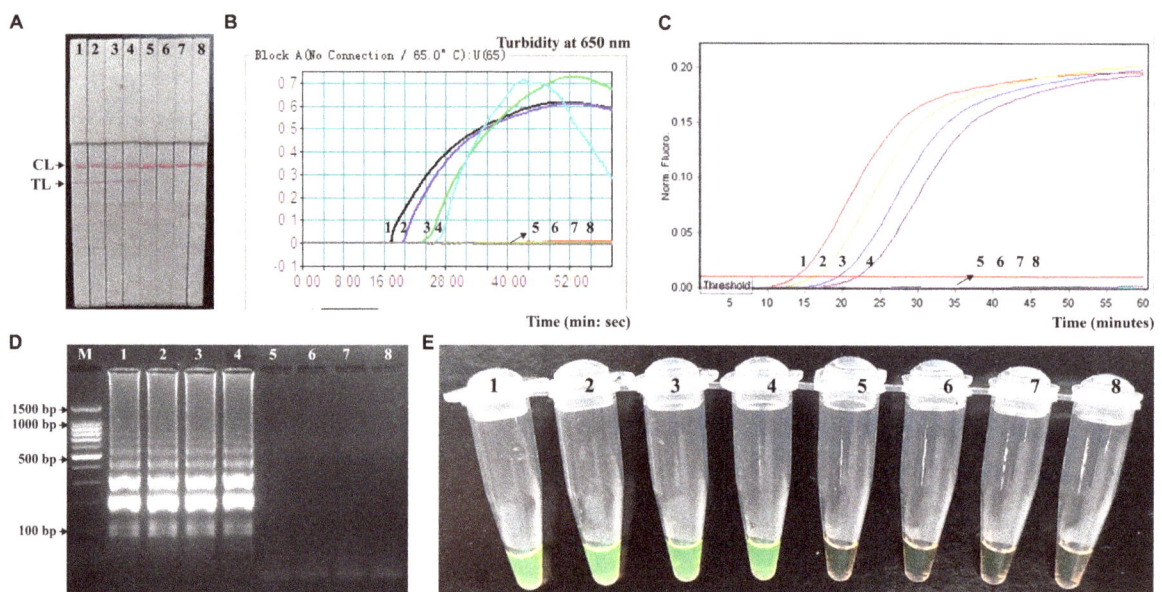

FIGURE 8 | Analytical sensitivity of MCDA-LFB for detecting target pathogens in artificially contaminated fecal samples. Five monitoring techniques, including lateral flow biosensor **(A)**, real time turbidity **(B)**, real time fluorescence **(C)**, gel electrophoresis **(D)**, and colorimetric indicator (FD) **(E)**, were applied for analyzing the amplification products. The serial dilutions of target templates were subjected to standard MCDA or ET-MCDA reactions. Strips **(A)**/Turbidity signals **(B)**/Fluorescence signals **(C)**/Lanes **(D)**/Tubes **(E)** 1–8 represented the DNA levels of 5860 CFU, 586 CFU, 58.6 CFU, 5.86 CFU, 0.586 CFU and 0.0586 CFU per reaction, negative control (non-contaminated fecal sample) and blank control (DW). The genomic DNA levels of 5860 CFU, 586 CFU, 58.6 CFU and 5.86 CFU, per reaction produced the positive reactions.

Due to negate the need for special reagents, electrophoresis and amplicon detection equipment, the MCDA-LFB assay was more suitable than other MCDA-based methods for simple, rapid and specific detection in a variety of fields with short turnaround times. Moreover, the use of the ten specific primers targeting the *ipaH* gene (*Shigella* spp.-specific gene) provides a high degree of specificity for nucleic acid amplification, and the analytical specificity was successfully assessed in this study (**Figure 7**). The detection was positive only for the four *Shigella* species, and was negative for non-*Shigella* species and blank control. Hence, the MCDA-LFB assay offered a high degree of selectivity for detecting *Shigella*.

CONCLUSION

A reliable MCDA-LFB technique was successfully devised for detection of *Shigella* app. causing severe diarrhea in both developed and developing countries, which could achieve the infection control, clinical care and epidemiologic investigations. The MCDA-LFB assay reported here is simple, sensitive and specific, and did not require special reagents and expensive apparatus. The use of the newly designed biosensor could offer a rapid, objective and easily interpretable readout of the assay's results. Therefore, the *Shigella*-MCDA-LFB assay was especially useful in field, point-of-care and resource-limited settings. Furthermore, the proof-of-concept technique (MCDA-LFB) may be reconfigured to detect a wide variety of nucleic acid sequences by re-designing the specific MCDA primers.

REFERENCES

Baker, K. S., Dallman, T. J., Ashton, P. M., Day, M., Hughes, G., Crook, P. D., et al. (2015). Intercontinental dissemination of azithromycin-resistant shigellosis through sexual transmission: a cross-sectional study. *Lancet Infect. Dis.* 15, 913–921. doi: 10.1016/S1473-3099(15)00002-X

Chua, A., Yean, C. Y., Ravichandran, M., Lim, B., and Lalitha, P. (2011). A rapid DNA biosensor for the molecular diagnosis of infectious disease. *Biosens. Bioelectron.* 26, 3825–3831. doi: 10.1016/j.bios.2011.02.040

Echeverria, P., Sethabutr, O., and Pitarangsi, C. (1991). Microbiology and diagnosis of infections with *Shigella* and enteroinvasive *Escherichia coli*. *Rev. Infect. Dis.* 13(Suppl. 4), S220–S225. doi: 10.1093/clinids/13.Supplement_4.S220

Ge, Y., Wu, B., Qi, X., Zhao, K., Guo, X., Zhu, Y., et al. (2013). Rapid and sensitive detection of novel avian-origin influenza A (H7N9) virus by reverse transcription loop-mediated isothermal amplification combined with a lateral-flow device. *PLoS ONE* 8:e69941. doi: 10.1371/journal.pone.0069941

Haley, C. C., Ong, K. L., Hedberg, K., Cieslak, P. R., Scallan, E., Marcus, R., et al. (2010). Risk factors for sporadic shigellosis, FoodNet 2005. *Foodborne Pathog. Dis.* 7, 741–747. doi: 10.1089/fpd.2009.0448

Khan, W. A., Griffiths, J. K., and Bennish, M. L. (2013). Gastrointestinal and extra-intestinal manifestations of childhood shigellosis in a region where all four species of *Shigella* are endemic. *PLoS ONE* 8:e64097. doi: 10.1371/journal.pone.0064097

Koh, X. P., Chiou, C. S., Ajam, N., Watanabe, H., Ahmad, N., and Thong, K. L. (2012). Characterization of *Shigella* sonnei in Malaysia, an increasingly prevalent etiologic agent of local shigellosis cases. *BMC Infect. Dis.* 12:122. doi: 10.1186/1471-2334-12-122

Liao, S.-C., Peng, J., Mauk, M. G., Awasthi, S., Song, J., Friedman, H., et al. (2016). Smart cup: a minimally-instrumented, smartphone-based point-of-care molecular diagnostic device. *Sens. Actuators B Chem.* 229, 232–238. doi: 10.1016/j.snb.2016.01.073

Mandal, P., Biswas, A., Choi, K., and Pal, U. (2011). Methods for rapid detection of foodborne pathogens: an overview. *Am. J. Food Technol.* 6, 87–102. doi: 10.3923/ajft.2011.87.102

McKillip, J. L., and Drake, M. (2004). Real-time nucleic acid–based detection methods for pathogenic bacteria in food. *J. Food Prot.* 67, 823–832.

Njuguna, H. N., Cosmas, L., Williamson, J., Nyachieo, D., Olack, B., Ochieng, J. B., et al. (2013). Use of population-based surveillance to define the high incidence of shigellosis in an urban slum in Nairobi, Kenya. *PLoS ONE* 8:e58437. doi: 10.1371/journal.pone.0058437

Nygren, B., Schilling, K., Blanton, E., Silk, B., Cole, D., and Mintz, E. (2013). Foodborne outbreaks of shigellosis in the USA, 1998–2008. *Epidemiol. Infect.* 141, 233–241. doi: 10.1017/S0950268812000222

Schroeder, G. N., and Hilbi, H. (2008). Molecular pathogenesis of *Shigella* spp.: controlling host cell signaling, invasion, and death by type III secretion. *Clin. Microbiol. Rev.* 21, 134–156. doi: 10.1128/CMR.00032-07

Thiem, V. D., Sethabutr, O., von Seidlein, L., Van Tung, T., Chien, B. T., Lee, H., et al. (2004). Detection of *Shigella* by a PCR assay targeting the ipaH gene suggests increased prevalence of shigellosis in Nha Trang, Vietnam. *J. Clin. Microbiol.* 42, 2031–2035. doi: 10.1128/JCM.42.5.2031-2035.2004

Vikesland, P. J., and Wigginton, K. R. (2010). Nanomaterial enabled biosensors for pathogen monitoring-a review. *Environ. Sci. Technol.* 44, 3656–3669. doi: 10.1021/es903704z

Villalobo, E., and Torres, A. (1998). PCR for detection of *Shigella* spp. in mayonnaise. *Appl. Environ. Microbiol.* 64, 1242–1245.

Von Seidlein, L., Kim, D. R., Ali, M., Lee, H., Wang, X., Thiem, V. D., et al. (2006). A multicentre study of *Shigella* diarrhoea in six Asian countries: disease burden, clinical manifestations, and microbiology. *PLoS Med.* 3:e353. doi: 10.1371/journal.pmed.0030353

Wang, Y., Wang, Y., Lan, R., Xu, H., Ma, A., Li, D., et al. (2015a). Multiple endonuclease restriction real-time loop-mediated isothermal amplification: a novel analytically rapid, sensitive, multiplex loop-mediated isothermal amplification detection technique. *J. Mol. Diagn.* 17, 392–401. doi: 10.1016/j.jmoldx.2015.03.002

Wang, Y., Wang, Y., Luo, L., Liu, D., Luo, X., Xu, Y., et al. (2015b). Rapid and Sensitive Detection of *Shigella* spp. and *Salmonella* spp. by multiple endonuclease restriction real-time loop-mediated isothermal amplification technique. *Front. Microbiol.* 6:1400. doi: 10.3389/fmicb.2015.01400

Wang, Y., Wang, Y., Ma, A. J., Li, D. X., Luo, L. J., Liu, D. X., et al. (2015c). Rapid and sensitive isothermal detection of nucleic-acid sequence by multiple cross displacement amplification. *Sci. Rep.* 5:11902. doi: 10.1038/srep11902

Wang, Y., Wang, Y., Zhang, L., Li, M., Luo, L., Dongxin, L., et al. (2016a). Endonuclease restriction-mediated real-time polymerase chain reaction: a novel technique for rapid, sensitive and quantitative detection of nucleic-acid sequence. *Front. Microbiol.* 7:1104. doi: 10.3389/fmicb.2016.01104

Wang, Y., Wang, Y., Zhang, L., Liu, D., Luo, L., Li, H., et al. (2016b). Multiplex, rapid and sensitive isothermal detection of nucleic-acid sequence by endonuclease restriction-mediated real-time multiple cross displacement amplification. *Front. Microbiol.* 7:753. doi: 10.3389/fmicb.2016.00753

AUTHOR CONTRIBUTIONS

Conceived and designed the experiments: YiW, JX and CY. Performed the experiments: YiW and YaW. Analyzed the data: YiW. Contributed reagents/materials/analysis tools: YiW, YaW, JX, and CY. Designed the software used in the analysis: YiW. Wrote the manuscript: YiW, JX, and CY.

FUNDING

We acknowledge the financial supports of the grant (Mega Project of Research on the Prevention and Control of HIV/AIDS, Viral Hepatitis Infectious Diseases 2013ZX10004-101 to CY) from the Ministry of Science and Technology, People's Republic of China, and grant (2015SKLID507 to CY) from State Key Laboratory of Infectious Disease Prevention and Control, China CDC.

Warren, B., Parish, M., and Schneider, K. (2006). *Shigella* as a foodborne pathogen and current methods for detection in food. *Crit. Rev. Food Sci. Nutr.* 46, 551–567. doi: 10.1080/10408390500295458

Zhao, Y., Chen, F., Li, Q., Wang, L., and Fan, C. (2015). Isothermal amplification of nucleic acids. *Chem. Rev.* 115, 12491–12545. doi: 10.1021/acs.chemrev.5b00428

Disclosures: YW and CY have filed for a patent from the State Intellectual Property Office of the People's Republic of China, which covers the novel assay and sequences included in this manuscript (Application number CN201610942289.8).

Conflict of Interest Statement: The authors declare that the research was conducted in the absence of any commercial or financial relationships that could be construed as a potential conflict of interest.

16

Power and Time Dependent Microwave Assisted Fabrication of Silver Nanoparticles Decorated Cotton (SNDC) Fibers for Bacterial Decontamination

Abhishek K. Bhardwaj[1,2]*, Abhishek Shukla[2], Rohit K. Mishra[3]*, S. C. Singh[2,4], Vani Mishra[3], K. N. Uttam[2], Mohan P. Singh[5], Shivesh Sharma[3] and R. Gopal[2]*

[1] Centre for Environmental Science, University of Allahabad, Allahabad, India, [2] Laser Spectroscopy and Nanomaterials Lab., Department of Physics, University of Allahabad, Allahabad, India, [3] Centre for Medical Diagnostic and Research, Motilal Nehru National Institute of Technology, Allahabad, India, [4] High-Intensity Femtosecond Laser Laboratory, The Institute of Optics, University of Rochester, Rochester, NY, USA, [5] Centre of Biotechnology, University of Allahabad, Allahabad, India

Edited by:
Jayanta Kumar Patra,
Dongguk University, South Korea

Reviewed by:
Shanmugharaj A. M.,
Kyung Hee University, South Korea
Amit Kumar Mandal,
Raiganj University, India
Ramaraju Bendi,
Nanyang Technological University,
Singapore
Pranjal Chandra,
Indian Institute of Technology
Guwahati, India

*Correspondence:
Abhishek K. Bhardwaj
bhardwajak87@gmail.com
Rohit K. Mishra
rohit_ernet@yahoo.co.in
R. Gopal
rgopal.prof@gmail.com

Plasmonic nanoparticles (NPs) such as silver and gold have fascinating optical properties due to their enhanced optical sensitivity at a wavelength corresponding to their surface plasmon resonance (SPR) absorption. Present work deals with the fabrication of silver nanoparticles decorated cotton (SNDC) fibers as a cheap and efficient point of contact disinfectant. SNDC fibers were fabricated by a simple microwave assisted route. The microwave power and irradiation time were controlled to optimize size and density of silver nanoparticles (SNPs) on textile fibers. As prepared cotton fabric was characterized for ATR-FTIR, UV-VIS diffuse reflectance, SEM and TEM investigations. Size of SNPs as well as total density of silver atoms on fabric gets increased with the increase of microwave power from 100 W to 600 W. The antibacterial efficacy of SNPs extracted from SNDC fibers was found to be more effective against Gram-negative bacteria than Gram-positive bacteria with MIC 38.5 ± 0.93 μg/mL against *Salmonella typhimurium* MTCC-98 and 125 ± 2.12 μg/mL against *Staphylococcus aureus* MTCC-737, a linear correlation coefficient with R^2 ranging from ~0.928–0.935 was also observed. About >50% death cells were observed through Propidium Iodide (PI) internalization after treatment of SNPs extracted from SNDC fibers with concentration 31.25 μg/mL. Generation of ROS and free radical has also been observed which leads to cell death. Excellent *Escherichia coli* deactivation efficacy suggested that SNDC fibers could be used as potentially safe disinfectants for cleaning of medical equipment, hand, wound, water and preservation of food and beverages.

Keywords: silver nanoparticles decorated cotton (SNDC) fibers, microwave irradiation, SNPs functionalized textile, antibacterial efficacy, cell death, ROS

INTRODUCTION

Silver nanoparticles (SNPs) are extensively employed in healthcare and pharmaceutical products that include coating material for medical devices, orthopedic or dental graft materials, air/water filters, food containers, topical aids for wound repair, clothing and textile fabrics (Piccinno et al., 2012; Bendi et al., 2015). Since ancient time silver is being used as an effective antimicrobial agent, the characteristic feature that has been accredited to SNPs with the advent of nanotechnology enhanced its antimicrobial activities by several orders (Davies and Etris, 1997; Swarnkar et al., 2016; Mahmoud et al., 2016). However, increasing interaction of SNPs with human and the environment has been an important issue of concern as it has been found to cause several undesired problems for the later. The NPs could migrate from textile to human sweat and increase its exposure to skin affecting human physiology (Windler et al., 2012). Their release into waterways could adversely affect the aquatic life (Mahmoudi and Serpooshan, 2012). Also, high surface energies of SNPs tend to agglomerate resulting in application difficulties (Grumezescu et al., 2013). This is an important element to immobilize SNPs on a matrix system that would allow an efficient and effective disinfectant by preventing its aggregation and thus reducing its migration plausible threat to the environment (Agnihotri et al., 2013; Zafar et al., 2014).

Researchers have rigorously employed flexible substrates such as fabrics, plastics, textiles, and papers for NPs immobilization. SNPs specifically have been incorporated into a range of cellulosic materials, such as bacterial cellulose, cotton fabric, filter paper, and cellulose gels (He et al., 2003; Maneerung et al., 2008; Ferraria et al., 2009; Dankovich and Gray, 2011; Bendi and Imae, 2013). In this context, cotton owing to its natural softness, permeability, high moisture absorbency, mechanical strength, renewable and heat retaining properties (Ravindra et al., 2010; Tang et al., 2012) has been recently employed as the most successful material. The super-hydrophobic surface on cotton fabric guarantees its dryness and cleanness which are considered as desired features, in particular on its outside facet (Hoefnagels et al., 2007; Xu and Cai, 2008; Bae et al., 2009; Gao et al., 2009; Hao et al., 2010; Xu et al., 2010; Berendjchi et al., 2011). Cotton fabric is an ideal place for settling and growing pathogenic bacteria because of its porous and hydrophilic structure.

To prepare antimicrobial silver-treated cotton fabrics, most of the recent research activities are concerned on preparations of SNPs with controlled size and developing routes to impart SNPs to cotton fabrics. Traditionally, in preparing SNP, numerous reducing agents, such as sodium borohydride (NaBH$_4$), formaldehyde, sodium citrate, hydrazine, ascorbic acid, glucose and γ-ray or UV irradiation, were utilized to reduce the silver cations, while some polymeric materials, such as poly(vinylpyrrolidone) (PVP), poly(ethylene glycol) (PEG), and some surfactants were used as stabilizers to prevent NPs agglomeration and precipitation (Luo et al., 2005; Stevenson et al., 2012). In addition, during the process of antimicrobial finishing on cotton fabrics with SNPs, a binders, such as dimethyloldihydroxyethylene urea (DMDHEU), polyurethane resin, poly(2-aminoethyl methacrylate) (PAEMA) Polyacrylic

esters (PALS), and PDDA is required to fix the SNPs on the fibers to provide durability of antimicrobial properties (Zhang et al., 2009; Liu et al., 2014). Nevertheless, the synthesis of a monodisperse and stable SNPs suspension is challenging and may go through tedious and complex procedures, which may hinder the practical applications of SNPs on textiles. The size of the SNPs can be controlled by the concentration of silver nitrate and reducing agent, temperature, and duration of reaction (Dankovich, 2014; Liu et al., 2014).

In this paper, we report *in situ* fabrication of SNPs impregnated cotton fiber by domestic microwave irradiation. The bactericidal efficacy of the SNDC fibers was tested against both Gram positive and Gram negative bacteria. This study reveals that the SNDC fibers might be successfully employed to small scale system for point of use water decontamination, surface sterilization of medical or other equipment, and wound healing. This is a cheap, portable, eco-friendly, and point of use system for contact killing of microbes and disinfection of drinking water.

MATERIALS AND METHODS

Procurement of Chemicals and Culture Media

Analytical grade chemicals such as Silver nitrate (AgNO$_3$ 99.98%), trisodium citrate (~99%) and poly (diallyldimethylammonium chloride) (PDDA) (M$_w$ = 200000–350000 g.mol^{-1}), and Milli-Q grade water were purchased from Merk, medical cotton and cotton bandage (made Krishna Handloom Pvt. Ltd, India) as well as nutrient agar (NA), Mueller Hinton Broth (MHB) and Mueller Hinton Agar (MHA) were procured from Himedia Pvt. Ltd for antibacterial assay. For cell permeability propidium iodide (PI) was purchased from Sigma Aldrich, Pvt. Ltd.

Selected Bacterial Pathogens

Three bacterial pathogens *Escherichia coli* (MTCC-723); *Staphylococcus aureus* (MTCC-737) and *Salmonella typhimurium* (MTCC-98) were procured from microbial type culture collection (MTCC), Chandigarh, India. The cultures were maintained on NA slants at 4°C throughout the study used as stock.

Fabrication of Silver Nanoparticles Decorated Cotton (SNDC) Fibers

In a typical synthesis procedure, cotton pieces (7 cm × 7 cm) were dipped in boiling double distilled water for 4 h to remove impurity and were transferred into the oven at 50°C for 6 h for drying. This cleaned cotton was immersed into the aqueous solution of 2.5 M PDDA for 12 h and dried at room temperature. After that PDDA modified cotton soaked in 2 M trisodium citrate. Wet cotton pieces were squeezed gently for the removal of extra trisodium citrate followed by addition of 6 ml, 0.01 M aqueous silver nitrate solution. For microwave assisted decoration of cotton fiber by SNPs, they were placed in domestic microwave oven (Sharp model no MW73V/XT, 2.45 GHz, 800 W) with rotating disk. The density of SNPs loaded on fiber and their sizes were varied with time (60–180 s) and power (100–600 W) of microwave irradiation. In order to avoid over heating/burning of cotton, the microwave was switched off after every 60 s of irradiation. In order to remove excess un-reacted silver precursor and loosely bound SNPs, treated cottons were kept in warm water for 2 h followed by drying in hot air oven at 45°C for 8 h. A mechanism for the fabrication of SNDC fibers is depicted in (**Figure 1**).

Characterization of Silver Nanoparticles Decorated Cotton Fibers

Qualitative imaging of SNDC fiber samples was done by standard photography and microscopy (magnification 200 X by digital microscope) which show the change in color from white to yellow/orange, an indication of SNPs adsorbed on the surface of the cotton fiber. The density of SNPs on cotton fibers was measured from diffused UV-VIS reflectance spectra recorded using PerkinElmer Lambda 35 double beam spectrophotometer equipped with Labsphere RSA-PE-20 diffused reflectance accessory with barium sulfate as white standard. ATR-FTIR spectra of SNDC fiber samples were recorded using ABB MB3000 series FTIR spectrometer (ABB, Bomem Inc. Canada), Tecnai G2-20 (FEI Company, Netherland), high resolution transmission electron microscope (HRTEM) operating at 200 kV and field emission scanning electron microscope (FESEM, Nova NanoSEM 450 series)was used for size and shape measurements of as produced SNPs.

Extraction of Silver Nanoparticles from SNDC Fibers

PDDA polymer is responsible to provide good binding affinity between cotton fibers and SNPs. We used normal SNDC fibers (PDDA untreated SNDC) for SNPs extraction, due to its lower binding affinity with SNPs and higher particle release capacity in compare to PDDA treated SNDC fibers. PDDA untreated cotton were used to prepare SNDC (microwave irradiation 600 W, 1 min) fallowed by their transfer into 100 mL warm deionized water (70°C) and continuous shaking for four hrs consequently SNPs was release in the water medium and form colloidal SNPs. Extracted colloidal solution of SNPs were used for UV-VIS absorption, TEM imaging and flowcytometry investigations.

Antibacterial Susceptibility of Extracted SNPs

Silver nanoparticless extracted from SNDC fibers were used for determining antibacterial activity as per broth microdilution method described by Mishra et al. (2016a). Briefly, SNPs solution extracted from the SNDC fibers is used as stock solution with 20 mg/ml, which was then serially diluted 1:10 in the medium in order to attain final concentration ranging from 500 to 3.9 µg/mL. Each well was subsequently filled with 100 µl of inoculum. The initial concentration of inoculum was 1×10^6 cells/ml (adjusted according to 0.5 McFarland) in MHB. The plates were stored at 35 ± 2°C in a wet chamber for 24 h and experiments were conducted in triplicate.

Quantification of Bacterial Growth

After 24 h of incubation, the optical density of the microtiter plates was recorded spectrophotometrically at 492 nm using SpectraMaxplus[384] (Molecular Devices, USA). The changes in OD over time were used to generate growth curves at each drug concentration against the control. The normalized OD of the SNPs treated wells (OD obtained after subtraction of the background OD) was used for the generation of turbidimetric growth curves. Percentage of growth inhibition at each drug concentration was calculated using the formula:

$$\text{Growth inhibition (\%)} = \frac{OD_{492} \text{ of wells containing the Ag NPs soln.}}{OD_{492} \text{ of well without treatment}} \times 100 \quad (1)$$

Determination of Minimum Inhibitory Concentrations (MICs)

Minimum inhibitory concentrations were determined spectrophotometrically using the software SoftMax® Pro-5 (Molecular Devices, USA). For the SNPs solution, the MIC was determined as the lowest nanoparticle concentration showing $\geq 70\%$ growth inhibition compared with the growth in the treatment-free well. Each test was performed in triplicates.

Determination of Minimum Bactericidal Concentration (MBC)

Minimum bactericidal concentration was defined as the lowest concentration of the SNPs at which 99.99% or more of the initial inoculum was killed. Hundred microliter aliquot of inoculum was taken aseptically from those wells that did not show turbidity and it was poured on MHA plates followed by incubation for 24 h at 35 ± 2°C. The absence of growth reflected that the concentration was lethal. The number of surviving organisms was determined by viability counts. All tests were performed in triplicates.

Flowcytometric Analysis for Plasma Membrane Permeabilization and ROS Detection

The measurement of membrane permeability of selected bacterial pathogens treated with SNPs were observed through flowcytometer using fluorescence dye i.e., PI and Protein leakage. The selected bacterial inocula 1×10^6 cells/ml were prepared and adjusted according to 0.5 McFarland. Each bacterial cell suspension was incubated with various concentrations 15.62 and 31.25 µg/mL of SNPs for 24 h at 35 ± 2°C. Following incubation, cells were washed and resuspended in PBS and subsequently stained with PI (20 µg/ml) for 20 min in a dark chamber. The cells were analyzed by BD Accuri C6 (Becton Dikinson, San Jose, CA, USA). Intrinsic parameter (SSC-A) and fluorescence in FL-2 channel for PI were acquired and recorded over the logarithmic scale. The changes in treated cells werc compared with untreated cells as well as with cells treated with SNPs.

For ROS detection, the bacterial cell pellet was suspended in a LB broth and incubated with DCFH-DA reagent at 37°C for 30 min in the dark. The fluorescence was measured as above at an

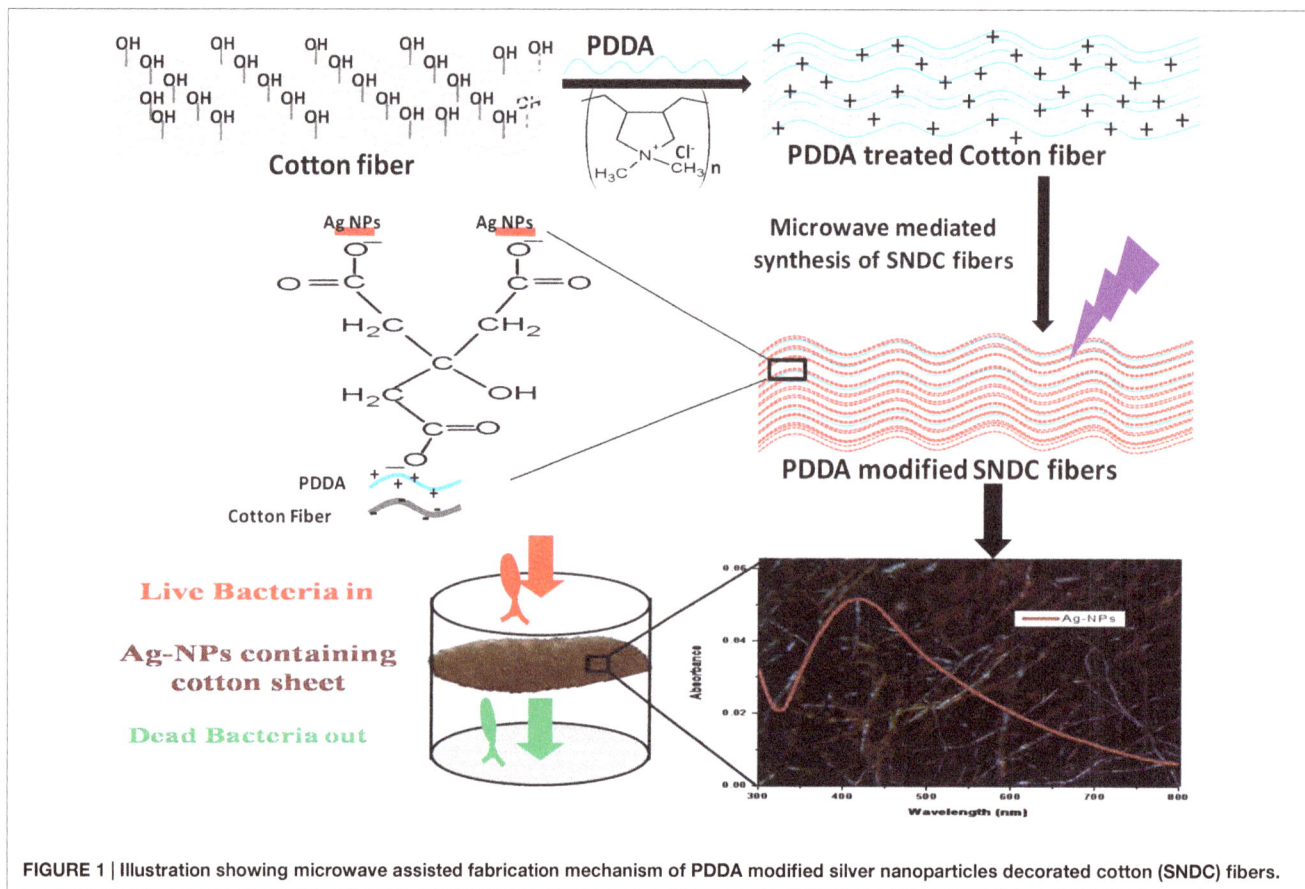

FIGURE 1 | Illustration showing microwave assisted fabrication mechanism of PDDA modified silver nanoparticles decorated cotton (SNDC) fibers.

excitation wavelength 485 nm and emission wavelength 528 nm. Protein leakage was analyzed in culture supernatant using Bradford assay as per the manufacturer's protocol. Ten microliter supernatant from each bacterial culture was transferred to 96 well plates followed by adding 250 µl of Coomassie Blue reagent. After mixing the plate on a plate shaker for 30 min and further incubation at room temperature for 10 min, the absorbance was measured at 595 nm using a spectrophotometer (SpectraMaxplus[384], Molecular Devices, USA). BSA was used as the standard for which a standard curve was drawn at each experiment to determine the protein concentration for each sample.

Antibacterial Efficiency of Silver Nanoparticles Decorated Cotton Fibers

The antibacterial susceptibility of SNPs-cotton and pure cotton fiber were carried out by disk diffusion assay. In this assay, E. coli, S. aureus and S. typhimurium were selected as the model bacteria. The microorganisms were cultured overnight at 37°C in NA. The final cell concentrations of bacterial inoculants were 10^6–10^7 colony forming unit (CFU)/ml. The fabricated samples were cut down into small pieces (0.5 cm × 0.5 cm) and delivered on the agar plates, incubated at 37°C for 24 h, for susceptibility study using modified disk diffusion assay technique (Hu et al., 2009; Ravindra et al., 2010). The culture plates were observed for the presence of the circular zone of bacterial

growth inhibition/clear zone around the SNDC fibers, expressed in terms of the average diameter of the zone of inhibition in millimeters.

Filtration Efficacy SNDC Fibers and Leaching of Ag+ Ions

The antibacterial efficacy of SNDC fibers was tested against E. coli (MTCC 723) because it is widely accepted as an indicator of fecal contamination in potable water. Cotton fiber with a thickness of 0.4 cm was used as a control and the same thickness of cotton fibers were fabricated with exposure to microwave irradiation (600 W) for 1 min.

The inocula suspension of E. coli was prepared and adjusted according to 1×10^8 CFU/mL in liquid media. This suspension was used as a model of contaminated water which permeates through SNDC fibers at the rate of 100 mL within 8 min. A small amount of E. coli bacteria were retained in the cotton filter and almost passed through SNDC fibers. These isolated bacteria from the effluent were centrifuged and analyzed for viability. The qualitative re-growth experiments were performed in MHB. Selected bacterial growth kinetics was analyzed with an optical density at 492 nm for every 2 h as compared with the positive control. Further, to identify bactericidal efficacy of SNDC fibers the effluent was placed on MHA and incubated overnight at 37°C for 24 h for observation of bacterial colonies. The Ag+ ions concentration leached from SNDC based filter in

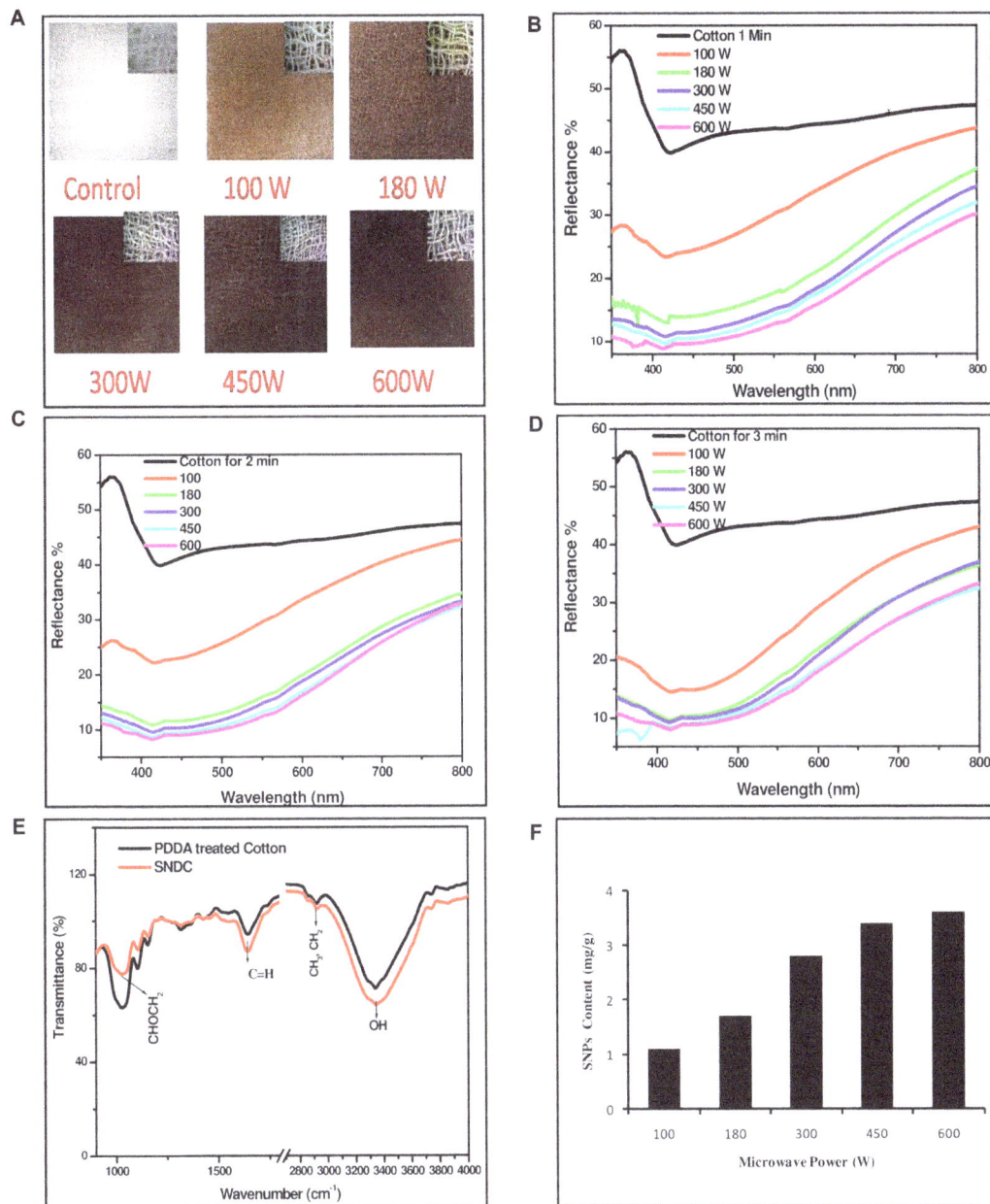

FIGURE 2 | Silver nanoparticles decorated cotton sheets prepared at different time and power of microwave irradiation, (A) SNDC fibers prepared at 100–600 W for 3 min microwave irradiation, **(B)** UV-VIS diffuse reflectance spectra of SNDC at 1 min, **(C)** 2 min, **(D)** 3 min, **(E)** ATR-FTIR spectra cotton and SNDC fibers, **(F)** Content of SNPs on different SNDC sheets.

the disinfection process was analyzed using Inductively Coupled Plasma Mass Spectrometer (ICP-MS) (Thermo Fisher Scientific, Germany).

Statistical Analysis

All experiments were independently repeated in triplicates. Results were expressed as mean ± standard error (SE). The statistical analysis was performed in MS Excel and Origin 6.1 software.

RESULTS AND DISCUSSION

Microwave irradiation is a one of the most promising method for rapid and green synthesis of SNPs (Raveendran et al., 2006; Dankovich, 2014). The SNPs were readily formed on cellulosic cotton sheet by microwave irradiation of the cotton saturated with precursor and reducing agent. This reaction does not occur at room temperature. Here reduction reaction catalyzed by the microwave irradiation results reduction of silver ions (Ag$^+$) into silver neutrals (Ag^0), which get start nucleation to form SNPs

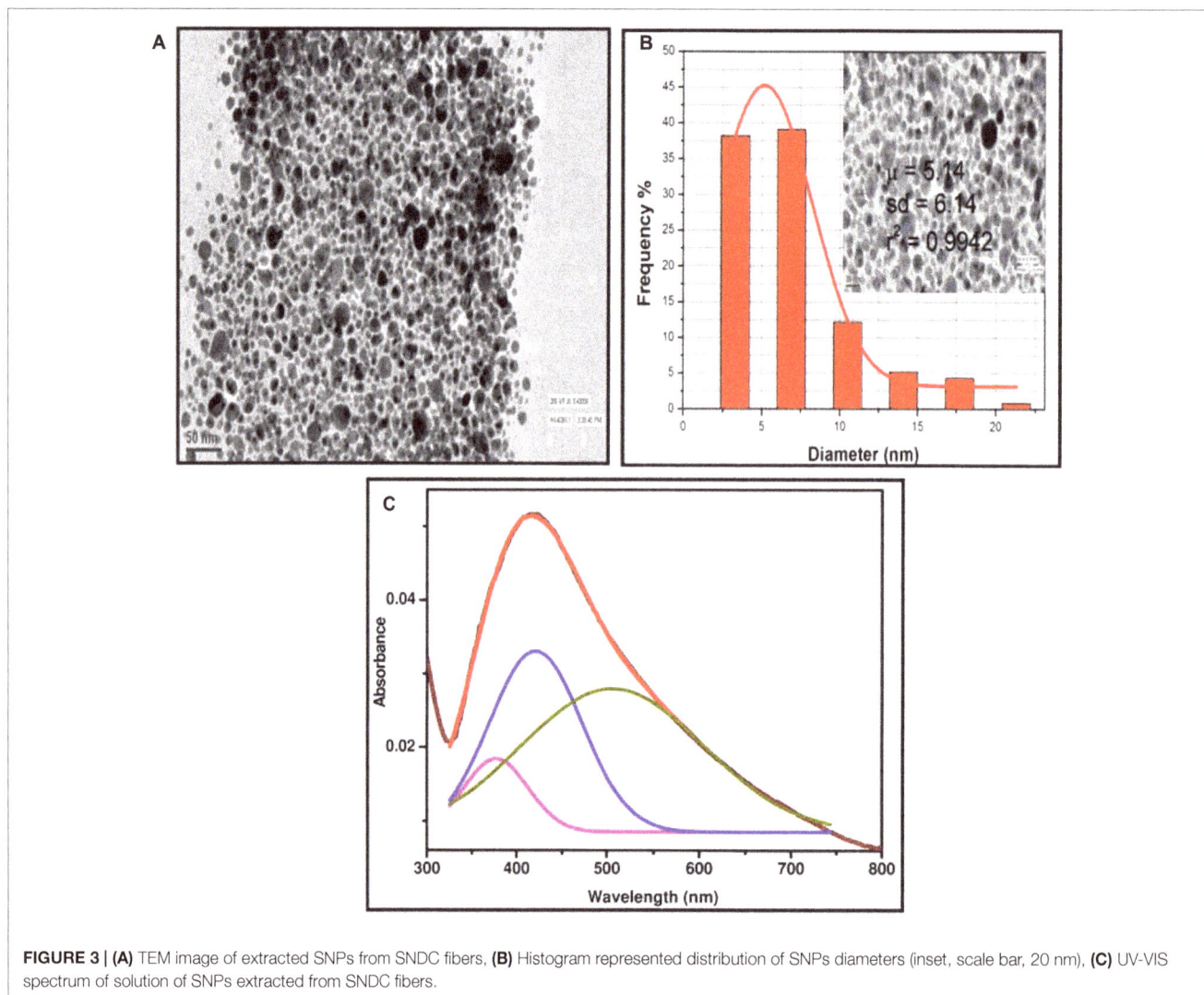

FIGURE 3 | (A) TEM image of extracted SNPs from SNDC fibers, **(B)** Histogram represented distribution of SNPs diameters (inset, scale bar, 20 nm), **(C)** UV-VIS spectrum of solution of SNPs extracted from SNDC fibers.

on the PDDA modified surface of cotton fibers to reduce their surface energy. Power and time of microwave irradiation are varied in order to increase density of SNPs on the surface of cotton fibers.

UV-Visible Diffuse Reflectance Spectroscopy for the Investigation of Loading Density of SNPs on Cotton Fibers

Color of cotton sheet soaked with aqueous solution of silver precursor changed from white to yellow/orange and finally brown with the increase in time of microwave irradiation at constant power or with the raise of power for given time. Change in the color of SNDC fibers can be directly correlated with the amount of SNPs loaded on them. UV-VIS diffuse reflectance spectra of SNDC fibers samples with variations in time of irradiation and microwave power are illustrated in (**Figure 2**). Valley of the reflectance spectra in the range of 430–450 nm is due to the out plane quadrupole plasmon resonance and some partial

color change are also due to localized surface plasmon resonance (LSPR) (Kelly et al., 2003; Sherry et al., 2006; Shopa et al., 2010; Dankovich and Gray, 2011; Tang et al., 2012; Dankovich, 2014).

UV-VIS diffuse reflectance spectra of SNDC fiber sheets fabricated with varying microwave power from 100 to 600 W for 1, 2, and 3 min of irradiations are shown in (**Figures 2B–D**). Optical photographs of undecorated and SNDC fiber at 100–600 W microwave irradiance for 3 min of irradiation are shown in (**Figure 2A**) with corresponding microscopic (20X) images in the inset. With the increase of microwave power or time of irradiation, reflectance in the complete spectral range decreases, which is an evidenced of increase in the density of SNPs on cotton fiber. Decrease in the reflectance is more pronounced at ~450 nm corresponding to the SPR absorption of silver. Reflectance corresponding to SPR decreases, while its position slightly shifted toward longer wavelength side with the increase of power or time of irradiation, which shows that density as well as size of the NPs increases with the increase in power or time of microwave irradiation (Tang et al., 2012). Consistent color change observed by necked eye further

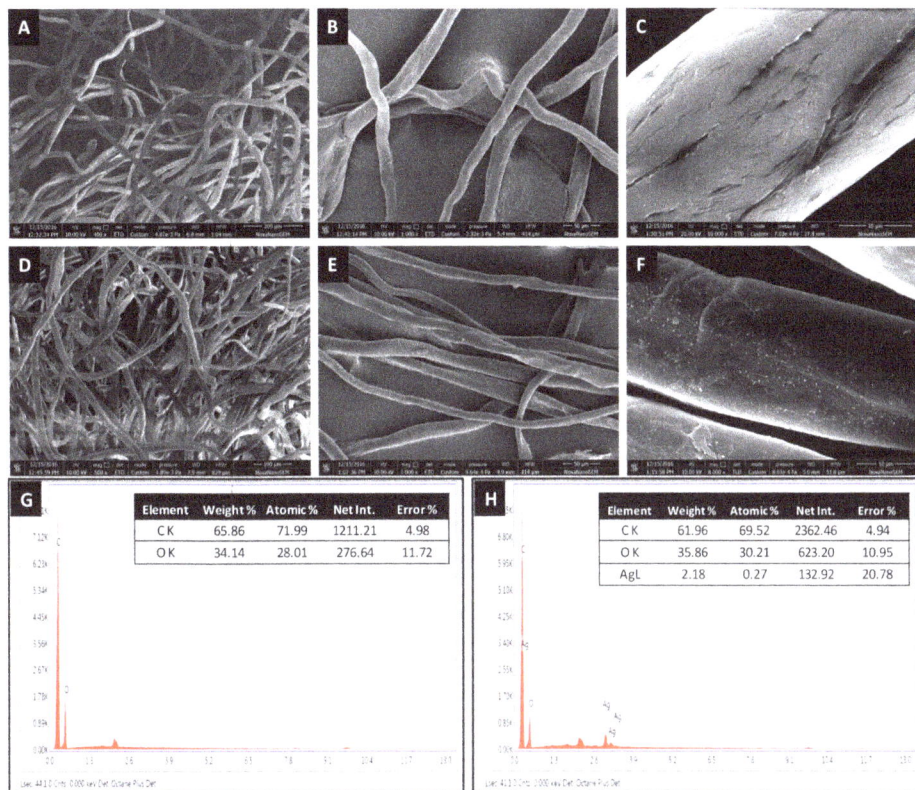

FIGURE 4 | Scanning electron micrograph (SEM) of SNDC fibers with different magnification **(A–C)** normal cotton fibers **(D–F)** SNDC fibers, **(G,H)** EDX of normal cotton fibers and SNDC fibers.

increase in the depth of valley at 450 nm in the reflectance spectra state that loading of SNPs increases with time as well as with power of microwave irradiation. This is further elaborated by calculated SNP content of SNDC fibers shown in **Figure 2F**. Presence of SNPs on the cotton fiber is verified by spectroscopic as well as microscopic investigation. UV-VIS absorption spectrum of colloidal solution of NPs obtained by washing of an SNDC fibers sample prepared by microwave irradiation (600 W, 1 min) of PDDA untreated cotton fiber has intense SPR peak at 418 nm (**Figure 3C**), which proves presence of SNPs on cotton fibers. Same extracted colloidal solution has been used for TEM investigation and flowcytometry. SPR absorption spectrum, deconvoluted into three Gaussian peaks centered at 377, 421, and 505 nm, shows trimodal distribution of SNPs, which is also verified by TEM (**Figure 3C**). The variations of thermodynamic environment like temperature are responsible to regulate morphology and stability of SNPs (Agnihotri et al., 2014; González et al., 2014). In this study interaction of water and microwave radiation creates high temperature with increasing the power may cause quick water evaporation from aqueous solution. In case of increasing power of microwave, SNPs loading was greater on cotton fiber but SPR peak was broaden which suggest wide particle size distribution. This broadening is due to quick synthesis of SNPs at higher temperature catalyzes the rate of reaction and faster growth of nucleation takes place at different layer of cotton fiber. The same tendency was found

TABLE 1 | Determination of minimum inhibitory concentration (MICs) of extracted SNPs from SNDC fibers against selected bacterial pathogens (μg/mL).

Selected parameters	Antibacterial efficacy of SNPs against selected bacterial pathogens (μg/mL)		
	E. coli (MTCC-723)	*S. aureus* (MTCC-737)	*S. typhimurium* (MTCC-98)
MICs	42.5 ± 1.33	125 ± 2.12	38.5 ± 0.93
IC_{50}	31.2 ± 0.58	72.4 ± 1.91	25.3 ± 0.37
MBC	62.5 ± 1.89	158 ± 3.19	62.5 ± 1.89

in (**Figures 2B–D**). However, the cellulose polymer begins to degrade at high temperatures (Madorsky et al., 1958) and SNPs formation at elevated temperatures lead to a weaker structure for the cotton filter, which shorten its lifespan as a water filter.

Here limited concentration of reducing agent and precursor are taken which is responsible to stop reaction automatically when reaction was complete. However, nucleation and growth of NPs continued. During observation of defuse reflectance we found that there is no SPR peak shift at particular power and different time, it is because of only continuous loading of SNPs on the surface of cotton fiber. This is confirmed by the spectra of different time of irradiation with observed dip SPR.

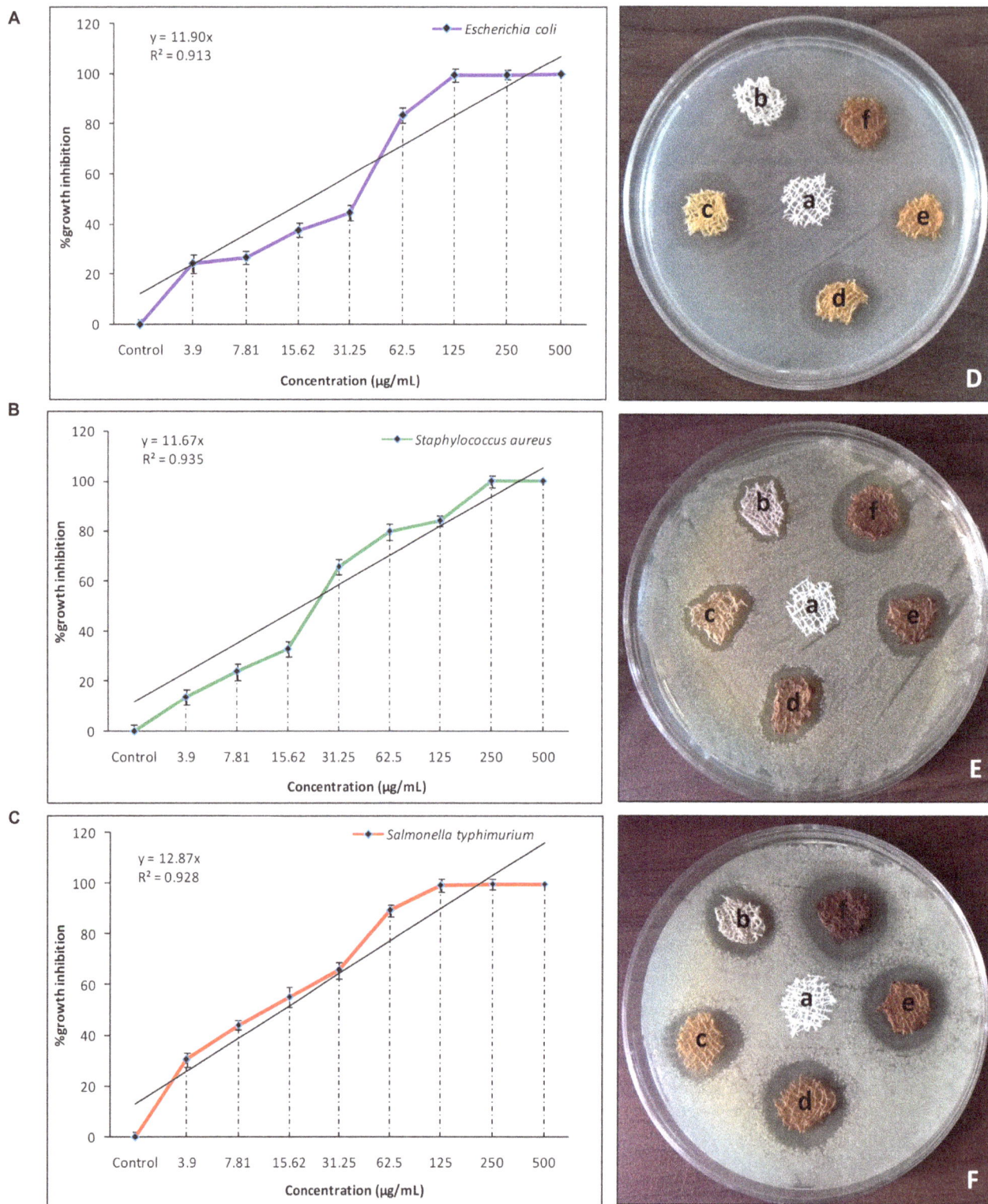

FIGURE 5 | Antibacterial activity of extracted SNPs and SNDC fibers (A) Graphical representation of %growth inhibition over concentration against *E. Coli* **(B)** *S. aureus* **(C)** *S. typhimurium* (Data represent mean ± standard error of three replicates), **(D)** Zone of inhibition of fabricated SNDC fiber at different power of microwave radiation for 1 min exposure against *E. coli* **(E)** *S. aureus* **(F)** *S. typhimuriume* (a) Untreated cotton served as control (b) Treated with 100 W (c) 180 W (d) 300 W (e) 450 W (f) 600 W.

Fourier-Transform Infrared Spectroscopy for the Diagnosis of Interaction between Cotton Fiber and SNPs

Fourier-transform infrared (FTIR) spectra of SNDC fibers in attenuated total reflectance (ATR) mode, illustrated (**Figure 2E**), have a broad band corresponding to the stretching frequency of the hydroxyl group around 3400 cm^{-1} and a peak around 1642 cm^{-1} corresponds group of the cotton fiber (Mahadeva et al., 2014), while peaks at 1028, 1106, 1156, and 1427 cm^{-1} can be identified as a characteristic peak of cotton/cellulose (Fakin et al., 2012; Peng et al., 2016). The three vibration peak of C-O locate within the spectral range of 1200–1000 cm^{-1} are 1156 cm^{-1}, 1106 cm^{-1}, and 1028 cm^{-1} suggest cellulosic content of cotton. The spectral peak observed at 2925 (CHn), 1642 (C = C) and 1427 cm^{-1} (CH$_2$) are characteristics of PDDA (Ding et al., 2014). The characteristic peaks position of PDDA treated cotton fiber after silver fabrication lying at 1642 cm^{-1} (C = C) and 1427 cm^{-1} (CH$_2$) are shifted (Xu et al., 2014). This shifting is with respect to their original value in cotton, clearly indicates the existence of SNPs over surface of PDDA modified cotton fiber. Pure cotton has less transmittance than decorated cotton at certain above given peaks. The intensities of the peaks is lower for pure cotton than decorated cotton fabrics indicating the hydroxyl groups of cellulose before silver coating, whereas after SNPs loading the intensities of the related peaks are increased. This might be due to decrease of this group on the cotton fiber surface. The decreasing of hydroxyl groups may demonstrate physical adsorption of silver ions to these groups (Nourbakhsh and Ashjaran, 2012).

TEM and SEM for Size, Shape and Distribution Measurements

Colloidal aqueous extracted SNPs synthesized by microwave-assisted reduction of SNPs over PDDA unmodified cotton has been taken for HRTEM characterization. The almost small, spherical, and monodispersed SNPs particles are shown in (**Figure 3A**), the average particle size of measured was found to be about 5 nm from TEM image (**Figure 3B**). The image is fully supported deconvoluted of SPR absorption spectra in to three Guassian peaks which shows the three different particles distribution pattern. SEM micrograph and EDX of PDDA treated cotton fiber and SNDC fiber (**Figure 4**) clearly reveals that small, spherical SNP are homogeneously deposited over the surface of cotton fibers. The EDX analysis (X-ray mapping) is also confirmed the amount and continuous distribution of SNPs (**Appendix A** and Supplementary Figure S1).

Mechanism of SNDC Fabrication

The interaction of SNPs and cotton fibers can be understood through electrostatic interaction force. It is believed that cotton carry negative surface charges due to the presence of hydroxyl and carboxyl groups and PDDA, as a strong cationic polyelectrolyte, carries positive charges (Ribitsch et al., 1996; Grancaric et al., 2005; Tang et al., 2012). Consequently cotton fibers with negative charges have strong possibility to coat by PDDA with positive charges by electrostatic force. After

reduction of Ag$^+$ using citrate carried negative charge over the citrate stabilize SNPs (Tang et al., 2015; Peng et al., 2016). Thus citrate stabilized SNPs having negative charges can be adsorbed on the PDDA modified cotton fabrics with positive charges through electrostatic interaction. Moreover, it is found that PDDA unmodified SNDC fibers were less dark than modified SNDC fibers at same energy and time of microwave radiation during SNDC fabrication. Which confirm PDDA is playing a significant role in better interaction cotton fiber and SNPs. The mechanism of microwave assisted SNDC fabrication on cotton fibers is discussed as shown in (**Figure 1**).

Antibacterial Susceptibility of Extracted SNPs Determination of Minimum Inhibitory Concentrations Determination of Minimum Bactericidal Concentration

The *in vitro* antibacterial susceptibility was measured in terms of growth rates of *E. coli, Salmonella typhimurium* and *S. aureus* using turbidimetric growth analysis over a concentration range of NP (0–500 μg/mL). The MICs of the NPs against all the three strains are represented in (**Table 1**; **Figures 5A–C**) with percentage growth inhibition curve against concentration range along with a linear regression coefficient between the two plotted parameters. All the three strains exhibited a significant correlation between concentration range and % growth inhibition with a maximum $R^2 = 0.935$ for *S. aureus,* followed by *Salmonella typhimurium* and *E. coli* with R^2 values equal to 0.928 and 0.913 respectively (**Figures 5A–C**). SNPs showed better efficacy against Gram negative bacteria than Gram positive bacteria (Pandey et al., 2014). The results are in agreement with the previously reported findings (Kim et al., 2007; Jung et al., 2008; Ruparelia et al., 2008). This could be attributed to the cell membrane structures possessing the different amount of lipid and peptidoglycan layer (Mishra et al., 2016b). Gram positive bacteria with high peptidoglycan may prevent the action of NPs across bacterial cell wall (Feng et al., 2000; Pal et al., 2007).

Flowcytometric Analysis for Plasma Membrane Permeabilization

The effects of NPs over the plasma membrane integrity of bacterial strains were evaluated in terms of PI internalization through flowcytometry and protein leakage analysis. Two median doses 15 and 30 μg/L was selected from the specified concentration range and evaluated the role of SNPs in affecting membrane permeability. A dose dependent depletion in cell survival % in all three strains as reflected through PI influx. *E. coli* over treatment with a higher concentration of NPs revealed ∼60.1% of cell death and lower concentration resulted in ∼28% cell death (**Figure 6**). Furthermore, a higher dose of NPs resulted in PI internalization in ∼61 and 53% of *Salmonella* and *S. aureus* respectively; and, a lower dose of NPs resulted in ∼49 and 38% of cell death of two bacterial strains respectively (**Figures 6A–C**).

Comparative overlay histogram sequence density plots clearly revealed that SNPs could induce dose dependent membrane permeability in bacterial cells (**Appendix A** and Supplementary Figure S2). The membrane permeability could further facilitate

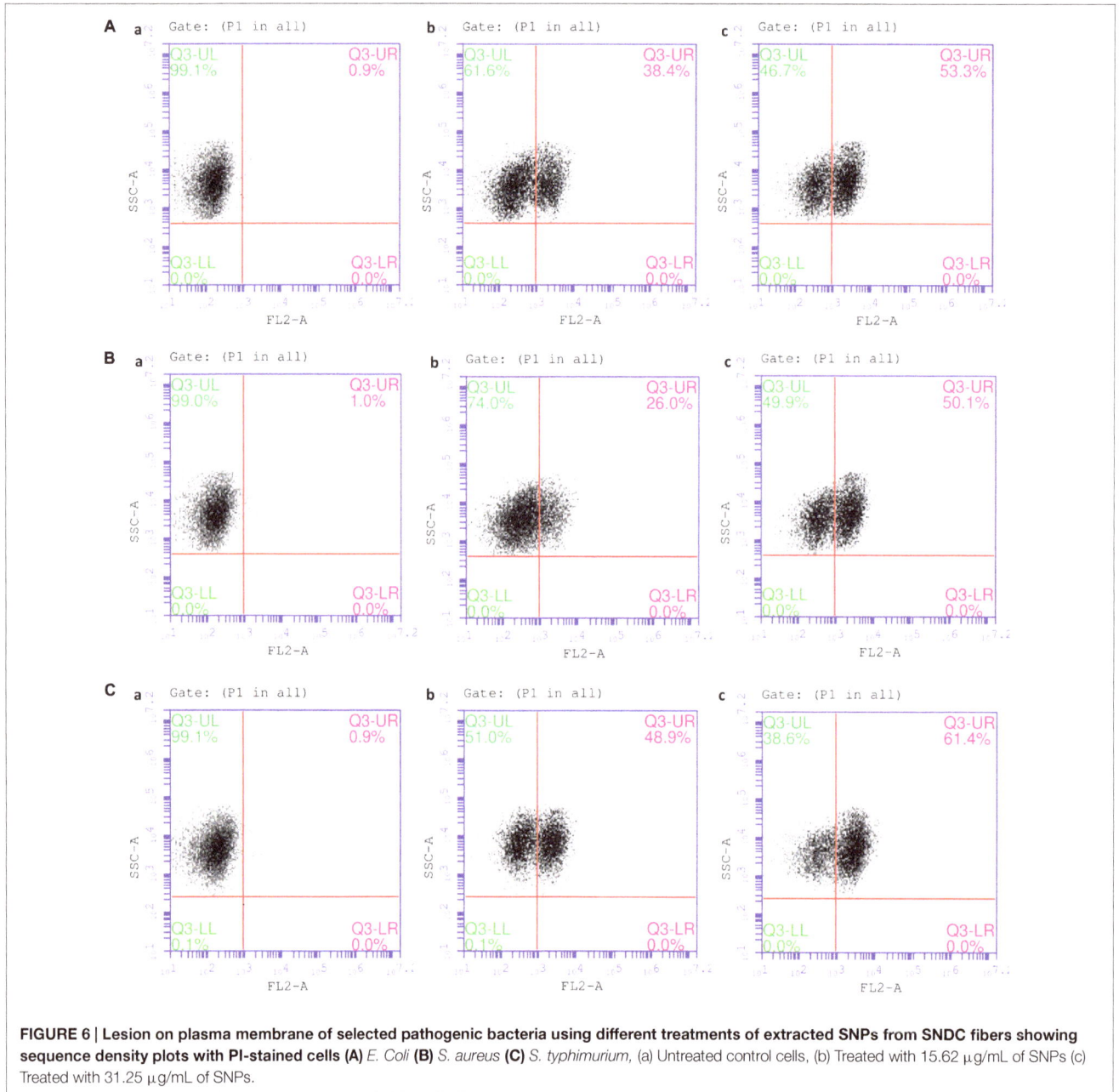

FIGURE 6 | Lesion on plasma membrane of selected pathogenic bacteria using different treatments of extracted SNPs from SNDC fibers showing sequence density plots with PI-stained cells (A) *E. Coli* **(B)** *S. aureus* **(C)** *S. typhimurium,* (a) Untreated control cells, (b) Treated with 15.62 μg/mL of SNPs (c) Treated with 31.25 μg/mL of SNPs.

SNPs penetration in bacterial cells and result in protein dysfunction (Hsueh et al., 2015). Protein leakage analysis further confirmed the loss of membrane integrity over NP-exposure (**Figure 7A**). A dose dependent increase in protein concentration was observed in culture supernatant of all the three bacterial strains as compared to control groups. The observations made through spectrophotometer and protein leakage analysis were found to be in line with the anti-bacterial susceptibility assay and previous research findings (Chakraborti et al., 2014). The bactericidal effect of SNPs was further studied in terms of Reactive Oxygen Species generation. **Figure 7B** reveal magnitude of ROS formation in terms of fluorescent counts due to DCA formation in bacterial cells exposed to SNPs. A dose dependent increase in ROS was observed against all bacterial species. *E. coli* and *S. typhimurium* showed a significant increase in oxidative stress level both at 15.62 μg/mL and 31.25 μg/mL, however, data remained insignificant in case of *S. aureus.* Although exact mechanism involving antibacterial efficacy of SNPs have not been clearly revealed ROS and free radicals have been variously documented to incur antibacterial potential to NPs (Sen et al., 2013; Dakal et al., 2016). The binding of Ag^+ over the microbial cell membrane induces cytotoxicity which further inhibits mitochondrial respiratory chain enzymes (Blecher and Friedman, 2012) and disrupts electron transport chain by uncoupling oxidative phosphorylation (Belluco et al., 2016) subsequently leading to cell death.

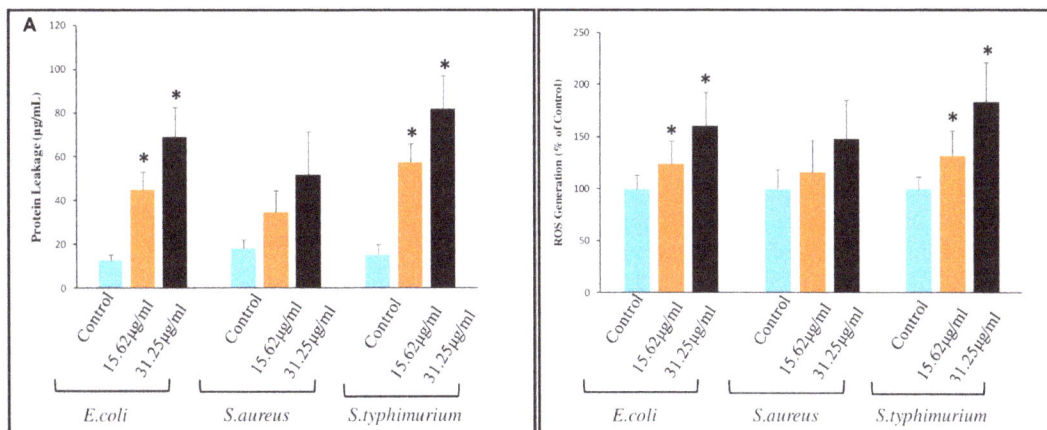

FIGURE 7 | Evaluation of membrane disruption by (A) Protein leakage and (B) Relative fluorescence intensity showing cellular ROS formation potential of SNPs. Data are represented as Mean ± SEM, *$p < 0.05$ compare to control.

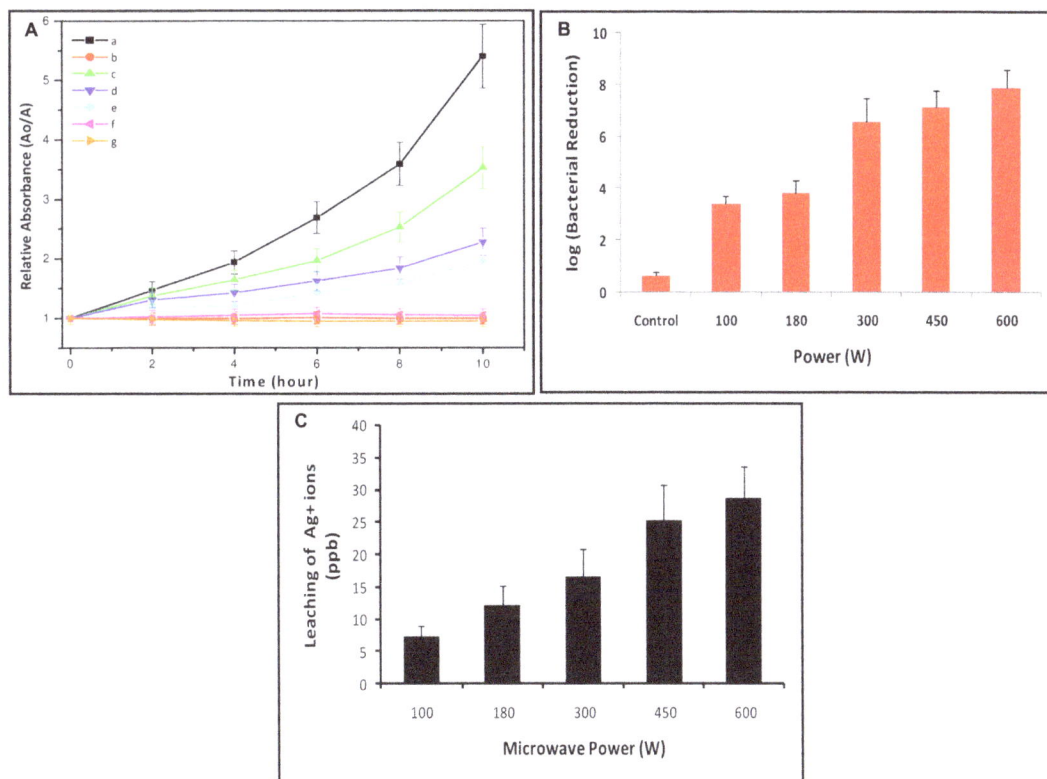

FIGURE 8 | (A) Relative absorbance after each 2 h interval of *E. coli* permeated with different SNDC fibers at 492 nm. (a) Positive control (without SNPs in cotton) (b) Negative control (without bacteria) (c) Treated with 100 (d) 180 (e) 300 (f) 450 and (g) 600 W, (B) Log reduction of *E. coli* (1×10^8 CFU/mL) count after permeation through SNDC sheets. (C) Leaching of Ag$^+$ obtained in the effluent measured by ICPMS.

Antibacterial Efficacy of SNPs Decorated Cotton Using disk Diffusion Assay

Zone of inhibition of SNDC fibers against some gram-positive (*S. aureus*) and gram-negative bacteria (*E. coli* and *Salmonella typhimurium*) strains are shown in (**Figure 5**).

The synthesized samples size (0.5 cm x 0.5 cm) are labeled as (a) Control (PDDA untreated cotton) and (b) treated with 100 W (c) 180 W (d) 300 W (e) 450 W (f) 600 W SNDC fiber at for 1 min exposure. Radiant zones of inhibition were observed due to SNPs loading dependency against tested bacterial pathogens. The results are clearly shown in (**Figures 5D–F**).

Filtration Efficacy of Silver NPs Decorated Cotton on *E. coli*

The SNDC fibers provides suitable and effective bactericidal activity as model *E. coli* suspensions were poured through the 0.4 cm thick dry consolidated SNDC fibers sheet which was put over filter paper. The average percolation time for 100 mL of bacterial suspension was 8 min. Some *E. coli* were retained in the cotton filter, but most of them passed through and were isolated from the effluent by centrifugation and analyzed for viability. The qualitative re-growth experiments in MHB showed inactivation of bacteria at the highest silver concentration (**Figure 8A**) shows the exponential growth curve in the positive control sample (undecorated cotton or without SNPs in cotton).

While the negative control sample (without bacteria) shows no growth. The bacteria growth after percolation through the SNDC fibers was almost completely deactivated for the cotton with the highest SNPs content (microwave irradiation 100–600 W for 1 min). The lower SNPs containing cotton showed a maximum bacterial growth reduction in comparing with the positive control. To check further the bactericidal effectiveness of SNDC fibers, the isolated effluent bacteria was added to NA plates after passing through the cotton. The plate count experiments showed maximum log 7.8 reductions of viable *E. coli* in the effluent, as compared to the initial concentration of bacteria (10^8 CFU/mL) (**Figure 8B**). The positive control also showed a reduction in bacteria by log 0.59, most likely due to some bacteria remaining on the surfaces of cotton fiber. The SNDC fibers prepared within one min with different microwave power at 100, 180, 300, 450, and 600 W, responsible to log bacterial reduction log 3.3 ($\pm$0.09), log 3.8 ($\pm$0.11), log 6.5 ($\pm$0.19), log 7.1 ($\pm$0.2) and log 7.8 ($\pm$0.22) respectively. This clear difference was observed in reduction potency of bacteria due to increasing concentration of SNPs over cotton that are also supported by UV-VIS reflectance spectra.

The antimicrobial nature of a few SNPs is believed to be only through the contact killing mechanism, which contributes an even greater potential lethality when bacteria come in contact with them (Li et al., 2009; Srinivasan et al., 2013). However, in this work contact mode are preferred for the testing of the bactericidal effectiveness of the SNDC fibers, we have approved model bacterial suspension through an SNDC fiber sheet and analyze bacterial viability in effluent water. In this purification, SNDC fibers was not used for removal of *E. coli* from effluent by filtration but rather the deactivation of bacteria as they percolate through the highly porous SNDC fibers structure. The large pore size of the cotton and base as filter paper allows good retention and contact killing of bacteria as well as allow reasonable balance flow by gravity, without the need for pressure or suction. Due to possible human health effects from silver exposure, we analyzed the silver content of the effluent water. We used centrifugation to separate bacteria from the silver leached out from the SNDC sheet. Highest amount of silver content of ~28.75 ppb was observed (SNDC prepared at 600 W for 1 min) using ICP-MS (**Figure 8C**). The amount of silver leaching from the SNDC filter thus meets the US-EPA guideline for drinking water of less than 100 ppb (Dankovich, 2014). This product can be easily prepared at home for multipurpose point of use decontamination of water, surface sterilization of medical equipment and human hand.

CONCLUSION

Microwave assisted SNDC fiber samples have been prepared through *in situ*, rapid, convenient, environmental friendly and cost effective method. The SNPs were well dispersed and stabilized on the surface of the cotton fiber. The fabricated NPs were found to be effective against both Gram positive and Gram negative bacteria as revealed through broth microdilution and disk diffusion assay. Cell death could also be credited to ROS generation. Furthermore, the PI influx and protein leakage studied also indicated toward membrane damage incurred due to NPs exposure. Thus the results of contact killing of microbes indicate that it can be applied as a point of use surface disinfectant of wounds, preservative as well as for the development of an effective filter for drinking water in developing countries.

AUTHOR CONTRIBUTIONS

AB, RG, and RM conceived and designed the experiments; KU, SCS, AB, AS, and MS carried out the synthesis, characterization and analysis of silver nanoparticles decorated cotton (SNDC) fibers; AB, RM, and SS performed the antibacterial experiments; VM, RM, and AB and performed and analyzed ROS and flowcytometry data. AB, RG, RM, VM, and SCS wrote the paper. All authors have read and approved the final manuscript.

ACKNOWLEDGMENTS

The authors are very much thankful to the University of Allahabad for providing UGC fellowship during D. Phil. Degree and Science and Engineering Research Board (SERB), Department of Science and Technology, New Delhi under the Fast Track Young Scientist Scheme (SB/YS/LS-44/2014). Also thankful to CMDR, MNNIT for flowcytometry, Dr. R. K. Kotnala NPL New Delhi for SEM and All India Institute of Medical Sciences (AIIMS) New Delhi for TEM facility.

REFERENCES

Agnihotri, S., Mukherji, S., and Mukherji, S. (2013). Immobilized silver nanoparticles enhance contact killing and show highest efficacy: elucidation of the mechanism of bactericidal action of silver. *Nanoscale* 5, 7328–7340. doi: 10.1039/C3NR00024A

Agnihotri, S., Mukherji, S., and Mukherji, S. (2014). Size-controlled silver nanoparticles synthesized over the range 5-100 nm using the same protocol and their antibacterial efficacy. *RSC Adv.* 4, 3974–3983. doi: 10.1039/C3RA44507K

Bae, G. Y., Min, B. G., Jeong, Y. G., Lee, S. C., Jang, J. H., and Koo, G. H. (2009). Superhydrophobicity of cotton fabrics treated with silica nanoparticles and water-repellent agent. *J. Colloid Interface Sci.* 337, 170–175. doi: 10.1016/j.jcis.2009.04.066

Belluco, S., Losasso, C., Patuzzi, I., Rigo, L., Conficoni, D., Gallocchio, F., et al. (2016). Silver as antibacterial toward *Listeria monocytogenes. Front. Microbiol.* 7:307. doi: 10.3389/fmicb.2016.00307

Bendi, R., and Imae, T. (2013). Renewable catalyst with Cu nanoparticles embedded into cellulose nano-fiber film. *RSC Adv.* 3, 16279–16282. doi: 10.1039/b000000x

Bendi, R., Imae, T., and Destaye, A. G. (2015). Ag nanoparticle-immobilized cellulose nanofibril films for environmental conservation. *Appl. Catal. A* 492, 184–189. doi: 10.1016/j.apcata.2014.12.045

Berendjchi, A., Khajavi, R., and Yazdanshenas, M. E. (2011). Fabrication of superhydrophobic and antibacterial surface on cotton fabric by doped silica-based sols with nanoparticles of copper. *Nanoscale Res. Lett.* 6, 1–8. doi: 10.1186/1556-276X-6-594

Blecher, K., and Friedman, A. (2012). Nanotechnology and the diagnosis of dermatological infectious disease. *J. Drugs Dermatol.* 7, 846–851.

Chakraborti, S., Mandal, A. K., Sarwar, S., Singh, P., Chakraborty, R., and Chakrabarti, P. (2014). Bactericidal effect of polyethyleneimine capped ZnO nanoparticles on multiple antibiotic resistant bacteria harboring genes of high-pathogenicity island. *Colloids Surf. B Biointerfaces* 121, 44–53. doi: 10.1016/j.colsurfb.2014.03.044

Dakal, T. C., Kumar, A., Majumdar, R. S., and Yadav, V. (2016). Mechanistic basis of antimicrobial actions of silver nanoparticles. *Front. Microbiol.* 7:1831. doi: 10.3389/fmicb.2016.01831

Dankovich, T. A. (2014). Microwave-assisted incorporation of silver nanoparticles in paper for point-of-use water purification. *Environ. Sci. Nano* 1, 367–378. doi: 10.1039/C4EN00067F

Dankovich, T. A., and Gray, D. G. (2011). Bactericidal paper impregnated with silver nanoparticles for point-of-use water treatment. *Environ. Sci. Technol.* 45, 1992–1998. doi: 10.1021/es103302t

Davies, R. L., and Etris, S. F. (1997). The development and functions of silver in water purification and disease control. *Catal. Today* 36, 107–114. doi: 10.1016/S0920-5861(96)00203-9

Ding, L., Liu, Y., Guo, S.-X., Zhai, J., Bond, A. M., and Zhang, J. (2014). Phosphomolybdate poly (diallyldimethylammonium chloride)-reduced graphene oxide modified electrode for highly efficient electrocatalytic reduction of bromate. *J. Electroanal. Chem.* 727, 69–77. doi: 10.1016/j.jelechem.2014.06.001

Fakin, D., Veronovski, N., Ojstršek, A., and Božiè, M. (2012). Synthesis of TiO2-SiO2 colloid and its performance in reactive dyeing of cotton fabrics. *Carbohydr. Polym.* 88, 992–1001. doi: 10.1016/j.carbpol.2012.01.046

Feng, Q., Wu, J., Chen, G., Cui, F., Kim, T., and Kim, J. (2000). A mechanistic study of the antibacterial effect of silver ions on *Escherichia coli* and *Staphylococcus aureus. J. Biomed. Mater. Res.* 52, 662–668.

Ferraria, A. M., Boufi, S., Battaglini, N., Botelho do Rego, A. M., and ReiVilar, M. (2009). Hybrid systems of silver nanoparticles generated on cellulose surfaces. *Langmuir* 25, 1996–2001. doi: 10.1021/la902477q

Gao, Q., Zhu, Q., Guo, Y., and Yang, C. Q. (2009). Formation of highly hydrophobic surfaces on cotton and polyester fabrics using silica sol nanoparticles and nonfluorinated alkylsilane. *Ind. Eng. Chem. Res.* 48, 9797–9803. doi: 10.1021/ie9005518

González, A., Noguez, C., Beránek, J., and Barnard, A. (2014). Size, shape, stability, and color of plasmonic silver nanoparticles. *J. Phys. Chem. C* 118, 9128–9136. doi: 10.1021/jp5018168

Grancaric, A. M., Tarbuk, A., and Pusic, T. (2005). Electrokinetic properties of textile fabrics. *Color. Technol.* 121, 221–227. doi: 10.1111/j.1478-4408.2005.tb00277.x

Grumezescu, A. M., Andronescu, E., Holban, A. M., Ficai, A., Ficai, D., Voicu, G., et al. (2013). Water dispersible cross-linked magnetic chitosan beads for increasing the antimicrobial efficiency of aminoglycoside antibiotics. *Int. J. Pharm.* 454, 233–240. doi: 10.1016/j.ijpharm.2013.06.054

Hao, L. F., An, Q. F., Xu, W., and Wang, Q. J. (2010). Synthesis of fluoro-containing superhydrophobic cotton fabric with washing resistant property using nano-SiO2 sol-gel method. *Adv. Mater. Res.* 121, 23–26. doi: 10.4028/www.scientific.net/AMR.121-122.23

He, J., Kunitake, T., and Nakao, A. (2003). Facile in situ synthesis of noble metal nanoparticles in porous cellulose fibers. *Chem. Mater.* 15, 4401–4406. doi: 10.1021/cm034720r

Hoefnagels, H., Wu, D., De With, G., and Ming, W. (2007). Biomimetic superhydrophobic and highly oleophobic cotton textiles. *Langmuir* 23, 13158–13163. doi: 10.1021/la702174x

Hsueh, Y.-H., Lin, K.-S., Ke, W.-J., Hsieh, C.-T., Chiang, C.-L., Tzou, D.-Y., et al. (2015). The antimicrobial properties of silver nanoparticles in *Bacillus subtilis* are mediated by released Ag+ ions. *PLoS ONE* 10:e0144306. doi: 10.1371/journal.pone.0144306

Hu, W., Chen, S., Li, X., Shi, S., Shen, W., Zhang, X., et al. (2009). In situ synthesis of silver chloride nanoparticles into bacterial cellulose membranes. *Mater. Sci. Eng. C* 29, 1216–1219. doi: 10.1016/j.msec.2008.09.017

Jung, W. K., Koo, H. C., Kim, K. W., Shin, S., Kim, S. H., and Park, Y. H. (2008). Antibacterial activity and mechanism of action of the silver ion in *Staphylococcus aureus* and *Escherichia coli. Appl. Environ. Microbiol.* 74, 2171–2178. doi: 10.1128/AEM.02001-07

Kelly, K. L., Coronado, E., Zhao, L. L., and Schatz, G. C. (2003). The optical properties of metal nanoparticles: the influence of size, shape, and dielectric environment. *J. Phys. Chem. B* 107, 668–677. doi: 10.1021/jp026731y

Kim, J. S., Kuk, E., Yu, K. N., Kim, J.-H., Park, S. J., Lee, H. J., et al. (2007). Antimicrobial effects of silver nanoparticles. *Nanomedicine* 3, 95–101. doi: 10.1016/j.nano.2006.12.001

Li, Y., Leung, W. K., Yeung, K. L., Lau, P. S., and Kwan, J. K. (2009). A multilevel antimicrobial coating based on polymer-encapsulated ClO2. *Langmuir* 25, 13472–13480. doi: 10.1021/la901974d

Liu, H., Lv, M., Deng, B., Li, J., Yu, M., Huang, Q., et al. (2014). Laundering durable antibacterial cotton fabrics grafted with pomegranate-shaped polymer wrapped in silver nanoparticle aggregations. *Sci. Rep.* 4:5920. doi: 10.1038/srep05920

Luo, C., Zhang, Y., Zeng, X., Zeng, Y., and Wang, Y. (2005). The role of poly (ethylene glycol) in the formation of silver nanoparticles. *J. Colloid Interface Sci.* 288, 444–448. doi: 10.1016/j.jcis.2005.03.005

Madorsky, S., Hart, V., and Straus, S. (1958). Thermal degradation of cellulosic materials. *J. Res. Nat. Bur. Stand.* 60, 343–349.

Mahadeva, S. K., Walus, K., and Stoeber, B. (2014). Piezoelectric paper fabricated via nanostructured barium titanate functionalization of wood cellulose fibers. *ACS Appl. Mater. Interfaces* 6, 7547–7553. doi: 10.1021/am5008968

Mahmoud, W. M., Abdelmoneim, T. S., and Elazzazy, A. M. (2016). The impact of silver nanoparticles produced by *Bacillus pumilus* as antimicrobial and nematicide. *Front. Microbiol.* 7:1746. doi: 10.3389/fmicb.2016.01746

Mahmoudi, M., and Serpooshan, V. (2012). Silver-coated engineered magnetic nanoparticles are promising for the success in the fight against antibacterial resistance threat. *ACS Nano* 6, 2656–2664. doi: 10.1021/nn300042m

Maneerung, T., Tokura, S., and Rujiravanit, R. (2008). Impregnation of silver nanoparticles into bacterial cellulose for antimicrobial wound dressing. *Carbohydr. Polym.* 72, 43–51. doi: 10.1016/j.carbpol.2007.07.025

Mishra, R. K., Mishra, V., Sharma, S., Pandey, A. C., and Dikshit, A. (2016a). Anti-dermatophytic potential of *Ajuga bracteosa* wall ex benth leaf extract mediated AgNPs with particular emphasis to plasma membrane lesion. *Mater. Focus* 5, 249–257. doi: 10.1166/mat.2016.1320

Mishra, R. K., Ramakrishna, M., Mishra, V., Pathak, A., Rajesh, S., Sharma, S., et al. (2016b). Pharmaco-phylogenetic investigation of methyl gallate isolated from *Acacia nilotica* (L.) delile and Its cytotoxic effect on NIH3T3 mouse fibroblast. *Curr. Pharm. Biotechnol.* 17, 540–548. doi: 10.2174/1389201017666160127110759

Nourbakhsh, S., and Ashjaran, A. (2012). Laser treatment of cotton fabric for durable antibacterial properties of silver nanoparticles. *Materials* 5, 1247–1257. doi: 10.3390/ma5071247

Pal, S., Tak, Y. K., and Song, J. M. (2007). Does the antibacterial activity of silver nanoparticles depend on the shape of the nanoparticle? A study of the gram-negative bacterium *Escherichia coli*. *Appl. Environ. Microbiol*. 73, 1712–1720. doi: 10.1128/AEM.02218-06

Pandey, J. K., Swarnkar, R., Soumya, K., Dwivedi, P., Singh, M. K., Sundaram, S., et al. (2014). Silver nanoparticles synthesized by pulsed laser ablation: as a potent antibacterial agent for human enteropathogenic gram-positive and gram-negative bacterial strains. *Appl. Biochem. Biotechnol*. 174, 1021–1031. doi: 10.1007/s12010-014-0934-y

Peng, L., Guo, R., Lan, J., Jiang, S., and Wang, X. (2016). Microwave-assisted coating of silver nanoparticles on bamboo rayon(fabrics)modified with poly (diallyldimethylammonium chloride). *Cellulose* 23, 2677–2688. doi: 10.1007/s10570-016-0931-0

Piccinno, F., Gottschalk, F., Seeger, S., and Nowack, B. (2012). Industrial production quantities and uses of ten engineered nanomaterials for Europe and the world. *J. Nanopart. Res*. 14, 1109–1120. doi: 10.1007/s11051-012-1109-9

Raveendran, P., Fu, J., and Wallen, S. L. (2006). A simple and "green" method for the synthesis of Au, Ag, and Au–Ag alloy nanoparticles. *Green Chem*. 8, 34–38. doi: 10.1039/B512540E

Ravindra, S., Mohan, Y. M., Reddy, N. N., and Raju, K. M. (2010). Fabrication of antibacterial cotton fibres loaded with silver nanoparticles via "Green Approach". *Colloid Surf. A* 367, 31–40. doi: 10.1016/j.colsurfa.2010.06.013

Ribitsch, V., Stana-Kleinschek, K., and Jeler, S. (1996). The influence of classical and enzymatic treatment on the surface charge of cellulose fibres. *Colloid Polym. Sci*. 274, 388–394. doi: 10.1007/BF00654060

Ruparelia, J. P., Chatterjee, A. K., Duttagupta, S. P., and Mukherji, S. (2008). Strain specificity in antimicrobial activity of silver and copper nanoparticles. *Acta Biomater*. 4, 707–716. doi: 10.1016/j.actbio.2007.11.006

Sen, I. K., Mandal, A. K., Chakraborti, S., Dey, B., Chakraborty, R., and Islam, S. S. (2013). Green synthesis of silver nanoparticles using glucan from mushroom and study of antibacterial activity. *Int. J. Biol. Macromol*. 62, 439–449. doi: 10.1016/j.ijbiomac.2013.09.019

Sherry, L. J., Jin, R., Mirkin, C. A., Schatz, G. C., and Van Duyne, R. P. (2006). Localized surface plasmon resonance spectroscopy of single silver triangular nanoprisms. *Nano Lett*. 6, 2060–2065. doi: 10.1021/nl061286u

Shopa, M., Kolwas, K., Derkachova, A., and Derkachov, G. (2010). Dipole and quadrupole surface plasmon resonance contributions in formation of near-field images of a gold nanosphere. *Opto Electr. Rev*. 18, 421–428. doi: 10.2478/s11772-010-0047-2

Srinivasan, N., Shankar, P., and Bandyopadhyaya, R. (2013). Plasma treated activated carbon impregnated with silver nanoparticles for improved antibacterial effect in water disinfection. *Carbon* 57, 1–10. doi: 10.1016/j.carbon.2013.01.008

Stevenson, A. P., Bea, D. B., Civit, S., Contera, S. A., Cerveto, A. I., and Trigueros, S. (2012). Three strategies to stabilise nearly monodispersed silver nanoparticles in aqueous solution. *Nanoscale Res. Lett*. 7:151. doi: 10.1186/1556-276X-7-151

Swarnkar, R. K., Pandey, J. K., Soumya, K. K., Dwivedi, P., Sundaram, S., Prasad, S., et al. (2016). Enhanced antibacterial activity of copper/copper oxide nanowires prepared by pulsed laser ablation in water medium. *Appl. Phys. A* 122, 1–7. doi: 10.1007/s00339-016-0232-3

Tang, B., Sun, L., Li, J., Kaur, J., Zhu, H., Qin, S., et al. (2015). Functionalization of bamboo pulp fabrics with noble metal nanoparticles. *Dyes Pigm*. 113, 289–298. doi: 10.1016/j.dyepig.2014.08.015

Tang, B., Zhang, M., Hou, X., Li, J., Sun, L., and Wang, X. (2012). Coloration of cotton fibers with anisotropic silver nanoparticles. *Ind. Eng. Chem. Res*. 51, 12807–12813. doi: 10.1021/ie3015704

Windler, L., Lorenz, C., Von Goetz, N., Hungerbuhler, K., Amberg, M., Heuberger, M., et al. (2012). Release of titanium dioxide from textiles during washing. *Environ. Sci. Technol*. 46, 8181–8188. doi: 10.1021/es301633b

Xu, B., and Cai, Z. (2008). Fabrication of a superhydrophobic ZnO nanorod array film on cotton fabrics via a wet chemical route and hydrophobic modification. *Appl. Surf. Sci*. 254, 5899–5904. doi: 10.1016/j.apsusc.2008.03.160

Xu, B., Cai, Z., Wang, W., and Ge, F. (2010). Preparation of superhydrophobic cotton fabrics based on SiO2 nanoparticles and ZnO nanorod arrays with subsequent hydrophobic modification. *Surf. Coat. Technol*. 204, 1556–1561. doi: 10.1016/j.surfcoat.2009.09.086

Xu, F., Deng, M., Liu, Y., Ling, X., Deng, X., and Wang, L. (2014). Facile preparation of poly (diallyldimethylammonium chloride) modified reduced graphene oxide for sensitive detection of nitrite. *Electrochem. Commun*. 47, 33–36. doi: 10.1016/j.elecom.2014.07.016

Zafar, S., Negi, L. M., Verma, A. K., Kumar, V., Tyagi, A., Singh, P., et al. (2014). Sterically stabilized polymeric nanoparticles with a combinatorial approach for multi drug resistant cancer: in vitro and in vivo investigations. *Int. J. Pharm*. 477, 454–468. doi: 10.1016/j.ijpharm.2014.10.061

Zhang, F., Wu, X., Chen, Y., and Lin, H. (2009). Application of silver nanoparticles to cotton fabric as an antibacterial textile finish. *Fibers Polym*. 10, 496–501. doi: 10.1007/s12221-009-0496-8

Conflict of Interest Statement: The authors declare that the research was conducted in the absence of any commercial or financial relationships that could be construed as a potential conflict of interest.

17

Antibacterial Activity and Synergistic Antibacterial Potential of Biosynthesized Silver Nanoparticles against Foodborne Pathogenic Bacteria along with its Anticandidal and Antioxidant Effects

Jayanta Kumar Patra[1] and Kwang-Hyun Baek[2]*

[1] Research Institute of Biotechnology and Medical Converged Science, Dongguk University-Seoul, Goyang-si, South Korea,
[2] Department of Biotechnology, Yeungnam University, Gyeongsan, South Korea

Edited by:
Abd El-Latif Hesham,
Assiut University, Egypt

Reviewed by:
Carmen Losasso,
Istituto Zooprofilattico Sperimentale
delle Venezie, Italy
Conrad Oswald Perera,
University of Auckland, New Zealand

*Correspondence:
Kwang-Hyun Baek
khbaek@ynu.ac.kr

Silver nanoparticles plays a vital role in the development of new antimicrobial substances against a number of pathogenic microorganisms. These nanoparticles due to their smaller size could be very effective as they can improve the antibacterial activity through lysis of bacterial cell wall. Green synthesis of metal nanoparticles using various plants and plant products has recently been successfully accomplished. However, few studies have investigated the use of industrial waste materials in nanoparticle synthesis. In the present investigation, synthesis of silver nanoparticles (AgNPs) was attempted using the aqueous extract of corn leaf waste of *Zea mays*, which is a waste material from the corn industry. The synthesized AgNPs were evaluated for their antibacterial activity against foodborne pathogenic bacteria (*Bacillus cereus* ATCC 13061, *Listeria monocytogenes* ATCC 19115, *Staphylococcus aureus* ATCC 49444, *Escherichia coli* ATCC 43890, and *Salmonella* Typhimurium ATCC 43174) along with the study of its synergistic antibacterial activity. The anticandidal activity of AgNPs were evaluated against *Candida* species (*C. albicans* KACC 30003 and KACC 30062, *C. glabrata* KBNO6P00368, *C. geochares* KACC 30061, and *C. saitoana* KACC 41238), together with the antioxidant potential. The biosynthesized AgNPs were characterized by UV-Vis spectrophotometry with surface plasmon resonance at 450 nm followed by the analysis using scanning electron microscope, X-ray diffraction, Fourier-transform infrared spectroscopy and thermogravimetric analysis. The AgNPs displayed moderate antibacterial activity (9.26–11.57 mm inhibition zone) against all five foodborne pathogenic bacteria. When AgNPs were mixed with standard antibacterial or anticandidal agent, they displayed strong synergistic antibacterial (10.62–12.80 mm inhibition zones) and anticandidal activity (11.43–14.33 mm inhibition zones). In addition, the AgNPs exhibited strong antioxidant potential. The overall results highlighted the potential use of maize industrial waste materials in the synthesis of AgNPs and their utilization in various applications particularly as antibacterial substance in food packaging, food preservation to protect against various dreadful foodborne pathogenic bacteria together with its biomedical, pharmaceutical based activities.

Keywords: antibacterial, anticandidal, antioxidant, foodborne bacteria, green synthesis, silver nanoparticles, *Zea mays*

INTRODUCTION

Nanotechnology is an emerging field of interdisciplinary research that includes all spheres of science starting from physics, chemistry, biology, and especially biotechnology (Natarajan et al., 2010). Nanoparticles (NPs) are a group of materials synthesized from a number of metals or non-metal elements with distinct features and extensive applications in different fields of science and medicine (Matei et al., 2008). Among them, silver nanoparticles (AgNPs) have been extensively studied because of their good electrical conductivity, as well as their potential for use in optical applications in nonlinear optics, as spectrally selective coatings for solar energy absorption, biolabeling, intercalation materials for electrical batteries as optical receptors, and catalysts in chemical reactions. Nanoparticles also have potential biological applications, such as biosensing, catalysis, drug delivery, imaging, nano device fabrication, and for use as antimicrobial agents and in medicine (Ghosh et al., 1996; Geddes et al., 2003; Nair and Laurencin, 2007; Jain et al., 2008; Sharma et al., 2009; Zargar et al., 2014). AgNPs release Ag$^+$ ions that interact with the thiol groups in bacterial proteins and affect the DNA replication, resulting in destruction of the bacteria (Marini et al., 2007). Additionally, nanoparticles have been shown to have potential anti-bacterial activity and significantly higher synergistic effects when applied with many antibiotics (Devi and Joshi, 2012).

Synthesis of AgNPs employing chemical and physical methods has been extensively studied throughout the world; however, these methods are often environmentally toxic, technically laborious and economically expensive (Gopinath et al., 2012). Accordingly, biological methods for synthesis of AgNPs using plants, microorganisms and enzymes have been suggested as possible eco-friendly alternatives (Mohanpuria et al., 2008). The synthesis of AgNPs using plants or plant extracts as reducing and capping agents is considered advantageous over other biological processes because they eliminate the need for the elaborate process of culturing and maintaining biological cells, and can be scaled up for large-scale nanoparticle synthesis (Saxena et al., 2012; Valli and Vaseeharan, 2012). Overall, plant-mediated nanoparticles synthesis is a cost-effective, environmentally friendly, a single-step method for biosynthesis process that is safe for various human therapeutic and food based uses (Kumar and Yadav, 2009). Generally, the AgNO$_3$ is reduced by the action of the reducing agents (plant extracts) to form silver nanoparticles which are further stabilized by the bioactive compounds from the biological extracts to form a stable silver nanoparticle.

During recent years, the use of agricultural and industrial wastes in the synthesis of different types of metal nanoparticles has been extensively investigated (Basavegowda and Lee, 2013; Ramamurthy et al., 2013; Nezamdoost et al., 2014) A number of food crops are industrially used for production of different types of food products and processed food. Among these, maize (*Zea mays*) is widely used throughout the world for production of popcorn, chips, corn oil, corn starch, and many other materials. Only the kernels of the corn plant are edible, while rest of the crop are occasionally used as animal feed or ingredients in beverages. Different parts of the maize plant have been effectively utilized in traditional medicines as strong therapeutic agents (Konstantopoulou et al., 2004; Ullah et al., 2010; Solihah et al., 2012). A number of bioactive compounds, such are polyphenols (chlorogenic acid, caffeic acid, rutin, ferulic acid, morin, quercetin, naringenin, and kaempferol), anthocyanins, flavonoids, flavonols, and flavanols have been reported to be present in the *Z. mays* plant and its various parts such kernel, leaves, roots etc., which are responsible for its antioxidant, antiinflammatory and other medicinal potential (Ramos-Escudero et al., 2012; Bacchetti et al., 2013; Pandey et al., 2013). Hence utilization of maize waste materials in the synthesis of nanoparticles would be a profitable approach in ecofriendly and cost effective nanoparticle synthesis.

Recently there is emergence of multi drug resistant pathogenic bacterial strains and most of the available antibiotics are not active against these pathogens (Andersson and Hughes, 2010; Huh and Kwon, 2011). These drug resistant pathogens are more pathogenic with high mortality rate than that of wild strain. The scientific community is continuously searching for a new classes of disinfection systems that could act efficiently against these pathogens. Silver-containing systems, and especially the AgNPs are these days one of the strong alternatives in search for various antibacterial drugs, as these nanoparticles have been reported previously to exhibit interesting antibacterial activities against a broad spectrum of pathogenic bacteria (Shahverdi et al., 2007; Lara et al., 2011; Guzman et al., 2012; Rai et al., 2012; Ouay and Stellacci, 2015). However, studies on AgNPs are still under investigation as antimicrobial and the studies performed since now demonstrated that a case by case evaluation have to be done for each nanoparticle and bacterial target. The bactericidal effects of ionic silver and the antimicrobial activity of colloidal silver particles is generally influenced by the size of the particles, i.e., the smaller the particle size, the greater the antimicrobial activity (Zhang et al., 2003).

There are many advantages of AgNPs to be used as an effective antimicrobial agents. They are highly effective against a broad range of microbes and parasites, even at a very low concentration with very little systemic toxicity toward humans (Ouay and Stellacci, 2015). AgNPs have been reported to be used and tested for several applications including prevention of bacterial colonization and elimination of microorganisms on various medical devices, disinfection in wastewater treatment plants, and silicone rubber gaskets to protect and transport food and textile fabrics (Guzman et al., 2012).

The present study investigated synthesis of AgNPs using the waste leaves of ears of corn following a green route and evaluate its potential application as antibacterial compound against a number of five foodborne pathogenic bacteria (*Bacillus cereus* ATCC 13061, *Listeria monocytogenes* ATCC 19115, *Staphylococcus aureus* ATCC 49444, *Escherichia coli* ATCC 43890, and *Salmonella* Typhimurium ATCC 43174) along with its anticandidal potential against five different *Candida* species (*C. albicans* KACC 30003 and KACC 30062, *C. glabrata* KBNO6P00368, *C. geochares* KACC 30061, and *C. saitoana* KACC 41238) and their antioxidant potentials. Utilization of these industrial waste materials in the synthesis of nanoparticles could add the values to the economy of industry.

MATERIALS AND METHODS

Sample Preparation

The whole corn of *Z. mays* L. (**Figure 1A**) was purchased from a local market located at Gyeonsan, Republic of Korea. The ear leaves (**Figure 1C**) were collected from the corn (**Figures 1A,B**) and cut into small pieces of approximately 1 cm. A total of 20 g of leaf pieces were then placed in a 250 mL conical flask, after which 100 mL of double distilled water was added and the samples were boiled for 15 min with continuous stirring. The aqueous extract of corn leaves (ACL) was cooled to room temperature, filtered and stored at 4°C before being used for the synthesis of AgNPs.

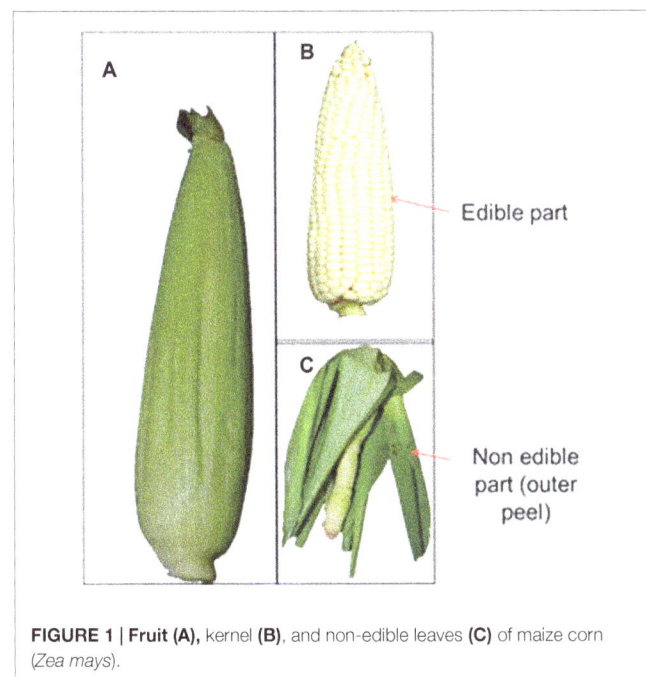

FIGURE 1 | Fruit (A), kernel (B), and non-edible leaves (C) of maize corn (*Zea mays*).

Biosynthesis of AgNPs

The synthesis of AgNPs was conducted by the green synthesis route using ACL. Briefly, 20 mL of ACL was added to 500 mL conical flasks containing 200 mL of 1 mM $AgNO_3$ and stirred continuously at room temperature until the solution became reddish brown. The concentration of ACL to $AgNO_3$ was maintained at 1:10 ratio with the use of less concentration of ACL and $AgNO_3$ in order to control the shape and size of the nanoparticles.

Characterization of AgNPs

The newly synthesized AgNPs were characterized by UV-VIS spectroscopy, scanning electron microscopy (SEM), energy-dispersive X-ray spectroscopy (EDS), Fourier-transform infrared spectroscopy (FT-IR), thermogravimetric and differential thermogravimetric (TGA/DTG) analysis, and X-ray powder diffraction (XRD) using standard analytical procedures (Basavegowda and Lee, 2013; Ramamurthy et al., 2013).

The synthesis of the AgNPs was monitored by the UV-Vis spectroscopy analysis by measuring the absorption spectra between 350 and 550 nm at a resolution of 1 nm using a microplate reader (Infinite 200 PRO NanoQuant, TECAN, Mannedorf, Switzerland). Changes in the color of the reaction mixture were observed every 3 h during incubation. The surface morphology of the AgNPs was analyzed using FE-SEM. The AgNPs were powdered using an agate mortar and pestle, then uniformly spread over the sample holder and sputter coated with platinum in an ion coater for 120 s, after which they were observed by FE-SEM (S-4200, Hitachi, Japan). Elemental composition analysis of the powdered AgNPs was conducted using an EDS detector (EDS, EDAX Inc., Mahwah, NJ, USA) attached to the FE-SEM machine. FT-IR analysis of the powdered AgNPs and the ACL extract was conducted using a FT-IR spectrophotometer (Jasco 5300, Jasco, Mary's Court, Easton, MD, USA) in the wavelength range of 400–4000 cm^{-1}. The powdered AgNPs sample was blended with potassium bromide (KBr) in a 1:100 ratio using an agate mortar and pestle, then compressed into a 2 mm semi-transparent disk using a specially designed screw knot, after which different modes of vibrations were analyzed for the presence of different types of functional groups in AgNPs and ACL extract.

Effects of high temperature on synthesized AgNPs were evaluated using a TGA machine (SDT Q600, TA Instruments, New Castle, DE, USA). For TGA analysis, powdered AgNPs (3.0 mg) were placed in an alumina pan and heated from 20 to 700°C at a ramping time of 10°C/min under a N_2 atmosphere in a specially designed heating chamber. The corresponding weight loss data were recorded using a computer attached to the TG/DTG machine with the SDT software. The AgNPs nanoparticles were analyzed by XRD (X'Pert MRD model, PANalytical, Almelo, The Netherlands). Prior to use, the AgNPs were dried at 60°C in a vacuum oven and ground to fine powder using an agate mortar and pestle. The samples were uniformly spread over the glass sample holder and subsequently analyzed at 30 kV and 40 mA with Cu Kα radians at an angle of 2θ. The average particle diameter of AgNPs was calculated from the XRD pattern, according to the line width of the maximum intensity reflection peak. The size of the nanoparticles was calculated through the Scherer equation (Yousefzadi et al., 2014).

Biological Activity of AgNPs
Antibacterial Activity of AgNPs

The antibacterial potential of AgNPs was determined against five different foodborne bacteria (*B. cereus* ATCC 13061, *L. monocytogenes* ATCC 19115, *S. aureus* ATCC 49444, *E. coli* ATCC 43890, and *S.* Typhimurium ATCC 43174) by the standard disk diffusion method (Diao et al., 2013). The bacterial pathogens were obtained from the American Type Culture Collection (ATCC, Manassas, VA, USA) and maintained on nutrient agar media (Difco, Becton, Dickinson and Company, Sparks Glencoe, MD, USA). Prior to use, the colloidal solution of the AgNPs was prepared by dissolving AgNPs in 5% dimethyl sulfoxide (DMSO, 1000 µg/mL) and sonicating the samples at 30°C for 15 min. Filter paper disks containing 50 µg of AgNPs/disk were used for the assay. Standard antibiotics, kanamycin and rifampicin, at 5 µg/disk were taken as positive controls, while 5% DMSO was used as the negative control. The overnight grown cultures of tested bacteria were diluted to 1×10^{-7} colony forming unit were used for the assay. The antibacterial activity of the AgNPs was determined by measuring the diameter of zones of inhibition after 24 h of incubation at 37°C. The minimum inhibitory concentration (MIC) and minimum bactericidal concentration (MBC) of the AgNPs were determined by the two-fold serial dilution method (Kubo et al., 2004). Different concentrations of AgNPs (100–3.12 µg/mL) were used for MIC test. Prior to test, initially 200 µg of the AgNP was added to initial tube containing, 2 mL of NB media, then 1 mL from it was transferred to next tube which contains 1 mL of only NB media and mixed properly, then the dilution was made till the concentration of the last tube was 3.12 µg/mL. The control tube contains only 1 mL of NB media. Then 10 µL of the tested pathogen was added to each tube. This procedure was repeated for all the tested pathogens. Then all the tubes were mixed properly and were incubated at 37°C overnight in a shaker incubator. The lowest concentration of AgNPs that did not show any visible growth of test organisms was determined as the MIC. Further, the MIC concentration and the next higher concentration were spread on NA plates and incubated for another 24 h at 37°C. The concentration that did not show any growth of a single bacterial colony on the NA plates was defined as the MBC value. Both MIC and MBC values were expressed as µg/mL.

Synergistic Potential of AgNPs

The synergistic activity of the AgNPs was determined with antibiotics (kanamycin and rifampicin) or anticandidal agent (amphotericin b).

Synergistic antibacterial activity of AgNPs

The synergistic antibacterial potential of AgNPs, as well as kanamycin and rifampicin as a standard antibiotics was determined against five foodborne pathogenic bacteria, *B. cereus* ATCC 13061, *E. coli* ATCC 43890, *L. monocytogenes* ATCC 19115, *S. aureus* ATCC 49444, and *S.* Typhimurium ATCC 43174

by the standard disk diffusion method (Naqvi et al., 2013). The bacterial pathogens were freshly cultured on nutrient broth media (Difco, Becton, Dickinson and Company, Sparks Glencoe, MD, USA). AgNPs (1 mg/mL) and the standard antibiotics (kanamycin or rifampicin at 200 µg/mL) were mixed properly at a 1:1 ratio and sonicated for 15 min at room temperature. Different antibiotic disks were prepared by adding 50 µl of the AgNPs/antibiotics mixture solution to a 6 mm filter paper disk that contains 25 µg AgNPs and 5 µg antibiotics together. The synergistic antibacterial activity of the AgNPs/antibiotics mixture was measured after 24 h of incubation at 37°C in terms of the diameters of the zones of inhibition around the filter paper disks.

Synergistic anticandidal activity of AgNPs

The synergistic anticandidal potentials of AgNPs and amphotericin b, a standard antifungal agent, were determined against five different pathogenic *Candida* species, *C. albicans* KACC 30003 and KACC 30062, *C. glabrata* KBNO6P00368, *C. geochares* KACC 30061 and *C. saitoana* KACC 41238, by the disk diffusion method (Murray et al., 1995). These *Candida* species were obtained from the Korean Agricultural Culture Collection (KACC, Suwon, Republic of Korea). AgNPs (2 mg/mL) and amphotericin b (200 µg/mL) were mixed in a 1:1 ratio and sonicated for 15 min at room temperature. Paper disks were prepared by adding 50 µL of the AgNPs/amphotericin b mixture solution to a 6 mm filter paper disk that contains 50 µg AgNPs and 5 µg amphotericin b. The *Candida* species in liquid media were spread uniformly on potato-dextrose agar (PDA) media (Difco, Becton, Dickinson and Company, Sparks Glencoe, MD, USA), after which the anticandidal disks were placed on the plates and samples were incubated at 28°C for 48 h. The synergistic anticandidal activity of the AgNPs/amphotericin b mixture solution was determined by measuring the diameters of the zones of inhibition around the paper disk.

Antioxidant Activity of AgNPs

The antioxidant potential of the AgNPs was determined by 1,1-diphenyl-2-picrylhydrazyl (DPPH) radical scavenging, nitric oxide (NO) scavenging, 2,2′-azino-bis(3-ethylbenzothiazoline-6-sulphonic acid) (ABTS) radical scavenging and reducing power assays.

DPPH Radical Scavenging Activity of AgNPs

The DPPH free radical scavenging potential of AgNPs was determined as previously described (Patra et al., 2015). Briefly, five different concentrations (20–100 µg/mL) of AgNPs and ascorbic acid (ASA) as the standard reference compound was assayed. The absorbance of the reaction mixtures was recorded at 517 nm using the microplate reader and the results were interpreted as the percentage scavenging according to the following equation:

$$\text{Percentage scavenging} = \frac{\text{Abs}_C - \text{Abs}_T}{\text{Abs}_C} \times 100$$

where, Abs_C is the absorbance of the control and Abs_T is the absorbance of the treatment.

NO Scavenging Activity of AgNPs

The NO scavenging potential of AgNPs was determined by the standard procedure (Makhija et al., 2011). Five different concentrations (20–100 µg/mL) of AgNPs and ASA as the standard reference compound were taken for the assay. The absorbance of the reaction mixtures was recorded at 546 nm using a microplate reader, after which the results were calculated as the percentage scavenging activity according to Eq. 1.

ABTS Radical Scavenging Activity of AgNPs

The ABTS radical scavenging potential of AgNPs was determined by the standard procedure (Thaipong et al., 2006). Briefly, five different concentrations (20–100 µg/mL) of AgNPs and ASA as the standard reference compound was taken for the assay. The absorbance of the reaction mixtures was recorded at 750 nm using a microplate reader, and the results were interpreted according to Eq. 1.

Reducing Power of AgNPs

The reducing power of AgNPs was determined by the standard procedure (Sun et al., 2011). Briefly, five different concentrations (20–100 µg/mL) of AgNPs and ASA as the standard reference compound were assayed. The absorbance of the reaction mixture was measured at 700 nm against an appropriate control and the results were expressed as OD values at 700 nm.

Statistical Analysis

The results of all the experiments were expressed as the mean value of three independent replicates ± the standard deviation (SD). Statistical analysis of the significance differences between the mean values of the results were identified by one-way analysis of variance (ANOVA) followed by Duncan's test at the 5% level of significance ($P < 0.05$) using the Statistical Analysis Software (SAS) (Version: SAS 9.4, SAS Institute Inc., Cary, NC, USA).

RESULTS

Synthesis of AgNPs

The industrial wastes from maize plants (**Figures 1A,C**) after utilization of the kernels (**Figure 1B**) were used in the present study for synthesis of AgNPs. Biosynthesis of AgNPs was indicated by gradual color development in the reaction solution after 1 h of incubation and subsequent increases in the intensity of the color during the course of reaction. The formation of AgNPs was monitored by a number of characterization techniques as described below.

Characterization of AgNPs

The UV-Vis spectra of the synthesized AgNPs recorded at different time intervals are presented in **Figure 2**. The absorbance peaks indexed as different colors indicated the reduction of AgNO₃ by ACL with different time intervals (0 min, 30 min, 1 h, 3 h, 6 h, 12 h, and 24 h) at room temperature (**Figure 2**). The UV-Vis spectra of the synthesized AgNPs were further recorded after 24 h, but the intensity of the color did not intensify after 24 h, confirming that the reaction was completed within 24 h.

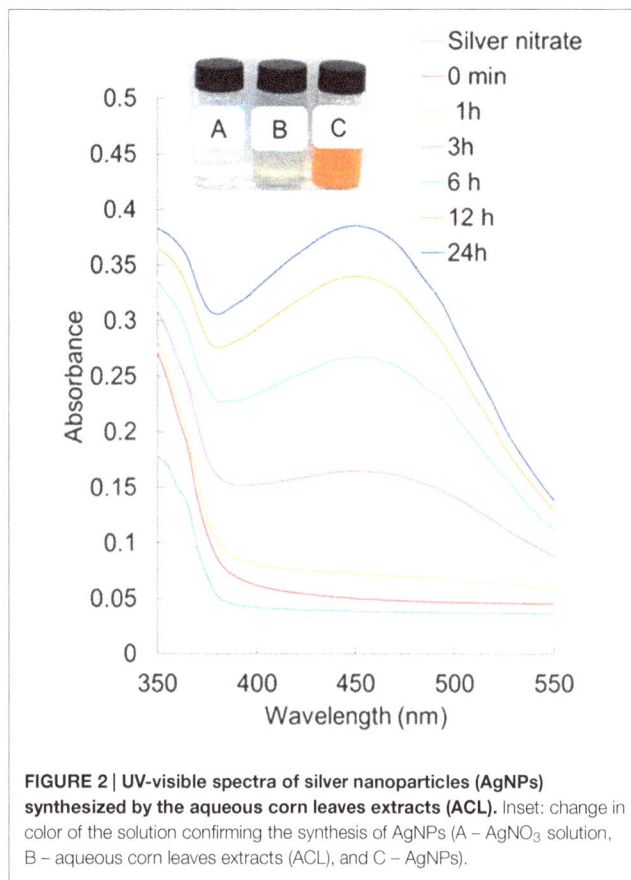

FIGURE 2 | UV-visible spectra of silver nanoparticles (AgNPs) synthesized by the aqueous corn leaves extracts (ACL). Inset: change in color of the solution confirming the synthesis of AgNPs (A – $AgNO_3$ solution, B – aqueous corn leaves extracts (ACL), and C – AgNPs).

The morphology of the synthesized AgNPs was revealed by FE-SEM analysis (**Figure 3A**). The FE-SEM image revealed the formation of a cluster of spherical beadlike structures of AgNPs that were strongly aggregated. The elemental composition of the synthesized AgNPs was determined by an EDS machine attached to the FE-SEM. The elemental composition confirmed that AgNPs were composed of 50.13% Ag, 31.20% C, 11.60% O, 6.20% Cl, and 0.87% Na (**Figures 3B,C**). FT-IR analysis of the ACL and AgNPs is shown in **Figure 4**. Absorption peaks located at 3438.38, 2353.05, 2167.81, 1645.40, 779.50, 664.44, and 564.98 cm^{-1} were observed upon ACL, whereas absorption peaks located at 3423.44, 2921.74, 2367.78, 1654.34, 1053.13, 671.22, and 527.40 cm^{-1} were observed for the AgNPs (**Figure 4**).

Thermogravimetric and differential thermogravimetric analysis of the synthesized AgNPs was conducted to show the nature of AgNPs at higher temperature (**Figure 5A**). A total of 44.01% weight loss was observed in three different phases when the AgNPs were heated to 700°C in a controlled N₂ atmosphere. The first phase of weight loss was observed between 30 and 150°C with a weight loss of 5.67%. In this phase, the water molecules that were attached to the AgNPs during the course of synthesis were degraded. The second phase of weight loss was observed between 150 and 470°C with a maximum weight loss of 33.77%. During this phase, organic molecules, such as alkanes, phenols, alkenes, proteins, and polysaccharides

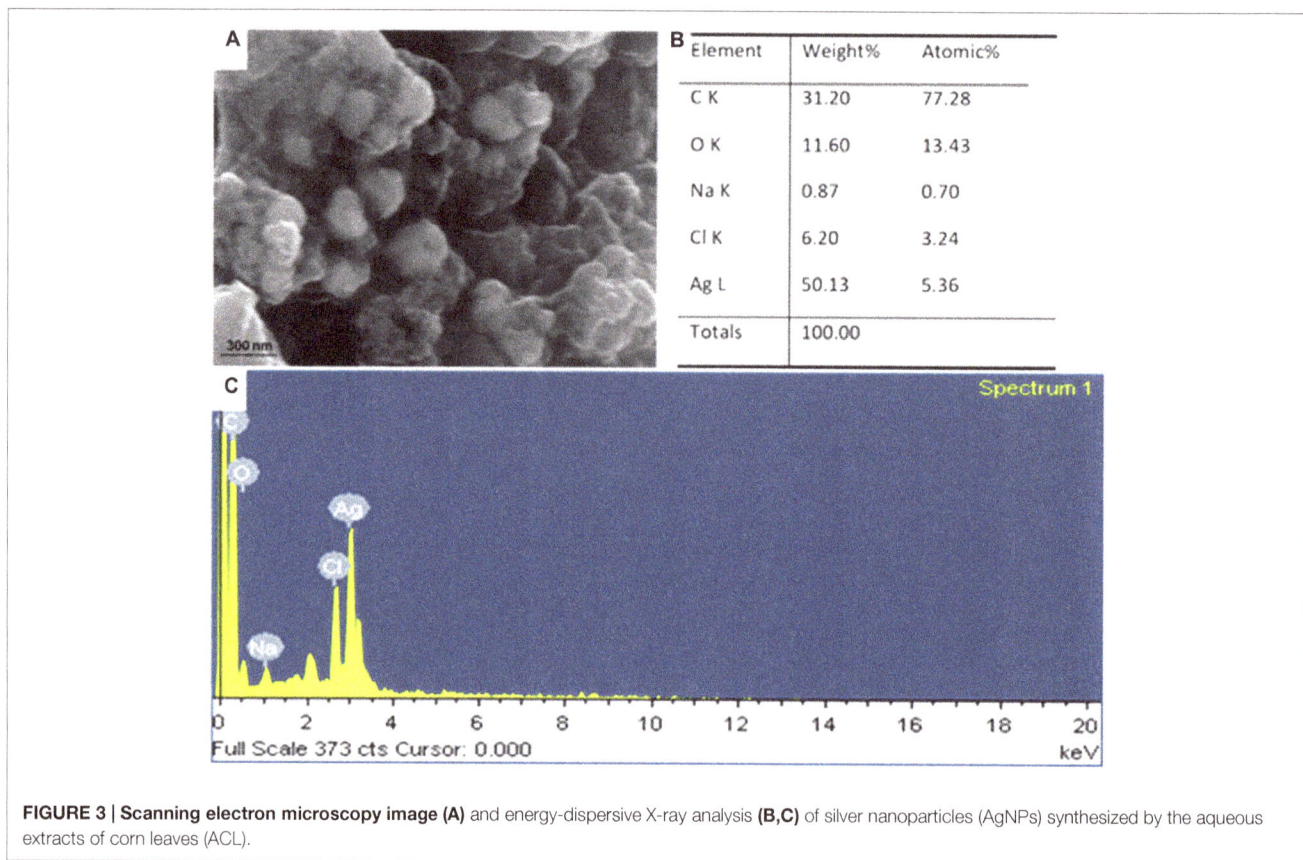

FIGURE 3 | Scanning electron microscopy image (A) and energy-dispersive X-ray analysis **(B,C)** of silver nanoparticles (AgNPs) synthesized by the aqueous extracts of corn leaves (ACL).

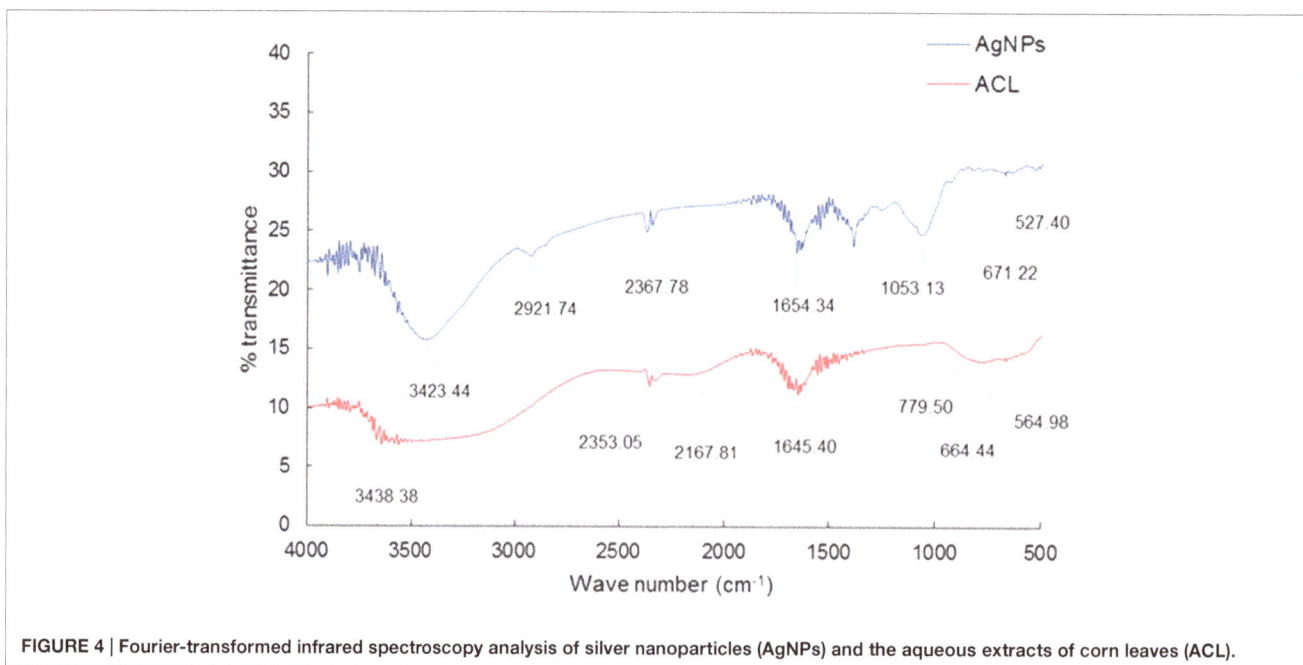

FIGURE 4 | Fourier-transformed infrared spectroscopy analysis of silver nanoparticles (AgNPs) and the aqueous extracts of corn leaves (ACL).

from the ACL that contributed to the reduction of the AgNPs as capping and stabilizing agents were degraded. The third phase extended from 480 to 700°C, during which time there was a weight loss of 4.57%. The nature of the synthesized AgNPs was analyzed by XRD (**Figure 5B**). The diffraction pattern showed six diffraction peaks at 27.78°, 32.04°, 38.41°, 46.12°, 54.95°, and 76.78°, which corresponded to (220), (122), (111), (231), (331), and (311) planes of silver, respectively.

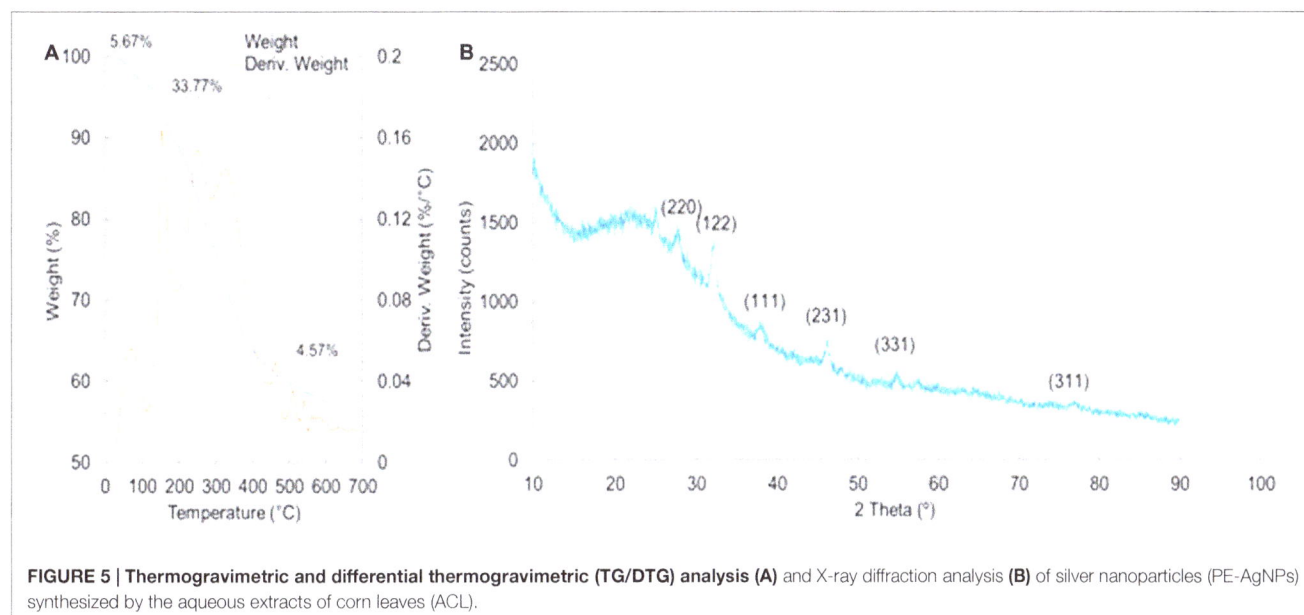

FIGURE 5 | Thermogravimetric and differential thermogravimetric (TG/DTG) analysis (A) and X-ray diffraction analysis **(B)** of silver nanoparticles (PE-AgNPs) synthesized by the aqueous extracts of corn leaves (ACL).

The average crystal size of the silver crystallites was calculated from the full width at half maximum (FWHMs) values of the diffraction peaks, using the Scherer equation. The estimated size of crystallite in different planes of silver was determined as 31.18, 35.74, and 69.14 nm with the mean value of all three peaks as 45.26 nm.

Biological Activity of AgNPs
Antibacterial Activity of AgNPs

The AgNPs at 50 μg/disk displayed moderate antibacterial activity against all five foodborne pathogenic bacteria, as indicated by diameter of inhibition zones of 9.26–11.57 mm (**Table 1**; **Figure 6**). The standard antibiotics, kanamycin, and rifampicin, at 5 μg/disk did not show any inhibitory activity against any of the five pathogens. Among the pathogenic bacteria, AgNPs were more active against *S. aureus* (11.57 mm inhibition zone) than *L. monocytogenes* (9.26 mm inhibition zone). The MIC and the MBC values of AgNPs against all five pathogenic bacteria ranged from 12.5 to 100 μg/mL (**Table 1**).

TABLE 1 | Antibacterial activity of AgNPs against five foodborne pathogenic bacteria.

Bacteria	AgNPs (50 μg/disk)	MIC (μg/mL)	MBC (μg/mL)
B. cereus ATCC 13061	11.39 ± 1.2[a]*	25	50
E. coli ATCC 43890	10.55 ± 0.27[b]	50	100
L. monocytogenes ATCC 19115	9.26 ± 0.31[c]	25	50
S. aureus ATCC 49444	11.57 ± 0.25[a]	12.5	25
S. Typhimurium ATCC 43174	11.22 ± 0.38[a]	50	100

Data are expressed as the mean zone of inhibition in mm ± SD. Values with different superscript letters are significantly different at P < 0.05.

Synergistic Antimicrobial Potentials of AgNPs
Synergistic antibacterial potential of AgNPs

The synergistic potential of the AgNPs together with the standard antibiotics, kanamycin and rifampicin, were evaluated against all five foodborne pathogenic bacteria and the results are presented in **Table 2** and **Figure 6**. At low concentrations (5 μg/disk), neither antibiotics exhibited any positive activity against any of the five pathogenic bacteria, which was also true for AgNPs at 25 μg/disk concentration. Thus, to study the synergistic antibacterial potential, both antibiotics and AgNPs were combined at this low concentration and their activities were tested against the five foodborne pathogens. When both antibiotic and AgNPs were mixed, they displayed strong antibacterial activity against all pathogens, with zones of inhibition ranging in diameter from 10.62 to 14.33 mm (**Table 2**).

Synergistic anticandidal potential of AgNPs

The synergistic anticandidal activities of the AgNPs are presented in **Table 3** and **Figure 7**. AgNPs at a concentration of 50 μg/mL did not exhibit any anticandidal activity against the five tested *Candida* species. However, when AgNPs (50 μg/disk) were combined with a standard anticandidal agent, amphotericin b (5 μg/disk), they displayed potent anticandidal activity against all five *Candida* species, with zones of inhibition ranging from 9.74 to 14.75 mm (**Table 3**; **Figure 7**).

Antioxidant Activity of AgNPs

The antioxidant potential of synthesized AgNPs was determined by *in vitro* assays of DPPH radical scavenging, NO scavenging, ABTS radical scavenging and reducing power. The DPPH radical scavenging potential of AgNPs is presented in **Figure 8A**. AgNPs displayed a moderate DPPH radical scavenging potential of 34.09% at 100 μg/mL, whereas ASA, which was taken as the reference standard, showed comparably high DPPH scavenging activity of 42.41% at 100 μg/mL (**Figure 8A**). AgNPs

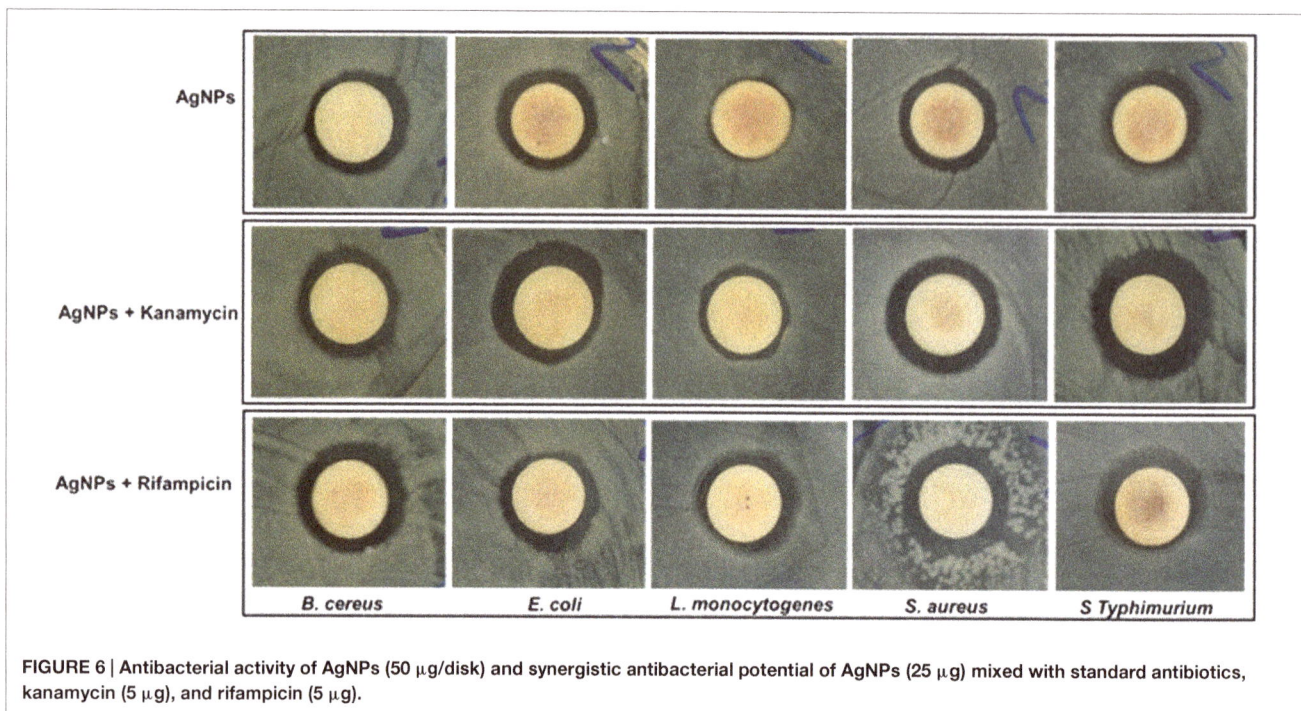

FIGURE 6 | Antibacterial activity of AgNPs (50 μg/disk) and synergistic antibacterial potential of AgNPs (25 μg) mixed with standard antibiotics, kanamycin (5 μg), and rifampicin (5 μg).

TABLE 2 | Synergistic antibacterial activity of AgNPs (25 μg) with standard antibiotics, kanamycin (5 μg) or rifampicin (5 μg), against foodborne pathogenic bacteria.

Bacteria	AgNPs + Kanamycin	AgNPs + Rifampicin
B. cereus ATCC 13061	$11.67 \pm 0.16^{c*}$	12.76 ± 0.44^{b}
E. coli ATCC 43890	12.45 ± 2.02^{b}	11.43 ± 0.21^{c}
L. monocytogenes ATCC 19115	10.62 ± 0.29^{d}	11.64 ± 0.23^{c}
S. aureus ATCC 49444	12.57 ± 0.20^{b}	14.33 ± 0.40^{a}
S. Typhimurium ATCC 43174	12.80 ± 0.31^{a}	11.45 ± 0.19^{c}

*Data are expressed as the mean zone of inhibition in mm ± SD. Values in each same column with different superscript letters are significantly different at P < 0.05.

TABLE 3 | Synergistic anticandidal activity of AgNPs (50 μg) with a standard antifungal agent, amphotericin b (5 μg), against pathogenic *Candida* species.

Candida species	Mean inhibition zone in mm ± SD
C. albicans KACC 30003	$10.34 \pm 0.29^{cd*}$
C. albicans KACC 30062	12.88 ± 0.15^{b}
C. glabrata KBNO6P00368	10.98 ± 0.71^{c}
C. glochares KACC 30061	14.75 ± 0.30^{a}
C. saitoana KACC 41238	9.74 ± 0.14^{d}

*Values with different superscript letters are significantly different at P < 0.05.

exerted comparably high NO scavenging potential of 82.63% at 100 μg/mL compared with that of 41.95% of ASA at 100 μg/mL (**Figure 8B**). The ABTS radical scavenging potential of AgNPs is presented in **Figure 8C**. AgNPs exhibited a moderate value of 49.29% ABTS radical scavenging potential relative to 82.20% by ASA at 100 μg/mL. AgNPs also displayed strong reducing power (**Figure 8D**).

DISCUSSION

The concentration of ACL to $AgNO_3$ was maintained at 1:10 ratio with the use of less concentration of ACL and $AgNO_3$ in order to obtain small and controlled size of nanoparticles. It is presumed that less the concentration of the plant extracts and $AgNO_3$ used, the smaller will be the size of the nanoparticles. This hypothesis has been proved by several researchers (Ghosh et al., 2011; Chandran et al., 2014) and can be confirmed again by

the present study. The appearance of brown color in the reaction solution (**Figure 2**, inset) was a clear indication of the formation of AgNPs in the reaction mixture (Wu et al., 2004; Kumar et al., 2008; Kumari and Philip, 2013). The characterization of the synthesized AgNPs was achieved using techniques, such as UV-Vis spectroscopy, FE-SEM, EDS, FT-IR, TG/DTG analysis, and XRD analysis (Zhang et al., 2006; Choi et al., 2007; Vilchis-Nestor et al., 2008). These techniques are used for determination of different parameters, such as nature, particle size, characteristics, crystallinity, and surface area.

Spectral analysis revealed that the surface plasmon resonance phenomena (SPR) absorption maxima peak of the synthesized AgNPs occurred at 450 nm with a high absorbance value specific for AgNPs (**Figure 2**) (Kelly et al., 2003; Nazeruddin et al., 2014). In general, typical AgNPs show characteristic SPR at wavelengths ranging from 400 to 480 nm (Pal et al., 2007; Nazeruddin et al., 2014), which was also observed in the present investigation. The SPR absorbance is sensitive to the shape, size and nature of particles present in the solution, and also depends upon inner particle distance and the surrounding media

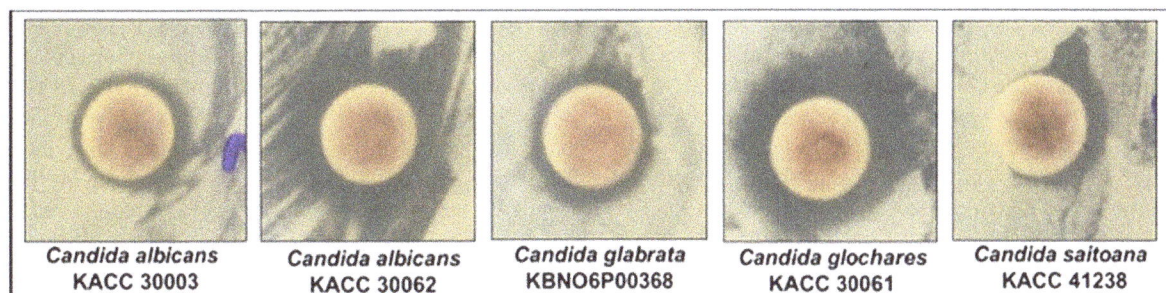

FIGURE 7 | Synergistic anticandidal potential of AgNPs (50 μg) mixed with a standard amphotericin b (5 μg).

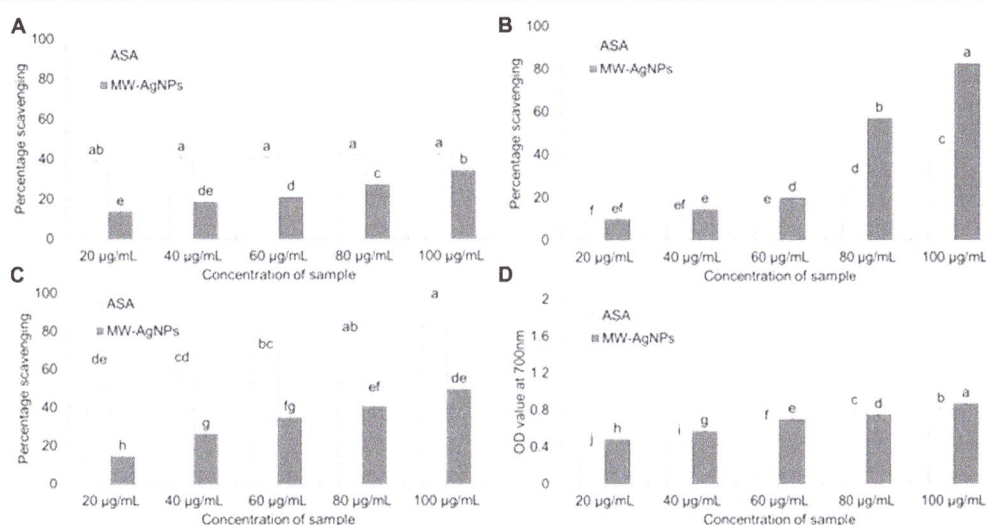

FIGURE 8 | Antioxidant potentials of AgNPs and ascorbic acid (ASA). (A) DPPH radical scavenging. **(B)** Nitric oxide scavenging. **(C)** ABTS radical scavenging. **(D)** Reducing power assay. Different superscript letters in each column indicate significant differences at $P < 0.05$.

(Nazeruddin et al., 2014). The surface morphology was seen by FE-SEM (**Figure 3A**) and its elemental composition by EDS analysis (**Figures 3B,C**). The EDS pattern indicates that the synthesized AgNPs were crystalline in nature, which is caused by the reduction of silver ions. A strong typical absorption peak was observed at 3.0 keV, which is typical of the absorption of metallic silver nanocrystallites due to SPR (Das et al., 2013) (**Figure 3C**). Similar results upon SEM and EDS analysis of different types of AgNPs were reported previously (Vijaykumar et al., 2013; Nazeruddin et al., 2014; Muthukrishnan et al., 2015; Velusamy et al., 2015).

The intense peaks in FT-IR spectra of both ACL and AgNPs (**Figure 4**) located at 3438 and 3423 cm^{-1} corresponded to O–H stretching of the alcohols and phenolic compounds, while the intense peaks at 1654 and 1645 cm^{-1} corresponded to – C=C-H stretching of the alkenes group (Ramamurthy et al., 2013; Rajeshkumar and Malarkodi, 2014). A major peak was observed at 2921 cm^{-1} in AgNPs, which could be assigned to the C-H stretching vibrations of methyl, methylene, and methoxy groups (Feng et al., 2009). The mechanism of adsorption and capping

of AgNPs by ACL can be explained through the coordination of carbonyl bonds (3423 cm^{-1}) and subsequent electron transfer from C=O to AgNPs (Qiu et al., 2006). The peak at 3423 cm^{-1} that corresponds to O-H stretching, 1654 cm^{-1} that corresponds to – C=C-H stretching of the alkenes group, and 1053 cm^{-1} might be contributed to by the C-O groups of the polysaccharides in the ACL extract that acted as reducing, capping, and stabilizing agents for the synthesis of AgNPs (Muthukrishnan et al., 2015). The slight shifting in the position of different peaks in the AgNPs from the ACL extract might have been due to progression of the reduction reaction with capping and stabilization of AgNPs by the various secondary metabolites present in the ACL. It is thus evident from the FT-IT spectra of the AgNPs that the bioactive compounds such are polyphenols (chlorogenic acid, caffeic acid, rutin, ferulic acid, morin, quercetin, naringenin, and kaempferol), anthocyanins, flavonoids, flavonols, and flavanols which were previously reported to be present in *Z. mays* (Ramos-Escudero et al., 2012; Bacchetti et al., 2013; Pandey et al., 2013) plays a vital role in the capping and stabilization of the AgNPs.

Thermogravimetric analysis of the AgNPs during this period (**Figure 5A**) indicated that the organic molecules from the ACL had mostly taken part in the synthesis and capping of the AgNPs, but were degraded at higher temperatures (Shaik et al., 2013). The nature of the synthesized AgNPs analyzed by XRD showed that the six diffraction peaks corresponded to (220), (122), (111), (231), (331), and (311) planes of silver, respectively, as per the standard FCC structures of Ag (JCPDS Card no. 04-0783) (Khurana et al., 2014; Roy et al., 2015). This structural characteristic pattern confirmed that the AgNPs had a crystalline structure. (**Figure 5B**).

The development of resistant pathogenic strains has recently affected healthcare systems worldwide (Rajeshkumar and Malarkodi, 2014). Therefore, the positive effects of AgNPs toward a number of foodborne pathogenic bacteria (**Table 1**) could be useful in formulation of new antibacterial drugs against resistant bacterial pathogens. Silver exhibits toxicity toward microorganisms but little or no toxicity toward animal cells; therefore, these properties of Ag particles could be more beneficial for AgNPs in the development of more potent drugs against the pathogens.

The exact cause of the antibacterial action of AgNPs against the pathogenic bacteria is not completely understood. Few studies have shown that the electrostatic attraction between negatively charged bacterial cells and the positively charged nanoparticles could be responsible for its bactericidal effects (Sondi and Salopek-Sondi, 2004). There are several possible proposed mechanisms for the positive antibacterial activity of AgNPs which includes the degradation of enzymes, inactivation of cellular proteins and breakage of DNA (Jiang et al., 2004; Sharma et al., 2009; Guzman et al., 2012). It is presumed that due to smaller size the AgNPs might have attached to the surface of the bacterial cell membrane and disturbed its power functions, such as permeability and respiration and then it could have easily penetrate to the inside of the bacteria and could have caused further damage, possibly by interacting with sulfur- and phosphorus-containing compounds, such as DNA resulting in cell lysis (Gibbins and Warner, 2005; Guzman et al., 2012; Singh et al., 2014; Yousefzadi et al., 2014; Ramesh et al., 2015; Swamy et al., 2015). Ovington (2004) reported the potential antimicrobial activity of nanocrystalline silver products by the process of releasing a cluster of highly reactive silver cations and radicals inside the pathogen body or the cell surface, this could be a possible reason for the antibacterial activity of AgNPs in the present study. The main possible mechanism of antimicrobial action of AgNPs could be that, due to the dissolution of AgNPs, antimicrobial Ag^+ ions are released which can interact with sulfur-containing proteins in the bacterial cell wall, which may lead to compromised functionality (Levy, 1998; Lansdown, 2004; Ovington, 2004; Reidy et al., 2013).

Similarly, many authors have also reported that due to the smaller size, the interaction of the nanoparticles is better with the targeted pathogen and thus these were more effective (Panacek et al., 2006; Suriya et al., 2012). In addition to this, it is also believed that the AgNPs after penetration into the bacteria membrane might have inactivated their enzymes, generating hydrogen peroxide that ultimately resulted in the death of the bacteria (Raffi et al., 2008). Furthermore, it is also believed that the high affinity of silver for sulfur or phosphorus compounds could be a possible reason for its antibacterial activity against the pathogenic bacteria because as sulfur and phosphorus are abundantly found throughout the cell membrane, the AgNPs could have reacted with sulfur-containing proteins inside or outside the cell membrane and in turn affected the cell viability causing leakage of bacteria leading to lysis (Hamouda and Baker, 2000). Morones et al. (2005) have reported the presence of AgNPs not only at the surface of cell membrane, but also inside the bacteria using the scanning tunneling electron microscopy (STEM), this proves that due to their smaller size, AgNPs have penetrated inside the bacteria and fungi, causing damage by interacting with phosphorus- and sulfur-containing compounds such as DNA. It is evident that with the decrease in the size of the particles to the nanoscale range, the specific surface area of a dose of nanoparticles increases, which allows for greater material interaction with the surrounding environment such as the cell membrane of the targeted pathogenic bacteria. Thus for the inherently antibacterial materials, such as zinc and silver, increasing the surface to volume ratio enhances the antibacterial effect that results in positive antimicrobial activity due to a number of reasons, such as the release of antibacterial metal ions from the particle surface and the antibacterial physical properties of a nanoparticle related to cell wall penetration or membrane damage (Seil and Webster, 2012). It has also been reported that the crystallographic structure surface and high surface-to-volume ratio increase the contact area of metallic nanoparticles with the body of the microorganism that influences the antibacterial activity of nanosized silver particle (Kora and Arunachalam, 2011). These properties of AgNPs made it a potential candidate for the industries in development of modern antimicrobial products. Moreover, the AgNPs could be useful in the formulation of polymer materials for packaging of food items and other durable materials that could be affected by microorganisms (Rhim and Ng, 2007; Finnigan, 2009).

Further study on the synergistic antibacterial activity of AgNPs with the antibiotics that showed as strong positive result (**Table 2**) could be due to the easy penetration of the mixture solution into the bacterial cell membrane, causing serious damage to the cells and death of the bacteria. This synergistic potential of AgNPs with the antibiotics could help minimize the extensive use of antibiotics that has resulted in development of many antibiotic resistant strains. Previously, the positive synergism impact of nanoparticle-antibiotics combination at significantly low concentration have been demonstrated against a number of dreadful multi-drug resistant pathogenic bacteria (Li et al., 2005; Birla et al., 2009; Ruden et al., 2009; Fayaz et al., 2010; Hwang et al., 2012; McShan et al., 2015; Barapatre et al., 2016; Deng et al., 2016; Panacek et al., 2016). All these authors have proposed that such positive results of nanoparticle-antibiotics combination might be due to the differences in size of prepared Ag-NPs the bonding reaction between them which enable the mixture to better interact with the pathogen.

Similarly, the synergistic anticandidal activities of the AgNPs as presented in **Table 3** and **Figure 7** confirmed that the use of AgNPs together with lower concentrations of anticandidal agents could be beneficial in clinical applications by enabling the application of lower amounts of anticandidal agents to avoid adverse effects and development of drug resistant pathogens (Gajbhiye et al., 2009). Candida are among the most common pathogenic yeast influencing humans, but treatments for Candida infections are limited because of development of resistant *Candida* sp., limited availability of antifungal drugs and high costs (Khan et al., 2003). Thus, the use of AgNPs together with low concentrations of amphotericin b has the potential for improved treatment of Candida related diseases. There have been only several reports on the antibacterial effect of AgNPs against *Candida* sp. (Panacek et al., 2009; Vazquez-Munoz et al., 2014; Artunduaga Bonilla et al., 2015; Lara et al., 2015), and the present study also corroborates with previous claim. As a previous report of the effective synergistic effects of AgNPs combined with antibiotics (Monteiro et al., 2013), the present study also corroborates with their findings.

The strong antioxidant potential of the AgNPs (**Figure 8**), could also make it a good source of natural agent for antioxidant. The strong DPPH and NO scavenging potential of AgNPs could make it a potential candidate for drug delivery. The moderate ABTS scavenging potential of AgNPs compared to ASA might be due to the different types of functional groups from the ACL that have attached to the surface of AgNPs during synthesis and capping of AgNPs (Adedapo et al., 2008). AgNPs also displayed strong reducing power, which might be attributed to the presence of phenolic compounds from ACL extract on the surface of AgNPs as surface stabilizers and capping agents. It is observed in the present investigation that the antioxidant activity of the AgNPs was higher than that of the standard ASA which might be possible due to the size and crystalline nature of the AgNPs as reported by various authors (El-Rafie and Abdel-Aziz Hamed, 2014; Bhakya et al., 2016; Gopukumar et al., 2016).

CONCLUSION

In the present study, AgNPs were synthesized using the aqueous extract of *Z. mays* corn leaves, which is a novel approach of waste utilization in nanoparticle synthesis. The results revealed that the synthesized nanoparticles are within the nanometer range, with an SPR of 450 nm based on UV-Vis spectroscopy. The elemental composition and crystallinity structure confirmed the synthesized particles to be Ag. The AgNPs displayed positive antibacterial activity against different foodborne pathogenic bacteria, as well as strong synergistic antibacterial and anticandidal activity with low concentrations of antibiotics and anticandidal agents. The AgNPs exhibited strong antioxidant potential. Based on these results, this approach of utilization of industrial wastes in nanoparticle synthesis can be beneficial in large scale fabrication of nanomaterials. The synergistic study of AgNPs with common antibiotics and anticandidal agents could be beneficial in formulation of antibacterial products and anticandidal drugs to be used in various food, agricultural, cosmetic, and pharmaceutical industries and future platforms for preparing nano-medicines, and targeted drug delivery.

AUTHOR CONTRIBUTIONS

JP carried out all the experiment and wrote the manuscript. JP and K-HB designed and edited the manuscript.

ACKNOWLEDGMENT

This work was supported by grants from the Systems and Synthetic Agro-biotech Center through the Next-Generation Bio-Green 21 Program (PJ011117), Rural Development Administration, Republic of Korea.

REFERENCES

Adedapo, A. A., Jimoh, F. O., Afolayan, A. J., and Masika, P. J. (2008). Antioxidant activities and phenolic contents of the methanol extracts of the stems of *Acokanthera oppositifolia* and *Adenia gummifera*. *BMC Complement. Altern. Med.* 8:54. doi: 10.1186/1472-6882-8-54

Andersson, D. I., and Hughes, D. (2010). Antibiotic resistance and its cost: Is it possible to reverse resistance? *Nat. Rev. Microbiol.* 8, 260–271. doi: 10.1038/nrmicro2319

Artunduaga Bonilla, J. J., Paredes-Guerrero, D. J., Sanchez-Suarez, C. I., Ortiz-Lopez, C. C., and Torres-Saez, R. G. (2015). In vitro antifungal activity of silver nanoparticles against fluconazole resistant *Candida* species. *World J. Microbiol. Biotechnol.* 31, 1801–1809. doi: 10.1007/s11274-015-1933-z

Bacchetti, T., Masciangelo, S., Micheletti, A., and Ferretti, G. (2013). Carotenoids, phenolic compounds and antioxidant capacity of five local Italian corn (*Zea mays* L.) kernels. *J. Nutr. Food Sci.* 3:6. doi: 10.4172/2155-9600.1000237

Barapatre, A., Aadil, K. R., and Jha, H. (2016). Synergistic antibacterial and antibiofilm activity of silver nanoparticles biosynthesized by lignin-degrading fungus. *Bioresour. Bioprocess.* 3:8. doi: 10.1186/s40643-016-0083-y

Basavegowda, N., and Lee, Y. R. (2013). Synthesis of silver nanoparticles using Satsuma mandarin (*Citrus unshiu*) peel extract: a novel approach towards waste utilization. *Mater. Lett.* 109, 31–33. doi: 10.1016/j.matlet.2013.07.039

Bhakya, S., Muthukrishnan, S., Sukumaran, M., Grijalva, M., Cumbal, L., Franklin-Benjamin, J. H., et al. (2016). Antimicrobial, antioxidant and anticancer activity of biogenic silver nanoparticles – an experimental report. *RSC Adv.* 6, 81436–81446. doi: 10.1039/C6RA17569D

Birla, S. S., Tiwari, V. V., Gade, A. K., Ingle, A. P., Yadav, A. P., and Rai, M. K. (2009). Fabrication of silver nanoparticles by *Phoma glomerata* and its combined effect against *Escherichia coli*, *Pseudomonas aeruginosa* and *Staphylococcus aureus*. *Lett. Appl. Microbiol.* 48, 173–179. doi: 10.1111/j.1472-765X.2008.02510.x

Chandran, K., Song, S., and Yun, S. I. (2014). Effect of size and shape controlled biogenic synthesis of gold nanoparticles and their mode of interactions against food borne bacterial pathogens. *Arabian J. Chem.* doi: 10.1016/j.arabjc.2014.11.041

Choi, Y., Ho, N., and Tung, C. (2007). Sensing phosphatase activity by using gold nanoparticles. *Angew. Chem. Int. Ed. Engl.* 46, 707–709. doi: 10.1002/anie.200603735

Das, J., Paul Das, M., and Velusamy, P. (2013). *Sesbania grandiflora* leaf extract mediated green synthesis of antibacterial silver nanoparticles against selected human pathogens. *Spectrochim. Acta A Mol. Biomol. Spectrosc.* 104, 265–270. doi: 10.1016/j.saa.2012.11.075

Deng, H., McShan, D., Zhang, Y., Sinha, S. S., Arslan, Z., Ray, P. C., et al. (2016). Mechanistic study of the synergistic antibacterial activity of combined silver nanoparticles and common antibiotics. *Environ. Sci. Technol.* 50, 8840–8848. doi: 10.1021/acs.est.6b00998

Devi, L. S., and Joshi, S. R. (2012). Antimicrobial and synergistic effects of silver nanoparticles synthesized using soil fungi of high altitudes of eastern Himalaya. *Mycobiology* 40, 27–34. doi: 10.5941/MYCO.2012.40.1.027

Diao, W. R., Hu, Q. P., Feng, S. S., Li, W. Q., and Xu, J. G. (2013). Chemical composition and antibacterial activity of the essential oil from green huajiao (*Zanthoxylum schinifolium*) against selected foodborne pathogens. *J. Agric. Food Chem.* 61, 6044–6049. doi: 10.1021/jf4007856

El-Rafie, H. M., and Abdel-Aziz Hamed, M. (2014). Antioxidant and anti-inflammatory activities of silver nanoparticles biosynthesized from aqueous leaves extracts of four *Terminalia* species. *Adv. Nat. Sci.* 5:035008.

Fayaz, A. M., Balaji, K., Girilal, M., Yadav, R., Kalaichelvan, P. T., and Venketesan, R. (2010). Biogenic synthesis of silver nanoparticles and their synergistic effect with antibiotics: a study against gram-positive and gram-negative bacteria. *J. Nanomed. Nanotechnol.* 6, 103–109. doi: 10.1016/j.nano.2009.04.006

Feng, N., Guo, X., and Liang, S. (2009). Adsorption study of copper (II) by chemically modified orange peel. *J. Hazard. Mater.* 164, 1286–1292. doi: 10.1016/j.jhazmat.2008.09.096

Finnigan, B. (2009). "Barrier polymers," in *The Wiley Encyclopedia of Packaging Technology*, ed. K. L. Yam (New York, NY: John Wiley & Sons), 103–109.

Gajbhiye, M., Kesharwani, J., Ingle, A., Gade, A., and Rai, M. (2009). Fungus-mediated synthesis of silver nanoparticles and their activity against pathogenic fungi in combination with fluconazole. *Nanomedicine* 5, 382–386. doi: 10.1016/j.nano.2009.06.005

Geddes, J. R., Carney, S. M., Davies, C., Furukawa, T. A., Kupfer, D. J., Frank, E., et al. (2003). Relapse prevention with antidepressant drug treatment in depressive disorders: a systematic review. *Lancet* 361, 653–661. doi: 10.1016/S0140-6736(03)12599-8

Ghosh, P. K., Saxena, T. K., Gupta, R., Yadav, R. P., and Davidson, S. (1996). Microbial lipases: production and applications. *Sci. Prog.* 79, 119–157.

Ghosh, S., Patil, S., Ahire, M., Kitture, R., Jabgunde, A., Kale, S., et al. (2011). Synthesis of gold nanoanisotrops using *Dioscorea bulbifera* tuber extract. *J. Nanomater.* 45, 1–8. doi: 10.1155/2011/354793

Gibbins, B., and Warner, L. (2005). The role of antimicrobial silver nanotechnology. *Med. Device Diagn. Ind. Mag.* 1, 1–2.

Gopinath, V., MubarakAli, D., Priyadarshini, S., Priyadharsshini, N. M., Thajuddin, N., and Velusamy, P. (2012). Biosynthesis of silver nanoparticles from Tribulus terrestris and its antimicrobial activity: a novel biological approach. *Colloids Surf. B Biointerfaces* 96, 69–74. doi: 10.1016/j.colsurfb.2012.03.023

Gopukumar, S. T., Sana-Fathima, T. K., Alexander, P., Alex, V., and Praseetha, P. K. (2016). Evaluation of antioxidant properties of silver nanoparticle embedded medicinal patch. *Nanomed. Nanotech. Op. Acc.* 1:NNOA-MS-ID-000101.

Guzman, M., Dille, J., and Godet, S. (2012). Synthesis and antibacterial activity of silver nanoparticles against gram-positive and gram-negative bacteria. *Nanomed. Nanotechnol. Biol. Med.* 8, 37–45. doi: 10.1016/j.nano.2011.05.007

Hamouda, T., and Baker, J. R. Jr. (2000). Antimicrobial mechanism of action of surfactant lipid preparations in enteric gram-negative bacilli. *J. Appl. Microbiol.* 89, 397–403. doi: 10.1046/j.1365-2672.2000.01127.x

Huh, A. J., and Kwon, Y. J. (2011). "Nanoantibiotics": a new paradigm for treating infectious diseases using nanomaterials in the antibiotics resistant era. *J. Control. Release* 156, 128–145. doi: 10.1016/j.jconrel.2011.07.002

Hwang, I., Hwang, J. H., Choi, H., Kim, K. J., and Lee, D. G. (2012). Synergistic effects between silver nanoparticles and antibiotics and the mechanisms involved. *J. Med. Microbiol.* 61, 1719–1726. doi: 10.1099/jmm.0.047100-0

Jain, P. K., Huang, X., El-Sayed, I. H., and El-Sayed, M. A. (2008). Noble metals on the nanoscale: optical and photothermal properties and some applications in imaging, sensing, biology, and medicine. *Acc. Chem. Res.* 41, 1578–1586. doi: 10.1021/ar7002804

Jiang, H., Manolache, S., Lee-Wong, C., and Denis, F. S. (2004). Plasma-enhanced deposition of silver nanoparticles onto polymer and metal surfaces for the generation of antimicrobial characteristics. *J. Appl. Polymer Sci.* 93, 1411–1422. doi: 10.1002/app.20561

Kelly, K. L., Coronado, E., Zhao, L. L., and Schatz, G. C. (2003). The optical properties of metal nanoparticles: the influence of size, shape, and dielectric environment. *J. Phys. Chem. B* 107, 668–677. doi: 10.1021/jp026731y

Khan, Z. U., Chandy, R., and Metwali, K. E. (2003). *Candida albicans* strain carriage in patients and nursing staff of an intensive care unit: a study of morphotypes and resistotypes. *Mycoses* 46, 476–486. doi: 10.1046/j.0933-7407.2003.00929.x

Khurana, C., Vala, A. K., Andhariya, N., Pandey, O. P., and Chudasama, B. (2014). Antibacterial activities of silver nanoparticles and antibiotic-adsorbed silver nanoparticles against biorecycling microbes. *Environ. Sci. Proc. Impact* 16, 2191–2198. doi: 10.1039/c4em00248b

Konstantopoulou, M. A., Krokos, F. D., and Mazomenos, B. E. (2004). Chemical composition of corn leaf essential oils and their role in the oviposition behavior of *Sesamia nonagrioides* females. *J. Chem. Ecol.* 30, 2243–2256. doi: 10.1023/B:JOEC.0000048786.12136.40

Kora, A. J., and Arunachalam, J. (2011). Assessment of antibacterial activity of silver nanoparticles on *Pseudomonas aeruginosa* and its mechanism of action. *World J. Microbiol. Biotechnol.* 27, 1209–1216. doi: 10.1007/s11274-010-0569-2

Kubo, I., Fujita, K., Kubo, A., Nihei, K., and Ogura, T. (2004). Antibacterial activity of coriander volatile compounds against *Salmonella* choleraesuis. *J. Agric. Food Chem.* 52, 3329–3332. doi: 10.1021/jf0354186

Kumar, A., Vemula, P. K., Ajayan, P. M., and John, G. (2008). Silver-nanoparticle-embedded antimicrobial paints based on vegetable oil. *Nat. Mater.* 7, 236–241. doi: 10.1038/nmat2099

Kumar, V., and Yadav, S. K. (2009). Plant-mediated synthesis of silver and gold nanoparticles and their applications. *J. Chem. Technol. Biotechnol.* 84, 151–157. doi: 10.1002/jctb.2023

Kumari, M. M., and Philip, D. (2013). Facile one-pot synthesis of gold and silver nanocatalysts using edible coconut oil. *Spectrochim. Acta A. Mol. Biomol. Spectrosc.* 111, 154–160. doi: 10.1016/j.saa.2013.03.076

Lansdown, A. B. G. (2004). A review of the use of silver in wound care: facts and fallacies. *Br. J. Nurs.* 13, 6–19. doi: 10.12968/bjon.2004.13.Sup1.12535

Lara, H. H., Garza-Trevino, E. N., Ixtepan-Turrent, L., and Singh, D. K. J. (2011). Silver nanoparticles are broad-spectrum bactericidal and virucidal compounds. *J. Nanobiotechnol.* 9:30. doi: 10.1186/1477-3155-9-30

Lara, H. H., Romero-Urbina, D. G., Pierce, C., Lopez-Ribot, J. L., Arellano-Jimenez, M. J., and Jose-Yacaman, M. (2015). Effect of silver nanoparticles on *Candida albicans* biofilms: an ultrastructural study. *J. Nanobiotechnol.* 13:91. doi: 10.1186/s12951-015-0147-8

Levy, S. B. (1998). The challenge of antibiotic resistance. *Sci. Am.* 3, 32–39.

Li, P., Li, J., Wu, C., Wu, Q., and Li, J. (2005). Synergistic antibacterial effects of β-lactam antibiotic combined with silver nanoparticles. *Nanotechnology* 16:1912. doi: 10.1088/0957-4484/16/9/082

Makhija, I. K., Aswatha-Ram, H. N., Shreedhara, C. S., Vijay Kumar, S., and Devkar, R. (2011). In vitro antioxidant studies of sitopaladi churna, a polyherbal ayurvedic formulation. *Free Radic. Antioxid.* 1, 37–41. doi: 10.5530/ax.2011.2.8

Marini, M., De Niederhausern, N., Iseppi, R., Bondi, M., Sabia, C., Toselli, M., et al. (2007). Antibacterial activity of plastics coated with silver-doped organic-inorganic hybrid coatings prepared by sol–gel processes. *Biomacromolecules* 8, 1246–1254. doi: 10.1021/bm060721b

Matei, A., Cernica, I., Cadar, O., Roman, C., and Schiopu, V. (2008). Synthesis and characterization of ZnO-polymer nanocomposites. *Int. J. Mater. Form.* 1, 767–770. doi: 10.1007/s12289-008-0288-5

McShan, D., Zhang, Y., Deng, H., Ray, P. C., and Yu, H. (2015). Synergistic antibacterial effect of silver nanoparticles combined with ineffective antibiotics on drug resistant *Salmonella* typhimurium DT104. *J. Environ. Sci. Health. C Environ. Carcinog. Ecotoxicol. Rev.* 33, 369–384. doi: 10.1080/10590501.2015.1055165

Mohanpuria, P., Rana, N. K., and Yadav, S. K. (2008). Biosynthesis of nanoparticles: technological concepts and future applications. *J. Nanopart. Res.* 10, 507–517. doi: 10.1111/j.1460-9568.2009.06927.x

Monteiro, D. R., Silva, S., Negri, M., Gorup, L. F., de Camargo, E. R., Oliveira, R., et al. (2013). Antifungal activity of silver nanoparticles in combination with nystatin and chlorhexidine digluconate against *Candida albicans* and *Candida glabrata* biofilms. *Mycoses* 56, 672–680. doi: 10.1111/myc.12093

Morones, J., Elechiguerra, J., Camacho, A., Holt, K., Kouri, J., Ramirez, J., et al. (2005). The bactericidal effect of silver nanoparticles. *Nanotechnology* 16, 2346–2353. doi: 10.1088/0957-4484/16/10/059

Murray, P. R., Baron, E. J., Pfaller, M. A., Tenover, F. C., and Yolke, R. H. (1995). *Manual of Clinical Microbiology*, 6th Edn. Washington, DC: ASM Press.

Muthukrishnan, S., Bhakya, S., Senthil Kumar, T., and Rao, M. V. (2015). Biosynthesis, characterization and antibacterial effect of plant-mediated silver nanoparticles using *Ceropegia thwaitesii* – an endemic species. *Ind. Crop. Prod.* 63, 119–124. doi: 10.1016/j.indcrop.2014.10.022

Nair, L. S., and Laurencin, C. T. (2007). Silver nanoparticles: synthesis and therapeutic applications. *J. Biomed. Nanotechnol.* 3, 301–316. doi: 10.1166/jbn. 2007.041

Naqvi, S. Z. H., Kiran, U., Ali, M. I., Jamal, A., Hameed, A., Ahmed, S., et al. (2013). Combined efficacy of biologically synthesized silver nanoparticles and different antibiotics against multidrug-resistant bacteria. *Int. J. Nanomed.* 8, 3187–3195. doi: 10.2147/IJN.S49284

Natarajan, K., Selvaraj, S., and Ramachandra, M. V. (2010). Microbial production of silver nanoparticles. *Dig. J. Nanomater. Biostruct.* 5, 135–140.

Nazeruddin, G. M., Prasad, N. R., Prasad, S. R., Shaikh, Y. I., Waghmare, S. R., and Adhyapkca, P. (2014). *Coriandrum sativum* seed extract assisted in situ green synthesis of silver nanoparticle and its anti-microbial activity. *Ind. Crop. Prod.* 60, 212–216. doi: 10.1016/j.indcrop.2014.05.040

Nezamdoost, T., Bagherieh-Najjarn, M. B., and Aghdasi, M. (2014). Biogenic synthesis of stable bioactive silver chloride nanoparticles using *Onosma dichroantha* Boiss. root extract. *Mater. Lett.* 137, 225–228. doi: 10.1016/j.matlet. 2014.08.134

Ouay, B., and Stellacci, F. (2015). Antibacterial activity of silver nanoparticles: a surface science insight. *Nano Today* 10, 339–354. doi: 10.1016/j.nantod.2015. 04.002

Ovington, L. G. (2004). The truth about silver. *Ostomy Wound Manage.* 50, 1–10.

Pal, S., Tak, Y. K., and Song, J. M. (2007). Does the antibacterial activity of silver nanoparticles depend on the shape of the nanoparticle? A study of the Gram-negative bacterium *Escherichia coli. Appl. Environ. Microbiol.* 73, 1712–1720.

Panacek, A., Kolar, M., Vecerova, R., Prucek, R., Soukupova, J., Krystof, V., et al. (2009). Antifungal activity of silver nanoparticles against *Candida* spp. *Biomaterials* 30, 6333–6340. doi: 10.1016/j.biomaterials.2009.07.065

Panacek, A., Kvitek, L., Prucek, R., Kolar, M., Vecerova, R., Pizurova, N., et al. (2006). Silver colloid nanoparticles: synthesis, characterization, and their antibacterial activity. *J. Phys. Chem. B* 110, 16248–16253. doi: 10.1021/ jp063826h

Panacek, A., Smekalova, M., Kilianova, M., Prucek, R., Bogdanova, K., Vecerova, R., et al. (2016). Strong and nonspecific synergistic antibacterial efficiency of antibiotics combined with silver nanoparticles at very low concentrations showing no cytotoxic effect. *Molecules* 21:26. doi: 10.3390/ molecules21010026

Pandey, R., Singh, A., Maurya, S., Singh, U. P., and Singh, M. (2013). Phenolic acids in different preparations of Maize (*Zea mays*) and their role in human health. *Int. J. Curr. Microbiol. App. Sci.* 2, 84–92.

Patra, J. K., Kim, S. H., and Baek, K. H. (2015). Antioxidant and free radical-scavenging potential of essential oil from *Enteromorpha linza* L. prepared by microwave-assisted hydrodistillation. *J. Food Biochem.* 39, 80–90. doi: 10.1111/ jfbc.12110

Qiu, L., Liu, F., Zhao, L., Yang, W., and Yao, J. (2006). Evidence of a unique electron donor-acceptor property for platinum nanoparticles as studied by XPS. *Langmuir* 22, 4480–4482. doi: 10.1021/la053071q

Raffi, M., Hussain, F., Bhatti, T. M., Akhter, J. I., Hameed, A., and Hasan, M. M. (2008). Antibacterial characterization of silver nanoparticles against *E. coli* ATCC-15224. *J. Mater. Sci. Technol.* 24, 192–196.

Rai, M. K., Deshmukh, S. D., Ingle, A. P., and Gade, A. K. (2012). Silver nanoparticles: the powerful nanoweapon against multidrug-resistant bacteria. *J. Appl. Microbiol.* 112, 841–852. doi: 10.1111/j.1365-2672.2012.05253.x

Rajeshkumar, S., and Malarkodi, C. (2014). In vitro antibacterial activity and mechanism of silver nanoparticles against foodborne pathogens. *Bioinorg. Chem. Appl.* 2014, 1–10. doi: 10.1155/2014/581890

Ramamurthy, C. H., Padma, M., Daisy mariya samadanam, I., Mareeswaran, R., Suyavaran, A., Suresh Kumar, M., et al. (2013). The extra cellular synthesis of gold and silver nanoparticles and their free radical scavenging and antibacterial properties. *Colloids Surf. B Biointerfaces* 102, 808–815. doi: 10.1016/j.colsurfb. 2012.09.025

Ramesh, P. S., Kokila, T., and Geetha, D. (2015). Plant mediated green synthesis and antibacterial activity of silver nanoparticles using *Emblica officinalis* fruit

extract. *Spectrochim. Acta A Mol. Biomol. Spectrosc.* 142, 339–343. doi: 10.1016/ j.saa.2015.01.062

Ramos-Escudero, F., Munoz, A. M., Alvarado-Ortiz, C., Alvarado, A., and Yanez, J. A. (2012). Purple corn (*Zea mays* L.) phenolic compounds profile and its assessment as an agent against oxidative stress in isolated mouse organs. *J. Med. Food.* 15, 206–215. doi: 10.1089/jmf.2010.0342

Reidy, B., Haase, A., Luch, A., Dawson, K. A., and Lynch, I. (2013). Mechanisms of silver nanoparticle release, transformation and toxicity: a critical review of current knowledge and recommendations for future studies and applications. *Materials* 6, 2295–2350. doi: 10.3390/ma6062295

Rhim, J. W., and Ng, P. K. W. (2007). Natural biopolymer-based nanocomposite films for packaging applications. *Crit. Rev. Food. Sci.* 47, 411–433. doi: 10.1080/ 10408390600846366

Roy, K., Sarkar, C. K., and Ghosh, C. K. (2015). Photocatalytic activity of biogenic silver nanoparticles synthesized using potato (*Solanum tuberosum*) infusion. *Spectrochem. Acta A Mol. Biomol. Spectrosc.* 146, 286–291. doi: 10.1016/j.saa. 2015.02.058

Ruden, S., Hilpert, K., Berditsch, M., Wadhwani, P., and Ulrich, A. S. (2009). Synergistic interaction between silver nanoparticles and membrane-permeabilizing antimicrobial peptides. *Antimicrob. Agents Chemother.* 53, 3538–3540. doi: 10.1128/AAC.01106-08

Saxena, A., Tripathi, R. M., Zafar, F., and Singh, P. (2012). Green synthesis of silver nanoparticles using aqueous solution of *Ficus benghalensis* leaf extract and characterization of their antibacterial activity. *Mater. Lett.* 67, 91–94. doi: 10.1016/j.matlet.2011.09.038

Seil, J. T., and Webster, T. J. (2012). Antimicrobial applications of nanotechnology: methods and literature. *Int. J. Nanomed.* 7, 2767–2781. doi: 10.2147/IJN.S24805

Shahverdi, A. R., Fakhimi, A., Shahverdi, H. R., and Minaian, M. S. (2007). Synthesis and effect of silver nanoparticles on the antibacterial activity of different antibiotics against *Staphylococcus aureus* and *Escherichia coli. Nanomed. Nanotechnol. Biol. Med.* 3, 168–171. doi: 10.1016/j.nano.2007.02.001

Shaik, S., Kummara, M. R., Poluru, S., Allu, C., Gooty, J. M., Kashayi, C. R., et al. (2013). A green approach to synthesize silver nanoparticles in starch-co-poly(acrylamide) hydrogels by *Tridax procumbens* leaf extract and their antibacterial activity. *Int. J. Carbohydrate. Chem.* 2013:539636. doi: 10.1155/ 2013/539636

Sharma, V. K., Yngard, R. A., and Lin, Y. (2009). Silver nanoparticles: green synthesis and their antimicrobial activities. *Adv. Colloid Interface. Sci.* 145, 83–96. doi: 10.1016/j.cis.2008.09.002

Singh, K., Panghal, M., Kadyan, S., Chaudhary, U., and Yadav, J. P. (2014). Antibacterial activity of synthesized silver nanoparticles from Tinospora cordifolia against multi drug resistant strains of *Pseudomonas aeruginosa* isolated from burn patients. *J. Nanomed. Nanotechnol.* 5:192. doi: 10.4172/2157-7439.1000192

Solihah, M. A., Wan Rosli, W. I., and Nurhanan, A. R. (2012). Phytochemicals screening and total phenolic content of Malaysian *Zea mays* hair extracts. *Int. Food Res. J.* 19, 1533–1538.

Sondi, I., and Salopek-Sondi, B. (2004). Silver nanoparticles as antimicrobial agent: a case study on *E. coli* as a model for gram-negative bacteria. *J. Colloid Interface. Sci.* 275, 177–182. doi: 10.1016/j.jcis.2004.02.012

Sun, L., Zhang, J., Lu, X., Zhang, L., and Zhang, Y. (2011). Evaluation to the antioxidant activity of total flavonoids extract from persimmon (*Diospyros kaki* L.) leaves. *Food Chem. Toxicol.* 49, 2689–2696. doi: 10.1016/j.fct.2011.07.042

Suriya, J., Bharathi-Raja, S., Sekar, V., and Rajasekaran, R. (2012). Biosynthesis of silver nanoparticles and its antibacterial activity using seaweed *Urospora* sp. *Afr. J. Biotechnol.* 11, 12192–12198. doi: 10.5897/AJB12.452

Swamy, M. K., Mohanty, S. K., Jayanta, K., and Subbanarasiman, B. (2015). The green synthesis, characterization, and evaluation of the biological activities of silver nanoparticles synthesized from *Leptadenia reticulata* leaf extract. *Appl. Nanosci.* 5, 73–81. doi: 10.1007/s13204-014-0293-6

Thaipong, K., Boonprakob, U., Crosby, K., Cisneros-Zevallos, L., and Byrne, D. H. (2006). Comparison of ABTS, DPPH, FRAP, and ORAC assays for estimating antioxidant activity from guava fruit extracts. *J. Food Comp. Anal.* 19, 669–675.

Ullah, I., Ali, M., and Farooqi, A. (2010). Chemical and nutritional properties of some maize (*Zea mays* L.) varieties grown in NWFP, Pakistan. *Pakistan J. Nutr.* 9, 1113–1117. doi: 10.3923/pjn.2010.1113.1117

Valli, J. S., and Vaseeharan, B. (2012). Biosynthesis of silver nanoparticles by *Cissus quadrangularis* extracts. *Mater. Lett.* 82, 171–173. doi: 10.1016/j.matlet.2012.05.040

Vazquez-Munoz, R., Avalos-Borja, M., and Castro-Longoria, E. (2014). Ultrastructural analysis of *Candida albicans* when exposed to silver nanoparticles. *PLoS ONE* 9:e108876. doi: 10.1371/journal.pone.0108876

Velusamy, P., Das, J., Pachaiappan, R., Vaseeharan, B., and Pandian, K. (2015). Greener approach for synthesis of antibacterial silver nanoparticles using aqueous solution of neem gum (*Azadirachta indica* L.). *Ind. Crop. Prod.* 66, 103–109. doi: 10.1016/j.indcrop.2014.12.042

Vijaykumar, M., Priya, K., Nancy, F. T., Noorlidah, A., and Ahmad, A. B. A. (2013). Biosynthesis, characterization and anti-bacterial effect of plant-mediated silver nanoparticles using *A. Nilgirica*. *Ind. Crops Prod.* 41, 235–240. doi: 10.1016/j.indcrop.2012.04.017

Vilchis-Nestor, A. R., Sanchez-Mendieta, V., Camacho-Lopez, M. A., Gomez-Espinosa, R. M., Camacho-Lopez, M. A., and Arenas-Alatorre, J. A. (2008). Solvent less synthesis and optical properties of Au and Ag nanoparticles using *Camellia sinensis* extract. *Mater. Lett.* 62, 3103–3105. doi: 10.1016/j.matlet.2008.01.138

Wu, N., Fu, L., Aslam, M., Wong, K. C., and Dravid, V. W. (2004). Interaction of fatty acid monolayers with cobalt nanoparticles. *Nano Lett.* 4, 383–386. doi: 10.1021/nl035139x

Yousefzadi, M., Rahimi, Z., and Ghafori, V. (2014). The green synthesis, characterization and antimicrobial activities of silver nanoparticles synthesized from green alga *Enteromorpha flexuosa* (wulfen) J. Agardh. *Mater. Lett.* 137, 1–4. doi: 10.1016/j.matlet.2014.08.110

Zargar, M., Shameli, S., Reza Najafi, G., and Farahani, F. (2014). Plant mediated green biosynthesis of silver nanoparticles using *Vitex negundo* L. extract. *J. Ind. Eng. Chem.* 20, 4169–4175. doi: 10.3390/molecules16086667

Zhang, L. Z., Yu, J. C., Yip, H. Y., Li, Q., Kwong, K. W., Xu, A. W., et al. (2003). Ambient light reduction strategy to synthesize silver nanoparticles and silver coated TiO2 with enhanced photocatalytic and bactericidal activities. *Langmuir* 19, 10372–10380. doi: 10.1021/la035330m

Zhang, W., Qiao, X., Chen, J., and Wang, H. (2006). Preparation of silver nanoparticles in water-in-oil AOT reverse micelles. *J. Colloid Interface Sci.* 302, 370–373. doi: 10.1016/j.jcis.2006.06.035

Conflict of Interest Statement: The authors declare that the research was conducted in the absence of any commercial or financial relationships that could be construed as a potential conflict of interest.

Permissions

List of Contributors

Ronaldo Bertolucci Jr. and Rodrigo T. Ribeiro
Laboratório de Biocolóides, Departamento de Bioquímica, Instituto de Química, Universidade de São Paulo, São Paulo, Brazil

Jorge L. M. Sampaio
Laboratório de Microbiologia, Departamento de Análises Clínicas e Toxicológicas, Faculdade de Ciências Farmacêuticas, Universidade de São Paulo, São Paulo, Brazil

Ana M. Carmona-Ribeiro and Letícia Dias de Melo Carrasco
Laboratório de Biocolóides, Departamento de Bioquímica, Instituto de Química, Universidade de São Paulo, São Paulo, Brazil,
Laboratório de Microbiologia, Departamento de Análises Clínicas e Toxicológicas, Faculdade de Ciências Farmacêuticas, Universidade de São Paulo, São Paulo, Brazil

Meerambika Mishra
School of Life Sciences, Sambalpur University, Burla, India

Ananta P. Arukha
Department of Infectious Diseases and Pathology, University of Florida, Gainesville, FL, United States

Tufail Bashir
School of Biotechnology, Yeungnam University, Gyeongsan, South Korea

Dhananjay Yadav
Department of Medical Biotechnology, Yeungnam University, Gyeongsan, South Korea

G. B. K. S. Prasad
School of Biochemistry, Jiwaji University, Gwalior, India

Dinesh K. Dahiya
Advanced Milk Testing Research Laboratory, Post Graduate Institute of Veterinary Education and Research – Rajasthan University of Veterinary and Animal Sciences at Bikaner, Jaipur, India

Renuka
Department of Biochemistry, Basic Medical Science, South Campus, Panjab University, Chandigarh, India

Monica Puniya
Food Safety Management System Division, Food Safety and Standards Authority of India, New Delhi, India

Umesh K. Shandilya
Animal Biotechnology Division, National Bureau of Animal Genetic Resources, Karnal, India

Tejpal Dhewa
Department of Nutrition Biology, Central University of Haryana, Mahendergarh, India

Nikhil Kumar
Department of Life Sciences, Shri Venkateshwara University, JP Nagar, India

Sanjeev Kumar
Department of Life Science, Central Assam University, Silchar, India

Anil K. Puniya
College of Dairy Science and Technology, Guru Angad Dev Veterinary and Animal Sciences University, Ludhiana, India
Dairy Microbiology Division, ICAR-National Dairy Research Institute, Karnal, India

Pratyoosh Shukla
Enzyme Technology and Protein Bioinformatics Laboratory, Department of Microbiology, Maharshi Dayanand University, Rohtak, India

Shruti Shukla, Vivek K. Bajpai and Young-Kyu Han
Department of Energy and Materials Engineering, Dongguk University, Seoul, South Korea

Yuvaraj Haldorai
Department of Nanoscience and Technology, Bharathiar University, Coimbatore, India

Yun Suk Huh and Seung Kyu Hwang
Department of Biological Engineering, Biohybrid Systems Research Center (BSRC), World Class Smart Lab (WCSL), Inha University, Incheon, South Korea

Eleni N. Gkana and George-John E. Nychas
Laboratory of Microbiology and Biotechnology of Foods, Department of Food Science and Human Nutrition, Faculty of Foods, Biotechnology and Development, Agricultural University of Athens, Athens, Greece

Nikos G. Chorianopoulos
Institute of Technology of Agricultural Products, Hellenic Agricultural Organization-DEMETER, Athens, Greece

Agapi I. Doulgeraki
Laboratory of Microbiology and Biotechnology of Foods, Department of Food Science and Human Nutrition, Faculty of Foods, Biotechnology and Development, Agricultural University of Athens, Athens, Greece
Institute of Technology of Agricultural Products, Hellenic Agricultural Organization-DEMETER, Athens, Greece

Ram Prasad
Amity Institute of Microbial Technology, Amity University, Noida, India

Atanu Bhattacharyya
Department of Entomology, University of Agricultural Sciences, Gandhi Krishi Vigyan Kendra, Bengaluru, India

Quang D. Nguyen
Research Centre of Bioengineering and Process Engineering, Faculty of Food Science, Szent István University, Budapest, Hungary

Ruby Yadav, Puneet K. Singh and Pratyoosh Shukla
Enzyme Technology and Protein Bioinformatics Laboratory, Department of Microbiology, Maharshi Dayanand University, Rohtak, India

Anil K. Puniya
Division of Dairy Microbiology, Indian Council of Agricultural Research (ICAR) – National Dairy Research Institute (NDRI), Karnal, India,
College of Dairy Science and Technology, Guru Angad Dev Veterinary and Animal Sciences University, Ludhiana, India

Joana Oliveira
School of Microbiology, University College Cork, Cork, Ireland

Laurens Hanemaaijer and Thijs R. H. M. Kouwen
DSM Biotechnology Center, Delft, Netherlands

Horst Neve
Max Rubner-Institut, Kiel, Germany

John MacSharry
APC Microbiome Institute, University College Cork, Cork, Ireland

Douwe van Sinderen and Jennifer Mahony
School of Microbiology, University College Cork, Cork, Ireland
APC Microbiome Institute, University College Cork, Cork, Ireland

Mi-Ran Go, Song-Hwa Bae, Hyeon-Jin Kim, Jin Yu and Soo-Jin Choi
Department of Applied Food System, Major of Food Science and Technology, Seoul Women's University, Seoul, South Korea

Uma Singhal, Ram Prasad and Ajit Varma
Amity Institute of Microbial Technology, Amity University, Noida, India

Manika Khanuja
Centre for Nanoscience and Nanotechnology, Jamia Millia Islamia, New Delhi, India

Kanchan Vishwakarma, Neha Upadhyay and Jaspreet Singh
Department of Biotechnology, Motilal Nehru National Institute of Technology Allahabad, Allahabad, India

Vijay P. Singh
Government Ramanuj Pratap Singhdev Post Graduate College, Baikunthpur, India

Sheo M. Prasad
Ranjan Plant Physiology and Biochemistry Laboratory, Department of Botany, University of Allahabad, Allahabad, India

Devendra K. Chauhan and Shweta
D D Pant Interdisciplinary Research Lab, Department of Botany, University of Allahabad, Allahabad, India

Shiliang Liu
College of Landscape Architecture, Sichuan Agricultural University, Chengdu, China
College of Agriculture, Food and Natural Resources, University of Missouri, Columbia, MO, United States

Durgesh K. Tripathi
Centre for Medical Diagnostic and Research, Motilal Nehru National Institute of Technology Allahabad, Allahabad, India

Shivesh Sharma
Department of Biotechnology, Motilal Nehru National Institute of Technology Allahabad, Allahabad, India
Centre for Medical Diagnostic and Research, Motilal Nehru National Institute of Technology Allahabad, Allahabad, India

Durgesh K. Tripathi
Centre of Advanced Study in Botany, Banaras Hindu University, Varanasi, India,
Center for Medical Diagnostic and Research, Motilal Nehru National Institute of Technology Allahabad, Allahabad, India

Ashutosh Tripathi, Swati Singh, Yashwant Singh, and Devendra K. Chauhan
D. D. Pant Interdisciplinary Research Laboratory, Department of Botany, University of Allahabad, Allahabad, India

Kanchan Vishwakarma
Department of Biotechnology, Motilal Nehru National Institute of Technology Allahabad, Allahabad, India

Gaurav Yadav
Center for Medical Diagnostic and Research, Motilal Nehru National Institute of Technology Allahabad, Allahabad, India
Department of Biotechnology, Motilal Nehru National Institute of Technology Allahabad, Allahabad, India

Vivek K. Singh
Department of Physics, Shri Mata Vaishno Devi University, Katra, India,
Lawrence Berkeley National Laboratory, Berkeley, CA, USA

Rohit K. Mishra
Center for Medical Diagnostic and Research, Motilal Nehru National Institute of Technology Allahabad, Allahabad, India

R. G. Upadhyay
Veer Chand Singh Garhwali Uttarakhand University of Horticulture and Forestry, Tehri Garhwal, India

Nawal K. Dubey
Centre of Advanced Study in Botany, Banaras Hindu University, Varanasi, India

Yonghoon Lee
Department of Chemistry, Mokpo National University, Mokpo, South Korea

Hyeon-Jin Kim, Mi-Ran Go, Jin Yu, Song-Hwa Bae and Soo-Jin Choi
Division of Applied Food System, Major of Food Science and Technology, Seoul Women's University, Seoul, South Korea

Hyoung-Jun Kim, Kyoung-Min Kim, Jae Ho Song and Jae-Min Oh
Department of Chemistry and Medical Chemistry, College of Science and Technology, Yonsei University, Wonju, South Korea

Hongxia Liu, Ling Gao, Jinzhi Han, Zhi Ma, Zhaoxin Lu, Chong Zhang and Xiaomei Bie
College of Food Science and Technology, Nanjing Agricultural University, Nanjing, China

Chen Dai
College of Life Sciences, Nanjing Agricultural University, Nanjing, China

Yi Wang, Yan Wang, Jianguo Xu and Changyun Ye
State Key Laboratory of Infectious Disease Prevention and Control, Beijing, China
National Institute for Communicable Disease Control and Prevention, Beijing, China
Collaborative Innovation Center for Diagnosis and Treatment of Infectious Diseases, Beijing, China
Chinese Center for Disease Control and Prevention, Beijing, China

Jayanta Kumar Patra
Research Institute of Biotechnology and Medical Converged Science, Dongguk University-Seoul, Goyang-si, South Korea

Kwang-Hyun Baek
Department of Biotechnology, Yeungnam University, Gyeongsan, South Korea

Index